CONTENTS

NEW DEVELOPMENTS IN BIOSCIENCES:
THEIR IMPLICATIONS FOR LABORATORY ANIMAL SCIENCE

CONTRIBUTORS*

P. Agrawala, Abteilung für experimentelle Fortpflanzungsbiologie, Klinik für Gynäkologie des Rindes, Tierärtzlichen Hochschule Hannover, Hannover, Federal Republic of Germany (203)

E. Ahonen, Department of Clinical Physiology, Kuopio University Central Hospital, SF-70210 Kuopio, Finland (413)

A.F. Angulo, National Institute of Public Health and Environmental Hygiene, P.O. Box 1, 3720 BA Bilthoven, The Netherlands (443, 447)

C. Bacchus, Institut für Humangenetik und Anthropologie, Universität Heidelberg, Im Neuenheimer Feld 328, D-6900 Heidelberg, Federal Republic of Germany (405)

A.K. Banerjee, Laboratory Animal Centre, Erasmus University, P.O. Box 1738, 3000 DR Rotterdam, The Netherlands (437, 443, 447)

B.R. Barber, Farmitalia Carlo Erba, Istituto Ricerche Nerviano, Milano, Italy (191)

F. Bartnik, Henkel KGaA, Postfach 1100, D-4000 Düsseldorf 1, Federal Republic of Germany (215)

A. Baskerville, PHLS, Centre for Applied Microbiology and Research, Porton Down, Salisbury, Wiltshire, SP4 0JG, United Kingdom (99)

V. Baumans, Department of Laboratory Animal Science, Veterinary Faculty, State University, P.O. Box 80.166, 3508 TD Utrecht, The Netherlands (295, 305, 431)

E. Baunack, Zentrales Tierlaboratorium, Medizinische Hochschule Hannover, Konstanty-Gutschow-Strasse 8, D-3000 Hannover 61, Federal Republic of Germany (203)

D.W. van Bekkum, Radiobiological Institute TNO, P.O. Box 5815, 2280 HV Rijswijk, The Netherlands (7)

A. Berns, Division of Molecular Genetics, Netherlands Cancer Institute, Plesmanlaan 121, 1066 CX Amsterdam and Department of Biochemistry, University of Amsterdam, Amsterdam, The Netherlands (175)

A.C. Beynen, Department of Laboratory Animal Science, Veterinary Faculty, University of Utrecht, P.O. Box 80.166, 3508 TD Utrecht and Department of Human Nutrition, Agricultural University, De Dreijen 12, 6703 BC Wageningen, The Netherlands (279, 431)

M.G.C.W. den Bieman, Department of Laboratory Animal Science, Veterinary Faculty, University of Utrecht, P.O. Box 80.166, 3508 TD Utrecht, The Netherlands (197)

R. Boot, National Institute of Public Health and Environmental Hygiene, P.O. Box 1, 3720 BA Bilthoven, The Netherlands (71)

The number in parentheses is the opening page number of the contributor's article.

T. Bowman, Smith Kline and French Laboratories, 709 Swedeland Road L620, Swedeland, PA 19479, U.S.A. (449)

G. Brem, Institut für Tierzucht und Tierhygiene, Ludwig-Maximilians-Universität, Veterinärstrasse 13, D-8000 München 22, Federal Republic of Germany (93, 331, 337, 343)

B. Brenig, Institut für Tierzucht und Tierhygiene, Ludwig-Maximilians-Universität, Veterinärstrasse 13, D-8000 München 22, Federal Republic of Germany (331, 337)

K. Burow, Kali-Chemie Pharma GmbH, Hans-Böckler-Allee 20, 3000 Hannover, Federal Republic of Germany (185)

W. Buselmaier, Institut für Humangenetik und Anthropologie, Universität Heidelberg, Im Neuenheimer Feld 328, D-6900 Heidelberg, Federal Republic of Germany (405)

I. von Butler, Institut für Tierzucht der technischen Universität München, D-8050 Freising-Weihenstephan, Federal Republic of Germany (373)

S. Campbell, Smith Kline and French Laboratories, 709 Swedeland Road L620, Swedeland, PA 19479, U.S.A. (449)

B.P. Cats, Wilhelmina Children's Hospital, P.O. Box 18009, 3501 CA Utrecht, The Netherlands (47)

A. Clement, Institut für Tierzucht und Tierhygiene, Ludwig-Maximilians-Universität, Veterinärstrasse 13, D-8000 München 22, Federal Republic of Germany (343)

G. Clough, Alanann Consultancy Services, P.O. Box 230, York, YO1 1GG, United Kingdom (239)

V. Cools, Laboratory of Veterinary Physiology, State University Centre of Antwerp, Slachthuislaan 68, 2008 Antwerp, Belgium (393)

E. Cox, Laboratory of Veterinary Physiology, State University Centre of Antwerp, Slachthuislaan 68, 2008 Antwerp, Belgium (393)

F. Dagnæs-Hansen, Bomholtgaard Breeding and Research Center Ltd., Bomholtvej 10, DK-8680 Ry, Denmark (375)

F. Deerberg, Central Institute for Laboratory Animal Breeding, Hermann-Ehlers-Allee 57, D-3000 Hannover 91, Federal Republic of Germany (425)

A.J.M. Degen, Organon International B.V., Drug Safety Research and Development Laboratories, P.O. Box 20, 5340 BH Oss, The Netherlands (221)

A.M.L. van Delft, Scientific Development Group, Organon International B.V., P.O. Box 20, 5340 BH Oss, The Netherlands (59)

P. Demant, The Netherlands Cancer Institute, Antoni van Leeuwenhoekhuis, Plesmanlaan 121, 1066 CX Amsterdam, The Netherlands (209)

K.M. Dhasmana, Department of Anaesthesiology, Medical Faculty, Erasmus University, P.O. Box 1738, 3000 DR Rotterdam, The Netherlands (437)

L. Duzzi, Farmitalia Carlo Erba, Istituto Ricerche Nerviano, Milano, Italy (191)

P.M.C.A. van Eerd, Primate Center TNO, P.O. Box 5815, 2280 HV Rijswijk, The Netherlands (63)

M. Ferrandi, Farmitalia Carlo Erba, Istituto Ricerche Nerviano, Milano, Italy (191)

P. Ferrari, Farmitalia Carlo Erba, Istituto Ricerche Nerviano, Milano, Italy (191)

L. Francis, PHLS, Centre for Applied Microbiology and Research, Porton Down, Salisbury, Wiltshire, SP4 0JG, United Kingdom (99)

M.R. Gamble, The Boots Company PLC, Research Department, Nottingham, NG2 3AA, United Kingdom (117)

R.L. Gardner, Imperial Cancer Research Fund, Developmental Biology Unit, Department of Zoology, South Parks Road, Oxford, OX1 3PS, United Kingdom (147)

K. Gärtner, Medizinische Hochschule Hannover, Zentrales Tierlaboratorium, Konstanty-Gutschow-Strasse 8, D-3000 Hannover 61, Federal Republic of Germany (121, 359)

C. Goosen, Primate Center TNO, P.O. Box 5815, 2280 HV Rijswijk, The Netherlands (67, 305)

P. de Greeve, Veterinary Public Health Inspectorate, Section Animal Experimentation, P.O. Box 5406, 2280 HK Rijswijk, The Netherlands (255)

T. Haaranen, National Laboratory Animal Center, University of Kuopio, P.O. Box 6, 70211 Kuopio, Finland (409)

J.W.M. Haas, Small Animal Center, Agricultural University, De Dreijen 12, 6703 BC Wageningen, The Netherlands (431)

J. Hahn, Abteilung für experimentelle Fortpflanzungsbiologie, Klinik für Gynäkologie des Rindes, Tierärztlichen Hochschule Hannover, Hannover, Federal Republic of Germany (203)

M. Hakumäki, Department of Physiology, University of Kuopio, SF-70211 Kuopio, Finland (413)

C.J.A. van den Hamer, Department of Radiochemistry, Interuniversity Reactor Institute, Mekelweg 15, 2629 JB Delft, The Netherlands (419)

H. Hanhijärvi, Oy Star Ab Pharmaceutical Co., P.O. Box 33, 33721 Tampere, Finland (409)

W. Hardegg, Institut für Versuchstierkunde der Universität Heidelberg, Im Neuenheimer Feld 347, D-6900 Heidelberg, Federal Republic of Germany (317, 365)

A.A.M. Hart, The Netherlands Cancer Institute, Antoni van Leeuwen-hoekhuis, Plesmanlaan 121, 1066 CX Amsterdam, The Netherlands (209)

J. Hartikainen, Department of Physiology, University of Kuopio, SF-70211 Kuopio, Finland (413)

J. Hau, Department of Pathology, Laboratory Animal Unit, Royal Veterinary and Agricultural University, Copenhagen, Denmark (81, 87)

C. Heckl-Ensslin, Institut für Tierzucht der technischen Universität München, D-8050 Freising-Weihenstephan, Federal Republic of Germany (373)

H.J. Hedrich, Zentralinstitut für Versuchstierzucht, Hermann-Ehlers-Allee 57, D-3000 Hannover 91, Federal Republic of Germany (163)

H. Heinecke, Central Institute of Microbiology and Experimental Therapy, Academy of Sciences of the GDR, Beutenbergstrasse 11, DDR-6900 Jena, German Democratic Republic (225)

K.K. van Hellemond, Small Animal Center, Agricultural University, De Dreijen 12, 6703 BC Wageningen, The Netherlands (431)

H. van Herck, Division of Laboratory and Special Animal Diseases, Department of Veterinary Pathology and Department of Laboratory Animal Science, State University, P.O. Box 80.166, 3508 TD Utrecht, The Netherlands (295)

E. Hirsjärvi, Central Finlands Association for Laboratory Animal Welfare, Tikantie 23, 40520 Jyväskylä, Finland (29)

P.A. Hirsjärvi, Department of Applied Zoology, University of Kuopio, P.O. Box 6, 70211 Kuopio, Finland (399)

J.A.P. van Hoof, Vakgroep Bestuur en Beleid, P.O. Box 9108, 6500 HK Nijmegen, The Netherlands (75)

J.A.R.A.M. van Hooff, Laboratory of Comparative Physiology, Jan van Galenstraat 40, 3572 LA Utrecht, The Netherlands (41)

M.C. Horzinek, Institute of Virology, Veterinary Faculty, State University, P.O. Box 80.164, 3508 TD Utrecht, The Netherlands (11)

A. Houvenaghel, Laboratory of Veterinary Physiology, State University Centre of Antwerp, Slachthuislaan 68, 2008 Antwerp, Belgium (393)

F. Iglauer, Medizinische Hochschule Hannover, Zentrales Tierlaboratorium, Konstanty-Gutschow-Strasse 8, D-3000 Hannover 61, Federal Republic of Germany (121)

H.F.P. Joosten, Organon International B.V., Drug Safety Research and Development Laboratories, P.O. Box 20, 5340 BH Oss, The Netherlands (221)

N.-C. Juhr, Institut für Versuchstierkunde und Versuchstierkrankheiten, Freie Universität Berlin, Krahmerstrasse 6, D-1000 Berlin 45, Federal Republic of Germany (127)

M.A. Junnila, Institute of Occupational Health, Helsinki, Finland (399)

J. Kamerman, Organon International B.V., Drug Safety Research and Development Laboratories, P.O. Box 20, 5340 BH Oss, The Netherlands (221)

J. Kaspareit-Rittinghausen, Central Institute for Laboratory Animal Breeding, Hermann-Ehlers-Allee 57, D-3000 Hannover 91, Federal Republic of Germany (425)

P. Klír, Institute of Physiology, Czechoslovak Academy of Sciences, Prague, Czechoslovakia (35, 379)

R. Kluge, Abteilung für Versuchstierkunde des Klinikums der RWTH Aachen, Pauwelsstrasse, D-5100 Aachen, Federal Republic of Germany (105, 185)

J. Kong-A-San, Laboratory Animal Centre, Erasmus University, P.O. Box 1738, 3000 DR Rotterdam, The Netherlands (443)

G. Kuhn, Institut für Versuchstierkunde der Universität Heidelberg, Im Neuenheimer Feld 347, D-6900 Heidelberg, Federal Republic of Germany (317, 365)

K. Künstler, Henkel KGaA, Postfach 1100, D-4000 Düsseldorf 1, Federal Republic of Germany (215)

I. Kunstýř, Medizinische Hochschule Hannover, Institut für Versuchstierkunde, Postfach 610180, D-3000 Hannover, Federal Republic of Germany (111)

M. Landi, Smith Kline and French Laboratories, 709 Swedeland Road L620, Swedeland, PA 19479, U.S.A. (449)

E. Länsimies, Department of Clinical Physiology, Kuopio University Central Hospital, SF-70210 Kuopio, Finland (413)

J. Leuenberger, Institut für biologisch medizinische Forschung AG, Wölferstrasse 4, CH-4414 Füllinsdorf, Switzerland (353)

J.A. Loopstra, Division of Laboratory and Special Animal Diseases, Department of Veterinary Pathology, State University, Yalelaan 1, 3584 CL Utrecht, The Netherlands (311)

H. van Loveren, Laboratory for Pathology, National Institute of Public Health and Environmental Hygiene, P.O. Box 1, 3720 BA Bilthoven, The Netherlands (17)

M. Luukkainen, Central Finlands Association for Laboratory Animal Welfare, Tikantie 23, 40520 Jyväskylä, Finland (29)

P. Mäenpää, Department of Biochemistry, University of Kuopio, Finland (325)

U. Märki, Institut für biologisch medizinische Forschung AG, Wölferstrasse 4, CH-4414 Füllinsdorf, Switzerland (353)

K. Mensing, Medizinische Hochschule Hannover, Zentrales Tierlaboratorium, Konstanty-Gutschow-Strasse 8, D-3000 Hannover 61, Federal Republic of Germany (121)

J. Meyer, Institut für Tierzucht und Tierhygiene, Ludwig-Maximilians-Universität, Veterinärstrasse 13, D-8000 München 22, Federal Republic of Germany (337, 343)

R.T. Moore, PHLS, Centre for Applied Microbiology and Research, Porton Down, Salisbury, Wiltshire, SP4 0JG, United Kingdom (99)

H. Morse, Hope Farms B.V., P.O. Box 85, 3440 AB Woerden, The Netherlands (389)

J.R. Needham, The Microbiology Laboratories, 56 Northumberland Road, North Harrow, Middlesex, United Kingdom (117)

T. Nevalainen, National Laboratory Animal Center, University of Kuopio, Savilahdentie 11 M, 70210 Kuopio, Finland (325, 409, 413)

K.J. Öbrink, Institute of Physiology, Uppsala University, Biomedical Center, Box 572, S-75123 Uppsala, Sweden (31)

I. Ortlepp, Abteilung für Labortierkunde, Universität Zürich, Winterthurerstrasse 190, CH-8057 Zürich, Switzerland (359)

A.D.M.E. Osterhaus, The National Institute of Public Health and Environmental Hygiene, P.O. Box 1, 3720 BA Bilthoven, The Netherlands (143)

P. Parenti, Farmitalia Carlo Erba, Istituto Ricerche Nerviano, Milano, Italy (191)

W. Pitterman, Henkel KgaA, Postfach 1100, D-4000 Düsseldorf 1, Federal Republic of Germany (215)

A.A. Polak-Vogelzang, National Institute of Public Health and Environmental Hygiene, P.O. Box 1, 3720 BA Bilthoven, The Netherlands (447)

T.B. Poole, Universities Federation for Animal Welfare, 8 Hamilton Close, South Mimms, Potters Bar, Herts EN6 3QD, United Kingdom (231)

O.M. Poulsen, Institute of Obstetrics and Gynaecology, University of Odense, Denmark (81, 87)

M. Přibylová, Institute of Physiology, Czechoslovak Academy of Sciences, Videnska 1083, 14220 Prague, Czechoslovakia (379)

K.G. Rapp, Zentralinstitut für Versuchstierzucht, Hermann-Ehlers-Allee 57, D-3000 Hannover 91, Federal Republic of Germany (105, 185)

B. Ratcliffe, Polytechnic of North London, Holloway Road, London, N7 8DB, United Kingdom (267)

W. Rating, Department of Anaesthesiology, Medical Faculty, Erasmus University, P.O. Box 1738, 3000 DR Rotterdam, The Netherlands (437)

R.T. Raymond, PHLS, Centre for Applied Microbiology and Research, Porton Down, Salisbury, Wiltshire, SP4 0JG, United Kingdom (99)

J.M. van Ree, Rudolf Magnus Institute for Pharmacology, State University, Vondellaan 6, 3521 GD Utrecht, The Netherlands (53)

I.C. Reetz, Zentralinstitut für Versuchstierzucht, Hermann-Ehlers-Allee 57, D-3000 Hannover 91, Federal Republic of Germany (163)

K. Ross, Institut für Tierzucht und Tierhygiene, Ludwig-Maximilians-Universität, Veterinärstrasse 13, D-8000 München 22, Federal Republic of Germany (337)

W. Rossbach, Institut für biologisch medizinische Forschung AG, Wölferstrasse 4, CH-4414 Füllinsdorf, Switzerland (353)

H.C. Rowsell, Canadian Council for Animal Care, 1105-151 Slater Street, Ottawa, Ontario, K1P 5H3, Canada (289)

S. Salardi, Farmitalia Carlo Erba, Istituto Ricerche Nerviano, Milano, Italy (191)

R.J. Samsom, Director-in-Chief of Health Protection Branche, Ministry of Welfare, Health and Cultural Affairs, Rijswijk, The Netherlands (1)

H. Schötz, Abteilung für Labortierkunde, Universität Zürich, Winterthurerstrasse 190, CH-8057 Zürich, Switzerland (359)

H.J. Schuurman, Division of Immunopathology, Department of Internal Medicine and Pathology, University Hospital, Catharijnesingel 101, 3511 GV Utrecht, The Netherlands (17)

H. Sigg, Nestlé Research Center, Nestec Ltd., Vers-chez-les-Blanc, 1000 Lausanne 26, Switzerland (349)

F.R. Stafleu, Department of Laboratory Animal Science, State University, P.O. Box 80.166, 3508 TD Utrecht, The Netherlands (431)

T. Svoboda, Institute of Physiology, Czechoslovak Academy of Sciences, Videnska 1083, 14220 Prague, Czechoslovakia (379)

K. Tahvanainen, Department of Clinical Physiology, Kuopio University Central Hospital, SF-70210 Kuopio, Finland (413)

P. Tamborini, Institut für Toxicologie, ETH und Universität Zürich, 8603 Schwerzenbach, Switzerland (349)

G. van Tintelen, Small Animal Center, Agricultural University, De Dreijen 12, 6703 BC Wageningen, The Netherlands (431)

L. Torielli, Farmitalia Carlo Erba, Istituto Ricerche Nerviano, Milano, Italy (191)

M. Veldhuizen, Department of Radiochemistry, Interuniversity Reactor Institute, Mekelweg 15, 2629 JB Delft, The Netherlands (419)

R. Velleuer, Medizinische Hochschule Hannover, Zentrales
Tierlaboratorium, Konstanty-Gutschow-Strasse 8, D-3000 Hannover 61,
Federal Republic of Germany (121)

T.S. Veninga, Department of Radiopathology, State University of
Groningen, Bloemsingel 1, 9713 BZ Groningen, The Netherlands (389)

R. Verhoeff - de Fremery, Hubrecht Laboratory, Netherlands Institute for
Developmental Biology, Uppsalalaan 8, 3584 CT Utrecht, The Netherlands
(311)

F.J.M. Vervoordeldonk, Hubrecht Laboratory, Netherlands Institute for
Developmental Biology, Uppsalalaan 8, 3584 CT Utrecht, The Netherlands
(311)

J. Vijg, TNO Institute for Experimental Gerontology, P.O. Box 5815, 2280
HV Rijswijk, The Netherlands (133)

H.-M. Voipio, National Laboratory Animal Center, University of Kuopio,
Savilahdentie 11 M, 70210 Kuopio, Finland (325)

J.G. Vos, Laboratory for Pathology, National Institute of Public Health
and Environmental Hygiene, P.O. Box 1, 3720 BA Bilthoven, The
Netherlands (17)

M. Waller, Uppsala Biomedical Center, Box 570, S-75123 Uppsala, Sweden
(31)

T. Waller, Pharmacia AB, 75182 Uppsala, Sweden (385)

R. Wanke, Institut für Tierpathologie, Ludwig-Maximilians-Universität,
Veterinärstrasse 13, D-8000 München 22, Federal Republic of Germany (93)

L. Wass, National Board of Universities and Colleges, Box 45501, 10430
Stockholm, Sweden (31)

W.H. Weihe, Biologisches Zentrallaboratorium, Universitätsspital,
Universität Zürich, Rämistrasse 100, 8091 Zürich, Switzerland (245)

D. Wesseling, Department of Laboratory Animal Science, Veterinary
Faculty, State University, P.O. Box 80.166, 3508 TD Utrecht, The
Netherlands (305)

J.P. van Wouwe, Department of Pediatrics, University Hospital Leiden,
P.O. Box 9600, 2300 RC Leiden and Department of Radiochemistry,
Interuniversity Reactor Institute, Mekelweg 15, 2629 JB Delft, The
Netherlands (419)

T.D. Yih, Organon International B.V., Drug Safety Research and Development Laboratories, P.O. Box 20, 5340 BH Oss, The Netherlands (221)

M. Ylinen, National Laboratory Animal Center, University of Kuopio, P.O.
Box 6, 70211 Kuopio, Finland (409)

A. Zogbaum, Abteilung für Labortierkunde, Universität Zürich,
Winterthurerstrasse 190, CH-8057 Zürich, Switzerland (359)

<u>L.F.M. van Zutphen</u>, Department of Laboratory Animal Science, Veterinary Faculty, University of Utrecht, P.O. Box 80.166, 3508 TD Utrecht, The Netherlands (197, 209)

<u>P. Zwart</u>, Division of Laboratory and Special Animal Diseases, Department of Veterinary Pathology, University of Utrecht, Yalelaan 1, 3584 CL Utrecht, The Netherlands (299, 311)

A.C. Beynen & H.A. Solleveld (Eds), New developments in biosciences: their implications for laboratory animal science, ISBN 978-94-010-7973-0
© 1988 Martinus Nijhoff Publishers, Dordrecht.

OPENING ADDRESS: THE FUTURE OF LABORATORY ANIMAL USE

R.J. SAMSOM

Ladies and Gentlemen,

   I should like to use this occasion to make a few remarks about the future of research using laboratory animals. It is a subject which calls for consideration at a time of accelerating scientific progress, apparently limitless technical possibilities and changing views about the place of animals in our society.
   This of course raises the question of whether we can envisage what the future holds in store, and that is no easy matter. At the end of the last century a drawing competition was organized with a science fiction-type theme, in which artists were asked to depict the world as it would look in fifty years' time. The average result was really nothing more than an extravagant version of the contemporary nineteenth-century world. And then again, some of us here graduated in the nineteen fifties, and the science of immunology which was taught to students then was scarcely more advanced than it had been back in 1900, certainly when one compares it with the way in which the subject has taken off since the 1950s. With these cautionary remarks in mind I shall look ahead only as far as the year 2000.
   What has happened in immunology illustrates the acceleration in scientific progress to which I have referred. Most of you will be able to add your own examples to the list. I shall mention a couple more to clarify my point. Molecular techniques such as RNA and DNA hybridisation and the generation of monoclonal antibodies open up new possibilities and mean, for example, that in diagnosing the causes of disease it is possible to abandon complicated techniques involving culture mediums, cell cultures, laboratory animals and incubated birds' eggs. The development of entirely new types of vaccine is just around the corner. Ultrasonic equipment has already made its appearance in medical and veterinary clinics. Biology is turning into biological engineering; medicine and veterinary science are evolving in the direction of health technology. Expectations are running high. Writing in the Journal of the American Veterinary Medical Association (1986, vol. 189, pp. 172-177), Pritchard confidently proclaims: "There is absolutely no scientific or technical reason why diseases worldwide cannot be brought under effective control during the early 21st century". Fortunately he does temper his euphoria by adding "That should be the goal (of the veterinary profession)". For the time being I shall translate this into my first prediction which is that we can expect an enormous amount of research to be carried out in the future, and this will include research involving animal experimentation.

My second prediction is this: given the increasing sophistication of research, there will be an increase in the stringency of the requirements which those who commission research expect animal experiments and laboratory animals to meet. There are already signs of this, for example in the genetic requirements which are made of laboratory animals and which are reflected in a shift from outbred to congenic strains. However, there are other factors, some of them well known, others still requiring further research, which have so far tended to be overlooked when setting up experiments, but which can influence test results and should thus not be neglected. In future more attention will have to be paid to these factors, which include aspects of the accommodation and care of laboratory animals: their spatial requirements, the effects of confinement, the type and design of boxes and cages and the materials used in their construction, the effects of bedding, light, temperature, ventilation, noise, the rhythm of day and night, the supply and constituents of feed, transport, animal handling, the age at which the animals are weaned, the extent of crowding, social interaction and disease. And I have not even mentioned all the types of biotechnics and animal models which already exist or are likely to be developed. The conclusion here is once again that more research is needed.

My third prediction is that there will be increasing interest in the welfare of laboratory animals. In the case of most laboratory animals we do not know enough about their normal behaviour and the factors which influence this, for better or for worse. There is a need for a better understanding of the ethology of laboratory animals. We must learn a great deal more about suffering, and particularly pain, as experienced by animals, and ways of preventing it. We know that alternatives must be found. There is no need in an audience such as this to list all of the promising areas to be explored here. One point that is worth noting is that so far the discovery of alternatives has tended to be a by-product of ongoing research. However, there is a growing realization that research specifically geared to the development of alternatives can also be valuable in its own right. The Dutch government is involved in setting up a platform to stimulate research into alternatives by providing financial and other support. A notable initiative here is the literature survey into the possibilities of replacement, reduction and refinement in the area of vaccine production, which was funded by the Ministry of Welfare, Health and Cultural Affairs in conjunction with five animal welfare and anti-vivisection organizations. The research report by Hendriksen, which is being translated into English, contains 56 recommendations covering the following subjects:
- amending Dutch guidelines;
- amending foreign or international guidelines;
- adapting production techniques and planning;
- corroborative research;
- innovatory research.

One noteworthy result here is that part of the project, which involved a computer study relating to research into tetanus and diphtheria, has already led to a saving of roughly 3000 laboratory animals per year in the Netherlands. It is gratifying to note that various organizations, including the WHO, have shown an interest in this study. Our experience so far suggests that there are plenty of ideas in the air, but shortage of resources and manpower means that many of these never get taken up. The platform I mentioned earlier can perform a useful role in this respect. I expect that the platform, which is essentially an artificial

creation, will not prove to be the only source of funding for research
into alternative methods in the future, and that provisions will
increasingly be made for such research within regular research budgets.

My fourth prediction concerns public interest in animal
experimentation. There is growing interest in this subject, as we can see
from the newspapers. The press is sometimes criticised for bias in its
approach to gathering information and for the bad habit of attributing
certain attitudes and beliefs to the public on the basis of pure
hypothesis. Dutch survey results suggest that the public attitude is one
of critical tolerance. Opinion polls in Switzerland and California
present the same picture.

What values are rated as important in life by the silent majority? In
1979 Goddijn et al. asked a sample of Dutch respondents which of 25
possible goals and values they regarded as the most important in their
lives and which they would place in second and third place. The results,
in simplified form, were as follows.

Survey question

"State the aim or value most important for your personal life; also state
your second and third choice".

Result (%)

	total number of mentions	aim/value first mentioned
health	52	23
happiness	38	17
peace	31	16
19 others		
exciting life	2	0
beauty	1	0
capability	1	0
	(300%)	(100%)

One could devote an entire symposium to these findings, but for our
purposes I should like to point to the discrepancy between the values
which feature at the top and the bottom of the list: health scored
highest, capability lowest. Other findings which emerged from the same
piece of research, which are not reproduced here, show that people relate
the important things in life primarily to their own material
circumstances, their own situation and that of their family. This
suggests that people attach a high value to their own welfare but fail to
appreciate fully what this necessitates or what price has to be paid for
it, particularly where more complex processes are involved. Public
information to educate people about science should take account of such
factors, which, though not the only ones, are nevertheless important.

The theme of this address is the future of animal experimentation;
this is not to say that the past should not be discussed. I shall
therefore allow myself a short historical digression. One's first
inclination is then to refer to the famous pioneers of the turn of the
century - names such as Von Behring, Koch and Ehrlich in Germany, Finsen

in Denmark, Pavlov and Metchnikoff in Russia, Golgi in Italy, Ramón in Spain and Laveran in France, the first winners of the Nobel Prize for physiology and medicine. Such names remind us of the successful leap forward made in biology, medicine and veterinary science. This sort of historiography is like a *petitio principii*, leading inevitably to the confirmation of what we set out to prove, namely that animal experimentation has been enormously valuable.

But an alternative view of natural science is also conceivable. One can get an idea of this by examining other scientific literature dating from the period in which the scientists to whom I have referred were receiving their Nobel prizes. One sees then that a great deal of work was done which led nowhere. This is not necessarily to belittle the work that was done; biomedical research is a painstaking business and the secrets of nature are not easily uncovered. But honesty compels us to admit that much of what was published, let alone the work that was never published, should not have passed scrutiny.

This pattern is likely to repeat itself in future. We shall see examples of successful research, but also of research which leads nowhere or is badly designed. Fortunately we can say that the likelihood of poorly-designed research is considerably lower now than it used to be, thanks to mechanisms such as conditional funding which have become widely used. This does not change the fact that we must continue to be alert to the dangers.

There is also a third way of looking at the past, and that is to look at the intellectual beliefs which underlie people's approach to reality and the judgements they make. This leads us into the realm of ethics and the philosophy of science. In fact both of the approaches to history which I have discussed derive from a particular philosophy, namely utilitarianism, which maintains that human beings should act in such a way that good prevails over evil. Without going into the value of this philosophy I should like to note that opinion is likely to be divided about what constitutes "good" and "evil" and how we are meant to decide how to act if what is good for one person is bad for another.

Utilitarianism is just one example. In general an understanding of people's teleological or deontological principles is extremely useful in the public debate about experiments using laboratory animals. A knowledge of history is an integral part of this, and such a view of history will or should increasingly be the one we adopt in future, for it is important not only to know what we are doing, but also why we are doing it.

To summarize my argument so far: I have made a number of predictions about the use of laboratory animals. These related to:
- the need for experimental research using animals;
- requirements for animal experiments and laboratory animals;
- the issue of alternatives;
- public information;
- people's sense of responsibility.

What role can the government play in all this? The policy which the Dutch government wishes to pursue in the field of animal experimentation may be summed up as follows: "While recognizing that animal experimentation is essential for certain purposes, the government is committed to keeping the number of experiments as low as possible and ensuring that these tests are carried out with the utmost care".

Naturally the first instrument one thinks of here is legislation. In many countries we are indeed witnessing the introduction of new legislation and the redrafting of old statutes, partly under the

influence of European integration. This is all to the good. However, a
number of comments are in order.

Legislation is important in itself, but it is not the whole story. One
might even venture to state that it doesn't really matter all that much
exactly what the law on animal experimentation says. Dutch experience
since the first serious discussion in the nineteen sixties about the
introduction of legislation in this field has been that the discussion in
itself had a positive effect on ways of thinking about the issue of
laboratory animals. The most important effect of legislation of any sort
is the stimulus it provides in increasing public awareness and
heightening people's sense of responsibility.

My second comment is that legislation is not the only instrument in
this process. The most important investment is probably in education. The
object of this education must be to cultivate a constantly critical
attitude to all actions and all experiments so that this in the end
becomes an automatic reflex. Awareness of this issue should be based on
five elements:
- the realization that animals are complex and in many cases social
  creatures and that account should be taken of individual differences;
- a knowledge of what constitutes normal and abnormal behaviour in
  animals;
- an understanding of the care and accommodation of laboratory animals;
- a knowledge of the technology used in experiments;
- a concept of responsibility based on an understanding of ethical
  precepts and the interests of society.
The fifth element in particular deserves attention in education on
training in laboratory animal science. Dutch law lays down standards of
competence required of researchers, animal technicians, animal
caretakers, veterinarians and other experts responsible for the welfare
of laboratory animals. The curricula for these staff, which Professor Van
Zutphen's group has been instrumental in drawing up, rightly contain
explicit references to the ethical and social aspects of animal
experimentation.

My third comment relates to the role of animal experimentation
committees. By this I mean in very general terms committees of experts
which examine experimental research involving laboratory animals in which
animals might be at risk. Dutch law makes no provision for such bodies as
yet, but it has effectively been decided that there is a need for them in
that the State Secretary has promised Parliament that statutory provision
for such committees will be made in a forthcoming amendment. Here, too,
the increased interest in this discussion has had an effect: since the
publication in 1985 of a favourable report on the matter by the Advisory
Committee on Animal Experimentation the percentage of institutions which
have voluntarily introduced such a committee has risen from 17 to 95%. I
cannot read you the Committee's report in full, but I shall quote a few
lines from it:
Discussing the objectives of such committees, the report states that one
aim would be "to ensure that before any animal experiment is begun a
conscious calculation is made of the importance of the research in
relation to the suffering of the laboratory animals used". The
committees' tasks would include "advising the licence-holder and staff on
all aspects of animal experiments, including design, implementation and
admissibility". On the composition of the animal experimentation
committees the Advisory Committee advocates "a balanced and diverse
composition, such that members include not only experts in research and

research policy, biotechnics and animal care experts, but also persons
with a more specialized interest in the ethical aspects of animal
experimentation. Those appointed should preferably be able to provide
impartial advice."

It is on these three instruments - the law, education and the
establishment of animal experimentation committees - that the Dutch
Government is basing its policy on animal experimentation for the present
and the future.

Thank you.

A.C. Beynen & H.A. Solleveld (Eds), New developments in biosciences: their implications for laboratory animal science, ISBN 978-94-010-7973-0

THE ACQUIRED IMMUNEDEFICIENCY SYNDROME IN MAN

D.W. VAN BEKKUM

Since AIDS was recognized first in patients in 1981, more than 30,000 cases have been registered in the U.S. and about 4,500 in Western Europe *) where the epidemic began later. In certain parts of Africa large proportions, e.g. up to 10% of the general population are infected. The exact incidence of the disease in those areas is unknown. In countries where the disease is registered, the number of cases is increasing exponentially with a doubling time of between 5 and 15 months (average 10 months) (1). So far, patients with AIDS have not recovered and about 50% of the registered cases have died. Cases with the full blown disease form the tip of an iceberg (2). AIDS is caused by infection with a retrovirus HIV (Human Immunedeficiency Virus) which is transmitted sexually and by transfer of blood or blood products. For each individual with full blown AIDS there are now 5 infected persons who have signs and symptoms such as malaise, weight loss, immunological abnormalities (the AIDS-Related Complex - ARC). Furthermore, in the U.S. alone between 1 and 1.5 million people are infected with the virus, but are still aSymptomatic. According to recent projections, 20-30% of the infected people will develop AIDS over a 5 years' period, but this percentage may be much higher in the long run (3). Accordingly, it has been projected that by the year 1991, approximately 270,000 individuals will have developed AIDS in the U.S. alone.

The ethiological agent of AIDS was described as Lymphadenopathy-Associated-Virus (LAV) by Montagnier (4), as Human T Lymphotropic Virus type III (HTLV-III) by Gallo (5) and as AIDS associated retrovirus (ARV) by Levy (6). These various isolates proved to be essentially the same virus, which for purposes of uniformity has been renamed the Human Immunedeficiency Virus (HIV) by a subcommittee of the Committee on the Taxonomy of Viruses.

HIV is a retrovirus that displays heterogeneity among various isolates, which may have implications for the development of effective vaccines. A large body of data has emerged on the molecular and biological properties of HIV. The virus has the flanking long terminal repeats typical of retroviruses as well as the pol, gag and env genes, coding for the polymerases such as reverse transciptase, the core proteins and the envelope, respectively. In addition, four unusual genes have been identified: tat, trs, sor and 3' orf, which encode proteins that help to control the expression of viral genes (7). Recently, a number of retroviruses closely resembling HIV, perhaps to be considered as substrains of HIV, have been identified, some of them in the serum of

HIV-negative individuals. These have been termed LAV-2, SBL virus, HTLV-IV and seem to be closely related to or perhaps identical with the Simian Immunodeficiency Virus (SIV) that has been isolated from healthy wild caught African Green monkeys and from diseased captive Macacus rhesus monkeys (8-9). It is highly probable that SIV-infected rhesus monkeys will be the animal model of choice for experimental studies on AIDS.

Following HIV-infection of the target cell, a DNA copy of the viral RNA is produced via the enzyme reverse transcriptase and this DNA (provirus) integrates into the genome of the host cell or can exist as an unintegrated DNA. The predominant target cell for HIV is the helper/inducer subset of T-lymphocytes, known as the T4 cells. Since this cell is killed following infection by HIV, the result is Severe Immune Deficiency. The T4 (CD4) molecule on the surface of this subset of human T-lymphocytes is the receptor of the HIV-virus. Following the binding of the virus, the virus gains entry to the cell by a process called endocytosis. Inside the cell, virus replication may take place and this leads to the death of the cell. The mere presence of T4 on the surface of a cell does not render these cells permissive to HIV-infection as was demonstrated with mouse cells provided with a functional T4 gene (10). On the other hand, when HIV DNA-copies were introduced into mouse fibroblasts by DNA-mediated gene transfer, the virus replicated and infectious virions were produced (11). Apparently, another step is required which can only be provided in human cells. It is assumed that a variety of as yet unknown cofactors cause the conversion of latent or low level infection to full blown disease. Once the immune deficiency has fully developed, opportunistic infections occur, one or more of which will eventually kill the patient.

HIV also infects other cell types such as monocytes, macrophages, hemopoietic stem cells and cells of the central nervous tissue. This explains several other symptoms observed in AIDS-patients, e.g. thrombocytopenia and severe neurological syndromes related to brain damage. The fact that an infected cell remains a potential source of virus due to the presence of genomic provirus, implies that radical treatment requires not only effective antiviral measures but also perhaps eradication of all infected cells, as is the present philosophy in the treatment for cancer: eradication of all malignant cells.

AIDS has several other aspects in common with cancer. AIDS-patients are at an increased risk for certain tumors, e.g. Kaposi's sarcoma and lymphomas (Table 1). A highly increased risk for lymphomas and other tumors occurs in many hereditary immune deficiencies as well as in transplant recipients who are treated with immunesuppressive drugs for prolonged periods of time (12-14) (Table 2). A number of different malignancies in animals are induced by retroviruses, some of which exist in a latent form, others causing immunesuppression under certain condition .

Because of the extreme complexity of the infection, spread of AIDS virus in the body and its cytopathogenic action , the development of vaccines and treatment modalities cannot proceed well without the use

TABLE 1.

EXCESS TUMORS IN AIDS PATIENTS

KAPOSI'S SARCOMA

LYMPHOMA

BURKITT RETICULO SARCOMA

CA. MOUTH, HEAD & NECK

CLOACOGENIC CA.

(HEPATOMA)

TABLE 2.

RISK OF CANCER IN IMMUNODEFICIENCY

HEREDITARY I.D.	increased
e.g. Wiscott-Aldrich	128 x exp.
RENAL TRANSPLANT RECIPIENTS	
reticulum cell sarcoma	350 x exp.
lymphoma	35 x exp.
skin cancer	4 x exp.
other cancers	2.5 x exp.
DIALYSIS PATIENTS	
non-Hodgkin's lymphoma	26 x exp.
AUTOIMMUNE DISEASES	variable increases

of animal models. Recently, two different virologically transmitted diseases have been discovered in rhesus monkey which induce diseases very similar to AIDS. Obviously, experiment, with these agents in rhesus monkeys require animals that are free of infectious diseases. The importance of pathogen-free monkeys for the study of other immune-deficiency conditions has been demonstrated previously (15).

*) Per December 1986 according to WHO-figures.

REFERENCES

1. May RM, Anderson RM: Nature 326, 1987.
2. Fauc AS: Proc. Natl. Acad. Sci. USA, 83, 1986.
3. US Public Health Service, 101, 1986.
4. Barré-Sinoussi F, Chermann JC et al: Science 220, 1984.
5. Gallo RC, Salaluddin SZ et al: Science 224, 1984.
6. Levy JA, Hoffman AD: Science 225, 1984.
7. Gallo RC: Scientific Am. 256, 1987.
8. Kanki PJ, Kurtl R: The Lancet 1330, 1985.
9. Kanki PJ, Alroy J, Essex M: Science 230, 1985.
10. Maddon PJ, Dalgleish AG et al: Cell 47, 1986.
11. Levy JA, Cheng-Mayer C et al: Science 232, 1986.
12. Penn I: Clin. Exp. Immunol. 46, 1981.
13. Hoover R, Fraumeni JF: The Lancet II, 1780 (1973).
14. Kimlen LJ, Eastwood JB et al: Br. Med. J. 1, 1980.
15. Balner H and Beveridge WIB(ed): Infections and Immunosuppression in Subhuman Primate . Copenhagen: Munksgaard, Denmark, 1970.

A.C. Beynen & H.A. Solleveld (Eds), New developments in biosciences: their implications for laboratory animal science, ISBN 978-94-010-7973-0
© 1988 Martinus Nijhoff Publishers, Dordrecht.

FELINE ACQUIRED IMMUNODEFICIENCY SYNDROMES

M.C.HORZINEK

1. INTRODUCTION

Virus infections with immunosuppressive consequences are well-known in veterinary medicine. Persistent as well as acute infections may temporarily suppress the host's immune reactivity, e.g. by causing necrosis of lymphoid tissue. The present review will focus on the most prominent of acquired immunodeficiency syndromes (AIDS) – namely those caused by retroviruses. Like in human AIDS, in a similar condition in macaques (SAIDS) and in equine infectious anemia there is a D-type retrovirus also in cats which is responsible for one form of the feline acquired immunodeficiency syndrome (FAIDS). The causative agent, feline T-lymphotropic retrovirus (FTLV), has been discovered by Pedersen et al. (20) only very recently. Far more information is available on infections caused by a C-type retrovirus, feline leukemia virus (FeLV), which has become the classic example of a horizontally transmitted immunosuppressive retrovirus.

2. FELINE LEUKEMIA – A FORM OF FELINE AIDS

From these remarks it is obvious that D-type retroviruses are not the only immunosuppressive representatives of the family; murine C-type retroviruses have been known to impair immune reactivity (1) and since its discovery more than two decades ago (13,14), rapid progress has been made in the elucidation of the natural history of FeLV. Its oncogenic properties were the subject of first studies (8) and an indirect immunofluorescence test was developed which allowed detection of antigen in peripheral blood cells of viremic cats (9). From detailed clinical studies it appeared that FeLV
* is transmitted horizontally from cat to cat (10,11),
* is harbored and excreted by healthy cats long before disease symptoms appear (4)
* may induce a variety of disease manifestations in addition to lymphosarcoma (2); chronic carriers frequently develop immune deficiencies which involve virtually all defense mechanisms of the host (6). This observation has led to comparisons with human AIDS (25) and to the designation feline AIDS (FAIDS; 7)

2.1. Pathogenesis of FeLV infection

Natural infection usually occurs after oronasal contact with virus-containing saliva. FeLV subsequently replicates in oropharyngeal lymphatic tissue (23) and can be demonstrated in blood monocytes some days later. These cells presumably carry the virus into the target organs, i.e. spleen, lymph nodes, gut, bladder and epithelium of salivary glands. At about the same time with the appearance of virus in the secretions it reaches the bone marrow an may be detected in peripheral blood leukocytes and in platelets. Viremia and virus excretion continue for 1 to 16 weeks (in rare cases for 18 months) in 70 to 90% of the cats. In kittens at weaning age (about 4 weeks) viremia is accompanied by general lymphadenopathy, fever and apathy. Anemia, leukopenia and thrombopenia are additional observations which may favour secondary infections leading to enterocolitis, pyothorax, septicemia, epistaxis, hematuria and melaena. Animals with these severe symptoms usually remain viremic and succumb to the infections. Most cats, however, recover within 1 to 16 weeks after FeLV infection and appear clinically and haematologically normal (22).

Virus-neutralizing antibodies are formed, viremia ceases and the cats are' protected against a new infection (5,22). However, many animals continue to harbour FeLV for some 6 to 8 months as a latent infection of the bone marrow and lymph nodes (21,24). Latency may therefore be regarded as an extension of the postviremic reconvalescence process. The immune mechanisms which eventually lead to virus elimination are cellular rather than humoral. Excretion usually stops together with the viremia but sometimes urine, saliva and tears may contain virus for some more weeks (12; Pedersen, 1984, personal communication).

2.2. Disease signs

During the early phase of the infection symptoms are usually absent. Only months or even years later, when persistent viremia has developed clinical signs appear. In persistently infected cats mortality is relentless and progressive, with about 50% succumbing per year. Depending upon the size of the population most animals will have died after three years; 83% of infected healthy pet cats died within 3.5 years compared to only 16% of the uninfected animals living in the same households (19).

There may be no other virus infection in cats where the clinical picture is as variable as that caused by FeLV. The infection is considered the principal cause of death in the domestic cat (5). The virus can cause degenerative (blastopenic) and proliferative (neoplastic) responses, the most important of the former being anemias, while lymphoid tumors predominate in the latter. In the context of this review, only conditions will be discussed which are an indirect consequence of the infection, namely those facilitated by the FeLV-induced immune deficiency.

The most important virus infections in the wake of FeLV is feline infectious peritonitis (FIP), which is caused by a coronavirus. Early observations had shown that about 40% of the cats suffering from FIP were also viremic with FeLV (2); more recent estimates report 46% (7). At times, FIP was even regarded as another expression of FeLV infection. The mechanism of this correlation has not been elucidated so far, but neither depletion of phagocytes in the blood nor depression of humoral immunity seems to play a role. It is rather the impaired cellular immune response that determines the exacerbation of a latent or sequestered FIP virus, eventually leading to the clinical picture and death 2 to 6 months after the start of persistent FeLV viremia (21).- Other viral infections which occur at high frequency (55%) in FeLV infected cats are those of the respiratory tract (7).

Also protozoan diseases such as toxoplasmosis and hemobartonellosis are intensified by FeLV infection. The same is true for cryptococcosis where about one third of the victims are latent FeLV carriers. Acute and chronic atypical or opportunistic bacterial infections occur very frequently (6), with stomatitis and gingivitis, oral ulcers (48%), recurrent nonhealing skin sores and abscesses (34%) and chronic generalized infections (57% FeLV infected), being predominant (7). Purulent otitis externa and the panleukopenia-like syndrome (a peracute enterocolitis) should also be mentioned in this context, the latter being a frequent cause of death in animals with myeloproliferative disease or hypoplastic anemia.

Most conditions have been observed in cats with reduced leukocyte counts and, consequently, diminished phagocytic activity.
However, also deficiencies in humoral immunity can be demonstrated (Pedersen 1984, personal communication).

2.3.Possible mechanisms of FeLV-induced immunodeficiency

In contrast to the human immunodeficiency virus where the target cell population has been identified and the pathogenesis – at least in part-explained, the functions governing FeLV immunosuppression are unknown so far. Hypothetic mechanisms may be one or several of the following, all of them depending upon the persistent viremia. Many infected cats have decreased levels of plasma complement which is known to play an essential role in the immune defense (15,16). Hypocomplementemia may have been caused by the circulating immune complexes – which are immunosuppressive in their own right (3). Alteration of cellular functions – e.g. inhibition of lymphocyte blastogenesis caused by the p15E protein of the FeLV envelope (18) – or depletion of the populations of lymphocytes, granulocytes and macrophages by FeLV-induced lysis may all contribute to the effects observed.

2.4. Prophylactic measures

Continuous surveillance by serologic testing for viremia (immunofluorescence, ELISA) of confined cat populations (catteries) and elimination of viremic animals has proven to be very effective in reducing the incidence of the infection (5). These measures will have no effect on the field population.

A few years ago the first FeLV vaccine has reached the market. It is an inactivated preparation produced by the Norden Laboratories, Lincoln, Nebraska, and is based on studies performed at the Ohio State University (17). Mass vaccination must show whether the convincing effects obtained in defined cat populations (about 70% protection against persistent viremia) are reproducible under the less than optimal conditions in the field.

3. DISCOVERY OF A FELINE T-LYMPHOTROPIC LENTIVIRUS

Although FeLV infection has been repeatedly compared with AIDS, the very recent description (february 1987) of a highly T-lymphotropic virus in cats may prove of greater significance for AIDS research. The new cytopathogenic virus had been isolated from buffy coat cells by cocultivation with peripheral blood leukocytes and stimulation with concanavalin A and human interleukin-2; it was named feline T-lymphotropic lentivirus (FTLV) on the basis of its morphology and the metal ion requirement of its reverse transcriptase.

In a FeLV-negative cattery where a number of cats had died and some had shown an immunodeficiency-like syndrome, only 6% of the normal cats showed serologic evidence of infection with the new agent, whereas 40% of cats with ill health were seropositive. After inoculation, kittens developed fever, leukopenia and a generalized lymphadenopathy which persisted for 5 months (20). It appears that FTLV strikes in an older age bracket than FeLV. This may be because of the long latent period between infection and the AIDS portion of the disease. The primary stage of the infection goes unnoticed in most cases or is manifested by generalized lymphadenopathy that persists for many months before abating. Much like human AIDS, the immunodeficiency stage follows lymphadenopathy; immunodeficient cats usually do not show pronounced lymphadenopathy. The alimentary tract from the mouth to the colon is most severely affected, and many cats suffer from diarrhea and subsequent emaciation. Chronic rhinitis and periodontitis are also frequent accompanying features. Some cats become very anemic, sometimes showing the picture of a myeloproliferative-type disorder characterized by pancytopenia. A frequently fatal panleukopenia-like syndrome of sudden enteritis is observed in the older (vaccinated) cats (Pedersen 1987, personal communications).

The FTLV is prevalent in the Western part of the USA; about 50% of the field serum samples from cats with oral cavity affections, chronic enteric and other types of disease have been found antibody-positive (Pedersen 1987, personal communication). The virus probably occurs world-wide; a collaborative study with Dr.Hans Lutz, Zurich, is presently being performed and involves several hundred sera from different European countries.

In view of the difficulty to obtain sufficient primates for AIDS research, the naturally occurring FTLV in cats may provide an even better model for pathogenesis studies than FeLV. In addition, FTLV should be incorporated into the list of agents whose absence in cats is required for achievement of the specified pathogen-free status.

REFERENCES

1. Ceglowski W. S., and Friedman H. (1968) Immunosuppression by leukemia virus. I. Effect of Friend disease virus on cellular and humoral hemolysin responses of mice to a primary immunization with sheep erythrocytes. J. Immun. 101, 594-603.
2. Cotter S. M., Hardy W. D., and Essex M. (1975) The association of the feline leukemia virus with lymphosarcoma and other disorders. J. Amer. Vet. Med. Assoc. 168, 449-454.
3. Day, N.K., O'Reilly-Felice, C., Hardy, W.D., Good, R.A., and Witkin, S.S. (1980) Circulating immune complexes associated with naturally occurring lymphosarcoma in pet cats. J.Immun. 126, 2363-2366.
4. Essex M., Hardy W. D., Cotter S. M., Jakowski R. M., and Sliski A. (1975) Naturally occurring persistent feline oncornavirus infection in the absence of disease. Infect. Immun. 11, 470-475.
5. Hardy W. D. (1980) Feline leukemia virus disease; in Hardy et al. (eds.) Feline leukemia virus, Elsevier/North Holland, New York.
6. Hardy W. D. (1982) Immunopathology induced by the feline leukemia virus. Springer Seminars Immunpath. 5, 75-106.
7. Hardy W.D., and Essex, M. (1986) FeLV-induced feline acquired immune deficiency syndrome. A model for human AIDS. Prog.Allergy 37, 353-376.
8. Hardy W. D., Gearing G., and Old L. J. (1969) Feline leukemia virus: occurrence of viral antigen in the tissues of cats with lymphosarcoma and other diseases. Science 166, 1019-1021.
9. Hardy W. D., Hirshaut Y., and Hess P. (1973) Detection of the feline leukemia virus and other mammalian oncornaviruses by immunofluorescence. In: Unifying Concepts of Leukemia Dutcher R.M. and Chieco-Blanchi, L. (eds.) Karger, Basel pp 773-799.
10. Hardy W. D., Hess P. W., and Essex M. (1975) Horizontal transmission of feline leukemia virus in cats. In: Comparative Leukemia Research 1973 , 67-74.
11. Hardy W. D., Old L. J., Hess P. W., Essex M., and Cotter S. (1973) Horizontal transmission of feline leukemia virus. Nature 244, 266-269.
12. Jarrett O., Golder M. C., and Stewart M. F. (1982) Detection of transient and persistent feline leukemia virus infections. Vet. Rec. 110, 225-228.

13. Jarrett W. F. H., Crawford E. M., Martin W. B., and Davie F. (1964) A virus-like particle associated with leukemia (lymphosarcoma). Nature 202, 567-569.
14. Kawakami T. G., Theilen G. H., Dungworth D. L., Munn R. J., and Beall S. G. (1967) "C" type viral particles in plasma of cats with feline leukemia. Science 158, 1049-1050.
15. Koblinsky L., Hardy, W.D., and Day, N.K. (1979a) Hypocomplementemia associated with naturally occurring lymphosarcoma in pet cats. J.Immun. 122, 2139-2142.
16. Koblinsky L., Hardy, W.D., Ellis, R., and Day, N.K. (1979b) Activation of feline complement by feline leukemia virus. Fed.Proc. 38, 1089.
17. Mathes L. E., Lewis M. G., and Olsen R. G. (1980) Immunoprevention of feline leukemia: efficacy testing and antigenic analysis of soluble tumour--cell antigen vaccine. Feline Leukemia Virus , 211-216.
18. Mathes , L.E., Olsen, R.G., Hebebrand, L.C., Hoover, E.A., and Schaller, J.P. (1978) Abrogation of lymphocyte blastogenesis by a feline leukemia virus protein. Nature 274, 687-689.
19. McClelland A. J., Hardy W. D., and Zuckerman E. E. (1980) Prognosis of healthy feline leukemia virus infected cats. Feline Leukemia Virus , 121-126.
20. Pedersen, N.C., Ho, E.W., Brown, M.L., and Yamamoto, J.K. (1987) Isolation of a T-lymphotropic virus from domestic cats with an immunodeficiency-like syndrome. Science 235, 790-793
21. Pedersen N. C., Meric S. M., Ho E., Johnson L., Plucker S., and Theilen G. H. (1984) The clinical significance of latent feline leukemia virus infection in cats. Feline Pract. 14(2), 32-48.
22. Pedersen N. C., Theilen G., Keane M. A., Fairbanks L., Mason T., Orser B., Chen C., and Allison C. (1977) Studies of naturally transmitted feline leukemia virus infection. Amer. J. Vet. Res. 40, 1523-1531.
23. Rojko J. L., Hoover E. A., Mathes L. E., Haese W. R., Schaller J. P., and Olsen R. G. (1978) Detection of feline leukemia virus in tissues of cats by paraffin embedding immunofluorescence procedure. J. Nat. Cancer Inst. 61, 1315-1321.
24. Rojko J. L., Hoover E. A., Quackenbush S. L., and Olsen R. G. (1982) Reactivation of latent feline leukemia virus infection. Nature 298, 385-388.
25. Trainin, Z., Wernicke, D., Ungar-Waren, H., and Essex, M. (1983) Suppression of the humoral antibody response in natural retrovirus infections. Science 220, 858-859.

A.C. Beynen & H.A. Solleveld (Eds), New developments in biosciences: their implications for laboratory animal science, ISBN 978-94-010-7973-0
© 1988 Martinus Nijhoff Publishers, Dordrecht.

IMMUNE DEFICIENCY SYNDROME IN RODENTS: THE NUDE RAT

H. VAN LOVEREN[1], H.J. SCHUURMAN[2], J.G. VOS[1]

1. Laboratory for Pathology, National Institute of Public Health and Environmental Hygiene, P.O. Box 1, 3720 BA Bilthoven, The Netherlands, and
2. Division of Immunopathology, Department of Internal Medicine and Pathology, University Hospital, Catharijnesingel 101, 3511 GV Utrecht, The Netherlands

INTRODUCTION

The athymic nude mutation was first recorded in 1953 in a colony of outbred hooded rats maintained at the Rowett Research Institute, Bucksburn, Aberdeen, U.K., and the symbol rnu for Rowett nude has been assigned to the recessive mutant gene (1). We have introduced the nude mutant in our institute and backcrossed it into the WAG rat strain for 12 generations. Previous studies (1-3) suggest that the athymic nude rat is in many ways very similar to the nude mouse, but the numbers of animals and parameters investigated were limited. Therefore the morphologic characteristics in the nude rat were further defined (4). We here present a survey of our studies of these nude animals during the last 10 years.

I MORPHOLOGY OF LYMPHOID AND ENDOCRINE ORGANS

At 3 weeks of age, the rudimentary thymus of the nude rat consists of undifferentiated epithelial cells and small cysts or ducts. At 7 weeks most epithelial cells have differentiated into serous and mucous acini and cysts are increased in size. Lymphocytes were not observed in the rudiment. In these aspects the morphology during postnatal development of the thymus rudiment resembles that in the nude mouse (6).

Mesenteric and popliteal lymph nodes in young (6-10 week old) nude rats exhibit a profound lymphocyte depletion of the thymus-dependent paracortex. These findings are in accordance with observations in the nude mouse (6.7). Germinal centers are only incidentally seen in the nude rat, while they were not observed in the nude mouse (6). In the spleen of nude rats the reticular framework around the central artery was virtually free of lymphocytes.

The morphology of the pituitary gland, thyroid, adrenals, ovaries, and testes in nude rats did not differ from that in their heterozygous littermates. In contrast, changes do occur in the endocrine organs of the nude mouse (8-11). For example , Pierpaoli and Sorkin (8) observed in the pituitary gland that granules were absent and that endoplastic reticulum in somatotropin producing cells of 12-week old nude mice was enlarged, whereas Ruitenberg and Berkvens (10) found a decrease in size and number of acidophilic (presumably STH) cells in 8- and 12-week old male nude mice. Also slightly reduced pituitary STH concentrations have been measured in the nude mouse, although no consistent differences in plasma STH levels were observed (12).

The absence of morphologic defects in ovary and testis may explain that homozygous nude rats of both sexes are fertile (3), while the reproductive systems of nude mice are underdeveloped (9,10) leading to poor fertility (9,1). The nude rat also showed the normal sexual difference in body weights, which has been reported to be absent in the nude mouse (9). The normal sexual dimorphism in the submandibulary salvary glands is less evident in the nude rat, while in the mouse the submadibulary glands of eihter sex are similar (9).

II IMMUNOLOGICAL CHARACTERISTICS

In the nude rats the number of blood neutrophils, eosinophils, and monocytes is increased (14). The number of lymphocytes of 3-week-old nudes was equal to that found in heterozygous littermates. In 16-week old nudes a pronounced lymphopenia was present. These data differ from results obtained in nude mice, where granulocyte counts are the same as in +/nu littermates (15), and lymphopenia is observed at a younger age (15). For serum immunogobulin estimations, it can be concluded that IgM levels in nude rats are within the normal range of heterozygous littermates. An age-related reduction occurs in the level of IgG. A similar age-related decrease of serum IgG (in particular of IgG$_1$ subclass) has been demonstrated in the nude mouse (16). This phenomenon can be ascribed to the transfer of IgG from mother to fetus via placenta and colostrum, as the data mentioned concern animals, being the offspring of heterozygous females. Skin allografts from inbred rats, rejected in approximately 10 days by +/rnu rats, were not rejected by nude animals, which confirms the results of Festing et al. (3). This indicates the absence of T-cell mediated alloreactivity. Also other measures of T-cell mediated responses are deficient. For instance, nude rats do not respond to ovalbumin or toxoid sensitization, assessed by either delayed-type hypersensitivity (ear challenge) or circulating anti-ovalbumin antibodies (17). Using the sensitive enzyme-linked immunosorbent assay (ELISA) to detect antibodies, 1:2 diluted serum samples from nude rats at day 21 after ovalbumin immunization revealed negative reactions, whereas sera of +/rnu rats could be diluted up to 30,000 fold to still yield a positive reaction. Also the antibody response to sheep red blood cells was virtually absent in nude rats (18), whereas nude mice synthesized antibodies after primary immunization with sheep red blood cells (15,19), albeit that a substantial decrease was found for 19 S antibody and a profound decrease was found for 7 S antibody (19). The ability of nude rats to maintain normal concentrations of serum IgM strongly contrasts with their absent IgM response to tetanus toxoid and ovalbumin; a contrast that is more pronounced in the nude rat than in nude mice (19). The nude rat exhibits a normal IgM-antibody response to E. coli lipopolysaccharide (LPS), being thymus-independent in the rat (17), and a normal in vitro response to the B-cell mitogen LPS. Thus it is reasonable to conclude that IgM antibodies present at normal levels in the nude rat have been produced in response to bacterial polysaccharides, probably of enteric bacterial origin (20). In vitro lymphocyte stimulation studies showed an absent response in spleen cells of nude rats to the T-cell mitogens PHA and Con A (17). Similar results were reported by Festing et al. (3). Also spleen cells from athymic mice were not stimulated _in vitro_ in the presence of these T-cell mitogens. We were not able to note any proliferative response in spleen cells from nude rats to pokeweed mitogen (PWM), which is in accord with results of Watson et al (21), as obtained in the nude mouse, but is at variance with data of Janossy and Greaves (22) and of Thurman et al. (23), who found good responses of spleen cells of nude mice upon stimulation with PWM. According to these authors, PWM is therefore considered a B-cell mitogen.

In conclusion, the immunological parameters assessed all do indicate, that the B-lymphocyte system is present in nude rats as in normal animals, while (functionally active) T-lymphocytes are absent. The thymus-dependent immune response of nude rats seem more deficient than of nude mouse.

III NATURAL CELL MEDIATED CYTOTOXICITY

Natural killer (NK) cell activity is a unique cell-mediated cytotoxicity phenomenon that requires no presensitization. It has been found to exist in man, and in various animals, including mice, rats, hamsters and guinea pigs (24-26). Natural killer cell activity, expressed by non-adherent lymphoid cell populations, were established in both euthymic +/rnu and nude rnu/rnu rats (27). Nude rats showed a significantly higher cytotoxic activity than their thymus-bearing +/rnu littermates. This phenomenon has also been observed in nude mice (27,28). Also, in thymectomized rats increased NK cell activity was reported (29). Thus, NK-cell activity is generated independently of a thymic influence. The higher levels of congenitally athymic or thymectomized rats can be related on one hand to a naturally existing mechanism compensating the deficient thymus-dependent cytotoxicity, on the other hand to the absence of a hypothetical feedback activity of the thymus on NK-cell genesis. In this respect it should be noted that NK-cell cytotoxicity may be closely related to the cytotoxic activity of T-cells with membrane expression of the gamma chain of the T-cell receptor (this cytotoxicity is not major histocompatibility-complex restricted, thus differing from that by T-cells with a functionally active alpha-beta heterodimeric T-cell receptor) (30). For athymic nude mice the functional expression of gamma gene transcripts on T-cells was reported (31).

IV MACROPHAGE ACTIVITY

Rat NK-cells aside, also rat macrophages, e.g. from the peritoneal cavity can exhibit cytotoxicity towards tumor target cells. For the development of macrophage-mediated cytotoxicity generally activation by lymphocyte products is required. In addition macrophages can express spontaneous cytotoxicity. This spontaneous cytotoxicity apparently is independent from the thymus, since it is manifested in macrophage populations in the peritoneal cavity of athymic rnu/rnu rats (27). Similar to NK activity, this spontaneous cytotoxicity is higher in rnu/rnu compared to +/rnu littermates. Macrophages from the peritoneal cavity of nude rats exhibit in addition an enhanced anti-microbial activity. For instance the ability to ingest Listeria monocytogens in vitro was significantly enhanced. Also Corynebacterium parvum elicited peritoneal macrophages of rnu/rnu origin phagocytose and lyse Staphylococcus albus at a faster rate compared to macrophages from +/rnu littermates (unpublished data). Since C. parvum, presented to evoke peritoneal macrophages, acts directly on macrophages without help of T-cells (30), we conclude from these experiments that macrophages from rnu/rnu rats show an enhanced ability to phagocytose and kill bacteria.

V RESPONSE TO INFECTIOUS DISEASES

Host-resistance to Listeria monocytogens depends on non-specific phagocytosis and on T-lymphocyte mediated cellular immune mechanisms. During the first day after infection protection is effected by natural anti-microbial activity of phagocytes. Subsequently immunity develops, in which sensitized T-cell activate macrophages to kill or inhibit the growth of the organisms (33-35). In nude rats the T-cell mediated component of resistance to L. monocytogenes is deficient, e.g. immunization does not render nude rats resistant to challenge with bacteria (36). Since the non-specific component of resistance operates at a higher level in nude rats, clearance of L. monocytogenes one or two days after infection is enhanced. By day 5 after infection, when in euthymic rats T-cell mediated reactions developed, clearance in athymic rats is lower than that in euthymic rats. The enhanced phagocytotic and lytic activity of macrophages in nude rats seems biologically very significant, since the number of bacteria required to kill the rats after infection is significantly higher in rnu/rnu compared to +/rnu rats. Athymic rats have with athymic mice in common that they show an increased non-specific defense to Listeria monocytogenes and a diminished T-cell mediated immune response (32,37,39). However, in contrast to nude rats, nude mice are not able to ultimately efficiently control and terminate an intravenous infection with Listeria; e.g. bacterial counts from the spleen of nude mice remain stable during observation periods of 3 weeks (2,8), or even 5 weeks (39). This difference between nude rats and nude mice may be explained by the fact that rats are generally more resistant to bacterial infections. Thus, in nude mice the severe deficiency in T-cell dependent resistance may result in a more dramatic change in the ultimate control of infection with Listeria monocytogenes than in nude rats.

As in true for resistance to the bacterial infection with Listeria monocytogenes, also the resistance to the infection with the parasite Trichinella spiralis is different in nude rats compares to +/rnu littermates (40). Nude rats are unable to mount IgM, IgG and IgE-class antibodies against T. spiralis after oral infection with muscle larvae. The expulsion of adult worms from the intestine is also strongly retarded in athymic rats; the numbers of worms recovered from rnu/rnu rats at days 91 after infection are approximately equal to those from +/rnu rats at days 14 after infection. During the 91-day experimental period the adult worms remained productive in rnu/rnu rats. On the other hand, the yield of muscle larvae in +/rnu rats was already maximal at day 35 after oral infection of the animals. Remarkably nude rats manifested virtually no inflammatory reaction around muscle larvae, while a strong reaction, consisting predominantly of mononuclear cells, was seen in thymus bearing littermates. Also inflammatory responses in the intestine of rats infected with T. spiralis, comprising mucosal mast cells, were absent in nude rats. These serological and parasitological findings are similar to those in athymic mice (41-44). Moderate inflammatory reactions can occur around muscle larvae of T. spiralis-infected nude mice (43), the peak response being day 28 after infection in both nu/nu and +/nu mice, indicating that the kinetics of the response was the same. Thus, also in this in vivo infection model the nude mice manifest to some extent a thymus-dependent immunity which does not occur in nude rats.

VI T-CELL DEVELOPMENT IN NUDE RATS DURING AGEING

The data mentioned thusfar concern nude animals at young ages (6-10 weeks). It is well known that nude mice generate to some extent thymus-dependent immune reactions during their lifespan (45,46). We analysed this development with aging in nude rats (47). For this study we kept the animals under strict pathogen-free conditions before analysis to exclude extensive stimulation by the environment and to increase the lifespan.

Some authors have been able to find germinal centers in the nude rat (48), although these structures are generally completely absent in the nude rats, as in nude mice (49). However, thymus-dependent areas in peripheral lymphoid organs do manifest an increased cellularity with age. Cells at these locations bear T-cell markers and may be capable of mitogen responsiveness in vitro. This T-cell reactivity develops from almost absence at 2 and 3 months to levels of about half the value in euthymic animals at 17 months. Using double marker analysis indications were obtained for changes in the T-cell population of the spleen with regard to phenotype.

There is a decrease in cells expressing a precusor T-cell phenotype (bearing the ER-4 marker) and an increase in cells with a mature phenotype (bearing the OX 19 and either ER 2 or OX 8 markers, i.e. cells with a helper /inducer or suppressor/ cytoxic phenotype). The development of cells with a mature T-cell phenotype was paralleled by the capacity of cells of mitogen responsiveness in vitro. The major mitogen-responsive cell is a T-cell with helper phenotype (with the ER-2 marker), which produces interleukin-2 after stimulation (P.

Joling, personal communication). This capacity develops with age in nude animals, as confirmed by interleukin-2 assessment in culture supernatants. These data were recently extended with immunocyto-chemical data using an antibody detecting subpopulations of T-cells that express the antigen receptor (J. Kampinga, personal communication). Such cells were absent in young nude rats, but appeared in low density in nude animals at 10 months of age or older ages. However, the presence of such did not parallel with the development of in vivo respondes to ovalbumin sensitization. In this respect it remains to be established whether nude rats develop thymus-dependent immune responses of a restricted antigen recognition repertoire during ageing.

The data on T-cell development in nude rats confirm and extend similar studies in nude mice (46,49-52). It is unknown whether an extrinsic stimulus is required for this development; local stimulation, e.g. at mucosal surfaces by food components is an attractive hypothesis in this respect (53), as T-cells were observed in peritoneal organs especially in mesenteric lymph nodes draining the intestine.

VII CONCLUDING REMARKS

Nude rats lack a functional thymus, and as a consequence they lack functional T-cell activities. They are unable to mount immune responses that depend on functional T-cells. Thymus independent responses appear to be normal. Non-specific defense mechanisms, i.e. cytotoxicity by natural killer cells, or macrophage activity develop to higher levels than in euthymic animals, apparently independently of the presence of the thymus. The activity of these cell types is however regulated by the thymus, and thus the spontaneous activity of natural killer cells and macrophages in nude rats may represent absence of a negative feedback regulation by a functional thymus. Also numbers of granulocytes, another cell type involved in non-specific defence mechanisms, are increased in nude rats.

Because of these differences in defense systems between nude rats and euthymic rats, the way these animals cope with infections are also different. In the case that the defense mechanism depends on functionally active T-cells nude rats can not cope with such an infection. Examples are the defense to the parasite <u>Trichinella spiralis</u> and the failure of nude rats to become sensitized to <u>Listeria monocytogenes</u>. However, in the case that the defense depends on non-specific resistance, such as is the case with primary <u>Listeria monocytogenes</u> infections early on after infection, nude rats actually perform better than euthymic rats.

Nude rats seem to be very similar to nude mice. However, the nude rat seems to be more severely immune deficient in T-cell dependent immune responses than nude mice are. Remarkably nude rats do not show alterations in endocrine systems which are apparent in nude mice.

In toxicology and pharmacology the most commonly used experimental animal is the rat. This species becomes increasingly important as a laboratory animal in immunologic investigations (54-57). Along with other tools in immunological investigations in the rat, the nude mutant represents a poweful tool to further unravel the immune system such as extra-thymic T-cell maturation (47), and the biologic functions of thymic components in the processing of immature lymphocytes to become mature and functionally active in antigen responses (58).

VIII <u>REFERENCES</u>

1. D. May. Rat News Lett. <u>2</u>, 14, 1977.

2. W.L. Ford, M.E. Smith. Rat News Lett <u>2</u>, 15, 1977.

3. M.F. Festing, D. May, T.A. Connors, P. Lovell, S. Sparrow. Nature <u>274</u>, 365, 1978.

4. J.G. Vos, J.M. Berkens, B.C. Kruyt. Clin. Immunol. Immunopathol. <u>15</u>, 213, 1980.

5. P. Groscurth, M. Müntener, G. Töndury. Beitr. Path. Bd. <u>154</u>, 125, 1975.

6. M. Müntener, P. Groscurth, G. Kistler. Beitr. Path. Bd. <u>155</u>, 56, 1975.

7. M.A.B. DeSousa, D.M.V. Parroth, E.M. Pantelouris. Clin. Exp. Immunol. <u>4</u>, 637, 1967.

8. W. Pierpaoli, E. Sorkin. Nature, <u>238</u>, 282, 1972.

9. J.G.M. Shire, E.M. Pantelouris. Comp. Biochem. Physiol. 47a, 93, 1984.

10. E.J. Ruitenberg, J.M. Berkvens. J. Pathol. 121, 225, 1977.

11. E.M. Pantelouris, S. Lintern-Moore. In "The nude Mouse in Experimental and Clinical Research, Eds. J. Fogg and B. Giovanella, pp. 29-49, Academic Press, New York, 1978.

12. N. Oshawa, F. Matsusaki, K. Esaki, T. Nomura. In "Proc. Ist. Int. Workshop on nude mice". Eds. J. Rijgaard and C.O. Povlsen, pp. 221-226. Gustav Fischer-Verlag. Stuttgart, 1974.

13. S.P. Flanagan. Genet.Res. 8, 295, 1966.

14. J.G. Vos, J.G. Kreeftenberg, B.C. Kruyt, W. Kruizinga, P. Steerenberg. Clin. Immunol. Immunopathol. 15, 229, 1980.

15. H.H. Wortis. Clin. Exp. Immunol. 8, 305, 1971.

16. J. Bloemen, H. Eyssen. Eur.J. Immunol. 3, 117, 1973.

17. J.G. Vos, J. Buys, P. Beekhof, A.M. Hagenaars. Am. N.Y. Acad. Sci. 320, 518, 1979.

18. E. Terada, T. Nakayama, K. Hioko, M. Saito, N. Okudaira. Exp. Animals 29, 365, 1980.

19. H. Pritchard, J. Riddaway, H.S. Micklem. Clin. Exp. Immunol. 13, 125, 1973.

20. E.M. Pantelouris. In "The nude mouse in experimental and Clinical Research", Eds. J. Fogh and B.C. Giovanella, pp. 51-73, Academic Press, New York, 1978.

21. J. Watson, R. Epstein, I. Nakoinz, P. Ralph. J. Immunol. 110, 43, 1973.

22. G. Janossy, M.F. Greaves. Clin. Exp. Immunol. 9, 483, 1971.

23. G.B. Thurmann, B.B. Silver, J.A. Hooper, B.C. Giovanella, A.L. Goldstein. In "Proc. Ist. Int. Workshop on nude mice". Eds. J. Rijgaard and C.O. Povlsen. pp. 105-117, Gustav Fischer Verlag, Stuttgart, 1974.

24. R.B. Herbermann, H.T. Holden. Adv. Cancer Res. 27, 305, 1978.

25. A. Altma, H.J. Rapp. J. Immunol. 121, 2244, 1978.

26. S.K. Dalta, M.T. Gallagher, J.J. Trentin. Int. J. Cancer 23, 728, 1979.

27. W.H. De Jong, P.A. Steerenberg, P.S. Ursem, A.D.M.E. Osterhaus

J.G. Vos, E.J. Ruitenberg. Clin. Immunol. Immunopathol. 17, 163, 1980.

28. R. Kiessling, E. Klein, H. Wigzell. Eur. J. Immunol. 5, 112, 1975.

29. G.R. Shellam. Int. J. Cancer 19, 225, 1977.

30. E.L. Reinhertz. Nature 325, 660, 1987.

31. Y. Yoshikai, M.D. Reis, T.W. Tale. Nature 324, 482, 1986.

32. E.J. Ruitenberg, L.M. Van Noorle-Jansen, W. Kruizinga, P.A. Steerenberg. Br. J. Exp. Pathol. 57, 310, 1976.

33. C. Cheers, I.F.C. McKenzie, H. Pavlov, L. Waid, J. York. Infect. Immun. 19, 763, 1978.

34. G.B. Mackanes. S.J. Exp. Med. 129, 973, 1969.

35. D.D. McGreogor, E.D. Crum, T.W. Jungi, R.G. Bell. Infect. Immun. 22, 209, 1978.

36. H. Van Loveren, G.W. Postma, D. Van Soolingen, W. Kruizinga, D.G. Groothuis, J.G. Vos. In "Proc. of the 5th Int. Workshop of Immune-deficient animals", Ed. J. Rijgaard. p.p. 108-111 Karger, Basel, 1987.

37. P. Emmerling, H. Finger, J. Bochemuhl. Infect. Immun. 12, 437, 1975.

38. P. Emmerling, H. Finger, H.Hof. Infect. Immun. 15, 382, 1977.

39. K. Takeya, S. Shimotori, T. Tanaguchi, K. Nomoto. J. Gen. Microbiol. 100, 373, 1980.

40. J.G. Vos, E.J. Ruitenberg, N. Van Basten, J. Buys, A. Elgersma, W. Kruizinga. Parasite Immunol. 5, 195, 1983.

41. A. Perrudet-Badoux, Y. Boussac-Aron, E.J. Ruitenberg, A. Elgersma. J. Parasitol. 66, 671, 1980.

42. E.J. Ruitenberg, A. Elgersma, W. Kruizinga and F. Leemstra. Immunol. 33, 581, 1977.

43. L. Gustowska, E.J. Ruitenberg, A. Elgersma. Parasite Immunol 2, 133, 1980.

44. H.K. Parmentier, E.J. Ruitenberg, A. Elgersma. Int. Arch. Allergy Appl.Immunol. 68, 260, 1982.

45. J.P. Lamelin, B. Lisowska-Bernestein, A. Matter, J.E. Ruyser, P. Vasalli. J. Exp. Med. 136, 984, 1972.

46. M.C. Raff. Nature 246, 350, 1973.

47. L.M.B. Vaessen, R. Broekhuizen, J. Rozing, J.G. Vos,

H.J. Schuurman. Scand. J. Immunol. _24_, 223, 1986.

48. J. Fossum. Transpl. Proc. _15_, 1638, 1983.

49. P. Nieuwenhuis, D. Opstelten. Am. J. Anat. _170_, 421, 1984.

50. H. Cantor, E. Simpson, V.L. Sato, C.G. Fathmann, L.A. Herzenberg. Cell. Immunol. _15_, 180, 1975.

51. H.R. McDonald, R.K. Lees, A.L. Glasbrook, B. Sordat. J. Immunol. _129_, 521, 1982.

52. M.E. Gershwin, B. Merchant, A.D. Steinberg. Immunol. _32_, 327, 1977.

53. H.J. Schuurman, J.G. Vos, R. Broekhuizen, C.J.W.M. Brandt, L. Kater. Scand. J. Immunol. _21_, 21, 1985.

54. G.J. Van Der Brugge-Gamelkoorn, M.B. Van Den Ende, T. Sminia. Cell Tissue Res. _239_, 117, 1986.

55. J.G. Hall, E. Orlans, J. Reynolds, C.J. Dean, J. Peppard, L.A. Guyre, S.M. Hobs. Int. Arch. Allergy Appl. Immunol. _39_, 75, 1979.

56. W. Van Eden, J. Holoshitz, Z. Nero, A. Frenkel, A. Klajman, I.R. Cohen. Proc. Nat. Acad. Sci. USA. _82_, 5064, 1985.

57. G.D. Majoor, P.J.C. van Breda Vriesman. Immunol. Today _7_, 98, 1986.

58. H.J. Schuurman, L.M.B. Vaessen, J.G. Vos, A. Hertogh, J.G.N. Geertzema, C.J.W.M. Brandt, J. Rozing. J. Immunol, _137_, 2440, 1986.

A.C. Beynen & H.A. Solleveld (Eds), New developments in biosciences: their
implications for laboratory animal science, ISBN 978-94-010-7973-0
© 1988 Martinus Nijhoff Publishers, Dordrecht.

ANIMAL WELFARE - COMMON INTEREST

MARIA LUUKKAINEN AND EVA HIRSJÄRVI

Biomedical research has made tremendous improvements during the last
few decades. The apparatus and methods have become more and more sophisticated, the analyses more and more precise, and the data processing has
offered new possibilities for statistical evaluation of the results.

This is all very well - there seems to be only one weak link in the
excellent experimental chain - the test animal.

There has no doubt been development in this field, too. When I (E.H.)
was beginning my research some 40 years ago, we did not much bother about
our test animals. We performed our experiments on those animals there
happened to be, without any consideration of their genetic or environmental background. The scientist of this day could not claim to have used
"white rats" and still get his results published; he has to give detailed
facts on the background of his animals.

Yet, all too many scientists still consider their test animals as
"material" like the reagents and equipment, which can be stored and taken
out when needed. A good experimenter gives appropriate care to his
apparatus and reagents, but he often believes the same type of care is
sufficient also for his test animals: good housing and feeding,
appropriate health checks.

But it is not so. An animal, as a living being, is far more sensitive
and complicated than any apparatus. The animal is sensitive to many
factors: physical, as the temperature, relative humidity, light and noise
of its surroundings, social, as the contacts with its cage-mates or man,
and psychologic, as the pleasure, fear and stress experienced in its
everyday life or caused by experimental procedures. All these factors
also affect the physiology and responses of the animals, and if not
properly controlled, may distort the results.

Therefore, the scientist should be intimately interested in his test
animals, and seek knowledge of his animals welfare.

The scientists are often prevented from considering their test
animals' welfare because of the traditional hostility between animal
rightists and scientists. On the other hand, the interests of scientists
and animal welfarists should not be divergent. The welfarists, while
accepting the need of experimentation on animals, work for more refined
methods and better conditions for the animals.

In this task the welfarist needs a sound knowledge of laboratory
animals - in fact a laboratory animal scientist would make the best
welfarist.

In the Central Finlands Association for Laboratory Animal Welfare the
scientists and welfarists have come to work together. The association was
started a few years ago by students of biology, who had seen and done

some animal experiments. Fortunately, they did not join the Animal Rightists, but sought co-operation with the scientists. Now the Association has members in various parts of Finland, professors as well as laboratory animal technicians, students and laymen. The Association has organized seminars and symposia on refined techniques and handling of laboratory animals.

The good relations between animal welfarists and laboratory animal technicians is expressed by the fact that the Association has financially supported the preparation of the Finnish Association of Laboratory Animal Technicians Slide Series on modern laboratory animal techniques. Neither the animal rightists nor scientists showed any interest in the slide series. The Central Finlands Association for Laboratory Animal Welfare has also supported the publishing of the first Finnish Guide to the Care and Use of Laboratory Animals.

For the benefits of both scientific research and laboratory animals, the hostility between the scientists and animal welfarists should be replaced by real tolerance and effective co-operation.

A.C. Beynen & H.A. Solleveld (Eds), New developments in biosciences: their implications for laboratory animal science, ISBN 978-94-010-7973-0
© 1988 Martinus Nijhoff Publishers, Dordrecht.

LABORATORY ANIMAL .... SCIENCE - SERVICE - USE
Background for a FELASA Strategy

KARL JOHAN ÖBRINK, MARGARETA WALLER and LARS WASS

This is a presentation to discuss a background for a FELASA strategy.

FELASA is a federation of laboratory animal science associations, i.e. all member associations are expected to be involved in Laboratory Animal Science (LAS). But we have a feeling that there is no consensus about what LAS is. What is it? What does LAS include?

1. WHAT DOES LABORATORY ANIMAL SCIENCE INCLUDE?

Somebody asked: "What is pharmacology?" and got the answer: "Pharmacology is what a pharmacologist does", and a pharmacologist is to us someone working in a pharmacological department or section. Likewise one could perhaps hope for a similar definition of laboratory animal science by proposing that LAS is what a laboratory animal scientist is doing. However, we will immediately find that this would not clarify the situation, because there is no definition of a laboratory animal scientist. There are many, who consider themselves involved in LAS but there are very few departments of LAS.

The reason why we have such difficulties in definitions may be that LAS is an auxiliary science that has no justification for its existence in itself. If biological or medical scientists should discontinue to need animals for their research there would be no more use for LAS.

One possible definition could perhaps be that LAS is what the members of FELASA are doing. If we accept that definition - and we think that this is what we in fact have to do - we will find that LAS will cover a very wide spectrum of activities. This is because, as we believe, the individual member associations of FELASA have somewhat different policies as fas as their member recruitment is concerned even if they have a science orientated core in common.

If we list all the different types of people that in one way or another are involved in the field of laboratory animals we will get the following two groups: The first group will include breeders, lab.animal veterinarians, curators, animal technicians, suppliers of equipment, feed and beddings etc. and lab.animal administrators, all those that we may call the "Producers", because together they produce the scientific instrument, the laboratory animal, that is used in the laboratories by the second group, the biomedical research people and their staff or the "Users". The Users are primarily and mostly only interested in getting a reliable animal model. This second group is by far the largest in number.

2. PROFILES OF THE FELASA ASSOCIATIONS

We thus have a feeling that the individual FELASA associations show different member profiles and consequently put emphasis on different activity programs. Probably the majority seem to have a "Producer" approach, not because they contain all groups in that category but because they do not contain very many users. In fact several associations seem to be led by veterinarians. Maybe LAS is sometimes even considered a veterinary disciplin? Perhaps the sentence in the final announcement to this FELASA symposium, that reads "We trust that FELASA '87 will contribute to the international stimulation of laboratory animal science in the field of (veterinary) medicine and biology", justifies such an assumption.

3. SCAND-LAS POLICY

Scandinavian Federation of Laboratory Animal Science (Scand-LAS), that we represent, seems at least in its planning to differ in this respect. It was started and ever since headed by medical people, i.e. by users. Furthermore, in our member policy we accept and welcome all categories of both producers and users and our strategy is to eventually reach a member situation with many more users than today.

What reason do we have for such a strategy? Well, the people who started the whole business of laboratory animal science and formed our associations realized that the animals used in research very often were of unknown origin and of poor quality and that the researchers were unaware of that fact or even did not bother. So a double task was undertaken - one - to produce better animals and - two - to persuade the users that they needed better animals. Even if quite some progress can be registered this work is still going on. Although many users do realize the fundamental importance of using "analytical grade" animals there are still many who let the price determine the choice of animals. Thus there is still the era when the "producers" have the initiative. However, we must continue to get the users involved and not only have them accept the concept of quality but they must also become demanding. As this will occur the direction in which LAS will move will to a large extent be guided by the biomedical science itself.

That is why we consider it so important to engage the users in the work of our associations.

4. THE REASON FOR PRESENTING THIS PAPER

FELASA today comprices all categories of people involved in the work with laboratory animals, i.e. both "Producers" and "Users". That means that we have people engaged in lab.animal science proper, people that are engaged in giving service in several ways to animal production or to animal use and we have people that are only using animals.

We have brought this matter to your attention because if we are not aware of the different approaches of our different associations and in the different groups within them, we may run into confusions and misunderstandings in our cooperative efforts. FELASA is a fragile organisation and may easily fall apart if we do not realize the many diversified activities within its frame. All groups must feel at home in FELASA and have possibilities to a progress through international cooperation. We must continue the scientific work of importance for the development of animal models used in biomedical research, and that includes improvement of education at all levels. We must continue to give professional service to the biomedical laboratories and we must promote a proper use of animals under high ethical conscience. This could be effectively achieved if we also looked for cooperation with the biomedical scientific unions of different kinds and perhaps arranged symposia and congresses jointly, e.g. with unions of physiology, pharmacology, biochemistry, immunology, microbiology and even biophysics. A success along these lines will be possible only if we take part in the international social and political work and try to influence upon the policy in lab.animal questions. That is what we think should be included in the strategy of FELASA.

Consequently FELASA must act not only every third year at a symposium but also in between to promote the ideas of LAS. That means for instance that if a standing committee is formed by the European Council to follow up the new European Convention on laboratory animals we should see to it that we are permanently represented through FELASA. But, on the other hand, that necessitates a closer work between our associations to form a jointly accepted policy. And again we want to stress that this must be founded on the understanding that LAS is a service discipline to the biological and medical sciences. Consequently the program must include not only research concerning the biology of the laboratory animal but also teaching, husbandry development, animal welfare and ethics.

From our horizon we are prepared to discuss and cooperate on that policy.

Dept of Physiology and Medical Biophysics, Uppsala University, Uppsala Biomedical Center, Box 572, S-751 23 Uppsala, Sweden; the Lab.animal dept, Uppsala Biomedical Center, and the National Board of Universities and Colleges, Stockholm, Sweden.

A.C. Beynen & H.A. Solleveld (Eds), New developments in biosciences: their implications for laboratory animal science, ISBN 978-94-010-7973-0

LABORATORY ANIMAL SCIENCE IN CZECHOSLOVAKIA

P.KLÍR

After the II.World War central breedings of laboratory animals, mainly small laboratory rodents were established in Czechoslovakia. For the increasing needs of pharmaceutical industry there was as one of the first established the breeding facility of the Research Institute for Pharmacy and Biochemistry.
Further development of scientific-research basis led in 1953
to establishing the breeding units of Czechoslovak Academy of
Sciences (ČSAV) and in 1956 to an independent enterprise of
commercial character VELAZ, and in 1957 moreover is established
the department of gnotobiology in the Institute of Microbiology
ČSAV. Laboratory animal breeding in Czechoslovakia underwent
the same developmental characteristics as in the states with
advanced scientific research and pharmaceutical industry, i.e.
from the first demands for quantity of production to high quality of animals, definition and standardization. Which means
from the conventional breeding to isolators and barrier units
with a more effective laboratory control of animals and environmental factors. By establishing specialized breeding units
in research institutes which ensured the demands of their own
workplaces and developed research of the laboratory animals
and fulfilling the demands from the part of VELAZ enterprise
there was formed the production of usual species of laboratory
animals. The production was increasing till the end of the
sixties, when in consequence of the increasing quality of
laboratory animals and new research approaches on molecular
level it slightly decreased and stabilized with small variations up to the present time (between 8-9 hundred thousand per year).
Proportions of vertebrate species used in research (1985) are:
mice 59.6, rats 29.5, rabbits 4.5, guinea pigs 4.9, hamsters
0.7, others (dogs, cats, minipigs, primates, chicken) 0.8
per cent.
At the end of the seventies in ČSAV there was created a new
project forming conditions for increasing quality and further
research development of the laboratory animals. Besides forming
new barrier units mainly for long-term holding of laboratory
rodents the Institute of Physiology and Institute of Molecular
Genetics ČSAV are responsible for the tasks concerning breeding,
and long-term holding of animals for experiments; development
of new technology and gnotobiotechnology; collections of
informations from laboratory animal science (LAS); establishing
a strain bank and cryobank of rats and mice; establishing diagnostic-reference laboratories (rats and mice); education of
staff and research development.

Research work aims to a forming and characterization of new animal models, genetic mapping and monitoring, preparing chimeras and transgenic animals, precising diagnosis of sub-clinical forms of infections diseases, optimalization and standardization of laboratory rodents nutrition, behaviour of laboratory rodents.

In the scope of the project was worked out and accepted the categorization of laboratory rats and mice (1,2) which on the one side unifies and precises genetic nomenclature and control on the other hand limits from the point of health state demands for individual categories of animals, the extent and frequency of their control. Subsequently it specifies demands for basic zoohygienical conditions and technology of breeding in indivi-dual categories. In brief the demands for categories I-IV are in table 1-4. Animal examination is performed in diagnostic-reference laboratories ČSAV.

The bank of small rodent strains is ensured in the form of frozen embryos mainly of mice and in the form of breeding nuclei in rats. The collection of 47 mice strains and 24 rat strains is included in the International Index of Laboratory Animals. The bank enables not only to get new breeding material but also further investigation in genetic mapping and monitor-ing.

The present research also includes routine preparation of chimeric mice from embryos of the same developing stage (8-cell blastomere). Preference chimeras from 8- and 4-cell blastomeres are prepared as well. After an unsuccessful transplantation of nuclei, electrofuses of blastomeres from 2-,4-,8-cell stages with enucleated oocyte or zygota were worked out (3,4).

In connection to research focused to searching new models, the classification of biological and animal model was formed (5,6,7). Biological model can be defined as a living system the study which according to specific rules permits to repro-duce and from analogy to deduce behaviour and characteristics of the original object, that is of another living system. From the definition it is evident that any living system may be a model i.e. every system as a complex of elements in mutual relations and actions existing in the living nature, i.e. in vivo, as well as living systems in vitro. Another definitions are derivable. Animal model is a living organism the study of which according to specific rules permits to reproduce and from analogy to deduce behaviour and characteristics of the original object, that is of humans or other animals species. The concept of an animal model was further precised for spon-taneous and induced animal model, genetic and naturally acquired animal model and others. Precise classification of models and model systems has, according to our opinion, a significance for extrapolation of results as well as for ethic norms and principles which are being prepared in Czechoslovakia at the present time.

Demands on searching, forming and standardizing new animal models turn out from the present needs of research and the studies of pathogenesis of serious human diseases. Spontaneous

TABLE 1.

UNDESIRABLE BACTERIA AND PARASITE FOR INDIVIDUAL CATEGORIES (I-IV)RATS AND MICE

CATEGORY	BACTERIA	PARASITE
IV III II I	Brucella sp.	Sarcoptes sp.
	E.rhusiopathie	Cestodes
	Salmonella sp.	Toxoplasma gondi
	Mycobacterium tub.	
	Y.pseudotuberculosis	
	Leptospira (some serotypes)	
	L.monocytogenes	
	Trichophyton sp.	
	Bacillus piliformis	Zoopat.protozoa
	Bordetella pronchisep.	(Spironucleus muris
	Cor.kutscheri	Giardia muris
	Strep.moniliformis	Eimeria sp.)
	Zoopath.fungi	Zoopath.nematoda
		(without spec.belong
		to category IV)
		all ectoparasites
	Pasterella sp.	Encephalitozoon cunic.
	Zoopath.mycoplasmas	Pneumocystis carinii
	Pseud.aeruginosa	Syphacia sp.
	Strep.pyogenes alfa	Aspiculuris tetraptera
	Staph.aureus	
	Proteus sp.	
	Klebsiella sp.	
	Aerobacter sp.	
	Citrobacter sp.	
	Pneumococcus sp.	

TABLE 2.

UNDESIRABLE VIRUSES FOR INDIVIDUAL CATEGORIES (I-IV)

CATEGORY

IV III II I	RATS	MICE
	LCM	LCM
	–	Ektromelie[+]
	Theiler-GD VII	Theiler-GD VII
	Sendai	Sendai
	Reo-3	Reo-3
	PVM	PVM
	Toolan-H1	MVM
	Kilham rat virus	Polyoma
	Mouse adenovirus	K virus
	RCV	Mouse adenovirus
	SDA	MHV
	LDH	LDH

[+]in unvaccinated breeding colonies only

38

TABLE 3.

V. AND VI. CATEGORY OF LABORATORY MICE AND RATS

CATEGORY V - only known microorganisms can be present with
which the animals were associated in a normal
or an arteficial way

CATEGORY VI- all microorganisms which can be ascertained by
the present accesible examination methods are
excluded

TABLE 4.

HYGIENIC AND ZOOHYGIENIC DEMANDS FOR CATEGORIES

CATEGORY	I	II	III	IV	V	VI
	b r e e d i n g f a c i l i t i e s				i s o l a t o r s	
ANIMALS source	I-V	II-V	III-VI	IV-VI	VI hysterectomy	
entry	through zone	box with filter	via I or C		hysterectomy	
exit	"	"	via A,DT,LF		via tunnel	
diet	natural	P,S	S	S	S	S
water	a	S,F,o	S	S	S	S
bedding	natural	S	S	S	S	S
cages and material	D	D or S	S	S	S	S
staff	change of dress	personal lock, D change of dress			change of dress	
AIR filtration	against dust	85 %/o 0,5 /um	99,97°/o		0,3 /um	
regulation	mechanical	automatical			mechanical	
pressure	normal	normal	>6mm wat.column.			
exchange	rat 1,4 m³/h		mouse 0,25 m³/h			
No animals	rat 10-12/m³		mouse 42 /m³			
°C	21 + 1					
RH°/o	55 + 5					
LIGHT	350 lx - 1 m from floor					
NOISE	recomm.max 60 dB					

S-sterilization (steam autoclave, radiation), P-pasteurization,
SA-peracetic acid, D-desinfection, a-acidification (2.3-2.5 pH)
f-filtration, o-ozonization, I-isolator, C-container,
A-autoclave, DT-dip tank, LF-laminar flow

mononuclear cell leukemia with lymphoid character was detected
and standardized in isohistogenic strain SD/ Ipcv rats (8).
The strain of laboratory rat with congenital hypercholesterolemia was selected and standardized (9). For cardiovascular
research are prepared RI strains from inbred strains SHR x BNlx.
At present there are 36 strains in F 16 to F 19 generations
of b x s inbreeding. Blood pressure was measured by direct

method in individual strains. The values were continual in the
scope of values between the both progenitore strains (10).

Another, nowadays, a traditional animal model used in institutes of ČSAV is ontogenetic pig model. This model is useful
especially in immunological studies because of six-layer
epitheliochorial placenta of diffuse type preventing immunoglobulin transfer from the mother to fetuses, which is convenient
situation for analyses of fetal immune reaction mechanisms
without the influence of passively transferred maternal antibodies. Newborn minipiglets are obtained by caesarian section
and reared under germ-free conditions. GF minipiglets are used
in developmental studies of immunocompetence. Also GF rats and
another techniques for rearing GF rabbits and GF chickens are
available in the Institute of Microbiology ČSAV (11).

It is necessary to mention here that the ontogenetic aspect
and multigeneration experiments are emphasized mainly when
studying physiologic specifities. We suppose the animal properties in the experiment to be predetermined by its ontogenetic development in the breeding and may be particularly influenced in sensitive life periods i.e. perinatal period,
weaning and puberty. This view becomes more important especially in studying aimed to nutrition and behaviour of laboratory
rodents.

Nutritional studies are focused to experiments with usual and
newly composed diets with the aim of optimalization and standardization of diets and their attendance. In this way hypercholesterolemic and lowered nutritional effects on the development of young after heat attendance of casein type of diet were
proved (12). As to the composition of semipurified experimental
diets attention is paid to the amount of fatty acids in diets.
In feeding the diets with a lower concentration of linoleic
acid (18:2n6) showed a slight decrease in weight gains. Feeding
hydrogenated fat resulted in decreasing litter size and increasing incidence of morphological abnormality of sperm (13).

The findings from behavioural studies are also used in breeding and holding animals. Besides the influence of the amount
of animals in cage to their physiological specifities and
health status (14) attendance was given to rat weaning. Premature weaning in the female laboratory rat on the 15th day
affected negatively some important components of their maternal
behaviour in adulthood. The pattern of this abnormal or defective
behaviour is transferred into consecutive generations (15).

LAS in Czechoslovakia is in its areas animal breeding,
husbandry and technology, animal facilities design and construction managed by own workplaces respectivelly resorts, which
as advisory organs form their commissions. The activities in
this field are evolved in ČSAV, Ministry of Health and Ministry
of Agriculture and Nutrition. Research is centrally managed and
coordinated in the scope of the planned research. Czechoslovak
Scientific Technical Society associates research workers
and ensures education of staff, seminaries, conferences and
symposia with national and international participation.

REFERENCES

1.Klír P, Bondy R, Přibylová M, Jelen P, Svoboda T, Pospíšil M, Pravenec M, Boubelík M, Čapková J: Categorization of labora- tory rats and mice on the basis of microbiological status. In XXII.Int.Symp. on Biological Models,Hrubá Skála,ČSSR, April 1984, 39.

2. Bondy R, Klír P, Přibylová M, Svoboda T, Jelen P, Pravenec M: Health quality control of laboratory animals of the Czechoslovak Academy of Sciences (ČSAV) Scand.J.Lab.Anim. Sci.14, 1987:15-24

3. Landa V: Enucleation and electrofusion of blastomeres from 2-cell mouse embryos. Gamete Res.1987(in press)

4. Landa V, Stevens LC: Transfer of nuclei from 4-,8- and 16- cell embryos into enucleated blastomeres of 2-cell mouse embryos. Nature 1987 (in press)

5. Klír P:What is a biomodel? In XIII Symp.on Biological Models Špindlerův Mlýn, 1986, 44.

6. Klír P: What is a biological model? Acta Univ.Carol.Med. 1987 (in press).

7. Klír P, Pravenec M: Models in biomedical research. In Methods in animal physiology and biomedical research,Eds. Deyl Z, Zicha J, CRC Press 1987 (in press).

8. Klír P, Svoboda T, Přibylová M, Pravenec M, Sladká M: Spontaneous mononuclear cell leukemia in Sprague-Dawley rats. 24th Scient.Meet.SOLAS,Heidelberg, Sept.9-12,1986.

9. Poledne R: Effect of diet on cholesterol metabolism in the Prague Hereditary hypercholesterolemic rat. In Nutritional effects on cholesterol metabolism. Ed.Beynen AC,Vorthuizen- Holland, 1986,112.

10.Pravenec M, Kuneš J: 1987 (unpublished data)

11.Kovářů F, Kovářů H, Fišar Z: Ontogenet ic analysis of some surface markers on pig lymphocytes using fluorescence activated cell sorter. Folia Microbicl. 30,1985,277-290.

12.Pospíšil M, Klír P: Influence of diets DOS 2bST and ST1 on the physiological properties of rats in SPF breeding colony. Physiol.bohemoslov. 34, 1985,273.

13.Haniš T, Zídek V, Klír P: Suitability of semipurified diet "SED" with different fats for laboratory rat. In XIII Symp. on Biological Models, Špindlerův Mlýn, 1986, 25.

14.Klír P, Bondy R, Lachout J, Haniš T: Physiological changes in laboratory rats caused by different housing. Physiol. bohemoslov. 33, 1984, 111-121.

15. Nováková V: Maternal behaviour in normally and prematurely weaned female laboratory rats. Physiol.bohemoslov.23,1975, 73.

A.C. Beynen & H.A. Solleveld (Eds), New developments in biosciences: their implications for laboratory animal science, ISBN 978-94-010-7973-0
© 1988 Martinus Nijhoff Publishers, Dordrecht.

ON THE ETHOLOGY OF PAIN, ITS EXPERIENCE AND EXPRESSION

J.A.R.A.M. VAN HOOFF

The question whether animals experience pain has found different answers. The question cannot be answered in as far as it deals with the aspect of subjective experience; this is unaccessible to empirical investigation, for fundamental reasons. Most scientists, however, will accept a 'postulate of homology': because sensory and neural processes and structures are organised in essentially similar ways, at least in mammals, they feel bound to assume that these animals know experiences that are homologous to ours, also with respect to the sensation of pain (20). A more modest approach is to investigate the behavioural responses to damaging influences.

In this contribution I shall view the phenomenon of pain, its behavioural manifestations and their adaptive significance from an ethological perspective. It is remarkable that pain has attracted only little attention in the ethological literature (13). In the quest for principles of behavioural organization and of the adaptive significance of behaviour in all its diversity, attention has not been drawn towards 'disrupted systems', i.e. animals injured or ill. In learning psychology the phenomenon has commanded much more attention. This is not astonishing since research concerning the conditioning of behaviour has typically used two categories of reinforcers, rewards and punishments. The latter mostly took the form of pain stimuli, namely electric shocks, administered in rather artificial experimental situations. Because of this a notion about the nature and functional significance of pain has become established, which is rather one-sided and narrow (4).

THE ROLE OF PAIN AS A CAUSE OF FEAR

Pain has always been regarded as a mechanism that warns when there is (the danger of) injury, and that releases behaviour averting the danger and preventing further damage by withdrawal and/or defense (5). The generation of fear (readiness to flee) of things or circumstances that were associated with the infliction of damage has also been regarded as an important function. Pain then figures as the unconditional stimulus in a Pavlovian conditioning scheme. This has led to the idea that fear is the conditioned response to pain. Pain causes fear and this motivates behaviour directed at the avoidance of danger (18).

In this conception of pain and its relation to fear the diversity of stimuli which in nature generate and reinforce fear has gone lost from sight. Ethologists in particular have realized that there is a variety of flight releasing stimuli which do not derive their effectiveness from an association with pain (21). Fear of predators, for instance, cannot primarily be based on pain conditioning, because the first confrontation with the supposed unconditional stimulus is often the last (3). Other mechanisms, for instance instigation by social example, must be postulated.

TWO KINDS OF PAIN

Experiments with electric shock have one more limitation: they do not lead
to injury. We know, however, that man experiences two kinds of pain. First,
there is an immediate, sharp and well localizable pain, the 'acute' or 'primary' pain, which triggers fast withdrawal reflexes. These reflexes remain
after the brain connections have been severed. So, even though we may be
aware of this pain, the reactions are not dependent on such awareness. This
implies that the existence of such reflexes in animals does not necessarily
imply that they sense pain. Later there is a less localized, slow growing
and long lasting dull pain, the 'chronic' or 'secondary' pain (5). There are
detailed ideas about the physiological mechanisms concerned. Well known is
the "gate theory" of Melzack & Wall (15), which postulates a peripheral
mechanism by which the primary pain process temporarily inhibits the transmission of secondary pain signals to central neural stations. In addition
there are central mechanisms of inhibition, and of late we know that endorphines play a role in these.

All this raises questions about the functional relations between these
phenomena. Recently Bolles & Fanselow (4) have brought the different data
within one explanatory frame-work, their so-called "perceptive-defensive-
recuperative" (PDR) model of pain, in which they distinguish three phases
of reaction to injury. In the perceptive phase pain stimuli alert the animal
to the source of pain, and the animal may learn about its characteristics.
These stimuli also immediately trigger reactions meant to escape from and
avert the danger: the defensive phase.Essential in the PDR model is the recuperative phase; now the injury really starts to hurt and this resets motivational priorities; the pain now motivates behaviour conducive to recovery
and the healing of the injury: withdrawal to a secluded and safe spot, lick-
ing of the wounds and refraining from other activities, e.g. a loss of appetite and a decreased motivation to feed.

The traditional notion is that pain strengthens the fear of the dangerous
situation and that fearfulness is thus connected with pain. The PDR conception, on the other hand, regards pain as a motivational antagonist not only
of fear, but also of aggression and other motivations. In the defensive
phase, when the organism has to deal with the situation he is involved in,
including the inherent danger, the pain hardly hurts him, if he senses it at
all. The pain is inhibited and endorphines may play a role in this. Only
after "the heat of the fight", after things have been settled, the pain
swells and now, in turn, inhibits all other motivations conflicting with the
organism's urge to 'withdraw in suffering'. A depressed and apathetic attitude and a lack of interest in all but the pain itself and the circumstances
associated with it, is characteristic for the human chronic pain patient
(16, 19). Similar behaviour occurs during illness; then it is a response to
the sensations of nausea and sickness, rather than of pain. This apathetic
behaviour undoubtedly is a biologically adaptive response, which like depres-
sions in other circumstances (e.g. when an organism repeatedly experiences
negative effects of its actions) need not be meaningless and pathological;
on the contrary: a curbing of motivations can enable the organism to sit and
wait for times in which action and initiative may prove fruitful again (25).
Similarly, pain-induced depression is to be regarded as an adaptive aspect
of the recuperative phase and not as a pathological effect.

The PDR model views pain as an element in a distinct motivational system.
Thus there is a connection with ethological theory, which sees behaviour as
a hierarchical structure of behavioural systems (13). These are separate
motivational systems, each tuned to specific functions. Each commands a more
or less extensive choice of behavioural routines and subroutines. When a par-

ticular system is activated, it inhibits other systems to a large extent, thus ensuring that its functional program can be executed without disturbance (unless a system with a much higher motivational urgency announces itself) (1).

The PDR conception has not remained undisputed (see the 'peer commentary' following the article). One objection is that fear really does increase pain sensitivity on occasion (think of the patient in the dentist's chair, cringing at the slightest touch). Then the acute, primary pain is feared. A refinement involving the distinction of primary and secondary pain is obviously needed (6). The model should be amended as follows: secondary pain generally inhibits fear and other motivations; and vice versa: strong other motivations can inhibit the perception of pain. However, the fearful expectation of a particular (primary?) pain heightens the sensitivity for that pain, and vice versa: such pain may generate fear of that pain (13).

FUNCTIONS OF PAIN BEHAVIOUR

With this model concerning the motivational structure of the pain system and its functional context in mind, we can investigate what behaviours are to be expected when the pain behaviour system is activated.

1. The primary pain elicits immediate adaptive reactions, such as withdrawal and protective reflexes (e.g. a reflex-like attack on a nearby animal which might have been responsible for the pain, i.e. "pain-induced aggression"). Such vehement reactions can easily be recognized as indications of pain.

2. In the subsequent recuperative phase, the response to secondary pain, can manifest itself in different ways.
a) The reprogramming of motor patterns to the effect that the damaged structures are spared. In principle such modifications (e.g. limping) are recognizable, and the veterinarian uses these to identify a source of pain (10, 17).
b) Withdrawal to a secluded and safe place and abstinence from all but certain most urgent activities. Apathy, depressivity and unresponsiveness are mirrored in dejected movements and posture.

It is remarkable that depressivity is not recognized as easily as an indication of pain. In everyday life pain is associated primarily with the primary reflexes, such as cringing, and with specific emotional expressions, such as screaming, yelping and moaning.

3. The latter form a functionally distinct class of pain indicators, because these are expression movements in the true sense. These are social signals, specifically evolved to inform others, in particular conspecifics, about the condition of the "sender". Whereas the condition of an animal can, in principle also be read from the behaviours of class 1 and 2, these are not "evolutionarily intended" to convey information. Since evolutionary theory tells us that signals will only evolve when the sender is benefitted by informing others about his state or intentions, the question rises what advantage an animal might have from informing others that it is suffering pain.

COMMUNICATION ABOUT PAIN

The expression of pain can have different social consequences:
1. A cry of pain can, just as a cry of fear, alarm conspecifics and thus warn them for the danger; it can also help them to learn about the characteristics of possibly dangerous situations. Even if the sender can profit no more from this, natural selection will nevertheless favour the evolution of such signals, as long as relatives of the sender profit from these; according to kin selection theory, the 'inclusive fitness' of the sender will be benefit-

ted by such 'altruistic' signal specialization (7, 11).

2. An animal may release <u>support</u> or <u>consideration</u> from its group members:

a) These may come to defend a victim of an attack, either by a hostile conspecific or by a predator.

b) During the recuperative phase some species give signals that originally
belong to the infantile repertoire; these are signals of helplessness
(such as yelping and moaning) that release parental protection and care.
Although the documentation on this is still fragmentary, there are a few
species in which even fully grown animals can release and receive such
care. Thus dolphins carry sick or wounded herd members at the sea surface, just as mother and 'aunts' would do a newly born (22). There are
exceptional examples of nearly adult handicapped monkeys in feral groups;
their plaintive calls brought group members, especially near relatives,
to collect them and carry them (2, 8). I saw hyaena's in Masai Mara wait
for a yelping pack member, limping behind, to catch up with them.

Clearly there are great differences between species in these respects.
Whether aid-seeking behaviours have evolved will depend on two factors:
1) Even simple forms of altruistic support will evolve rather in groups of
relatives (where kin-selection can operate) than in anonymous associations
such as a herd of wildebeest.
2) Extensive care-giving is to be expected in species which have developed
cooperative behaviours, for instance, in hunting and, above all, in brood
care. Such behaviour is to be expected in pack-living carnivores, such as
wolves and cape hunting dogs (cf.14) rather than in antelopes.
Some species appear to be over-sensitive and touchy in comparison with
others. Thus pigs cry blue murder when they get jammed. These are contact
animals and, particularly for the piglets, an immediate and clear signal of
pain is of great importance to prevent them from being squashed. The fact
that a wildebeest remains mute while being torn to pieces by cape hunting
dogs does not imply that it does not suffer from pain; during its evolution
the living conditions may never have made it profitable to raise an outcry.
So, expression movements of pain are closely linked to the possibility to
provoke an adaptive social response. This is also demonstrated by the fact
that the expression of pain seems to be dependent on the nature of the injury or disturbance. Thus animals such as horses and pigs remain mute when
struck by internal affections, which we may assume to be painful on the basis of the 'postulate of homology' (17).

CONCEALING PAIN

Just as the possibility of help may promote the expression of pain, there
may be circumstances promoting the inhibition of any sign that an animal is
injured or ill and, therefore, vulnerable. Thus an animal in a threatened
position of high rank might conceal strong suffering, lest its challengers
notice it. If predators look for easy prey showing signs of weakness, then
there must be a strong premium on concealment in such prey species. Systematic investigations of this aspect are still lacking, but anecdotal evidence
suggests its importance.

SOCIAL MODULATION OF PAIN EXPRESSION

Companion animals, such as dogs, are known to exaggerate fearful behaviour
if it is reinforced, for instance, by much wanted attention (12). In an analogous manner the expression of pain, and secondarily, the selective awareness of pain, can be reinforced if this expression leads to positive social
experience or to a relief from demands and obligations. In man much psycho-

genic chronic pain may be explainable in such terms (9, 16, 23, 24). Both
the expression and the tolerance of pain may thus be modified. Again little
systematic research on these processes has been done in animals, but there
are indications that particularly our companion animal, the dog, is prove
to such hypochondria, e.g. sympathetic lameness, asthmatic symptoms, anorexia (12).

THE MEASUREMENT OF PAIN

The complex relation between the supposed sensation of pain and its behavioural manifestations renders the objective measurement of pain problematic
but not impossible. The question of pain in fish offers an illustrative
example. In the Netherlands this question has become the subject of a hot
dispute between the society for the prevention of cruelty to animals and organizations of anglers. The latter expose the view that fish may not suffer
from pain, since they are cold-blooded animals.

Both organizations supported a research project intended to reveal in
which respects and to what extent angling causes suffering in fish.(26). A
hooked carp shows vehement reactions, such as jumping, struggling, head-
shaking, spitting air from the swimbladder, etc. What is the functional
significance of these reactions? A hooked carp may be severely injured in
the mouth and gill cavity. The line heavily restricts its freedom of movement. To establish what is the relative aversive value of these influences,
they were experimentally separated. In carps damage to the mouth, due to
biting a hook, evoked but mild reactions such as shaking and spitting. Often
the carp continued to forage and eat within minutes, with the hook still in
its mouth; bait avoidance was weak. When the carp was kept on a tight line
it showed vehement behaviours; these were also shown when the animal was
stimulated with the species-specific alarm substance or confined in a narrow
container. There is strong evidence that these behaviours are motivated by
fear. This condition led to strong hook avoidance. This leads to the conclusion that the fear provoking aspects of the angling process are more aversive than the pain provoking ones.

Note that not the intensity of the reaction is the decisive criterion (see
above), but rather the consequences for subsequent behaviour. They reveal
to what extent an animal takes trouble to avoid the potentially aversive
situation. In general, this criterion can be systematically operationalized
in an active avoidance conditioning set-up, in which the energy invested to
achieve avoidance is taken as a relative measure, as well as the speed with
which an association is established and the length of time it is maintained.

In an analogous fashion the degree to which normal behavioural routines
are disrupted, and the animal abstains from need reducing activities during
the recuperative phase, can be a relative measure of its suffering. Thus an
investigation about the aversive effects of the routine practice of removing
the testicles of male piglets suggested that the primary pain of the operation is largely subdued by the fear evoked by the procedure. However, the
conspicuous recuperative behaviour during more than 5 days indicates that
there must be appreciable secondary pain (27).

REFERENCES

1. Baerends GP: The functional organization of behaviour. Anim. Behav. 27, 726-38, 1979.
2. Berkson G: The social ecology of defects in primates. In Primate bio-social development. New York, Garland, 189-204, 1977.
3. Blanchard DC & Blanchard RJ: PDR theory - a psychological approach to biological questions. Behav. Brain Sci. 3, 302-3, 1980.

4. Bolles RC & Fanselow MS: Perceptual-defensive-recuperative model of fear and pain. Behav. Brain Sci. 3, 291-323, 1980.
5. Bond MR: Pain, its Nature, Analysis and Treatment. Edinburgh: Churchill Livingstone, 1979.
6. Bowsher D: Dual mechanism of pain. Behav. Brain Sci. 3, 303-4, 1980.
7. Brown, JL: The evolution of behavior. New York: Norton, 1975.
8. Fedigan LM & Fedigan L.: The social development of a handicapped infant in a free-living troop of Japanese monkeys. In Primate Bio-social Development (S Chevalier-Skolnikoff e.a.,eds) Garland,New York,205-22,1977.
9. Fordyce WE: Learning, processes in pain. In The Psychology of Pain (RA Sternbach, ed). New York: Raven, 49-72, 1978.
10. Grauvogel A: Zum Begriff des Leidens. Der prakt. Tierarzt 1, 36-44, 1983.
11. Hamilton WD: The genetic evolution of social behaviour. J. theor. Biol. 7, 1-52, 1964.
12. Hart B.: Canine Behavior. Santa Barbara: Vet. Practice Publ., 1980.
13. Hooff JARAM van: Pijn, gewaarwording en expressie: een ethologische beschouwing. Tijdschr. Diergeneesk. 110, 59-68, 1985.
14. Lawick-Goodall H van & Lawick-Goodall J van: Innocent Killers. London and Glasgow: Collins Clear-Type Press, 1976.
15. Melzack R & Wall PD: Pain mechanisms: a new theory. Science 150, 971-9, 1965.
16. Menges LJ: De psychologie van de patiënt met pijn. In Pijn bij Mens en Dier (P van Duijn et al., eds). Wageningen: Pudoc, 79-80, 1980.
17. Mickwitz G von: Schmerz und Schmerzreaktionen beim Tier. Der prakt. Tierarzt 1, 26-36, 1983.
18. Mowrer OH: On the dual nature of learning: a reinterpretation of conditioning and problem-solving. Harvard Educ. Rev. 17, 102-47, 1947.
19. Pilowsky I: Psychodynamic aspects of the pain experience. In The Psychology of Pain (RA Sternbach, ed.). New York: Raven, 203-18, 1978.
20. Putten G van: Pijn bij dieren. In Pijn bij Mens en Dier (P.van Duijn et al., eds). Wageningen: Pudoc, 12-25, 1980.
21. Salzen EA: The ontogeny of fear in animals. In Fear in Animals and Man. New York: Van Nostrand Reinhold, 125-63, 1979.
22. Siebenaler JB & Caldwell DK: Cooperation among adult dolphins. J. Mammol. 37, 126-8, 1956.
23. Sternbach RA: The Psychology of Pain. New York: Raven, 1978.
24. Sternbach RA: Psychogenic pain. Schmerz 2, 17-23, 1981.
25. Thierry B, Steru L, Chermat R & Simon T: Searching-waiting strategy: a candidate for an evolutionary model of depression. Behav. Neur. Biol. 1984.
26. Verheijen FJ: Pijn en angst bij een aan de haak geslagen vis. Biotechn. 25, 77-81, 1986.
27. Wemelsfelder F: Gedrag als mogelijke indicator voor pijn bij biggen. Zeist: IVO 'Schoonoord', 1982.

A.C. Beynen & H.A. Solleveld (Eds), New developments in biosciences: their implications for laboratory animal science, ISBN 978-94-010-7973-0
©1988 Martinus Nijhoff Publishers, Dordrecht.

PAIN IN NEONATES

B.P. Cats

> Understanding of pain in another starts with each person's
> isolated internal reference to his or her own experience
> (Bakan 1968)

Pain has a number of personal/perceptual aspects based upon
physiological, emotional, behavioral and developmental variables.
Developmental variables are probably the most important with regard to
the problems of defining, assessing and treating pain in the preverbal
human being. In this respect a premature baby differs only gradually
from a term newborn, as does a sick neonate from a healthy one,
regardless gestational age.
Although the concept of a neonate as a non-hearing, non-perceiving and
non-feeling being (which in the past has led to newborns undergoing
surgery without proper analgesia[1]) has largely been abandoned, the
evaluation of pain in neonates still lacks appropriate, easily
applicable criteria; it depends largely on clinical experience and
current analgetic therapy in neonates is mainly empirical. To a
certain degree this also applies to treatment of pain in older
children and adults: these, however, can verbally express their
experiences which has led to the development of a number of pain
inventories, scales and other assessment tools (as for instance the
pain-thermometer).
Reports on pain in neonates are extremely rare; as such the subject is
indexed in only one[2] out of the 119 books on neonatology or pediatric
surgery in our library. However, it is likely to get more attention
because investigators are increasingly focusing on the influence of
so-called "environmental factors" on neonatal development, especially in
neonates admitted to intensive care units.
Because of the lack of psychometric methods for pain assessment in the
newborn, the study of pain is restricted to the observation of
behavioral and physiological responses at different postconceptual
(p.c.) ages. Some endocrine-system responses may be useful as well.

BEHAVIORAL RESPONSES

The nervous systems of an immature baby and a term neonate differ more
from each other than the nervous systems of a term born and an adult.
Peripheral, spinal and central pathways for pain transmission and
perception may be immature at birth. This probably applies for pain
modulation as well and certainly applies for neonatal motor response

47

to painful stimuli, because myelinization of motor neurons is known to be incomplete at birth. Already in 1945 McGraw[3] demonstrated the evolution in motor responses to a pin-prick. Immediately after birth newborns react with gross body movement and occasionally reflex responses; the reaction increases in intensity during the first month and subsides during the second month. Subsequently their reactions change to 'purposeful withdrawal' of stimulated limbs around the age of one year. Anticipatory responses to avoid the pin prick stimulus do not occur until the age of one year. The phenomenon of inhibition of motor activity or "motor shut down", regularly seen in adults, is uncommon in neonates.

It is astonishing to note that, even in the seventies, suggestions were made about a possible lack of pain perception in human neonates, using the incomplete myelinization of neuronal pathways as an argument, because a substantial part of the afferent pathways involved in pain transmission are composed of unmyelinated (c) fibers. Although it may be assumed that differences exist in the processing of painful stimuli between the developing and the mature nervous system,
it seems beyond doubt that neonates, regardless gestational age at birth, do experience pain. The sociocultural and psychological factors known to modify pain perception in later life probably exert little or no influence in the neonatal period.

Evolution is also apparent in behavioral responses to pain such as facial distortion and grimacing: newborns react with eye-closure, brows-down and nasal root bulging whereas adults open their eyes widely, knot their brows and perspire. Although these responses can be measured objectively, results from studies fail to show any consistency with regard to the quantification of pain.

The most extensively studied behavioral response is without doubt the pain-cry. Auditory analysis of crying has already been described by Hippocrates and the topic of infant crying and screaming was reviewed in 1855 by Darwin. Sound spectrographic cry analysis is developed into a useful diagnostic tool in a variety of diseases[4] and has shown differences between healthy full-term and low birthweight infants. Sound spectrographic properties of pain cries in prematures of 31-36 wks. gestational age have recently been published by Thodén e.a.[5].

The impression, already described in the middle ages, that experienced "caregivers" are able to distinguish birth, pain, pleasure and hunger cries from each other has been substantiated by spectrographic analysis. Behavioral responses probably are reliable indices of emotional states only in neonates with adequate behavioral-state organization and control. Behavioral states already exist in utero but their final organization as well as circadian periodicity is developed only after birth. This development may be hampered by premature birth or perinatal illness. Because differential use of behavioral states determines a baby's possibility to communicate with his caregivers or to modulate incoming stimuli, a lacunar behavioral-state organization must inevitably lead to abnormal and/or inconsistent behavioral responses.

Physiologic responses

A number of physiologic responses may be used for the evaluation of pain (Table I).[6]

	SY ↑	PS ↑
a. Heart-rate	↑	↓
b. Respiratory rate	↑	↓
c. Blood pressure	↑	↓
d. Metabolic rate	↑	=
e. Peristalsis	↓	↑.
f. Emotional sweating	↑	=
g. Pupil size	dilatation	constriction
h. Urine production	↓	↑

Table 1: Physiologic responses occurring with sympathetic (SY↑) and
          parasympathetic (PS↑) stimulation.

These mainly originate from sympathetic nervous system stimulation but
parasympathetic responses may be encountered as well. The sympatheti-
cally mediated responses are common "acute-pain reactions", but even in
the youngest viable neonates they tend to be modified by adaptation in
case of recurrent or prolonged pain.
When using physiologic responses in pain evaluation, it has to be born
in mind that measurements of heart rate, respiratory rate, blood
pressure and metabolic reate show age-dependent alterations as well as
a physiologic intra-individual baseline variability.
The number of active sweat glands, for instance in the palm of the
hand, is influenced by emotionally arousing situations. Several
methods have been reported to measure the number of active sweat
glands, such as finger-printing with moisture-repellant solutions and
measuring galvanic skin resistance.
In 1982 Harpin and Rutter[7] were the first to identify a relationship
between emotional sweating and p.c. age. They documented increasing
palmar water loss as a result of a painful stimulus (Fig. 1) and
showed that distinct variations in palmar sweating occurred only from
the 36th p.c. week on.

Fig. 1: Recording of palmar water loss during heel prick in (a) a term
        baby aged 6 days and (b) a baby of 35 weeks' gestation aged
        4 days (reprinted with permission from Harpin and Rutter[7])

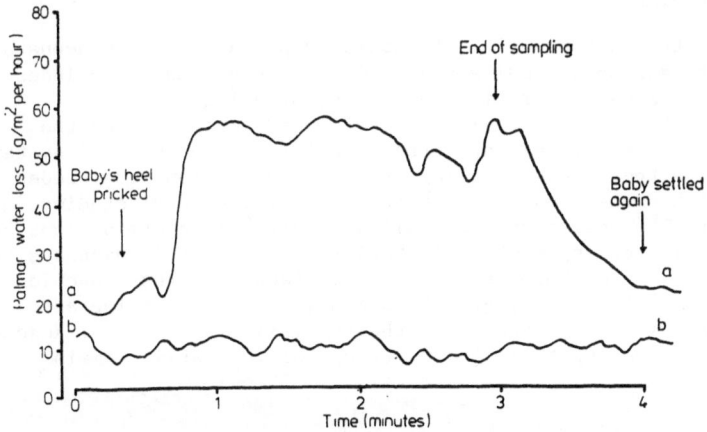

Owens and Todt, in 1984[8], showed that different arousing situations
(tactile stimulation vs a heel-lance) were followed by significantly
differing responses in heart rate and crying (Fig. 2).

Fig. 2: Mean heart rate and
percentage of crying
following different
arousing situations
(A = baseline, B =
tactile stimulation,
C = heel lance)
(reprinted with
permission from Owens
and Todt[8])

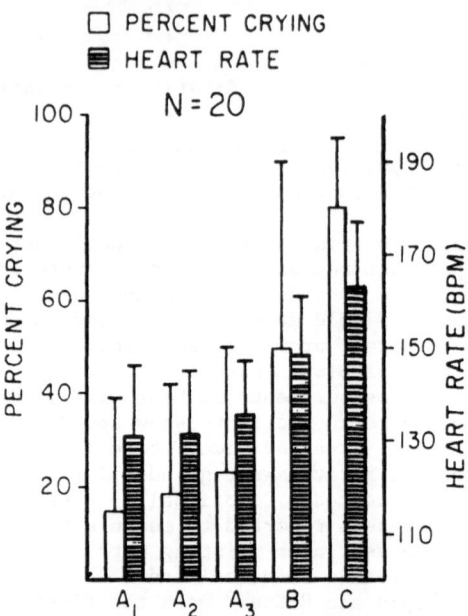

Changes in physiologic responses proportional to varying degrees of one
and the same arousing situation have never been demonstrated. Because a
variety of stimuli (such as handling, sucking, noise, non-nutritive
sucking etc.) induce changes in physiologic responses similar to those
occurring during pain, these responses can only serve as adjuncts in the
evaluation of pain.

Endocrine responses

Pain is one of the most powerful stressors, especially in the neonatal
period. Because adrenocortical activity is commonly used as an index of
stress, an increasing number of researchers, studying the
(inter)relationship between nociceptive stimuli, behavior, and the
endocrine system, have focussed on the pituitary-adrenocortical system.
Until recently either blood sampling or urine collection were needed for
measurement of hormone production, hampering developmental studies in
human neonates. The development of salivary cortisol determinations has
resulted in an increasing number of studies carried out in neonates and
infants. However, other hormones involved in (adult) stress-reaction
patterns - such as thyroxine, growth hormone, (nor)epinephrine and
insulin - cannot yet be determined without invasive procedures. Research
on pain, stress, and coping in the newborn therefore still remains
lacunar.

Conclusions

Evaluation of pain and analgesic therapy in the preverbal human being is
mainly based on clinical experience. All indicators of pain currently
available are non-specific: they signal distress-states caused by pain
as well as by other nociceptive stimuli. This also applies for
laboratory methods/biochemical determinations. The minute research in
this area must be seen as a consequence of the lack of appropriate
analytical tools rather than a result of the rarity of pain in infants.
At present the most promising adjuncts for evaluation seem the
determination of palmar sweating, registration of pain-induced
vocalization and salivary-cortisol measurements; additional instruments
may still be necessary to expand our knowledge about pain in the
preverbal human-being. Without doubt incomplete knowledge of the
behavioral evolution of sick vs healthy, term vs preterm neonates
restricts optimalization of pain evaluation and treatment as well as
it's research.

References

1. Swafford L.E., Allan D.: Pain relief in the pediatric patient. Med.
   Clin. N. Amer., 1968, 52, 131-136.
2. Holmes D.L., Reich J.N., Pasternak J.F.: The development of infants
   born at risk. Lawrence Erlbaum Ass., Hillsdale/London, 1984.
3. McGraw M.B.: The Neuromuscular Maturation of the Human Infant.
   Hafner, New York, 1945.
4. Michelsson K., Raes J., Thodén C.J. and Wasz-Höckert O.: Sound
   spectrographic cry analysis in neonatal diagnostics. An evaluative
   study. Journal of Phonetics, 1982, 10, 79-88.
5. Thodén C.J., Järvenpää A.L. and Michelsson K.: Sound spectrographic
   pain cry in prematures. In Infant Cry: Theoretical and research
   prospectives. Academic Press, New York, 1982.
6. Jacox A.K.: Assessing pain. Am. J. Nurs. 1979, 79, 895-900.
7. Harpin V.A., Rutter N.: Development of emotional sweating in the
   newborn infant. Arch. Dis. Child. 1982, 57, 691-695.
8. Owens M.E., Todt E.H.: Pain in Infancy: Neonatal Reaction to a Heel
   Lance. Pain 1984, 20, 77-86.

A.C. Beynen & H.A. Solleveld (Eds), New developments in biosciences: their
implications for laboratory animal science, ISBN 978-94-010-7973-0
© 1988 Martinus Nijhoff Publishers, Dordrecht.

ENDORPHINS AND PAIN

JAN M. VAN REE

1. INTRODUCTION

It is well known for many centuries already that opium, derived from
the poppy plant, causes pain relief and euphoria. This latter is thought
to be an important factor in the addictive properties of opium (Van Ree,
1979). After the discovery of morphine as the most effective analgesic and
addictive component of opium, attempts were made to prepare morphine related drugs in order to separate the desired analgesic and the undesirable
addictive properties. Although these attempts were not very succesful, the
structure activity relationship studies have revealed the concept that
specific opiate receptors are present in the body. In 1973 binding studies
with brain tissue indeed have suggested the existence of such receptors
(Terenius, 1973). Subsequently the presence of endogenous substances that
can activate these receptor systems was postulated. This suggestion accords
well with findings showing that animals and humans have to a certain extent
control over pain sensation. It was demonstrated in 1975 that brain tissue
contains two pentapeptides, called enkephalins, that have morphine-like
properties as assessed using isolated tissue preparations in vitro (Hughes
et al., 1975). Since then several peptides with morphine-like action have
been isolated from brain and other tissue. These substances are called endorphins (endogenous morphine). Soon after their discovery these peptides
have been implicated in pain related mechanisms, in chronic pain and
various psychopathological disorders such as psychosis, depression, mania
and addiction.

2. ENDORPHIN SYSTEMS

The several presently known endorphins belong to three major families,
arising from three distinct systems. Each system contains a big precursor
molecule of about 240 amino acids. Enzymatic processing can release from
this molecule peptide fragments with a certain biological function. These
fragments are however also precursor molecules for smaller peptides with
other biological activity. This may illustrate the neuropeptide concept,
that enzymatic processing of peptide molecules can evoke specific information enclosed in the molecule (De Wied, 1977). The three endorphin systems are designated as pro- opiomelanocortin, pro-enkephalin and pro-
dynorphin.

Pro-opiomelanocortin is located predominantly in the pituitary, but
also in neuronal pathways in the brain (Watson et al., 1979). From the
basal hypothalamus (nucleus arcuatus) and the brain stem (nucleus tractus
solitarius) these pathways spread to several structures of the limbic system and to the brain stem. In addition pro- opiomelanocortin is present in
the gut and peripheral nerves. Enzymatic processing can release from pro-
opiomelanocortin the hormones β-lipotropin (β-LPH) and adrenocorticotropin
(ACTH). β-LPH is further processed to the opioid peptide β-endorphin, which

is the precursor molecule for at least two opioid peptides, α- and γ-endorphin. The structure of the N-terminal part of these opioid peptides is identical to that of Met -enkephalin, but this enkephalin is not derived from these peptides.

Pro-enkephalin is located peripherally (e.g. in the adrenal medulla) and in widely distributed short neuronal pathways in the brain (Watson et al., 1979). Peptides belonging to this system have been demonstrated among others in the basal ganglia, hypothalamus, midbrain and spinal cord. Pro-enkephalin is processed to several opioid peptides, including the penta-peptides Met- and Leu-enkephalin, but also to longer sequences.

Pro-dynorphin is present in the hypothalamic-hypophyseal neuronal path-way, but also in other parts of the brain, such as the hippocampus and the spinal cord (Watson et al., 1981). The N-terminal part of the sequence of the opioid peptides derived from pro-dynorphin e.g. dynorphin and α- and β-neo-endorphin, is identical to that of Leu-enkephalin.

Several subpopulations of opioid receptors have been proposed. Some information is available that the opioid peptides belonging to a certain family can activate a specific subclass: β-endorphin may activate the μ, but also the δ receptor, enkephalins may activate the δ and to some extent the μ receptor, while dynorphin may activate predominantly the κ receptor (Ward, 1982). However, the exact relationship between these subclasses of opioid receptors and the physiological roles of the different opioid pep-tides have sofar not clearly been elucidated.

3. OPIATE EFFECTS OF ENDORPHINS

β-Endorphin is the most potent peptide in mimicking morphine-like ac-tion, at least when injected into body fluids. The other endorphins are also active, but more of these peptides is needed, presumably because of the more rapid degradation of the shorter endorphins. Injection of β-endor-phin into the cerebrospinal fluid of animals leads to a variety of behav-ioral changes which are also observed after administration of morphine, Thus, β-endorphin induces antinociception, hypothermia, hormonal changes, excessive grooming, and at higher dose levels profound immobilization and muscular rigidity. Most if not all of these effects are blocked by the specific opiate antagonists naloxone and naltrexone. β-Endorphin also shows common actions with morphine-like drugs after repeated administration. Thus, tolerance develops to the effect of β-endorphin on pain perception following repeated treatment (Van Ree et al., 1976). Chronic administration of β-endorphin induces physical dependence, characterized by withdrawal symptoms (Wei and Loh, 1976). Low doses of β-endorphin causes self-injec-ting behavior, indicating inherent addiction properties (Van Ree et al., 1979). Most if not all of these effects are however elicited by injecting rather high doses as compared to the amount of β-endorphin available in the brain, and can therefore be considered as pharmacological rather than physiological actions.

4. ENDORPHINS AND PAIN

Opioid receptors and various endorphins are present at different levels of the spinal cord and in brain pathways, activated by painful stimuli and involved in pain detection, pain sensation, tolerance to pain and the res-ponse to pain. The physiological role of endorphins can be tested by using opiate antagonists like naloxone, which block the opiate action of endor-phins, or specific antisera to endorphins which inactivate the physiologi-cally available endorphins. The studies with naloxone in both animals and humans do not indicate a major role of endorphins in pain perception. In general naloxone did not affect the pain threshold of animals and humans

without pain. However, when pain is present certain effects of naloxone
have been described, which may suggest that pain may activate endorphin
systems. Evidence is available suggesting that endorphins may be more concerned in motivational processes implicated in pain sensation and pain
tolerance and the integrated response to pain than in pain perception.

Several procedures exist to treat clinical pain. Most of them were
developed before information about endogenous systems involved in pain
control became available. Evidence has been presented for multiple endogenous analgesia systems (Watkins and Mayer, 1982). Both neural and hormonal pathways and both opioid and nonopioid substances play roles in the
complex modulation of pain transmission. These various systems can be selectively activated by different environmental manipulations. For example,
the neural-opioid system can be activated by morphine treatment and the
hormonal-opioid system by stress, classical acupuncture and variants of
acupuncture e.g. the twitch procedure in horses which calms down the animals and activates a pain-decreasing mechanism (Lagerwey et al., 1984).
Electrical brain stimulation can activate both neural-opioid and neural-
nonopioid systems, while transcutaneous nerve stimulation can enhance
activity in opioid or nonopioid systems, depending on the type of stimulation (low frequency, high intensity and high frequency, low intensity
respectively). That environmental manipulation activates opioid systems is
evidenced by increased levels of endorphins in plasma and/or cerebrospinal
fluid and by the blocking action of opiate antagonists. In general the
effects of activation of the hormonal systems can be prevented by removal
of the pituitary.

There are some reports showing that certain pathological conditions are
associated with pain as well as a deficiency in endorphin systems both
centrally and peripherally. Low levels of endorphin-like material have
been found in patients suffering from chronic pain, migraine or neurogenic
pain. However, more detailed studies are needed before definite conclusions
can be drawn about the involvement of endorphins in pathological conditions
related to pain.

5. ENDORPHINS AND EUPHORIA

Opium and morphine cause pain relief, but also enhance euphora, that is
thought to be a critical factor in the human abuse of opiates. Euphoria
is a feeling of well being. Physiologically such a feeling is present
during and/or after a delicious meal, enjoyable social contacts, a good
performance and sexual contacts, including orgams, among others. There are
some suggestions that endorphins are involved in physiological euphoria.
Release of hypothalamic β-endorphin has been reported when rats eat highly
palatable food. Manipulation of the opiate systems has marked effects on
social interactions (Panksepp et al., 1980). Interestingly, low doses of
β-endorphin increase social contacts of rats tested in dyadic encounters
(Van Ree and Niesink, 1983). It has been suggested that endorphins may be
release during sexual activity and may participate in orgasm, one positive
reward from sexual fulfilment (Henry, 1982). Thus, endorphins, especially
β-endorphin may be implicated in behaviors, which are accompanied by
euphoria, and it may be postulated that this euphoria is at least partly
mediated by the hedonistic action of these peptides.

However, the inherent reinforcing properties of endorphins as evidenced
by their selfadministration in experimental animals, may also lead to development of dependence on behaviors probably associated with release of
endorphins. This may be the underlying mechanism of certain addictive
habits, like gambling and jogging. Moreover, the endorphins have been implicated in dependence on drugs others than opiates e.g. cocaine and

ethanol. Thus, endorphins, particularly β-endorphin, may be a critical endogenous factor in addictive behavior in general. When a large amount of endorphins is released, euphoria may be changed to ecstasy, as is observed during or after some behaviors like ritual dancing ceremonies, which may induce a trance in which euphoria and analgesia are common symptoms (Henry, 1982). It has been proposed that a continuum "suffering - pain, including mental pain - euphoria - ecstasy" functions in the brain, in which endorphins are involved. Morphine and probably endorphins shift this continuum to the right leading to antinociception and increased euphoria. This concept links pain relief and addiction. Chronic (mental) pain may facilitate the release of endorphins in order to counteract the undesirable effects of the pain. When this release continues addictive behavior may develop, characterized by repeatedly performing the behavioral act(s) which is accompanied by or evoked by the release of endorphins. Such a process may be the underlying mechanism of stereotyped behaviors observed in single housed, neck tethered sows. These animals continuously perform stereotypies which may function as an effective strategy in order to cope with the conflict-inducing housing condition. But the ultimate behavior is narrowed and resembles addictive behavior in certain aspects. Interestingly, these stereotypies are naloxone reversible, implicating endorphins in the mentioned strategy (Cronin et al., 1985).

6. SYNOPSIS

Endorphin systems are implicated in several mechanisms controlling pain, which are present in the brain as well as in the spinal cord. These endorphin systems can be activated by procedures e.g. stress, electrical stimulation and acupuncture, resulting in relief of pain. Different endorphin systems and various types of opioid receptors are present in the brain, but their distinct role in processes involved in pain is still under debate. Endorphins have also been concerned in euphoria and addiction on various drugs and habits. The endorphin-induced decreased feeling of pain and increased euphoria may be somehow related and this relation could be of importance for the development of addictive behavior.

REFERENCES

1. Cronin, G.M., Wiepkema, P.R. and Van Ree, J.M. (1985). Endogenous opioids are involved in abnormal stereotyped behaviours of tethered sows. Neuropeptides 6, 527-530.
2. De Wied, D. (1977). Peptides and behavior. Life Sci. 20, 195-204.
3. Henry, J.L. (1982). Circulating opioids: possible physiological roles in central nervous function. Neurosci. Biobehav. Rev. 6, 229-245.
4. Hughes, J., Smith, T.W., Kosterlitz, H.W., Fothergill, L.A., Morgan, B.A. and Morris, H.R. (1975). Identification of two related pentapeptides from the brain with potent opiate agonist activity. Nature 258, 577-579.
5. Lagerweij, E., Nelis, P.C., Wiegant, V.M. and Van Ree, J.M. The twitch in horses: a variant of acupuncture. Science, in press.
6. Panksepp, J., Herman, B.H., Vilberg, T., Bishop, P. and DeEskinazi, F.G. (1980). Endogenous opioids and social behavior. Neurosci. Biobehav. Rev. 4, 473-487.
7. Terenius, L. (1973). Characteristics of the "receptor" for narcotic analgesics in synaptic plasma membrane fraction from rat brain. Acta Pharmacol. Toxicol. 33, 377-384.
8. Van Ree, J.M. (1979). Reinforcing stimulus properties of drugs. Neuropharmacology 18, 963-969.

9. Van Ree, J.M. and Niesink, R.J.M. (1983). Low doses of β-endorphin increase social contacts of rats tested in dyadic encounters. Life Sci. 33, 611-614.

10. Van Ree, J.M., De Wied, D., Bradbury, A.F., Hulme, E.C., Smyth, D.G. and Snell, C.R. (1976). Induction of tolerance to the analgesic action of lipotropin C-fragment. Nature 264, 792-794.

11. Van Ree, J.M., Smyth, D.G. and Colpaert, F. (1979). Dependence creating properties of lipotropin C-fragment (β-endorphin): evidence for its internal control of behavior. Life Sci. 24, 495-502.

12. Watkins, L.R. and Mayer, D.J. (1982). Organization of endogenous opiate and nonopiate pain control systems. Science 216, 1185-1192.

13. Watson, S.J., Akil, H., Berger, Ph.A. and Barchas, J.D. (1979). Some observations on the opiate peptides and schizophrenia. Arch. Gen. Psychiatry 36, 35-41.

14. Watson, S.J., Akil, H., Ghazarossian, V.E. and Goldstein, A. (1981). Dynorphin immunocytochemical localization in brain and peripheral nervous system: Preliminary studies. Proc. Natl. Acad. Sci. USA 78, 1260-1263.

15. Wei, E. and Loh, H. (1976). Physical dependence on opiate-like peptides. Science 193, 1262-1263.

16. Wood, P.L. (1982). Multiple opiate receptors: Support for unique mu, delta and kappa sites. Neuropharmacology 21, 487-497.

A.C. Beynen & H.A. Solleveld (Eds), New developments in biosciences: their implications for laboratory animal science, ISBN 978-94-010-7973-0
© 1988 Martinus Nijhoff Publishers, Dordrecht.

EFFECTS OF EXOGENOUS PAIN RELIEVING SUBSTANCES ON EXPERIMENTAL RESULTS

A.M.L. VAN DELFT
Scientific Development Group, Organon International B.V., Oss,
The Netherlands

Pain or injury, as inflicted upon an animal in an experimental setting, will disturb its homeostatic balance and change its behaviour and its affective state. These 3 changes may have far reaching consequences upon the outcome of experiments and may require the application of proper analgesia. I will discuss these physiological consequences of injury and the effects of analgesia, while assuming that the investigator is not interested in nociception, pain or pain control in itself, but in other biological questions which apparently cannot be answered with the help of completely intact and undisturbed animals.

Injury will not only activate all peripheral and central sensory systems involved in nociception and pain, but invariably also the sympathetic nervous system and the pituitary adrenal system. In parallel with the intensity of pain the levels of catecholamines, ACTH and adrenocortical hormones increase in an attempt by the animal to re-establish homeostasis. The behavioural consequences of injury may range from a mere reflex response to avoidance, flight or escape responses or aggression. One might describe the underlying affective state of the animal in pain as anxiety, discomfort, fear, anger or suffering.

If one now tries to comprehend the consequences of pain for the outcome of an experiment, it is clear that the activation of the sympathetic nervous system results in increased activity of many functions of the body including blood pressure and blood supply, the rate of cellular metabolism, blood glucose concentration and rate of blood coagulation. Higher levels of adrenocortical hormones markedly influence carbohydrate, protein and fat metabolism throughout the body, in addition to their effects on inflamma-tion and the immune system. Adding the behavioural consequences of pain to this already impressive list, one realizes that experiments in conscious animals, in which pain is a possible factor, are always biased by secondary influences on the parameters to be studied. If, on top of this, one adds the biological variation which exists even in inbred animals to a standardized pain stimulus, the logical conclusion is that blocking the pain of an animal in an experimental situation is a prerequisite for obtaining reliable answers to scientific questions.

The next part of this discussion concerns the side effects of pain relieving procedures as used in experimental animals. We will limit oursel-ves to pain relief obtained by analgesic drugs, since other methods like electrical analgesia or acupuncture are as yet not generally accepted. Three classes of analgesics are in vogue: major analgesics of the opiate/narcotic type, α-adrenergic-type analgesics and minor analgesics of the non-narcotic salicylic type. Introduction of analgesics in an experiment will also have far-reaching consequences on several body functions outside the sensory pain system.

Most widely used in animal experimentation are the major analgesics. Tens of thousand of papers have appeared dealing with the pharmacology of these compounds, including a wide range of actions in addition to pain relief. For this presentation we will limit ourselves to some aspects of their action relevant for the discussion. Opioids typically act after binding in the body to more or less well defined receptors. These actions have usually been interpreted with respect to the participation of 3 receptors: mu, kappa and sigma. Each drug may act as an agonist, a partial agonist or an antagonist on these receptors. With respect to side effects the mu receptor is thought to mediate respiratory depression, euphoria and physical dependence; the kappa receptor mediates miosis and sedation and the sigma receptor dysphoria, hallucinations and respiratory and vasomotor stimulation. Although this classification is an oversimplification, it indicates that the range of side effects of an individual opioid analgesic is for a larger part dependent on its receptor binding characteristics. Regional distribution in the central nervous system has been correlated with different effects like anticough, respiratory depression, nausea, vomiting or disturbed vision, while peripheral receptors throughout the gastro--intestinal tract participate in the regulation of gut functions, including motility. Acute or chronic administration of opioid-type analgesics may thus evoke a wide range of side effects which include respiratory depression, cardiovascular effects, sedation, excitation, emesis, constipation, blurred vision, mydriasis, miosis, spasmolysis, vagus stimulation, anorexia, stereotypies, catalepsia, etc. Marked differences exist with respect to side effects of individual opioid drugs. Xylazine is a presynaptic α-adrenergic drug similar to the well known antihypertensive clonidine. Presynaptic α-adrenergic agonists have marked analgesic properties in addition to their cardiovascular effects. Xylazine is used in veterinary practice as an analgesic and since adreno-receptors have wide spread occurrence in the body many side effects are inherent to its use. The most well known side effects are respiratory depression, heart rate and blood pressure decrease, gastric dilatation, paralytic ileus and sometimes CNS--stimulation. Its concommittant sedative and non-addictive properties contribute to its popularity.

The minor or antiphlogistic-, antipyretic analgesics usually act locally and in the periphery and do not penetrate the central nervous system to the same extent as the major analgesics. Their use has been limited to situations where mild analgesia is indicated. Since the mechanism of action of these drugs is antagonism of plasmakinins or prostaglandins, which are abundantly released at the site of injury, the side effect profile of these drugs is largely limited to functions where these tissue hormones play a role, such as platelet aggregation, mucus secretion, local blood flow regulation, bronchoconstriction and uterine contraction. Minor analgesics in general are less efficient in deep abdominal or central pain, but in sufficiently high doses do limit operative and postoperative pain to the same extent as the major analgesics. Since the side effect profile is a lot more favourable, more research into the efficacy of these drugs in experimental situations seems warranted.

In conclusion, pain has marked consequences for normal body function in animals. Experimental procedures, which cause pain, necessarily leave the investigator with a markedly disturbed animal and a small change for a proper answer to a scientific question. Analgesics take away the pain and its unwanted effects but introduce, albeit known, side effects. In order to make a proper choice for a good analgesic, the investigator should take into account the side effect profile of a particular analgesic.

Non-narcotic analgesics deserve more attention in replacing narcotic analgesics since they are cleaner in their pharmacological profile.

REFERENCES

1. Goodman and Gilman: The pharmacological basis of therapeutics, 7th edition, McMillan, NY 1985.
2. Wamsley JU: Opioid receptors: autoradiography. Pharmacol. Rev. 35 (1983) 69-83.
3. Martin WR: Pharmacology of opioids. Pharmacol. Rev. 35 (1984) 283-323.
4. Schmitt H, Le Douarec JC and Petillot N: Antinociceptive effects of some α-sympathomimetic agents. Neuropharmacol. 13 (1974) 289-294.
5. Oliverio A, Castellano C and Puglisi-Allegra S: Psychobiology of opioids. Int. Rev. Neurobiol. 25 (1984) 277-327.
6. Manara L and Bianchetti A: The central and peripheral influences of opioids on gastrointestinal propulsion. Ann. Rev. Pharmacol. Toxicol. 25 (1985) 249-273.
7. Kuhar MJ and Pasternak GW: Analgesics. Raven Press, N.Y. 1984.

A.C. Beynen & H.A. Solleveld (Eds), New developments in biosciences: their implications for laboratory animal science, ISBN 978-94-010-7973-0
© 1988 Martinus Nijhoff Publishers, Dordrecht.

VETERINARY CARE AT THE PRIMATE CENTER TNO

P.M.C.A. VAN EERD
Primate Center TNO, P.O. Box 5815, 2280 HV Rijswijk, The Netherlands

INTRODUCTION

From a medical point of view apes and the lower monkeys are undoubtedly more closely related to man than to animals.

Their high degree of susceptibility to human diseases requires knowledge in the human medical field. The transmission of infectious diseases from attending personnel to non-human primates and vice versa, must at all times be anticipated, although in the modern concept of primate housing all precautions are taken to avoid direct contact. In some cases the disease can be most dangerous for the human host, while the monkey has only little discomfort from this disease (for example herpes B = herpes simiae). The opposite situation can also take place, herpes simplex infection in human beings give only mild symptoms (vesiculae on mucous membranes) but causes high mortality among marmosets. Therefore it is very important that people, working with apes and monkeys have an up to date knowledge of these agents and the symptoms they can give in humans and sub-human primates. In short, be aware of the possible dangerous effects for man and animals.

In contrast to their natural habitat animals in modern facilities have constant exposure to their own discharges which with the humid conditions of the indoor environment promotes the occurrence of parasitic problems and infectious diseases. On the other hand, the new outdoor housing of our rhesus monkeys could lead to the occurrence of infections which we seldom have seen in the past such as avian tuberculosis, tetanus, pseudo-tbc and strongyloides. Overcrowding is an important factor in this matter. Some of the infections listed in Tables 1 and 2 were discussed in detail.

HEALTH SURVEILLANCE PROGRAM AT THE PRIMATE CENTER TNO

- Continuous veterinary surveillance on the health status of the animals together with a regular inventory of the circulating agents by means of bacteriological and parasitological examination.
- Mantoux test every half year on all animals and attending personnel.
- Different species are to be housed separately because of differences in individual sensibility to infectious agents.
- Good and adequate quarantinefacilities for sick animals from the colony have to be present.
- In garbage removal and cleaning procedures aerosolformation is to be avoided as much as possible. The disinfection procedures are to be carried out as strictly as possible.

64

- Autopsy material and other animal material must be burnt in plastic bags.
- Children under 15 years of age and pregnant women are <u>not</u> to be admitted to the monkey rooms.
- Infectious diseases among personnel and their relatives are to be reported.
- Preventive vaccinations of:
  apes   : Diphtheria, wooping cough, tetanus, poliomyelitis (4 times) (DKTP) in the 6th, 7th, 8th, 14th month of life
  measles in the 18th month of life
  pneumovax at an age of 3 years (pneumococci)
  rhesus monkeys: measles at an age of 1 year
  occasionally tetanus (toxoid).
- Special procedures:
  - TBC: special procedures are outlined in the TBC protocol applied in the Primate Center TNO
  - After a Shigella of Salmonella infection the animals are only to be returned to the colony after three consecutive bacteriological examinations, with an interval of 1 week.

TABLE 1. Major infectious diseases of apes and monkeys.

Systemic diseases	Skin and mucous membrane diseases
Virus: Marburgvirus hemorrhagic fever Ebola virus hemorrhagic fever Yellow fever (arbovirus) Rubeola (measles) Retrovirus: SAIDS (SIV)	**Virus:** Poxvirus: Monkey pox Yabavirus Molloscum contagiosum Herpesvirus Herpes B = simiae

Bacteria:
Clostridium tetani
Streptococcus pneumomiae
Pasteurella multocida septicaemia
Mycobacterium tuberculosis
    Humane
    Avian
    Bovine
	Bacteria:
	Streptococcus
	Staphylococcus

Parasites: Toxoplasma Oesophagostoma Strongyloides	**Parasites:** Mycoses: Microsporum Trichophytum Candida (moniliasis) Mouth and intestinal tube Scabies : sarcoptes Lice

TABLE 2.   Major infectious diseases of apes and monkeys.

Enteric diseases	Respiratory diseases
Virus:	**Virus:**
Rotavirus	Respiratory syncytial virus
Coronavirus	(coryza)
Reovirus	Parainfluenzavirus type 2 and 3
Hepatitis A	Rhinovirus
Hepatitis B	Rubeola (measles)
Hepatitis nonAnonB	
Bacteria:	**Bacteria:**
Shigella	Streptococcus pneumoniae
Salmonella	Staphylococcus
E. coli	Klebsiella
Campylobacter	Bordetella
Yersinia enterocolitica and	Pseudomonas
pseudotuberculosis	Haemophilus influenza
Proteus	Mycobacterium tuberculosis
	Corynebacterium
Parasites:	**Parasites:**
Strongyloides	Pneumonyssus simicola = lung mite
Oesophagostomum	Pneumocystiscarinii
Trichuris	Strongyloides (larva migrans)
Oxyuris	Toxoplasma
Ascaris	Ascaris (larva migrans)
Balantidium spp.: ileum +colon	
Giardia SPP: jejunum + ileum, both	
not primary pathogenic for monkeys	
Entamoeba spp.: E.coli, (E.histolytica)	
:ileum, caecum, colon: primary	
pathogenic	
Trichomonas	
Candida (moniliasis)	

Diseases of the nervous system	Reproduction diseases
Virus:	**Virus:**
Picornavirus (poliomyelitis)	Rubeola (measles)
Bacteria:	**Bacteria:**
Streptococcus pneumoniae	Mycoplasma
E. coli	Leptospira
Klebsiella	Trichomonas fetus
Pasteurella	Actinomyces
Clostridium tetani	Staphylococcus
	Corynebacterium
	Yersinia pseudotbc

CONCLUSIONS

The non-human primate has been shown to be responsible for a number of health hazards, both to humans and non-humans. It is claimed with justice that man himself apart, simian primates are the most dangerous animals with which he can associate. It is apparent that non-human primates will continue to play a role in various research efforts attempting to understand human biologic mechanisms and function. Their need and continued use in research is without question. A number of recommendations are made in the pamphlet "Health surveillance program and quarantine procedures at the Primate Center TNO" to minimize the health hazards from the use of these animals in research. This pamphlet, as handed out during the workshop held at the Primate Center TNO, is available on request. If interested, please direct your request to Dr. P.M.C.A. van Eerd at the Primate Center TNO.

A.C. Beynen & H.A. Solleveld (Eds), New developments in biosciences: their implications for laboratory animal science, ISBN 978-94-010-7973-0
© 1988 Martinus Nijhoff Publishers, Dordrecht.

DEVELOPING HOUSING FACILITIES FOR RHESUS MONKEYS: PREVENTION OF ABNORMAL BEHAVIOUR

C. GOOSEN
Primate Center TNO, P.O. Box 5815, 2280 HV Rijswijk, The Netherlands

If one is to conduct research in which experimental animals are used, one must keep the animals available for the research to be conducted. The care given to the animals before, during and also after such research, is referred to as animal husbandry. Although this area has received a great deal of attention over the years, it is currently one of the areas in experimental animal science which experiences new developments. The issues raised are so fundamental that animal husbandry might indeed become a focal point of scientific interest. The central point in animal husbandry is the concern for or protection of the well-being of the animal. Well-being then is taken in a broad sense, i.e. psychological or behavioural well-being in addition to the presence of adequate food and drinking water and the absence of diseases. This point has also become a concern to the general public. In many countries legislation on the protection of laboratory animal welfare has been drafted or become effective. In the Netherlands the act on the protection of experimental animals welfare is a framework law in which specific guidelines are to be provided. The law rules that special attention be given to certain mammalian species which are often used as household pets and to primates. The drafting of rules on husbandry of experimental animals takes place under the guidance of a committee of experts working with experimental animals (Goosen et al., 1984). This paper describes how the drafting of such rules on the housing of rhesus monkeys has incited new research and what developments in the near future can be expected in the practice of husbandry.

Housing facilities commonly in use for rhesus monkeys are based on single cage housing. This system stems from facilities used for sanitation of animals imported from countries of origin. After their trapping and before arrival at the laboratory, these animals often had contracted pathogens partly from human origin. In order to prevent spreading of such diseases, animals had to be housed separately in conditions which permitted rigorous cleaning. As research interests developed and laboratory breeding of rhesus monkeys was initiated, some larger cages were added. Such cages were used for introducing one or more females to a male to permit for mating, or for rearing young animals in peer groups after weaning. These housing conditions are obviously profoundly different from the conditions in a natural environment. Since they provide relatively little stimulation and permit for much fewer types of activity by the animals, there is good reason for concern about the state of well-being to the animals.

A grave difficulty in improving the well-being of animals is that the concept of well-being is still largely intuitive and subjective. Therefore, we have no good means to determine whether or not a state of

well-being is present in the animal and what measures should be taken to
institute or protect it. One could put forward (Fox, 1984) that the
presence of abnormal behaviour is to be interpreted as an indication of
disturbed wellbeing. Adjustment of the environment would lead to therapy
or prevention of abnormal behaviour and hence to institution of a state
of wellbeing. Although this view deserves sympathy, certainly in cases of
extremely abnormal behaviour, it does not solve the problem with respect
to well-being. One can always argue that an animal, however abnormal it
behaves that it does so because it likes doing that. Besides, there are
instances of clearly abnormal behaviour in which the indication of a
state of disturbed well-being is far from convincing (Wesseling et al.,
elsewhere this volume). Conversely, if there is no indication of any
abnormal behaviour, one cannot be confident that a state of well-being is
undisturbed. A tentative way out is to leave it as a problem and to concentrate on the prevention of abnormal behaviour. It is perfectly respec-
table to rule that housing conditions for experimental animals should not
induce abnormal behaviour, unless required by the research. Abnormal
behaviour is then defined as activities not normally seen in feral animals and which persist in the animal's behavioural repertoire.

The use of the word abnormal too may give rise to confusion. If a
phenomenon is called abnormal it often also means that the phenomenon is
not understood; scientifically the adjective is then a question in
disguise. When trying to answer such questions one seeks to find out how
the abnormal activity is derived from the normal. Once this is understood
one may be inclined to conclude that the activity is not abnormal after
all. This conclusion, however, is not justified, also when the origin of
an abnormal activity is understood, it still remains abnormal.

When working along the line of the treatment or prevention of abnormal behaviour, there are different possible approaches. One is to alter
the conditions rigorously and to try to provide a near-natural environment. The other is to be more conservative and to adjust the laboratory
environment. In the Primate Center we have attempted both these
approaches. The first approach which was meant to provide a near-natural
environment was the institution of large outdoor enclosures (10x27 m)
each for a group of monkeys. The enclosures also comprised small indoor
facilities to provide some shelter in case of extreme weather conditions.
The experience extending for over a period of three years so far has been
quite promising. However, it is doubtful that such a facility solves all
problems. The solution requires much space. Moreover, for many research
purposes, a more controlled environment is necessary. The second approach
therefore, was to determine which adjustments are required to avoid
various abnormal activities. To this end, we made an inventory of the
occurrence of various abnormal activities among individuals housed under
different conditions (Goosen, 1981). The abnormal activities of interest
comprised different categories: a) stereotyped locomotion, which consists
of making locomotory movement while the animal is not looking to an end-
pont of its walking path. b) Selfdirected infantile activities, such as
digit sucking and saluting, holding a hand near or against the brow, c)
selfdirected normally social activities, in this case selfaggression,
self sexual behaviour, and selfcaress which consists of adopting postures
seen in animals submitting to grooming while gazing away gently stroking
the fur, a behaviour interpreted as pretending to be groomed. While
following an individual throughout a certain observation period, the
moment of beginning and ending of the different activities were recorded
on a continuous time basis. Observations were made in the morning during

periods varying between 20 minutes and 3 hours. The variables investigated were: a) the size of the cage (160x80x80 cm), medium (160x80x80 cm), large (150x200x200 cm) or very large (10x27x2.5 m), b) the presence or absence of one or more cage mates, and c) the age of the subject, which could be an infant, 6 months or less, juvenile, between 1 and 2½ years, and adult, over 5 years. Comparison between the different conditions was a comparison partly of the same individuals under different conditions. The numbers of individuals per group varied between 5 and 20. The number of observations per individual also varied between 5 and 30.

Results showed that stereotyped locomotion was by far the most common type of abnormal behaviour. It was shown mostly under single cage conditions, for about 200 sec per thousand seconds of observation. There was no significant difference between juveniles and adults nor between a small and a medium cage. The presence of one or more social partners decreased the amount of time spent in stereotyped locomotion to about 40 seconds per thousand seconds of observation. There were no significant differences between the presence of one or more partners nor between a medium, large or very large cages. These results indicate that contrary to expectation (Paulk et al., 1977) cage size is probably not an important variable in the occurrence of stereotyped locomotion at least not among the investigated sizes. But the occurrence of the activity is clearly influenced by the presence of absence of social partners.

With regard to the ontogeny, it should be noted that stereotyped locomotion was not shown by infants nursed by their mothers. It was shown, however, in social or singly housed juveniles at levels similar to those in social or singly housed adults. This suggests that stereotyped locomotion emerges quickly already shortly after weaning. Subsequent qualitative observations confirmed this supposition; directly after weaning levels of locomotion were high. The locomotion was similar to stereotyped locomotion in the sense that it appeared agitated and that the animal did not look to a certain point it walked to; the locomotion was not as stereotyped in motor coordination or path as it is often in older animals. Stereotyped locomotion is rarely seen among wild caught animals or lab-born animals which remained in their natal social group for several years. This suggests that the ontogenesis of stereotyped locomotion is strongly promoted by a relatively early age at which the animals as infants were weaned from their mothers. In other words stereotyped locomotion is part of an early separation response which is shown especially under certain socially restrictive conditions.

Selfdirected infantile behaviour was shown by fewer individuals than was stereotyped locomotion. The amounts of time spent in these activities were highly variable but in the mean less than that spent in stereotyped locomotion (25 seconds per thousand seconds of observation). Ontogenetically selfdirected infantile behaviour is similar to stereotyped locomotion in so far as it emerges also shortly after weaning. However, the presence or absence of a partner nor cage size influenced the amount of time spent in these activities significantly.

The occurrence of selfdirected social activities were also quite variable and the mean amounts of time spent in selfaggression then relatively small: on the average about 1 second per 1000 seconds of observation in selfsex about 5 and in selfcaress about 10. These activities were on the average not significantly influenced by the presence or absence of social partners nor by cage size. Selfdirected social activities were seen, however, in adult animals only. The ontogeny of selfdirected social activities therefore, seems to result primarily from experiencing restricted physical access to social partners rather than from early weaning.

Returning to our goal of adjusting housing conditions, the recommendations to be proposed are firstly to wean the animals at ages later than 3 to 6 months and secondly, to house the animals together.

The recommendations seem rather modest and conservative but implementation requires major changes. Current facilities cannot accomodate social housing. Besides large scale changes have the risk of large scale introduction of new problems. For instance, leaving the infants with their mothers for longer periods of time may prolong lactation and thereby possibly decrease the colony's rate of reproduction. Institution of social housing might lead to the introdcution of serious problems due to social aggression. Proposed solutions therefore, are to be tested experimentally on a reasonable scale before they can be implemented. Nevertheless, testing of such solutions will cause the field of animal husbandry to further develop in the near future.

REFERENCES

Goosen, C., van der Gulden, W., Rozemond H., Balner H., Bertens A., Boot R., Brinkert J, Dienske H., Janssen A., Lammers A., Timmermans P. (1984) Recommendations for the housing of macaque monkeys. Laboratory Animals 18 (1984) 99-102.

Fox, M.W. (1986) Laboratory animal husbandry: ethology, welfare and animal husbandry. State University of New York Press, Albany, 1986.

Paulk H.H., Dienske H. and Ribbens L.G. (1977) Abnormal behaviourin relation to cage size in rhesus monkeys. Journal of Abnormal Psychology 86: 87-92.

A.C. Beynen & H.A. Solleveld (Eds), New developments in biosciences: their implications for laboratory animal science, ISBN 978-94-010-7973-0
© 1988 Martinus Nijhoff Publishers, Dordrecht.

MICROBIOLOGICAL QUALITY ASSURANCE AND QUALITY
ASSESSMENT OF LABORATORY ANIMALS

R.BOOT,
National Institute of Public Health and Environmental Hygiene
Bilthoven, the Netherlands.

In the National Institute of Public Health and Environmental Hygiene several animal species and strains are used in various fields of study. About 90% of the 125000 rodents and rabbits annually used are bred within the Institute (1986-87) Considering each combination of a species, a strain and a source as a population, 19 populations are produced and 27 are imported, qualitively ranging from germfree to conventional. For various reasons much attention is paid to the microbiological quality of experimental animals and animal experiments. Zoonotic infections, morbidity and mortality in experimental animals, as well as more subtle interference with experimental data have promoted the introduction of specified pathogen free (SPF) animals. According to international requirements, animals used for the production of live virusvaccins and monoclonal antibodies to be used in man, must be free of all potentially pathogenic micro-organisms and viruses.

Quality assurance of laboratory animals basically is a system of preventive hygienic measures controlling all sources and routes of microbiological and virological contamination of laboratory animals, including feral animals, pet animals and biological materials. One of the critical factors for the success of the quality assurance programme is the central ordening of animals, whether coming from internal or from external sources. At least even important is to have a Quality Control Officer (QCO) who is responsible for and has control of all animals and biological materials (tumors, parasites) entering the Institute. This QCO is supported by a Quality Control Group (QCG). Specialists in the field of virology, bacteriology, mycoplasmology, parasitology and pathology of laboratory animals perform the diagnostic examinations in animals and the screening of animals and biological material. Since they are experts in the epizootiology of the various groups of agents, they can provide information for management decisions which may reduce the risk of introducing disease into colonies or limit the spread of infections. It is important that all members of the QCG use animals themselves for various fields of study, hence quality assurance is in the interest of all involved. The QCO represents the QCG if the Board of Directors, the Committee on Animal Experiments or the Head of the Department of Animal Supply have to be advised on e.g. the implementation of hygienic measures and the consequences of contaminations for breeding and the research programme.

Basic for our quality assurance programme is the allocation of animals and biologicals to one of 5 classes of microbiological status (Fig.1).

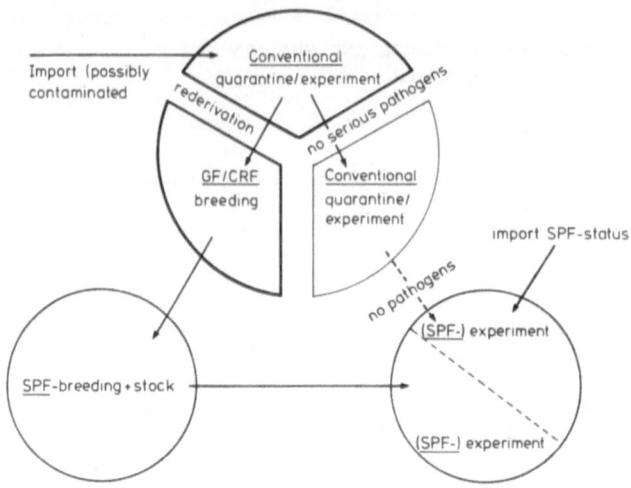

Fig.1: Microbiological Quality Assurance of laboratory animals needs separated housing and care according to microbiological status of animals and biologicals.

Most species and strains bred in the Institute are SPF animals, produced within classical SPF-type barrier systems (class II). All SPF-colonies were initially started as hysterectomy-derived germfree breeding nuclei (class I), deliberately contaminated with a strict anaerobic colonization resistant enteric microflora (CRF). Imported animals are separated into 2 groups. Animals that meet minimum microbiological requirements (SPF) are placed directly into the experimental facilities (class III), but are, if possible, kept separated from home bred SPF animals. The rationale for this approach is that quarantining and screening regular shipments from reliable sources is very laborious, yields only limited information additional to the health reports provided by the vendor and poses considerable logistic problems. Housing by source reduces the risk of mutual transmission of infectious agents that might be present. For short term experiments (e.g. up to 2 months) this approach is satisfactory if the supplier is reliable and immediately informs the customer of problems in his colony and if combined with a set of hygienic measures. To achieve workable conditions, the hygienic regimen for most experimental facilities (class III) is less strict than for SPF-breeding units, although in longterm toxicological experiments the regimen approaches the classical SPF-barrier concept.

Imported animals of unknown microbiological status and biologicals entering the Institute are presumed to be contaminated (class V) and therefore are kept strictly isolated. This means housing of animals or performing experiments with animals that are inoculated with tumorcells or parasites, in stainless steel isolator systems. Release from the absolute barrier system is only allowed if examination did not reveal the presence of pathogenic agents having zoonotic potential or probably causing considerable damage to the research programme.

Animals or biological material found to be free from serious pathogens can be housed in a "reversed" SPF type barrier-system (class IV) that is also used for experimental infections. Hygienic measures aim to keep infectious pressure on class II and III facilities as low as possible. Since the animals presumably will seldom meet SPF-requirements, introduction into class III facilities only seems a theoretical possibility.

Clearly animal technicians play a pivotal role in the quality assurance programme. Commonly they are the first to call attention to health problems in laboratory animals. Because of their intensive daily contacts with the animals, they must presumably be considered the key factor in the transmission of contaminations between animal populations. Therefore it is of fundamental importance that animals housed by status are attended by separate groups of animal technicians, as indicated by the circles in fig.1.

Assessment of the microbiological and virological quality of laboratory animals and animal experiments actually means monitoring the effect of preventive hygienic measures. If the breeding of SPF animals is considered a production proces that should yield as a final product, animals meeting high standards of quality, a separation can be made between proces-control and product-control. Examples of proces-control are the recording of physical parameters (time, temperature and pressure) during autoclaving and the determination of numbers of bacteria in diets. However with the present state of the art the contribution of each of the numerous hygienic measures taken to protect colonies is not known and hence requirements for some of the factors are not clear. For instance low numbers of bacteria in untreated diets do not exclude the introduction of contamination and high numbers do not per definition imply the presence of pathogens. Therefore more emphasis is placed on product-control, which is performed in two ways. Physiological parameters, e.g. growth and reproduction figures are used as indicators for the health of the breeding colony.

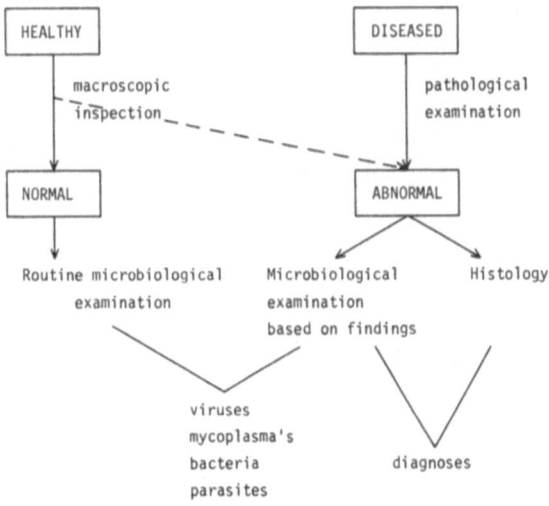

Fig.2: Microbiological and (histo)pathological examination of SPF-animals.

The microbiological, virological and (histo)pathological evaluation of these SPF-animals includes both periodic (routine) screening of healthy animals as well as the examination of diseased animals (Fig.2). Healthy animals are screened for the presence of a wide range of viruses, mycoplasma's, bacteria and parasites. Mostly these animals do not show clinical signs or gross abnormalities at macroscopic examinations. If clinical signs or lesions are suspected the animals are submitted for postmortem (histo)pathological examinations that will lead to a diagnosis or will guide further microbiological and virological examinations. Experiments, having limited risk to become contaminated are not monitored microbiologically on a routine basis. Information from short term experiments primarily comes from postmortem examination of animals showing clinical signs or dying unexpectedly. The status of long term toxicological experiments is assessed by screening sentinel animals for the presence of selected contaminating agents, postmortem examinations of these animals and by considering physiological data obtained in control-animals used in the toxicological studies.

A.C. Beynen & H.A. Solleveld (Eds), New developments in biosciences: their
implications for laboratory animal science, ISBN 978-94-010-7973-0
©1988 Martinus Nijhoff Publishers, Dordrecht.

CENTRALISATION - DECENTRALISATION: FAILURE - SUCCESS

Prof.dr. J.A.P. van Hoof

INTRODUCTION

When Wim van der Gulden invited me for a lecture at this meeting, I was
at first flabbergasted, as I was completely aware of my utter ignorance
of animal laboratories, as much as without any doubt you will be when you
heard my lecture.

I know of course that one should rather not talk about matters one does
not know about. But then, the invitor dressed his invitation so charmingly
that I could hardly resist. Moreover he used an interesting argumentation:
it might be refreshing to be exposed to the views of an outsider, more or
less in accordance with the dutch wisdom that children and foolish people
may tell truth.

Well, outsider I am, being a sociologist, more especially in the sociology of work and organisation, but without any experience or research in
animal laboratories. I had a couple of very stimulating talks in which I
could learn from the expertise of Wim van der Gulden and Marianne Kuipers,
but for the greater part I had to rely on my social sciences theory and
approach.

In sociology, the fundamental processes of organisation are considered
to be differentiation and integration, or put in other words: the division
of labor and the coordination. And that is where I started from.

What I will do in this lecture is to discuss three subjects along this
scheme:
1. Nature of work and characteristics of the operating core:
   - work
   - coordination
   - motivation
   - congruence
2. Organisation of animal laboratories:
   - internal organisation and leadership
   - external organisation and dependences
3. Culture and history.

Of course, I cannot and will not try to be exhaustive on these subjects.
I will highlight some points and leave other topics in darkness.

1. NATURE OF WORK AND CHARACTERISTICS OF THE OPERATING CORE

The operating core is of course of vital importance to any organisation: that is directly related to the production of the services.

The work of animal care-takers and animal technicians is in my opinion characterised by the following 6 properties:

- it is skilled work; it requires a specific training, ample experience and a special feeling, which cannot easily be transfered to other persons.
- the work is a very special combination of technical and social knowledge and skills.
- it is individual work, not directly connected with the performance of colleages as is the case with conveyor belts or group-work.
- it is service-work (at least the work of the technicians) in behalf of high-status researchers, with whom a relationship of mutual trust should be established; moreover the clients change over time.
- the quality of the performance cannot easily be assessed, leave alone be measured.
- within the organisation of the laboratory the operating core is in the position to spot problems and to initiate improvements in the actual work activities.

I hope this description is at least partly correct, and I will very much appreciate your comments on it.

Now the characteristics of the work have profound consequences for the kind of coordination and motivation that are needed and probably will occur in the laboratory if proper policy is followed.

As for coordination basically there are two types: direct control (or close supervision) and responsible autonomy. It may be expected that in a laboratory as I have in mind in my description of the work responsible autonomy is the most adequate type. More specifically: coordination should rely on mechanisms such as mutual adjustment and improvement of skills. Other types of coordination-mechanisms, such as standardisation of work-process, standardisation of output, direct control by supervisor are hardly applicable.

The nature of the work-task also requires a corresponding type of work motivation. As I see it the animal care-taker and animal technician should, because of the nature of his task, have a work-motivation in which the following 4 traits are well developed:

- appreciation of autonomy, which implies self-confidence and willingness to accept responsibility.
- identification with the task, by which I mean willingness to perform according to the norms of good work, without being controlled and sanctionised directly.
- a combination of service-mindedness and self-consciousness.
  On the one hand willingness to search for the best ways to meet the demands and the purposes of the client, on the other hand ability to maintain one's own insights and norms, and to deliberate on an equal footing with the client.
- social skills in the sense of ability and willingness to communicate and to establish good relations with changing partners, at the same time with also good but different relations with collegues and superiors.

Again: I would appreciate your comments on the statements. What I am trying to do actually is to apply the concept of congruence to the social system of animal laboratories. Congruence has to do with the social system character of organisations. Social system means: a whole of interdependent elements. This means that the state of one element or a change in one element affects the other elements or at least some of them One might think as a comparison of a mobile: a push or pull in one place activitates the steady state of the whole system.

Now congruence means that these interdependencies between elements are not at random: a specific change in one element requires and actually tends to involve a corresponding change in other elements. For example, when a work-task is defined in such a way that autonomy is important, then the workers should dispose of the skills needed and of a motivation to use them properly.

Congruence in social systems tends to occur because these systems move away from discongruence, which causes tensions that are expensive and ineffective. So that is why I suppose that the specific character of the work of animal care-takers and animal technicians require corresponding ways of coordination and corresponding types of motivation.

But: this is a rather deductive way of reasoning. It needs confrontation with the induction from empirical data, with facts, and these I hope to get from you.

If this treatise on congruence sounds too abstract and too theoretical to you, just skip it for the moment.

I think you are able to follow the rest of my story without the concept of congruence, and if you are interested in it, I will gladly try and repeat and explain it afterwards.

Nevertheless, I will go on using this way of reasoning in the next topic:

2. INTERNAL AND EXTERNAL ORGANISATIONS of animal laboratories.

Considering the nature of the work of the operating core and the congruence with coordination and motivation, the next step is the question of the effects of this pattern on the internal organisation. A number of 5 statements can be derived from the description above.

- leadership in a situation like this has as its main function: providing favorable conditions for the work by the operating core; these conditions imply the capacities and motivation of the workers but also the physical and social arrangements in the short run but especially on long term. Leadership should act indirectly: it advises, it gives support, it thinks ahead and provides for the necessary means and a favorable climate for its subordinates to work with.

- in the social sciences with regard to leadership a distinction is usual of social and instrumental leadership, or people-oriented and work- or task-oriented leadership.

Now to avoid misunderstanding I want to stress that social leadership does not mean a soft, submissive or permanent smile behaviour. Social leadership has to do with enabling the group and the individual workers to perform well. The opposite, instrumental leadership, is aimed at the same goal, but operates in a different way. Instrumental leadership influences the behaviour of subordinates in a direct way: by giving strict orders and instructions by maintaining close supervision and by aplying negative sanctions.

From the foregoing it may be obvious that in my opinion social leadership is required in animal laboratories, especially the centralised ones.

- The reverse of autonomy often is uncertainty and indistinctness. These two phenomena also are the reason sometimes why people refuse autonomy: it is unpredictable and dangerous.

I think that social leadership should concern itself especially with these aspects and provide for certaintly and clearness. The reason why I
stress this point is that an animal laboratory has to be flexible, has to
change according to the demands of the internal and external circumstances.
    Now one of the oldest topics in the social science of organisation is
the "resistance to change", which of course impedes flexibility. In my
opinion resistance to change often results from uncertainty and indistinc-
tion. In that case changes are threats to people or are very easily seen
as such, and of course people will then not be willing to cooperate and to
be creative. But most people are quite willing to cooperate with changes,
when these are interesting new experiences, when they are sure that the
boss will never let you down, and when they are allowed and stimulated
to learn from their mistakes and errors.
- Animal laboratories, even the centralised ones, will mostly be of a
  limited size. Together with the nature of work and leadership this re-
  inforces informal ways of interaction and organisation.
  Now informal organisation can be a powerful force in the pursuit of
  goals, but it can just as well be a powerful hindrance. Whether support
  or hindrance depends on the consensus or conflict between formal and informal organisation, or simply: between leader and the group of subordinates.
  Consensus is not a self-evidence. It has to be built and maintained very
  carefully, but it is a basic condition for coordinating activities by
  responsible autonomy.
- The organisation of the laboratory, because of its informal character,
  is very much dependent on individual persons and especially on the per-
  sonality of the leader or manager. Succes or failure are often determined by the capacities and the opinions of management.
  This should not be understood as a direct plea for the "strong man".
  The capacities and opinions of the manager I have in mind are related
  to the functions of management, especially those of motivating people,
  of providing the conditions for a good performance of work, and of maintaining the external relations of the organisation.
    My neat topic will discuss the external organisation. I should warn
  you that at this point the argument takes a turn. Whereas the internal
  organisation fairly well could be related to the characteristics of the
  core activities, the external organisation requires a different approach.
  Which goes to show that the concept of congruence cannot be stretched
  without restrictions.

    As I describe it, the centralised animal laboratory has many traits of
the type of organisation that is labeled a professional organisation.
Maybe the word "professional" sounds a bit too weighty, because of its
usual connotation with an academic training, but I think that the level
of formal education is in this respect of less importance than the nature of the task and the character of the organisation. For the sake of
distinction tel us call it a semi-professional organisation.
    Now this laboratory also is part of a larger organisation. This larger
unit itself also is a professional organisation in the sense of an
organisation where the core activities are performed by full-fledged
professionals.
    One might expect that the coexistance and cooperation between these
two professional organisations, the smaller semi-professional one and
the larger full-fledged professional organisations would hardly pose
any serious problem. Professionals tend to consider themselves as

reasonable and well-educated people and as fair colleages to other professionals.

However, from the viewpoint of the full-fledged professionals the animal laboratory is one of the service or support departments, part of the bureaucratic component of the overall organisation. These departments should be treated with suspicion, because they tend to suffer from some kind of goal-displacement: not the service-on-call is their factual main goal, but the growth, the glory and the independence of their department and its peculiar kind of rationality. Support-units - and why should animal laboratories be exempted - indeed display this tendency, when not kept in control. Centralised laboratories should be aware of this danger and of their foremost duty of providing services. Close relationship and serious dialogue with the client-departments can prevent the building of ivory towers and castles in the air.

However, the rational organisation is not only a matter of communication and mutual understanding. The external relations are primarily important as dependence-relations: who controls the input to the organisation (finances, people, knowledge, information, facilities), who profits by the output, how are the relations especially the power-relations between the parties involved.

It is not possible for me at this moment to draw a generalised picture of these dependence-relations. I suppose they are different for each case, but I have too little information about the different cases. Maybe this topic could be elaborated in the discussion.

An important means - although not the only one - is the financial budget or more generally the ways in which and the criteria according to which financial means are allocated; budgets, tasks and the appointment of top-management usually are the main mechanisms to control professional organisations.

It certainly would be worthwhile to reconnaitre these questions more thoroughly, but at this moment I will drop them and proceed to the last topic: culture and history.

3. CULTURE AND HISTORY

When considering the question what factors determine or influence success or failure of animal laboratories, it should be clear what is to be understood as success and failure.

However, I am afraid that often this is not clear at all: how many of the laboratories involved would have a clear-cut picture of the criteria for assessing their performance? How many of the external relations, namely client-departments and governing boards and councils have more elaborate instruments for evaluation than guess-work?

A few years ago Peters and Waterman published a book "In search of excellence" about the factors that influenced the obvious success of some 60 outstanding enterprises. The book was a definite hit: it has been sold in some millions of copies all over the world. For truth's sake, I have to add that it gained more approval from managers (and probably would-be managers and managers-to-be) than from social scientists, but nevertheless the main message of the book is important. This message is that the first and most important factor to success is to be found in the culture of the organisation, more specifically in the adherence especially by management of central values such as "customers really go first" or "spic-and-span cleanness pays". Actually not the content of these central values or goals is of most importance to success, but the fact that the organisation and especially management formulated these values and stuck to them with almost obsessive tenacity. The cultural identity of the organisation is by far more important than most organisations are aware of.

I would like to suggest that the same holds true for much smaller organisations such as animal laboratories. But because they are part of larger organisations the question now would be whether and which of the parties involved have a set of cultural values or goals they want to be realised through the laboratory, and whether the goals of these parties are in accordance with each other or conflict. The main parties involved are then:
- management of the laboratory
- the workers of the laboratory and their representatives
- the clients of the laboratory and their departments
- the different administrative levels of the overall organisation.

CONCLUSION

I do realise that I did not give a straightforward answer to the problem of centralisation-decentralisation as related to success or failure. But actually I do not believe that you would be interested in such an answer by an outsider. What I have been trying to do was to hand you some tools that might be of service to you if you put them into use.

A.C. Beynen & H.A. Solleveld (Eds), New developments in biosciences: their
implications for laboratory animal science, ISBN 978-94-010-7973-0
© 1988 Martinus Nijhoff Publishers, Dordrecht.

THE EFFECT OF HOMOGENIZATION AND PASTEURIZATION ON THE ALLERGE-
NICITY OF BOVINE MILK ANALYSED BY A MURINE ANAPHYLACTIC SHOCK
MODEL

O.M. POULSEN[1] AND J. HAU[2]

1. INTRODUCTION
A substancial literature deals with the diagnosis, characterization and treatment of cows milk allergy (1-3). Sera from milk
allergic patients have been analysed with respect to levels of
milk specific immunoglobulins, but no studies have shown any
significant diagnostic value of measurements of specific IgE
(or IgG) with respect to immediate hypersensitivity (3-5).
Despite the lack of correlation between the presense of circulating IgE and immediate hypersensitivity, in vitro measurements
of specific immunoglobulins produced in humans or in experimental animals have been one of the most widely used methods for
assessing the allergenicity of milk proteins (1-5).

The proteins of milk interact in various complexes, such as
protein-protein-complexes and protein-lipid-complexes, and the
composition of these complexes is markly rendered by the dairy
processing of milk (i.e. pasteurization and homogenization).
However, very little informations on the effects of milk processing on allergenicity of milk have been published (6,7). In
Scandinavia an increasing number of parents to cows milk allergic children claim that their children tolerate raw, untreated
milk, but the few available clinical reports indicate, that this
is the case only in a minor group of clinically well-documented
cows milk allergic children (6,7).

The aim of the present study was to examine in a dose-response
manner the effect of normal homogenization and pasteurization
on the allergenicity of cows milk, using systemic anaphylactic
reactions in mice as a model.

2. MATERIALS AND METHODS
2.1. Animals. The mice employed were inbred BALB/c Bom mice
maintained for several generations on pellets containing 15%
skimmed bovine milk proteins (The Panum Institute, University
of Copenhagen, Denmark). Mice of both sexes aged 5-8 weeks were
used and given water and feed ad libitum. As described previously (8), the mice were well sensitized by the oral route.

2.2. Preparation of milk samples for intraveneous challenge.
One batch of untreated milk was processed into various milk
samples different in fat content, extent of homogenization, and
extent of heat treatment on a large commercial dairy (Enigheden,
Copenhagen, Denmark) (Table 1).

TABLE 1. Different milk samples prepared by varying the fat
content, the pressure of homogenization and the degree
of heat treatment.

Nomenclature	Fat content %	Homogenization (kg/cm^2)	Heat treatment
0.05- 0 -0	0.05	0	None
0.05- 0 -LP	0.05	0	LP
0.05- 0 -B	0.05	0	B
0.05-175-LP[a]	0.05	175	LP
0.05-175-B	0.05	175	B
1.5 - 0 -LP	1.5	0	LP
1.5 - 0 -B	1.5	0	B
1.5 -175-LP[b]	1.5	175	LP
1.5 -175-B	1.5	175	B
3.5 - 0 -LP	3.5	0	LP
3.5 - 0 -B	3.5	0	B
3.5 -175-LP[c]	3.5	175	LP
3.5 -175-B	3.5	175	B

LP = Low pasteurization (73-76°C for 15 sec). B = Boiled for
20 min. Samples a,b and c are the milk types normally manifac-
tured in Denmark.

Each of the 13 different milk samples were diluted in 0.9% NaCl
to give three diluted samples with a protein content of 3.6 mg/ml
1.2 mg/ml and 0.4 mg/ml, respectively. The mice were challenged
intraveneously with 250 μl of diluted milk sample. Control mice
were injected with a solution of ovalbumin (Sigma) containing
12 mg/ml in 0.9% NaCl.

2.3. Anaphylactic shock score. The anaphylactic reactions, following intraveneous administration of antigen, were evaluated
and group into four scores:

0	No reaction observed
+	The animals are slow and mobile only if provoked
++	The animals are stationary and do not move if provoked
+++	The animals lie down and suffer from severe anaphylac-tic shock (i.e. whole body cramps).

The mice were observed during a 30 minutes periode following
intraveneous challange after which they were killed. However,
mice showing score +++ were killed immidiately.

3. RESULTS

The three-factorial experiment was carried out to study the effect of fat content, homogenization and pasteurization on the ability of milk to induce anaphylactic reactions in sensitized mice. Each of the 13 different milk types (table 1) were diluted to give challenge doses of 900 µg, 300 µg or 100 µg milk protein per mouse using intraveneous injections of 250 µl. Four mice fed skimmed milk powder containing pellets ad libitum were challenge at each dose level.

The homogenized milk samples induced strong anaphylactic reactions in the mice (Figure 1). The ability of homogenized milk to induce anaphylactic shock increased with increasing fat content. Boiling the milk samples with 0.05% fat (milk type 0.05-175-B) and 1.5% fat (milk type 1.5-175-B) reduced the ability of homogenized milk to induce anaphylactic reactions as compared with the low pasteurized, homogenized milk samples. However, no significant reduction was observed as a result of boiling the homogenized milk with 3.5% fat (milk type 3.5-175-B). Even at the highest dose (900 ug per mouse) no animals responded significantly upon challenge with the unhomogenized milk samples (0.05-0-0, 0.05-0-LP, 0.05-0-B, 1.5-0-LP, 1.5-0-B, 3.5-0-LP and 3.5-0-B) and, consequently, these results are not presented graphically in figure 1.

FIGURE 1. The effect of fat content, homogenization and pasteurization on the allergenicity of bovine milk.

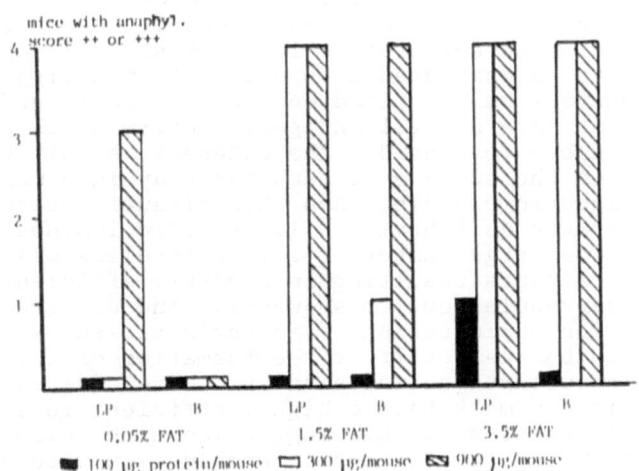

4. DISCUSSION

In the present study a murine anaphylactic shock model was used to demonstrate that homogenization of bovine milk strongly increased the ability of intraveneously injected milk to induce anaphylactic shock in sensitized mice. Inbred mice were chosen because of the minimal genetic variation, and among the inbred strains BALB/c Bom was found to be optimal due to the high concentration of mast cells in this strain (9). Consequently, high reproducibility and high sensitivity was obtained.

Sensitization by feeding orally on skimmed milk powder containing pellets was employed, since it was previously shown, that this mode of sensitization was as efficient as sensitization by intraperitoneal injections of milk with respect to the subsequent sensitivity to intraveneous challenge (8).

Anaphylactic shock could not be induced by administration of raw milk at the concentrations employed. Shock was, however, induced using homogenized milk, and there was increased effect correlated with increasing fat content. These results may be explained taking a colloid chemical viewpoint:

The concentration of surface exposed antigenic determinants is far higher in homogenized milk than in raw milk having the majority of the casein determinants hidden in the casein micelles. In raw, untreated milk the relatively large fat bodies are to some extent stabilized by a "membrane" composed of phospholipids, lipoproteins and other amphiphilic compounds. The caseins are associated in micelles with a kappa-casein core surrounding the alfa-casein and beta-casein molecules. The whey proteins (lactalbumin, lactoglobulin, bovine serum albumin, immunoglobulins, enzymes etc.) are found unbound in the liquid phase. Homogenization disrupts the large fat bodies into smaller bodies (< 0.8 um) resulting in a large increase in the surface area. The casein molecules and to a minor extent the whey proteins are absorbed by hydrophobic interaction to the exposed surfaces of the newly formed small fat bodies. The caseins and whey proteins bound to the fat bodies constitute a structurally well-organized , coherent protein layer. Mast cell degranulation is considered to be initiated by binding of an antigen molecule to several immunoglobulins fixed on the mast cell surface, i.e. the antigen must have several antigenic determinants (10, 11). It seems reasonable to consider the coherent, highly structured protein layer on the surface of each fat body in homogenized milk as one gigantic protein. Once this gigantic protein reaches contant with a mast cell having milk specific immunoglobulins on the surface, a large number of immunoglobulins will bind to antigenic determinants resulting in a highly efficient induction of mast cell degranulation. Consequently, the observed effect of homogenization on induction of anaphylactic shock in mice is suggested to be the result of the formation of colloid complexes, i.e. structures on which the proteins identical to the proteins in raw milk become highly efficient in induction of the allergic reactions. This suggestion might also explain the effect of increasing fat content in combination with homogenization.

Boiling for 20 minutes reduces the allergenicity of whey protein by complete denaturation of the heat labile proteins, whereas the heat stabile caseins are more or less unaltered by boiling (1,12,13). The observed reduction in the ability of boiled, homogenized milk with 0.05% and 1.5% fat content might in part de explained by the denaturation of the whey proteins situated

on the surface of the fat bodies. At 3.5% fat content the
amount of surface exposed casein in the homogenized milk might
be sufficiently high to induce anaphylactic shock at a challenge
dose of 300 µg protein per animal even though a part of the
determinants (i.e. the whey proteins) are lost due to boiling.

The present study has shown that the effect of milk allergens
is markly influenced by the structural association with the
milk fat. In general, it is not possible to analyse the effect
of the colloid chemical environment of a protein on the potential allergenicity of the protein by using in vitro methods
such as RAST or ELISA, since these methods, which measure presense of specific antibodies only, do not provide informations
on the cellular events in the allergic reactions. Thus in vivo
methods such as the presented murine anaphylactic shock model
are required in studies on the importance of the colloid chemical environment of allergens in connection with inhalation
allergy against house dust, fungal spores and pollen and with
food allergy against milk, fish, egg and gluten.

5. REFERENCES
1. Bahna SL, Gandhi MD. Milk hypersensitivity. I. Pathogenesis
    and Symptomatology. Ann. Allergy 50(4):218-223(1983).
2. Pearson JR, Kingston D, Shiner M. Antibody production to
    milk proteins in the jejunal mucosa of children with cow's
    milk protein intolerance. Pediatr. Res 17:406-412(1983).
3. Ford RPK, Hill DJ, Hosking CS. Cow's milk hypersensitivity:
    Immidiate and delayed clinical patterns. Arch. Disease
    Childhood 58:856-862 (1983).
4. Basuyau JP, Mallet E, Brunelle Ph, de Menibus CH. Les Into-
    lerances au Lait de Vache. La Presse Medicale 12(33):
    2041-43(1983).
5. Gjesing B, Østerballe O, Schwartz B, Wahn U, Løwenstein H.
    Allergen specific IgE antibodies against antigenic components
    in cow's milk and milk substitutes. Allergy (1986), in press.
6. Hansen LG, Høst A, Østerballe O. Allergiske reaktioner på
    uhomogeniseret mælk og råmælk. Ugeskrift for Læger.
    149(14):909-911(1987).
7. Juntunen K, Ali-yrkkö S. Goat's milk for children allergic
    to cow's milk. Short communication, Symposium on Role of
    Milk Proteins in Human Nutrition, Kiel 1983.
8. Poulsen OM, Hau J. The effect of homogenization and pasteu-
    rization on the allergenicity of bovine milk analysed by
    a murine anaphylactic shock model. Clinical allergy (1987),
    in press.
9. Poulsen OM, Hau J. Murine Passive Anaphylaxis Test (PCA)
    for the "all or none" determination of bovine whey proteins
    and peptides. Clinical Allergy 17:75-83(1987).
10. Brockhurst WE. Passive Cutaneous anaphylaxis. in Handbook
    of Experimental Immunology. Weir BM (ed). Beckwell Scienti-
    fic Publications Inc., Osney Mead, Oxford.
11. Lui CT, Das BR, Maurer PH. Immunochemical studies of the
    tryptic, chymotrypyic and peptic peptides of heat denatured
    Bovine Serum Albumin. Immunochemistry 4:1-10(1967).

12. McLaughlan P, Anderson KJ, Widdowson EM, Coombs RRA. Effect
    of heat on the anaphylactic-sensitising capacity of cows milk
    , goat's milk, and various infant formulae fed to guinea-
    pigs. Arch. Disease Childhood 56(3): 165-171(1981).
13 Heppell LM, Cant AJ, Kilshaw PJ. Reduction in the antigeni-
    city of whey proteins by heat treatment: A possible strate-
    gy for producing a hypoallergenic infant milk formula.
    British J. Nutrition 51:29-36(1984).

Authors: [1])O.M. Poulsen, Institute of Obstetrics and Gynaecology,
          University of Odense, Odense, Denmark.
          [2])J. Hau, Department of Pathology, Laboratory Animal
          Unit, Royal Veterinary and Agricultural University,
          Copenhagen, Denmark.

A.C. Beynen & H.A. Solleveld (Eds), New developments in biosciences: their implications for laboratory animal science, ISBN 978-94-010-7973-0

MURINE PASSIVE CUTANEOUS ANAPHYLAXIS TEST (PCA) FOR THE "ALL OR NONE" DETERMINATION OF ALLERGENICITY OF BOVINE WHEY PROTEINS AND PEPTIDES

O.M. POULSEN[1] AND J. HAU[2]

1. INTRODUCTION

The study of allergenicity of proteins and peptides has attracted considerable research activities in the past two decades, utilizing several in vitro and in vivo techniques. The usefulness of the in vitro techniques, such as immuno-precipitation tests (1), radio-allergo-sorbent-test (2) and enzyme-linked - immuno-sorbent-test (3), is limited due to a poor correlation between results obtained by these tests and provocation tests in the patients. Consequently, the in vivo assays, passive cutaneous anaphylaxis test (PCA) (4,5,6,) and anaphylactic shock models (3) using guinea pigs in both cases, are still widely employed and recommended when assessing the allergenicity of various compounds.

The present report describes the use of mice in PCA for the in vivo determination of the allergenicity of bovine whey proteins and peptides, and advocates for a more general replacement of guinea pigs by mice.

2. MATERIALS AND METHODS

2.1. Animals. The mice employed in the PCA test were as follows:

Group A: Inbred BALB/c Bom maintained at our laboratory for several generations on a diet without milk proteins (Ewos Brood stock feed-R-3,DK 8600, Silkebord, Denmark)

Group B: Inbred BALB/c Bom maintained for several generations on pellets containing 15% skimmed bovine milk proteins (The Panum Institute, University of Copenhagen, Denmark).

Group C: BALB/c Bom maintained by random breeding in a closed colony at the Royal Veterinary and Agricultural University, Copenhagen. These mice were maintained on the milk containing diet.

Mice of both sexes aged 5-8 weeks were used and given water and feed ad libitum.

2.2 Antibody preparation. Murine antiserum against whey proteins was prepared as described previously (7), using low doses of protein for high IgE production as suggested by Jarrett and collegues (8).

2.3. Antigen preparations. Concentrated whey proteins (CWP), hydrolyzed CWP and peptide fractions of partially hydrolyzed CWP (figure 1) were prepared as described previously. The pro-

FIGURE 1. Gelfiltration of hydrolyzed whey protein on Biogel P-30. Samples of hydrolyzed whey protein (5.0 ml) were separated and the eluted peptides pooled as indicated. Molecular weight markers were bovine serum albumin(BSA), lactoglobulin (LG), lactalbumin (LA), trypsin inhibitor (A), insulib chain B (B)and insulin chain A (C).

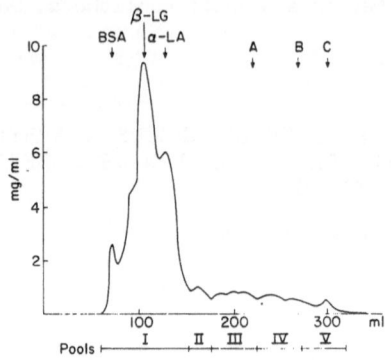

tein content of these preparations was measured with the micro Kjeldahl method or by measuring the absorbance at 280 nm using bovine serum albumin (Sigma) as standard (7).

2.4. Evans Blue preparation. A solution of 10 mg/ml of Evans Blue (Sigma) in isotonic saline was prepared and stored at -20°C in aliquots of 10 ml until use. Samples of proteins ranging from 0 to 4800 µg were mixed with 3.0 ml of the Evans Blue solution just prior to intraveneous administration.

2.5. Passive cutaneous anaphylaxis test. The mice were anaesthetized by intraperitoneal injection of propanidid (Sombrevin) (500 mg/kg), and the fur on a 3 x 4 cm square on the backside was removed using an electric razor. Samples of 100 µl, containing antibody preparation mixed with Freund's incomplete adjuvant (1:1), were injected intradermally in the center of the shaved area using 26-gauge needles. The use of FIA resulted in well-defined local colour reactions, whereas antiserum mixed with saline (1:1) gave more diffuse, less intensive blue colouration of the entire backside.

Following a period of 3 hours after administration of antibody, samples of 250 µl of the antigen-Evans Blue solutions were injected intraveneously in the tail vein. The blue skin reactions of the surface were recorded 30 min after the injection, at which time the mice were killed by dislocation of the neck and the skin removed to observe the colour reaction on the inside (corium) of the skin.

3. RESULTS
3.1. Sensitivity of the PCA method. Using an intraveneous administration of 160 µg whey protein in 250 µl Evans Blue solution the employed murine antibody preparation could sensitize the mice even at a 128-fold dilution.

The sensitivity of the PCA method was studied using the murine antibody preparation at a 1:8 dilution. The sensitivity was studied in three different groups of mice. A dilution series ranging from 0.04 µg/ml to 400 µg/ml of whey protein in Evans Blue solution was prepared. At each dilution step three mice of each grouppassively sensitized with the mice antiserum were challenged intraveneously with 250 µl antigen-Evans Blue Solution The sensitivity results are presented in table 1.

89

TABLE 1. Sensitivity of murine PCA-test.

	Group A	Group B	Group C
Detection limit. Smallest dose of whey protein capable of inducing PCA-reaction.	0.1 µg	2.0 µg	10 µg

At low concentrations of antigen (i.e. close to the detection limit) the reaction could be seen on the inside of the skin only. At higher concentrations of antigen the PCA reactions were also visible on the surface of the skin as dark blue circles. The highest sensitivity was obtained when using inbred BALB/c Bom mice fed the diet without milk proteins. This type of mice was used throughout the following experiments.

3.2. PCA-test for allergenicity of hydrolyzed bovine whey proteins and peptides. Peptides from partially hydrolysed CWP were separated by gel filtration (Figure 1). Pool I contained the unhydrolysed whey proteins, whereas pools II-V contained peptides of decreasing molecular weight (Table 2). The values of table 2 should merely be considered as the molecular weight range of the majority of peptides in the different pools, since there was a certain extent of overlapping.

TABLE 2. Molecular weight range of peptides in pools I-V.

Pool	Molecular weight
I	> 14 000
II	14 000 - 11 000
III	11 000 - 6 500
IV	6 500 - 3 400
V	< 3 400

Sixty-six mice of group A were passively sensitized by intradermal injection with 100 µl undiluted murine antibody:adjuvant (1:1), whereas nine control mice were given 100 µl saline:adjuvant intradermally. CWP, CWP-hydrolysate and peptide pools I-V were diluted in Evans Blue solution to give samples with a protein content of 100 µg/ml, 400 µg/ml and 1600 µg/ml, respectively. Three hours after the antibody injection the mice were challenged intraveneously. The control mice were given 250 µl CWP-Evans Blue solution. After 30 min the reaction on the surface of the skin was recorded. If no reaction was observed the mouse was killed and the skin inverted in order to inspect the inside of the skin. The results are presented in table 3. Reduced allergenicity was observed of CWP-hydrolysate and pool IV, and no allergenicity could be recorded of peptide pool V.

TABLE 3. Allergenicity of hydrolysed whey proteins and the resultant peptides separated by gel filtration.

Intradermal injection	Intraveneous injection antigen	PCA reaction (dose/animal)		
		25 µg	100 µg	400 µg
Ab	CWP	s	s	s
Ab	CWP-hydrolysate	None	w	w
Ab	Pool I	s	s	s
Ab	Pool II	s	s	s
Ab	Pool III	s	s	s
Ab	Pool IV	None	w	s
Ab	Pool V	None	None	None
Ab	Saline	None	None	None
Saline	CWP	None	None	None

Skin reaction: s: deep blue coloured zone larger than 10 mm,
w: blue coloured zone smaller than 3 mm easily visualized on the inside of the skin.

4. DISCUSSION

In the present study, the CWP-hydrolysate was able to induce a PCA reaction in mice, indicating that the hydrolysate contained peptides or proteins of sufficient size to induce a histamine release of the mast cells. Less than 5% of the peptides of the CWP-hydrolysate turned out to be larger than 6000 Da when analysed by gel filtration on Biogel P-6 (results not shown), explaining the weak PCA reaction induced by this preparation. This results and the result on pool III, are in agreement with Otani (9), who found that peptides larger than 10 000 Da, produced by proteolytic degradation of lactoglobulin, induced histamine release in passively sensitized guinea pigs. Pool IV, containing peptides in the range between 6500 - 3400 Da, induced weak PCA reactions in mice. Thus, this result is in disagreement with Otani (9), but confirms the observations of Lui et al. (5), who showed posetive PCA reactions in guinea pigs upon challenge with a 7200 Da peptide produced by hydrolysis of bovine serum albumin. Gelfiltration on Biogel P-6 showed, that the majority of peptides in pool IV were larger than 6000 Da (7). Pool V (<3400 Da) could not induce any PCA reactions in mice, indicating that the peptides had less that two epitopes per molecules (4).

A multitude of inbred mouse strains (which minimizes genetic variation) are available. In the present study, inbred BALB/c mice were chosen due to their relative high susceptibility to passive sensitization (10,11). Inbred BALB/c mice fed mlik-free pellets (Table 1, group A) showed a higher responsiveness than inbred BALB/c mice fed pellets containing milk p teins (Group B). It is well known that mice are immunized by the oral route (12,13), and circulating antibodies against whey proteins in these mice form complexes with the intraveneously injected whey proteins, thereby reducing the concentration of whey proteins available for induction of a PCA reaction. The outbred mice (Table 1, group C) showed a lower responsiveness compared with group B, probably indicating that the inherited high responsiveness of inbred BALB/c mice (11) was lost. At best, the PCA reaction could be induced in group A mice sensitized with murine antibodies by intraveneous challenge with 0.1 µg whey protein, a sensitivity very similar to the sensitivity reported in guinea pigs (4).

Several difficulties are associated with the use of guinea-pigs for PCA testing. Sedatives, anasthetics and stress tend to give a reduced or lacking response (4). Intraveneous injections are rather complicated, and since only one antigen/allergen can be tested per animal, the use of guinea-pigs also becomes a rather expensive, space- and time-consuming method. The general agreement between the results obtained on mice and the quoted results obtained on guinea-pigs indicates that mice should be considered to be of general interest for PCA tests applied to the assessment of allergenicity of proteins and peptides.

[1] O.M.Poulsen, Institute of Obstetrics and Gynaecology, University of Odense, Odense, Denmark.

[2] J. Hau, Department of Pathology, Laboratory Animal Unit, Royal Veterinary and Agricultural University, Copenhagen, Denmark.

5. REFERENCES

1. Løwenstein H, Krasilnikoff PA, Bjerrum OJ, Gudmund-Høyer E. Occurence of specific precipitins against bovine whey proteins in serum from children with gastrointerstinal disorders. Int Archs Allergy appl Immunol 1977;55:514-25.
2. Basayau JP, Mallet E, Brunelle Ph, de Minibus CH. Les intolérances au lait de vache. La Presse Mëdicale 1983;12:2041-43.
3. Zifshitz F. (ed). Clinical Disorders in Pediatric Nutrition. Vol. 4. Nutrition for Special Needs in Infancy. Protein Hydrolysates. Marcel Dekker, Inc. 1985.
4. Ovary Z. Passive Cutaneous Anaphylaxis.In:Ackroyd JF(ed) Immunological Methods. A Symposium. Oxford, Blackwell Scientific Publications 1964:259-83.
5. Lui CT, Das BR, Maurer PH. Immunochemical studies of the

tryptic,chymotryptic and peptic peptides of heat denatured bovine serum albumin. Immunochemistry 1967;4:1-10.
6. Takase M, Fukuwatari Y, Kawase K, Ogasa K, Suzuki S, Kuroume T. Antigenicity of Casein Enzymatic Hydrolysates. J. Dairy Sci 1979;62:1570-76.
7. Poulsen OM, Hau J. Murine passive cutaneous anaphylaxis test (PCA) for the "all or none" determination of allergenicity of bovine whey proteins and peptides. Clicical Allergy 1987; 17:75-83.
8. Jarrett EEE, Haig DM, McDougall W, McNulty E. Rat IgE production. II. Primary and booster reaginic antibody responses following intradermal or oral immunization. Immunology 1976; 30:671-77.
9. Otani H. Antigenicities of beta-lactoglobulin treated with proteolytic enzymes. Jpn J Zootech Sci 1981;52:47-52.
10.Levine BB, Vaz NM. Effect of combination of inbred strain, antigen, and antigen dose on immune responsiveness and reagin production in the mouse. A potential mouse model for immune aspects of human atopic allergy. Int Arch Allergy 1970;39:156-71.
12.Swarbrick ET, Stokes CR, Scoothill JF. Absorbtion of antigens after oral immunization and the simultaneous induction of systemic tolerance. Gut 1978;39:121-25.
13.André C, Bazin H, Heremans JF. Influence of repeated administration of antigen by the oral route on specific antibody-producing cells in the mouse spleen. Digestion 1973;9:166-75.

A.C. Beynen & H.A. Solleveld (Eds), New developments in biosciences: their
implications for laboratory animal science, ISBN 978-94-010-7973-0
© 1988 Martinus Nijhoff Publishers, Dordrecht.

PHENOTYPIC AND PATHO-MORPHOLOGICAL CHARACTERISTICS IN A HALF-
SIB-FAMILY OF TRANSGENIC MICE CARRYING FOREIGN MT-HGH GENES

G. Brem* and R. Wanke**

* Institut für Tierzucht und Tierhygiene
Lehrstuhl für Tierzucht: Prof. Dr. Kräußlich
** Institut für Tierpathologie
Lehrstuhl für Allgemeine Pathologie: Prof. Dr. v. Sandersleben
Ludwig-Maximilians-Universität München

1. INTRODUCTION

Transgenic mammals can be produced by microinjection of
foreign DNA into pronuclei of zygotes. Fusion genes where the
metallothionein I promotor has been connected to the structural
gene for rat or human growth hormone has been introduced in
mice by Palmiter et al. (1982, 1983). About 50 % of the animals
have expressed these additional genes, most of them grew
significantly larger than their littermates and enhanced growth
rate was shown to be heritable.
We have produced transgenic mice, rabbits and pigs carrying
foreign MTI-hGH fusion genes (Brem et al., 1985, Brenig and
Brem, 1987). We report here phenotypic and patho-morphological
characteristics of a transgenic half sib family in mice.

2. MATERIALS AND METHODS

A male transgenic mouse carrying 4 copies of the in vitro
recombinated MTI-hGH gene was bred to 5 normal females. When
offspring were weaned total nuclei acids were extracted from a
piece of tail tissue and used for DNA dot and Southern blot
hybridization to determine transgenic mice (for details see
Brenig and Brem, 1987). Both transgenic and non transgenic
siblings (controls) were reared together.
After weaning at an age of 3 weeks offspring were fed with
concentrate (Altrumin) ad libitum. They received water
supplemented with 0.7 % $ZnSO_4$ during the next 10 weeks. Body
weights of mice were recorded periodically. Starting at an age
of 5 month transgenic mice were mated permanent to transgenic
and normal mice.
After natural death or euthanasia the transgenic mice and
control siblings were necropsied. Histological preparations
were made from different organs (kidney and liver obligatory),
except when the tissue had become decomposed. Routine histology
was performed after fixation in 7% buffered formalin. Paraffin
sections were cut at 5µ and were stained as follows:
hematoxylin and eosin (HE), Masson's trichrome, periodic acid-
Schiff (PAS), Gomori's silver impregnation, alkaline Congo red.

Morphometric studies were performed on liver sections (5µ

thick stained with HE) of 4 transgenic mice and 3 controls. The size (area expressed in μ^2) of the hepatocytes and the nuclei was calculated by direct measuring under a microscope (magn. 400 x) using a Zeiss Morphomat 30; magnification was determined with the help of an object micrometer. A total of about 120 cells was measured per case. Ten test fields were selected at random in different sections of each liver.

3. RESULTS
3.1. Growth and fertility data of transgenic mice
The founder mouse was found to express hGH and was about 1.6 times heavier than control mice. Breeding of the founder male resulted in only three litters. 24 offspring were born, of which 20 survived (3 were eaten by their mother, 1 transgenic died one day after birth). 11 out of the surviving offspring were shown to be transgenic. They expressed the new gene and circulating hGH levels were between 185 and 1780 ng/ml. All of the transgenic offspring grew larger than non transgenic control siblings (table 1).

At weaning only small differences were observed between offspring that did and did not receive the transgene from their father. Growth rates of transgenic mice were accelerated most between 4th and 8th week.

TABLE 1. Average of body weights of transgenic mice and control siblings (4 month old)

	Sex	n	Body weight min.	max.	average (Std. dev.)	Growth ratio
Control	m	4	35	40	36.7 (2.4)	1
Transgenic	m	7	51	77	61.1 (8.4)	1.66
Control	f	5	30	32	30.5 (1.2)	1
Transgenic	f	4	56	76	63.5 (9.1)	2.08

Starting at an age of 5 months transgenic mice were mated to transgenic and to normal mice. None of these permanent matings resulted in offsprings although mating partners were changed several times.

The founder male died at an age of 243 days. Most transgenic offspring became ill and died. 6 mice died and 2 were killed because of illness between 138 and 192 days (∅ 168 days). One ill mouse was killed 316 days old and another one died at an age of 351 days. Only 1 mouse survived and was killed at an age of 436 days for further examination. Normally duration of life in the outbred strain used (NMRI) is more than 1.5 years.

The weights of the transgenic mice are compared to the weights of the littermates without MTI-hGH gene sequences. The transgenic animals are significantly heavier than sex-matched controls. In contrast to the latter non sex-dependant differences of the body weights are found in mice carrying MTI-hGH genes. The absolute and relative weights of the heart, the liver and the kidneys (males only) are significantly increased compared to controls of the same sex. No differences are found concerning the lungs (table 2).

TABLE 2. Absolute and relative organweights of transgenic and control mice

	male			female			Diff.	
	n	g	%	n	g	%	g	%
Bodyweight:								
Transgenic	8	47.1(11.6)	–	4	47.0 (5.8)	–	0.1	–
Control	3	36.2 (2.0)	–	5	30.6 (1.6)	–	5.6*	–
Difference		10.9*			16.4**			
Heart:								
Transgenic	7	.38(.16)	.80(.28)	3	.43(.06)	.93(.08)	-.06	-.13
Control	3	.16(.06)	.46(.16)	5	.13(.03)	.42(.09)	.03	-.04
Difference		.21*	.34*		.30**	.51***		
Liver:								
Transgenic	8	4.14(1.5)	8.68(1.6)	4	3.8(.4)	8.2(1.4)	.33	.77
Control	3	1.97(.45)	5.41(1.1)	5	1.2(.1)	4.0(.5)	.77	1.41
Difference		2.16*	2.02*		2.6**	4.2*		
Kidney:								
Transgenic	8	1.66(.53)	3.59(1.0)	3	.97(.3)	2.1(.4)	.7*	1.49**
Control	3	0.6 (0)	1.66(.09)	5	.48(.1)	1.57(.4)	.12	.09
Difference		1.07**	1.93**		.49	.53		
Lungs:								
Transgenic	7	.52(.11)	1.19(.34)	3	.47(.06)	1.01(.2)	.05	.18
Control	3	.43(.06)	1.2 (.17)	5	.36(.06)	1.17(.19)	.07	.03
Difference		.09	.01		.11	-.16		

3.2. Pathological findings

3.2.1. Necropsy findings. At necropsy, lesions of the urinary tract are the predominant findings in mice carrying MTI-hGH gene sequences. Grossly, the kidneys of 5 animals are indurated and of a pale yellowbrownish color. Small cysts can be seen on the surface. In these and two other mice the renal pelvis and the urinary bladder are dilated and white pulpy masses are found within the lumen of the bladder in some cases. The concrement consists of minerals and proteinaceous material.

Generally, the liver of the transgenic mice is enlarged without showing abnormalities in shape; a nodular surface of the liver is seen in one animal.

The postmortem examination of the controls reveals no lesions of the internal organs except for changes which are found usually as a consequence of euthanasia.

3.2.2. Histopathology.

- Kidney

Glomerular lesions are seen in all of the transgenic mice. In most cases the glomeruli are affected in a diffuse pattern. Different grades of hyalinization of the mesangium and thickening of the capillary wall are found. A few glomeruli are completely hyalinized. The hyalinized areas give a strongly positive reaction with PAS; silver impregnation shows an increased number and density of reticulin fibers. Congo red stain for demonstration of amyloid is negative. With increasing severity of the lesions, changes are seen in the other parts of the kidney: Cystic tubular atrophy is common and hyaline casts appear within dilated proximal tubules and collecting tubules.

The interstitial tissue shows focal fibrosis and lympho-plasmocellular infiltration predominantly in perivascular sites. Similar lesions are not detected in any of the controls.

- Liver

The hepatocytes of the transgenic mice are variable in size and shape. Most of them are distinctly enlarged. The hepatocytic nuclei are extremely pleomorphic. Homogeneous intranuclear inclusions, which are occasionally seen in the hepatocyte of normal mice, are found in a large number. As shown by electron microscopic studies these inclusions consist of cytoplasmic material and result from invagination of the nuclear membrane.

Necrosis of single hepatocytes and also of liver cell groups combined with inflammatory cellular infiltration is seen in many locations. Morphometric examination revealed that the average hepatocytic area is about four times enlarged compared to control livers (table 3). Liver cell nuclei are even six times larger resulting in a diminished hepatocyte:nucleus ratio. These differences are even more striking when calculated cell volumes respectively nuclear volumes are compared.

TABLE 3. Areas of hepatocyte cells and nuclei in transgenic mice

	Sex	Counts (n)	Hepatocyte (μm^2)		Nuclei (μm^2)		Relation H/N
Control	f	114	252	(75)	25	(9.3)	10.0
"	f	99	234	(68)	41	(14)	5.7
"	m	122	472	(138)	72	(22)	6.6
Average		335	327	(150)	47	(26)	7.0
Transgenic	f	131	2591	(1218)	503	(305)	5.2
"	f	121	1308	(569)	263	(143)	5.0
"	m	109	751	(281)	129	(57)	5.8
"	m	111	605	(269)	104	(66)	5.8
Average		472	1370	(1081)	261	(242)	5.2

4. DISCUSSION

The integration of the MTI-hGH fusion gene into the mouse genome results in accelerated growth rates and significantly increased body weights relative to control littermates.

The body weights of the transgenic mice show no sex dependant differences as found in controls and as can be usually observed in mice.

As to the increase in weight the internal organs are affected in a different manner. The heart, the liver and the kidneys (males only) are not only significantly heavier but also the organ:body weight ratio is elevated compared to the controls. In contrast to these organs the lung weights do not differ between the transgenic and the non transgenic animals.

The comparison of the hepatocytic enlargement to the increase in body weight indicates that the elevated liver weights of the transgenic mice is not a consequence of an increase in number of the hepatocytes but results from cellular hypertrophy. The liver cell nuclei are not only absolutely but also relatively enlarged as can be seen in a diminished hepatocyte:nucleus size ratio.

The hypertrophic liver cell changes may be understood as a consequence of elevated growth hormone expression according to Helweg-Larsen (1952), who found a relationship between the hepatocytic nuclear size and the presence of growth hormone. Degenerative changes and necrosis of liver cells are combined with the hepatocellular hypertrophy.

Obviously the lesions of the urinary tract are related to the integration of the MTI-hGH fusion gene. No similar pathologic changes are found in the controls. The renal glomeruli seem to be affected primarily as showing alterations in all the transgenic mice. The lesions of the other parts of the kidney seen in severely affected cases may be considered as secondary changes following primary glomerular dysfunction.

The pathogenesis of the glomerular injury is obscure. Similar glomerular alterations termed spontaneous, progessive glomerulosclerosis are described in the RF strain of mice by Gude and Upton (1962). There is also some resemblance to a form of so called diffuse intercapillary glomerulosclerosis found in aged mice of different strains and increased by X-ray exposure (Guttman and Kohn, 1960). No relation exists between the mice used in this study and the RF strain of mice. Severely affected kidneys are found in the transgenic mice at 5 months up to one year. Amyloidosis has been implicated in the pathogenesis of similar renal lesions but amyloid is not demonstrable in the kidneys of the transgenic animals.

REFERENCES
1. Brem G, B Brenig, HM Goodman, RC Selden, F Graf, B Kruff, K Springmann, J Hondele, J Meyer, EL Winnacker, H Kräußlich: Production of transgenic mice, rabbits and pigs by microinjection into pronuclei. Zuchthygiene 20, 251-252, 1985.
2. Brenig B, G Brem: Integration of hGH gene in transgenic mice and transmission to next generation. 3rd FELASA Symposium, Amsterdam, 1.-5.6.1987.
3. Gude WD, AC Upton: A histologic study of spontaneous glomerular lesions in aging RF mice. Am. J. Pathol. 40, 699-706, 1962.

4. Guttman PH, HJ Kohn: Progressive intercapillary glomerulosclerosis in the mouse, rat, and Chinese hamster, associated
   with aging and x-ray exposure. Am. J. Pathol. 37, 293-307,
   1960.
5. Helweg-Larsen: Nuclear Class Series. Copenhagen: Munhogaard,
   1952.
6. Palmiter RD, G Norstedt, RE Genlinas, RE Hammer, RL
   Brinster: Metallothionein-Human GH Fusion Genes stimulate
   growth of mice. Science 222, 809-814, 1983.
7. Palmiter RD, RL Brinster, RE Hammer, ME Trumbauer, MG Rosenfeld, NC Birnberg, RM Evans: Dramatic growth of mice that
   develop from eggs microinjected with metallothionein-growth
   hormone fusion genes. Nature 300, 611-615, 1982.

A.C. Beynen & H.A. Solleveld (Eds), New developments in biosciences: their implications for laboratory animal science, ISBN 978-94-010-7973-0
© 1988 Martinus Nijhoff Publishers, Dordrecht.

THE HOUSING AND HANDLING OF MARMOSETS AND TAMARINS INFECTED WITH AIDS AND OTHER RETROVIRUSES

LYNN FRANCIS, R.T. MOORE, R.T. RAYMOND, A. BASKERVILLE

1. INTRODUCTION

In the search for suitable animal models for AIDS and other retrovirus infections it was proposed to infect the common marmoset, Callithrix jacchus, and the cotton-topped tamarin, Saguinus oedipus, with the viruses HIV-1, HIV-2, HTLV-1 and HTLV-2. To comply with the Guidelines issued in the United Kingdom by the Advisory Committee on Dangerous Pathogens the work has to be undertaken at Animal Containment Level 3 (Health and Safety Executive, 1984; 1986). Scratches or bites from animals inoculated with AIDS and similar viruses must be regarded as a potential hazard to staff and the normal method of handling these animals was considered unacceptable. Normally, they would be caught by putting a leather-gloved hand into the cage and holding the animal around the shoulders. To avoid handling the animal in this manner a cage was designed to minimise risks from bites and scratches, and to enable the animal to be trapped safely in a nest box with a crush-back facility.

2. MATERIALS AND METHODS

Three points were considered when designing the cage.

a) The animal must be contained so that both normal husbandry and experimental manipulations can be safely carried out.

b) The cage environment must be optimal for the animal, within the above safety limitations.

c) The cage must be readily dismantled for cleaning in the animal room rather than by using a central cage washing facility.

Modifications to an existing design of marmoset cage (Modular Systems and Developments Limited, Woolwich, United Kingdom) carried out in collaboration with this company enabled these criteria to be met. The cage body is constructed of 16swg anodised aluminium with internal dimensions 500mm wide x 500mm deep x 750mm high (Fig. 1.). The 14 swg stainless steel mesh used for the front, roof and floor is 25mm x 13mm, which is too small for an adult marmoset or tamarin to put its hand through. The front section is removable, being suspended on two lift-off type hinges, and securely locked with a square key. A label holder, 50mm x 60mm, is welded to the back plate for the door lock. The vertical sliding door, 150mm x 300mm high is fastened by means of a dog-clip. The dirt tray (16 swg anodised aluminium) slides into runners formed from the extended body of the cage so that the cage components form an integral unit. Both the mesh floor and dirt tray are held in place by a swivel clip which locates in a slot in the rim of the tray. Either the grid or the tray may be removed without risk of the animal escaping. Two hardwood perches, 25mm diameter, are slotted into brackets on the cage sides and held by a locating pin fixed through the perch.

Food dishes placed on the grid are frequently pushed to the back of the

FIGURE 1. The cage with nest box in position.

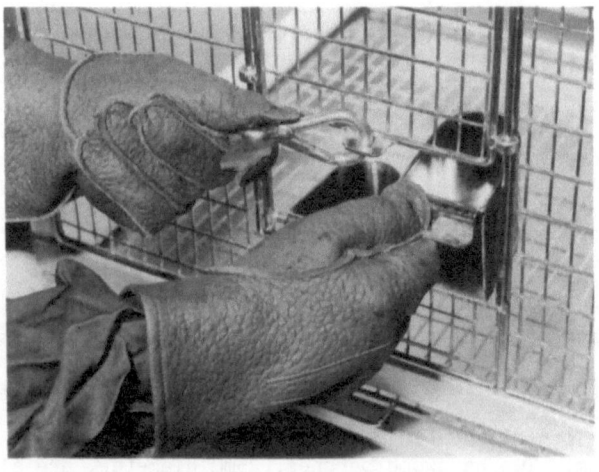

FIGURE 2. Insertion or removal of food hopper.

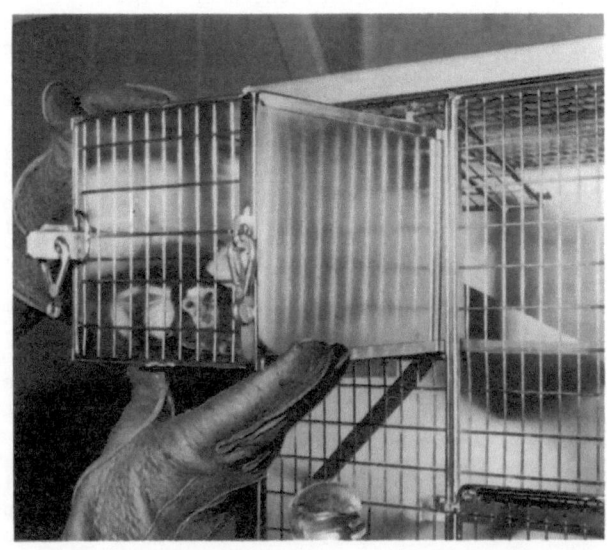

FIGURE 3.   Insertion or removal of nest box with marmoset locked inside.

FIGURE 4.   Crush back facility of nest box holding marmoset
            for intramuscular injection.

cage by the animal. To avoid having to reach into the cage to retrieve dishes, two food hoppers are hooked onto the bottom of the frame for the guillotine door, being held in place by the door (Fig. 2.).

The nest box (Fig. 3.) is constructed from 20 swg stainless steel except for the front, which has a mesh of the same dimensions as the cage. The internal measurements are 170mm wide x 250mm deep x 160mm high, with an opening on the side 100mm x 100mm. The nest box rests on a platform and is secured to the cage front by a dog-clip locked into two lugs. The animal can be trapped inside the nest box by means of a sliding panel inserted at the front, and this is locked in the closed position by a dog-clip. The nest box can be slid into position or removed either with an animal free in the cage or trapped in the box. When the box is removed the hinged door falls down and can be locked to the cage front with a dog-clip. The nest box is fitted with a crush back operated by a single side arm, (Fig. 4.) which can be locked into position with a dog-clip when the animal has been drawn up to the front.

The base of each cage slides onto runners on a rack, and a mobile rack holds four cages, two above the others. Eight units of four cages were constructed.

3. RESULTS

The cages have been used for housing marmosets and tamarins individually. Both species do utilise all the space available in the cage, including hanging from the mesh roof. Initially, animals were trained to enter the nest box by enticing them with food such as banana or malt loaf. When the animals began to associate trapping with the subsequent procedure of injection some of them became more wary of entering their nest box. These animals now have to be persuaded to enter their nest box by a technician wearing an elbow-length leather gauntlet inserting his forearm through the guillotine door. The animals are never actually caught by hand.

Trapping the animal safely in the nest box enables several procedures to be carried out. After ensuring the sliding door is locked the box can be removed from the cage so that the cage or its components can be cleaned in the animal room with 1% Tego MHG 103 (Th. Goldschmidt Ltd., Ruislip, United Kingdom). To facilitate animal weighing every nest box was initially pre-weighed and its weight recorded on the box. Animals are anaesthetised by using the crush-back to pull them up against the mesh front, and injecting them intramuscularly with doses of 0.15 - 0.18ml ketamine hydrochloride (Vetalar, Parke-Davis, Pontypool, United Kingdom). Only when an animal is fully anaesthetised is it removed from the nest box for experimental procedures, such as inoculation or femoral vein blood sampling.

4. DISCUSSION

The housing of marmosets and tamarins in these cages is one aspect of the containment facility in which animals are used for research with AIDS and other retroviruses. The animal rooms are part of a high containment area which is run under negative pressure with extracted air passing through High Efficiency Particulate Air Filters. All liquid effluent is sterilised and other waste material is autoclaved. Only designated staff are authorised to work in the suite with animals infected with AIDS viruses and they are all familiar with the code of practice which details the husbandry of the animals and the experimental procedures. Full protective clothing is worn, including a gown, rubber boots, two pairs of disposable gloves, a respirator and hood, and staff shower on exit from the suite.

The cages have proved satisfactory for the nine months which they have been in use. No animal has injured itself in the cage and there have been no escapes. There have been no incidents of staff receiving bite or scratch wounds from infected animals.

Staff are aware of the hazards of working with AIDS viruses and it is important that all possible safety precautions are taken to protect them. The design of these cages has given staff the confidence to work with marmosets and tamarins, knowing that, provided the code of practice is adhered to, the animals are safely contained .

REFERENCES
Health and Safety Executive, Advisory Committee on Dangerous Pathogens:
Categorisation of pathogens according to hazard and categories of
containment. London: HMSO, 1984.
Health and Safety Executive, Advisory Committee on Dangerous Pathogens:
LAV/HTLV III - the causative agent of AIDS and related conditions -
Revised guidelines. London: HMSO, 1986.

A.C. Beynen & H.A. Solleveld (Eds), New developments in biosciences: their implications for laboratory animal science, ISBN 978-94-010-7973-0
© 1988 Martinus Nijhoff Publishers, Dordrecht.

LONG-TERM OBSERVATION OF LITTER INTERVALS
IN PERMANENTLY MONOGAMOUS MATED Han:NMRI MICE

K. G. Rapp and R. Kluge

INTRODUCTION

Experimentators who work on reproduction performances of laboratory animals usually record the number of live and dead born pups. In addition the number of weaned pups per litter and the weaning weight is commonly included.

However, one character of reproduction performance which is as well important as the above mentioned is the time between mating and first litter and the intervals between the subsequent litters. Despite of its importance this trait is scarcely considered.

In the present study the results obtained from a long term observation concerning litter intervals of a defined mouse outbred population are discussed.

MATERIALS AND METHODS

Animals: The Han:NMRI mouse outbred population, maintained at the Central Institute for Laboratory Animal Breeding in Hannover (West Germany) since 1961 and bred permanently monogamous.

Population size included: 7 740 dams with a total of 47 236 litters.

Observation period: All breeding units which were set up between 11th of March 1979 and 24th of May 1983.

Breeding period of the units and particular steps: Permanent monogamous mating of animals having reached an age of 8 weeks. All litters were standardized to 10 pups; that is, the litters having less than 10 offspring were replenished by surplus young of other dams. According to the working time of the animal care takers the litters were weaned at an age of 20 - 22 days. The males were removed from breeding at an age of about 34 weeks. The dams finished their breeding time at an age of 37 - 41 weeks after weaning their last litter.

Breeding cages: Macrolon cages (type II) with a basal surface of 352 cm^2.

Animal house: Barriere type animal house [SPF] with excess pressure of 15mm H_2O in the animal rooms against the outside area, a room temperature of $22 \pm 2.5^{\circ}C$, a relative humidity of more than 40% and light-dark cycle of 10:14 hrs.

Nutrition: An autoclaved, complete diet provided as pellets, manufactured according to a formula of the Central Institute and drinking water acidified with HCl to pH 2.5. Access to food and water ad lib.

Data recording: Each working day all breeding cages without a litter were controlled. Estimation of the day of litter birth after duty off days according to the development of the pups, recording the reproduction data for subsequent data-processing.

Data analysis: Calculation of arithmetic means ($\bar{x}$), standard deviations (s), coefficients of variation (V), analyses of variance, and t-tests using a PDP 11/34 computer.

RESULTS

The main results of this study lasting more than 4 years (1979 - 1983) concerning the time between mating and first litter and the subsequent litter intervals in permanently monogamous bred Han:NMRI mice are summarized in Table 1. An uncritical preliminary review of the results show the NMRI mice to have an average time period between mating and first litter of nearly 25 days which is followed by mean litter intervals ($\bar{x}$) between 2nd and 8th litter decreasing from 31 to 24.4 days. The respective standard deviations (s) and coefficients of variation (V) were found to be comparatively high but decrease significantly after the 5th litter.

However, dividing the dams into 8 groups according to their maximum number of litters and evaluating the groups separately we get completely different results. Fig. 1 shows the development of the periods between mating and first litter and the subsequent intervals separated according to the 8 dam groups. By this figure it can be demonstrated that dams with a low total number of litters usually never reach intervals between 25 and 31 days; it becomes obvious that dams with a poor reproduction performance already show an above average time between mating and first litter having also extended subsequent intervals. Dams with 2 - 5 litters during the breeding period are characterized by permanent increasing litter intervals. In contrast, dams with a total of 6 - 8 litters express a comparatively long period only between the 1st and 2nd litter. This can be explained by the extreme stress during that time which results from the first lactation, the own body growth, and hormonal changes. Thus, the pregnancies after

Table 1. Intervals between litters in days in permanent monogamous bred Han:NMRI mice, 7740 mothers having a total of 47 236 litters

NUMBER OF LITTER	INTERVAL BETWEEN LITTERS			NUMBER OF ANIMALS	
	$\bar{x}$	s	V	n	%
1	24.87	10.32	41.5	7740	100.0
2	31.03	10.41	33.5	7670	99.1
3	30.37	10.93	36.0	7489	96.8
4	30.12	11.18	37.1	7193	92.9
5	29.28	10.24	35.0	6603	85.3
6	27.69	8.75	31.6	5319	68.7
7	26.09	6.86	26.3	3623	46.8
8	24.48	4.57	18.7	1599	20.7

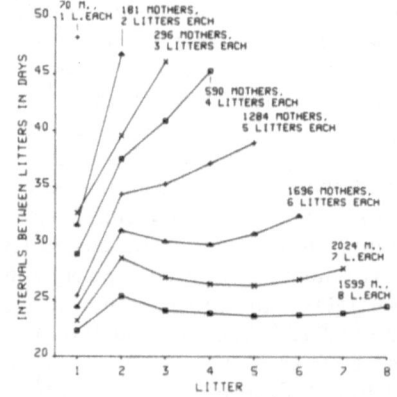

Fig. 1. Intervals between litters depending on number of litters per mother

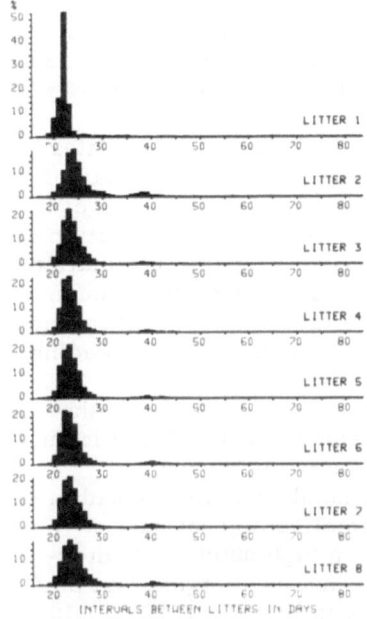

Fig. 2. Relative distribution
of intervals between litters,
1600 mothers with 8 litters
each = 12 800 litters

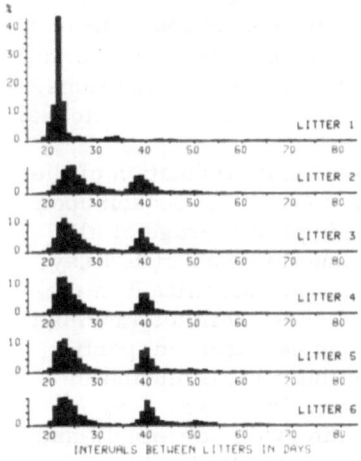

Fig. 3. Relative distribution
of intervals between litters,
1697 mothers with 6 litters
each = 10 182 litters

the first litter often result from later
occuring cycles but not from the post
partum estrus. The subsequent intervals
in the well reproducing female groups
decrease at first but increase later on.

Considering all female groups, of
course less time can be observed between mating and first litter than between
all subsequent litters. Despite of a great
variance most of the litter intervals show
highly significant differences within the
respective dam groups. However, the
differences between the dam groups concerning the intervals between litters 1 -
8 are even more distinct.

About two third of all 7740 females
included reach the 6th litter after 26
weeks of permanent monogamous breeding. Nearly the half has a total of 7
litters and about one fifth 8 litters
(Table 1). The total average of all
animals was calculated to be 6.1 litters
with a standard deviation of 1.58.

The relative distribution of the litter
intervals according to the maximum
number of litters can be demonstrated
by the female groups with a total of 8,
6, and 4 litters respectively. Their
behaviour is representative for the whole
population. Therefore the separate
demonstration of all other groups can be
abandoned.

The figures 2 - 4 show the relative
proportions (y-axis) of all intervals in
days (x-axis) for the litters of the respective dam groups. The sum of all
columns per female group and litter
number is 100% respectively.

Most intervals of the dams with 8 litters (Fig. 2) are found between 19 and 31
days with a mean peak at 23 days. These
short periods can only be explained
assuming most of the successful copulations to take place during the post partum
estrus. However, looking particularly at
the second and the last litter of this
group a small part of the intervals is
found with a peak at 39 days. This cumu-

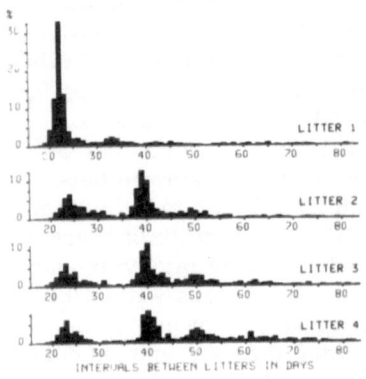

Fig. 4. Relative distribution
of intervals between litters,
590 mothers with 4 litters
each = 2 360 litters

lation becomes more distinct in the
groups having reached a total of 6 litters
during breeding (Fig. 3). The peak at 23
days significantly decreases in this
group.

Concerning the females with 4 litters
the number of intervals with a mean of
39 days exceeds those with a mean of 23
days. Fig. 4 shows a third distribution of
intervals having an average culmination
at 49 - 50 days. This third peak becomes
more and more distinct with decreasing
total litter numbers during the breeding
period.

Considering all dam groups together
having more than one litter the following
relations can be summarized:
An increasing number of litters within
a dam group implies a great part of
successful post partum copulations resulting in a high number of intervals of 23 days. Decreasing numbers of litters involve a significant shift
of interval distributions favouring the average culmination at 39 and 49
days. The time span of 16 days between the peaks of 23 and 39 days just
represents the mean lactation period. During this time the sexual cycle
is blocked. Thus, a successful copulation can either occur immediately
after birth during the post partum estrus or at the earliest after the main
lactation period when the sexual cycle is starting again.

However, not all of the litter intervals can be appointed to a distinct
distribution pattern. There is a great deal of influences as fetal atrophy,
abortions at each stage of pregnancy, and death of whole litters at birth,
which cannot be recorded and analysed under usual breeding conditions.

These influences may mask the real distribution pattern of the intervals
particularly in dams with a low number of litters.

Despite of an artificial climatisation and a total standardisation of the
environmental conditions within the animal house effects of season upon
the litter intervals could be observed. Considering the average of all
litters under investigation the intervals were found to be about 1.5 days
longer in winter than during the other seasons (highly significant in the
F-test). Dividing the dams into 8 groups according to their total number
of litters significantly longer winter intervals can be observed particularly in females with 2 - 3 litters concerning their 1st litter and in
females with 4 - 8 litters concerning their 2nd and 3rd litter.

In about 0.1% of all litters (41 out of 47 236) intervals between 15 and
18 days were recorded (26 x 18 days, 9 x 17 days, 4 x 16 days, and
2 x 15 days). These very short intervals only occurred after dead born
litters or after rearing only a very few or none pups in the preceding
litters.

In about 0.7% of the litters extended periods of pregnancy could be observed after the males had been removed from breeding. The longest time of pregnancy was found to be 41 days. These extended times of pregnancy without the presence of males might be caused by a retardation of nidation which cannot be explained sufficiently by the present material.

SUMMARY

During a period of several years the time span between mating and first litter and the subsequent litter intervals were recorded in Han:NMRI [SPF] outbred mice. A total of 7 740 females with 47 236 litters were included in this study.

The most important results can be summarized as follows:
- Dividing the females into 8 groups according to their total number of litters highly significant differences were observed between the dam groups concerning the time between mating and first litter and the following litter intervals.
- During the breeding period the litter intervals continuously increase in dams with a total of 2 - 5 litters.
- Females with a total of 6 - 8 litters have their maximum litter interval between the 1st and 2nd litter and before their last litter.
- The relative distribution of litter intervals in dams with a high reproduction shows an accumulation of intervals between 19 and 31 days with a peak at 23 days.
- A second peak at 39 days becomes more and more significant with decreasing numbers of litters during the breeding period.
- In dams with only a few litters there is a three peak distribution with the third peak at 49 - 50 days.
- A partly significant influence of season could be recorded despite of a complete standardisation and control of room climate. Considering the total number of litters under investigation the intervals were by 1.5 days shorter in summer than in winter.
- In 0.1% of all litters the intervals were found to be between 15 and 18 days. This was only observed when a very few or none pups were reared in the preceding litter.

A.C. Beynen & H.A. Solleveld (Eds), New developments in biosciences: their implications for laboratory animal science, ISBN 978-94-010-7973-0
©1988 Martinus Nijhoff Publishers, Dordrecht.

GENITAL INFLAMMATION IN MALE MICE. A MICROBIOLOGICAL STUDY

I. KUNSTÝŘ

1. INTRODUCTION

In a study on early embryo manipulation mice of the strains C57BL/6J and C57BL/6J^{awj} were used (Baunack et al. 1986). In order to obtain adequate numbers of early embryos in a short time the males were mated each weak with 1 or 3 females. This unusually heavy mating load often led to inflammation processes on the penis and prepuce causing impotence. As a result of this about 50 % of the 225 males had to be removed earlier from the breeding programme. No one male died. In a comparable group of normal breeding mice with permanent 1 + 1 mating the disease occured in about 6 % of the males. In order to confirm the hypothesis concerning the etiology of this condition, i.e. mating overload, minor genital injuries, and infection with ubiquitous bacteria, I examined 74 healthy and 81 ill males in a bacteriological study.

2. PROCEDURE
2.1. Materials and Methods
2.1.1. Animals. Between April 1985 and April 1987 81 males of the strain C57BL/6J and of the substrain C57BL/6J^{awj} showing acute or chronic inflammation of the preputial opening (Figs 1-3) were examined. According to the severity of the local pathological process the affected animals were divided in two groups: moderately ill (64) and considerably ill (17). 74 healthy males of the same genetic origin served as a control. All the animals were kept in two rooms under clean conventional conditions (room temperature 21+1 ^{o}C, humidity 50+5 %, artificial light between 7.00 - 19.00 h, polycarbonate cages "Macrolon I" with wooden granula bedding). A standard commercial diet (Altromin 1320) and tap water from plastic bottles were available ad libitum.

2.1.2. Examination. After euthanasia with CO_2, the animals were dissected and following 11 sites were cultivated on blood and Endo agar plates: penis, preputial gland, testis, epididymis, seminal vesicle, urinary bladder, kidney, spleen, liver, heart, and preputial concretion if present. Only one from each of the paired organs was cultivated. In order to obtain comparable numbers of bacteria the organ samples of about the same size were inoculated, i.e. pressed on the agar surface. Organ samples of smaller size were pressed two or three times on the agar surface. After 24 h aerobic incubation at 37 ^{o}C the kind and number of colonies (semiquantitative

evaluation) was read. My concern was with aerobic bacteria
only. In a few cases adspectory bacteriological diagnosis was
confirmed by additional biochemical examinations:mannit
fermentation [1] and plasma coagulase [2] in 28 Staphylococcus
aureus isolates, API 20 NE-Kit [3] in 3 so-called Pasteurella
pneumotropica isolates, and API 20 E-Kit [3] in 1 Proteus sp.
isolate. In one S. aureus isolate the lysotype was determined
in a special laboratory for fagotypisation. Protein A was also
examined in this isolate (Wullenweber-Schmidt 1987).

 2.1.3. <u>Statistical evaluation.</u> The Chi-square-test (Sachs
1969) was used to calculate the significance of the differences between the groups of animals. For this purpose the
bacterial infestation was expressed as none (sterile),
moderate ($\sim 10^1$) and heavy (10^2-10^3 or more) both for the
genital tract and for the internal organs. The influence of
pathogenic bacteria, S.aureus and P.pneumotropica, on the
severity of the disease was also calculated.

3. RESULTS
Figures 1 to 3 illustrate the <u>clinical manifestations</u> of the
disease.

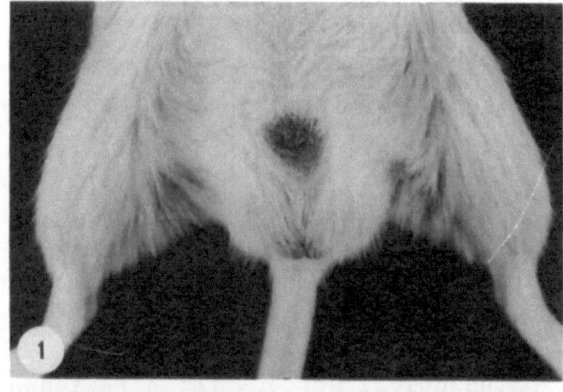

FIGURE 1.
Acute haemorrhagic
inflammation of the
genital opening in
a C57BL/6J[awj] male

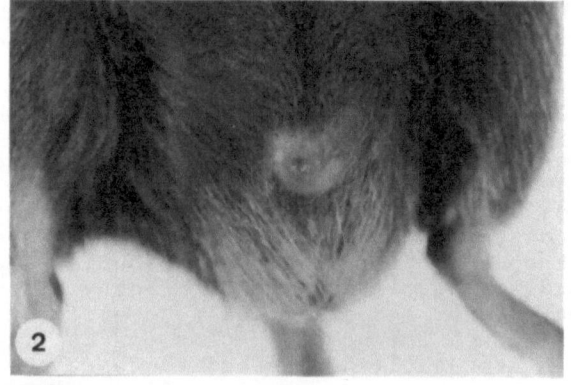

FIGURE 2.
Chronic proliferative
inflammation with
stricture of the
genital opening in
a C57BL/6J male

---

[1] Mannitol Salt Agar, CM55 Oxoid;
[2] Lyophilized rabbit plasma, 56352 Diagnostics Pasteur
[3] API 20 NE and API 20 E Kits, API System S.A.

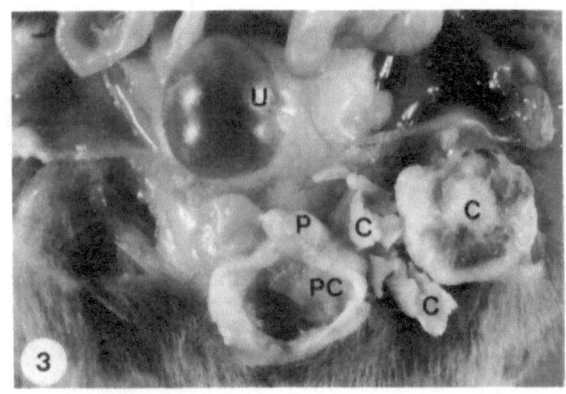

FIGURE 3.
Severe posthitis with
preputial concretion.
U = urinary bladder
P = penis
PC = enlarged
     preputial cavity
C = concretions from
    the enlarged
    preputial cavity

3.1. Bacterial species (or groups) detected were: E.coli, P.
mirabilis, enterococci, S.aureus, P.pneumotropica nomen
dubium, Corynebacterium sp., S.epidermidis. No significant
difference in the kinds of germs was found between the healthy
(group 1) and the ill (groups 2 and 3) mice. Besides the
presence of normal intestinal bacteria, two pathogenic
species, Staphylococcus aureus and so-called Pasteurella
pneumotropica, were often isolated in all groups. They
colonized mainly the genital tract occurring as hospitalism
iniciating germs (Staphylococcus in animal room 1, Pasteurella
in animal room 2) and having no influence on the severity of
the condition.
3.2. Bacterial numbers differed with high statistical
significance between the three groups of animals (Table 1 and
2). The number of isolated bacteria increased with increasing
severity of the disease.

TABLE 1.

Animal group	Location	n or %	Degree of bacterial infestation		
			sterile	moderate	heavy
1 (control) n = 74	genital tract	n	6	52	16
		%	8	70	22
	internal organs	n	48	26	0
		%	65	35	0
2 (moderate illness) n = 64	genital tract	n	0	17	47
		%	0	27	73
	internal organs	n	10	51	3
		%	16	80	4

x^2 (genitals)   = 38.5 (d.f. = 2) = p < 0.00001
   (int.organs) = 35.5 (d.f. = 2) = p < 0.00001

TABLE 2.

Animal group	Location	n or %	Degree of bacterial infestation		
			sterile	moderate	heavy
2 (moderate illness) n = 64	internal organs	n	10	51	3
		%	16	80	4
3 (severe illness) n = 17		n	1	11	5
		%	6	65	29

x^2 = 9.25 (d.f. = 2) = p < 0.01

4. DISCUSSION

The males with genital inflammation were given occasion to copulate 4 to 12 times more than the other males. Thus, it seems very probable that the higher copulatory frequency led to minor injuries on the penis and preputial opening and that the common and/or pathogenic bacteria, present in the environment and in the external genitals too, provoked the inflammation process. This pathogenic hypothesis is supported by bacteriological findings whereby not the quality but the quantity of bacteria plays a decisive role. The more severe the disease, the more bacteria were found. Our earlier experience with the disease, where the reduction of the number of mating females stopped genital inflammation supports also this hypothesis (Kunstýř et al. 1973).

Balanoposthitis in male dogs, especially those quickly sexually aroused, is also attributed to mixed infection with autochthonous bacteria present in the genital mucous membranes (Brass 1981).

The supposed minor injuries present were thought to cause the onset of ascending bacteremia. This bacteremia was more pronounced in males with preputial concretion, a site of high bacterial numbers. From the concretion, consisting of a highly proteinous secretion from the seminal vesicles (Kunstýř et al. 1982) desquamated epithelia, inflammation cells, spermias, urinary salts etc., a continous invasion of the blood stream with large numbers of different bacteria is possible.

Particularly surprising is the overall good general health of the males, despite the low level bacteremia present. It seems probable that, under good conditions, the animals can tolerate and devitalize small numbers of bacteria, even those of pathogenic nature, invading their blood stream, provided that these bacteria are of low virulence. This is the case for the Pasteurella pneumotropica in our house (Kunstýř et al. 1980). The Staphylococcus aureus, most probably of human origin (Wullenweber-Schmidt 1987), although coagulase and mannit positive, was obviously also of relatively low virulence for

our animals. Occasionally small abscesses caused either by
Pasteurella or Staphylococcus were found in the preputial
glands.
The same or a similar condition was observed, especially in
long term studies, on male mice of different strains
(references see Bendele & Carlton 1986). The latter call this
condition "obstructive uropathy" or "mouse uropathy syndrome".
I saw this condition years ago (unpublished) in a
gerontological study on 3000 C57BL/6J mice (Kunstyr &
Leuenberger 1975; Leuenberger & Kunstyr 1976) and interprete
it retrospectively in a similar manner: The urethral plug is
not a terminal but physiological phenomon; each healthy
sexually mature rodent male has an urethral plug (Kunstyr et
al. 1983). The secretion of seminal vesicles is the main
component of the ejaculate (Anderson et al. 1983) and the
urethral plug or coagulum. This is voided during copulation
and spontaneous ejaculation (Beach 1975; Huber & Bronson
1980). In long term studies, when both sexes are kept
separately and copulation con not occur, the second voidance
possibility, spontaneous ejaculation, may also be hampered
because of the animals' general weakness. As a consequence of
the senescent weakness and the presumed continuing activity of
the accessory glands the secretions accumulate in the
vesicles, in the urethra or the prepuce, and offer a breeding
place for ever present bacteria which, as a result, cause
local inflammation. Such a condition is perhaps better named
"mouse genito-urinary syndrom".~
I share the opinion of Militzer and Wecker (1986) that
C57BL/6J mice do not have a genetic disposition for skin
damages.

Acknowledgement. I am indebted to Prof. Dr. U. Ranft, Institut
für Biometrie, for his statistical evaluation of my findings
and to Priv. Doz. Dr. M. Wullenweber-Schmidt for his help in
identifying the Staphylococcus aureus isolate. My thanks for
technical help go to Mrs. Mahdi and Mrs. Neumann.

5. SUMMARY

Genital inflammation causing impotence but no death was
observed in about 50 % of the 225 C57BL/6J and C57BL/6J[awj]
males used in an intensive mating programme (1 or 3 new mating
females each week). No inflammations occured in females; they
occured in about 6 % of the males kept in comparable
conditions, as part of a permanent 1 + 1 breeding programme (4
to 12x lower copulation frequency). A semiquantitative
bacteriological study revealed the involvement of different,
mostly intestinal, in part pathogenic, bacteria
(Staphylococcus aureus, Pasteurella pneumotropica). It was not
the kind of bacteria invading the genitals and the internal
organs but their quantity which correlated (p < 0.00001 or
0.01) with the severity of the process. 16 out of 17 males
with severe inflammation, stricture and preputial concretion
had a bacteremia, i.e. 10^1 - 10^3 bacteria in each organ
sample. The protein-rich concretion, arising under voidance
blockage of the physiological urethral plug, contained the
highest numbers (>10^4) of bacteria. Thus, pathogenesis is very

probably: local inflammation and ascending infection with intestinal or pathogenic bacteria of low virulence from the environment which invade small injuries of the external genitals. The condition described as "mouse uropathy syndrome" (Bendele & Carlton 1986) can, together with senescent weakness, also be explained in this manner.

REFERENCES
1. Anderson RA, Oswald C, Willis BR, Zaneveld LJD: Relationship between semen characteristics and fertility in electroejaculated mice. J Reprod Fert, 68:1-7, 1983
2. Baunack E, Gärtner K, Schneider B: Is the "environmental" component of the phenotypic variability in inbred mice influenced by the cytoplasm of the egg? J Vet Med A, 33: 641-646, 1986
3. Beach FA: Variables affecting "spontaneous" seminal emission in rats. Physiol Behav, 15:91-95, 1975
4. Bendele AM, Carlton WW: Incidence of obstructive uropathy in male B6C3F1 mice on a 24 month carcinogenicity study and its apparent prevention by ochratoxin A. Lab Anim Sci, 36:282-285, 1986
5. Brass W: Kompendium der Kleintierkrankheiten. Hannover: Verlag M. & H. Schaper, 1981
6. Huber MHR, Bronson FH: Social modulation of spontaneous ejaculation in the mouse. Behav Neural Biol, 29:390-293, 1980
7. Kunstýř I, Hackbarth H, Naumann S, Rode B, Müller H: Zur Pasteurella-pneumotropica-Infektion in einer Barrieren-Ratten-Zuchteinheit. Z Versuchstierk, 22:303-308, 1980
8. Kunstýř I, Küpper W, Weisser H, Naumann S, Messow C: Urethral plug a new secondary male sex characteristic in rat and other rodents. Lab Anim, 16:151-155, 1982
9. Kunstýř I, Leuenberger HGW: Gerontological data of C57BL/6J mice. 1. Sex differences in the survival curves. J Gerontol, 30(2) 157-162, 1975
10. Kunstýř I, Odenthal KJ, Matthiesen T, Gärtner K: Venerische Infektion mit Impotentia coeundi bei der Maus. Z Versuchstierk, 15:371, 1973
11. Leuenberger HGW, Kunstýř I: Gerontological data of C57BL/6J mice. II. Changes in blood cell counts in the course of natural aging. J Gerontol, 31(6):648-653, 1976
12. Militzer K, Wecker E: Behaviour-associated alopecia areata in mice. Lab Anim, 20:9-13, 1986
13. Sachs L: Statistische Auswertungsmethoden. Berlin, Heidelberg, New York: Springer Verlag, 1969

A.C. Beynen & H.A. Solleveld (Eds), New developments in biosciences: their implications for laboratory animal science, ISBN 978-94-010-7973-0
© 1988 Martinus Nijhoff Publishers, Dordrecht.

MICROBIAL ASSESSMENT OF A SINGLE FUMIGATION BY FORMALDEHYDE OF A
MULTI-LEVEL ANIMAL FACILITY

M.R.GAMBLE & J.R.NEEDHAM *. The Boots Company PLC,Nottingham,UK & * The
Microbiology Laboratories,56 Northumberland Road,North Harrow,Mddsx.,UK.

1. INTRODUCTION
    A six level laboratory animal facility with basement and a plant room
on the roof had been vacated for a period of 18 months for complete
refurbishment. With the installation of new doors,floors, furniture, and
air conditioning systems, the building had been open to ingress by wild
birds,rodents and insect pests. Before reoccupation it had to be
rendered as microbiologically clean as possible. Immunologically
suppressed animals and medium term research projects might be compromised
if unappreciated levels of potentially pathogenic organisms were to remain
within the fabric of the building. It was decided to fumigate with
formaldehyde, to sterilise all surfaces as well as sub-surface structures,
ducts and conduits. After careful consideration it was decided to carry
this out as a single operation. Wind tunnel tests on a model indicated that
effluent air, discharged in a high velocity jet from roof level, did not
return to street level, even under a variety of wind conditions.
Calculations showed there would be an immediate and considerable dilution
factor on the discharged air so that it would be possible to generate
sufficient vapour to sterilise the 54 main areas involved and to remove the
vapours without risk to staff, operatives, passers-by, or the general
environment.
    This approach was only considered to be feasible because it was possible
to seal and completely isolate the building and to control the air
conditioning system from a remote computer terminal.

2. FORMALDEHYDE AND FUMIGATION
    The use of formaldehyde as an antimicrobial agent has been long
established (Hoffman,1971). Caution has been urged in its use because of its
reported toxic and potentially carcinogenic properties (Greenblatt,1983).
The traditional method of oxidation and vapourisation of formalin by
potassium permanganate is messy, hazardous to the operative and potentially
explosive (Robinson,1978). Atomisation of formaldehyde (a 40% solution) has
been shown to be less hazardous and equally effective (Ackland, Hinton, and
Denmeade,1980), but there are indications that sufficient concentrations can
only be maintained in smaller areas than that anticipated in this exercise
(Songer, Braymen, Mathis, and Monroe,1972).There is also the possibility of
formaldehyde reacting with any hypochlorites present (in disinfectants) to
form the potent lung carcinogen, bis-chlormethyl ether (Gamble,1977; Carson,
1978). A process of generating formaldehyde by heating paraldehyde granules
(a polymer containing 91% formaldehyde) was developed and shown to be
effective (Schilling, Weuffen, and Wigert,1978) and its use applied to the
poultry industry (Ide,1978) and to the beef industry (Scarlett and Mathewson,
1977). It was decided to modify this method for use in this exercise.

3. OPERATION

Thirty aluminium heating bowls were purchased and measured amounts of paraldehyde granules were placed in them at a rate of $300m^3$ per kilo.(Antec International Ltd., Windham Road, Sudbury, Suffolk.CO10 6XD.England). Each heater was fitted with a plug-in timeswitch accurate to $\pm$ 1 sec.(Davi 2000, Davi Marketing,Stamford, Lincs.PE9 2AS.England). They were then set to switch themselves on sequentially over a 2 min. period (each one being 1.5 kW) to reduce the switching load on the mains electricity supply. They were placed in corridors and rooms so that sufficient vapour could reach the 52 rooms it was intended to clean.

Two standard bacteriological swabs, moistened in sterile phosphate buffered saline solution A, were taken from each room from areas such as bench tops,inside cupboards, and walls and immediately plated on to a horse blood agar plate and a MacConkey agar plate. They were incubated at 37^oC. aerobically for 24hr.,after which they were inspected for signs of growth and assessed semi-quantitatively. Any plates with no growth were incubated for a further 48hr. and re-assessed.

In addition, two horse blood agar plates were left opened in each room. One plate had a confluent growth of Bacillus subtilis and the other a confluent growth of Staphylococcus aureus. They were placed in generally inaccessible places such as the back of cupboards and under sinks. The rooms and corridors were then lightly sprayed with water to increase the humidity.

The operation commenced on a Friday evening and involved six staff: the authors (an operator-in-charge and a microbiologist),an engineer, a building projects officer, a safety co-ordinator, and a technical services officer. Breathing apparatus was available for use and the UK Health & Safety Executive were informed of the proposals and the results.

All rooms were checked and the building was locked, the external doors were sealed with tape, and warning signs were erected around the building perimeter. The air conditioning was switched off but the background heating was left on. Each burner switched itself on and the building filled with vapour. This was confirmed by detector probes operated by the technical services officer. The perimeter of the building was checked for obvious signs of vapour and it was then left for 24hr. under the surveillance of Security staff.

At midnight on the Saturday the team reassembled and the engineer activated the air conditioning system. The perimeter and surrounding areas were again checked for the presence of vapour.

On the Sunday morning the air inside the building was checked by the technical services officer and declared safe. The microbiologist collected the agar plates from each room to incubate them at 37^oC. aerobically for seven days. They were then semi-quantitatively assessed. In addition a further two swabs were taken from each room as described above and treated in a similar manner. After incubation for 72hr. they were semi-quantitatively assessed as previously.

4. RESULTS

The spore forming Bacillus cultures were significantly reduced and all the Staphylococcus cultures were killed as shown in Table 1. The room burden of microbes was also significantly reduced, only six rooms having 11-25 colonies per plate and four rooms having 10 or less colonies per plate. A total of 94 plates had no growth at all. This demonstrates that in a large building the bacterial challenge can be reduced to insignificant levels in a single procedure with many areas appearing to have been sterilised.

	BEFORE						AFTER					
	‡‡	+‡	++	+	+-	−	‡‡	+‡	++	+	+-	−
Bacillus :	54	−	−	−	−	−	−	−	−	17	1	36
Staph. :	54	−	−	−	−	−	−	−	−	−	−	54
Swabs (2 per room):	1	6	16	37	21	23	−	−	−	6	4	94

‡‡   confluent growth   +   11-25 colonies

+‡   51-75% of plate covered   +-   10 or less colonies

++   26-50 colonies   −   no growth

Table 1 : Semi-quantitative assessment of microbial cultures and swabs before and after a single fumigation with formaldehyde.

REFERENCES

1. Ackland NR; Hinton MR; & Denmeade KR. Controlled formaldehyde fumigation system. Applied and Environmental Microbiology,39,480-487.(1980).
2. Carson FL.Formaldehyde and hydrochloric acid bis-CME. Fact or fantasy? J.Histotechnology,1,174-175.(1978).
3. Gamble MR. Hazard: Formaldehyde and hypochlorites. Laboratory Animals,11, 61.(1977).
4. Greenblatt M; Alpert LI; & Abraham JL. Update on formaldehyde. Pathologist,38,722-724.(1983).
5. Hoffmann RK. in Inhibition and Destruction of the Microbial Cell. WB Hugo (ed.),Academic Press, Lond.p225.(1971).
6. Ide PR. The sensitivity of some avian viruses to formaldehyde fumigation. Can.J.comp.Med.,43,211-216.(1979).
7. Robinson PJ. Fumigation incident. Chem.Ind.,18,723-724.(1978).
8. Scarlett CM; & Mathewson GK. Terminal disinfection of calf houses by formaldehyde fumigation. Veterinary Record,101,7-10.(1977).
9. Schilling B; Weuffen W; & Wigert H. Determination of gaseous formaldehyde from parformaldehyde tablets.2.Studies on the use of paraformaldehyde for bacterial count reduction, disinfection, cold sterilisation, and sterile storage of medical instruments. Pharmacie,33,103-104.(1978).
10.Songer JR; Braymen DT; Mathis RG; & Monroe JW. The practical use of formaldehyde vapor for disinfection. Health Lab.Sci.,9,46-55.(1972).

A.C. Beynen & H.A. Solleveld (Eds), New developments in biosciences: their implications for laboratory animal science, ISBN 978-94-010-7973-0

MECHANISMS OF NATURAL SELECTION MAINTAINING IN GROUPS OF HIGHLY STANDARDIZED MICE AND RATS, INFLUENCE ON INFECTION RESISTANCE

K. GÄRTNER, F. IGLAUER, K. MENSING, R. VELLEUER

Natural selection; i.e. mechanisms within animal collectives responsible for adaptation to altered environments is based on two main principles: (a) creation of individual variability and (b) elimination of particular individuals. Both principles - at least in part - persist in groups of highly standardized laboratory mice and rats. - The study reports on the maintainance of mechanisms creating inborn individual variability in inbred mice as well as on the persistance of sexual selection within groups of inbred male rats and the existence of three psychotypes within such animals. Finally the psychotype's different susceptibility against an infection with Mycoplasma arthritidis is investigated.

Maintenance of inborn mechanisms creating variability in inbred animals

The continuous variability of quantitative biological characteristics in natural populations can be attributed to genetical and non genetical causes.

In inbred strains the genetic variability is eliminated. But such elimination reduces to only a small extend the random distribution of various traits, such as body weight (Figure 1). The body weight's random distributions of four inbred mouse strains repeatedly estimated, were compared with that of a hybrid strain of extreme genetic diversity from which the four inbred strains could be derived. That is a mosaic hybrid produced by crossing the four inbred strains. A coefficient of variation for body weight of approximately 8.8 % was apparent in the hybrid strain. The genetically uniform inbred strains showed coefficients of variation ranging from 6 to 10 %. The elimination of genetic diversity through inbreeding has only a very small effect on the random distribution of the body weight (for details see 1, 2).

The extend of environmental component on the total varia- bility of the body weight can be assessed exactly by dividing blastocysts and preparing two monozygous twins which afterwards were transfered in different foster mothers and grew up in different litters. Diversities apparent between the two twins would be representative of the real environmental influences. A similar estimation was done in the two inbred strains AKR and C57BL6 as shown in Fig. 1. The two twin mates, however, were raised in the same foster mothers and litters (3, 4). Previous investigations in cattle show that the changed procedure makes no general differences.

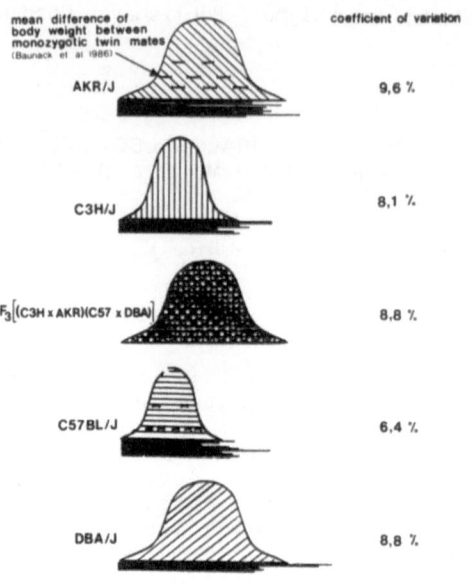

FIGURE 1. Distribution of
body weight in female inbred
mice (~100 days of age) in
four inbred strains and a
mosaic hybrid derived from
them (Gärtner et al. 1979,
Gärtner 1985)

Graphically the environmental component of variability is
represented by the length of the lines between the body
weights of the twin mates and amounts only 20 to 40 % of the
total variability (3). The remaining 60 %, represented by the
space between the lines, is inborn and caused by early
ontogenetic determinations (2), i.e. differences in plasma
components of the unfertilized oozytes. – In spite of the high
degree of standardization elimination of that very distinct
and most important type of inborn variability is not possible.
Population genetists use the term 'intangible variance', which
may cover also this component.
    The results support the hypothesis that the random distribu-
tion in inbred strains is inborn and not environmental. This
inborn variability implies the maintenance of a special
mechanism producing variability in order to support natural
selection. This mechanism determines for each individual
prospectively, prior to birth, the position in the distribution of the body weights in adults (for details see 2, 3, 4).
    For a better understanding of this observation from a
biological point of view it will be helpfull to think about a
group of animals like an European football team. Also their
pyramidal arrangement is similar to a random distribution. It
is a common knowledge that the players are notpositioned by
chance. Prior to the game, each player is selected for the
position best suited to his special capabilities. The similar
selection holds true for mice. Prior to birth, each individual
mouse receives signals from outside of the genom as to their
position in a range which we observe as a random distribution.
The random distribution we always find in highly standardized
inbred mice and rats is not the result of a large number of
trivial environmental conditions. On the contrary, this
distribution is an expression of an internal mechanism

prospectively establishing an optimal range for a living collective to win the game of life. Such a prospectively determined arrangement functions as a basal demand in evolutionary systems (2, 5, 6). That mechanism maintains in inbred strains.

Persistence of sexual selection in highly standardized rats

Similar to the football players' diversities other capabilities of the individual animals may vary according to their position in the body weight range. Investigated was the capability of the individual animals as potential fathers of the next generation.

We observed the copulatory behavioural patterns of male inbred rats. Special attention was given to the ejaculatory pattern, in a competitive design. Males were confronted with an estric female. Daily observations were performed for 100 minutes for a period of 10 or more days in randomly selected groups of four male rats of the same age and genotype. (for details see 7). The males discriminate into three psychotypes. Fig. 2 shows results of a comparable study with three inbred strains. The number of ejaculations an animal performed is expressed by the rank mean of the Friedmann rank-correlation test. The ordinate shows the frequency of animals with the same ranks of ejaculatory patterns. Type A-animals have a significantly high; type B intermediate; and type O significantly low ejaculatory performances. This pattern was similar in all three genotypes. Further investigations (7, 8) let suppose that type A animals father approx. 70 % of the subsequent generation. The type B males sire only 25 %; type O only 5 %.

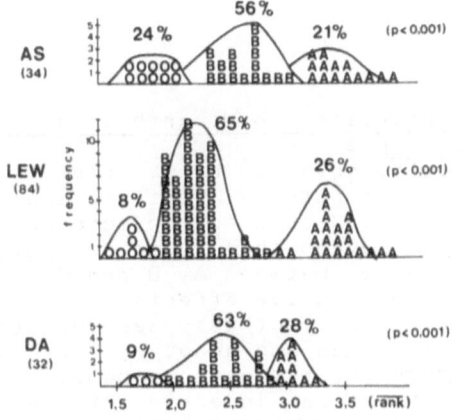

FIGURE 2. The frequency distribution of the ejaculatory behavioural patterns in male rats (four animals per cage of the inbred strains AS/Ztm, LEW/Ztm, DA/Ztm). Psychotype A many ejaculations; psychotype B intermediate ejaculations; psychotype O few ejaculations

With the football team analogy in mind and with some knowledge of the strategy of evolution (4, 5, 6), specific arrangement of the psychotypes was questioned in the distribution of body weight of the total collective. Results from another study (2) are shown in Fig. 3. The body weight distribution of 520 rats was cuttet into six slices or classes. In each class the frequency of psychotype A and

psychotype O was determined and compared. The frequencies are
different in each class. The O-animals are represented more
frequently in the outer classes. A-animals prefer placement
near the centre of the distribution. These results show that
the sexual selection expressed by the existence of the psychotypes A, B and O in groups of high standardized male rats is
linked to the body weight distribution. A similar centralizing
selection is known also from other species and described for
Drosophila (9), or quails (10), or in poultry (11). Hence,
this selection seems to be an epigenetic and general strategy
of internal selection expressed in animal population.

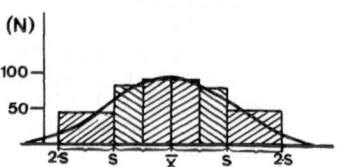

FIGURE 3. Centralizing
selection of type A-animals.
- Above: the body weight
distribution of 520 adult
male rats of the genotypes
DA, LEW and BH, cutted in six
slices of a dimension of "s"
or "s/2". - Middle: the
frequencies of 41 type O-rats
out of a sample of 164
animals in the six body
weight classes. - Below: the
frequencies of 130 type
A-rats out of a sample of 520
animals in the six body
weight classes.

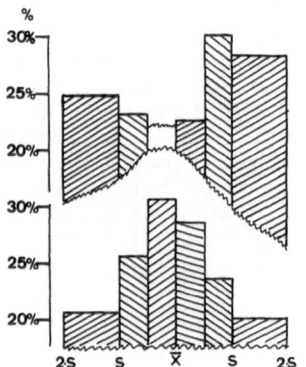

Susceptibility to M. arthritidis as influenced by the psychotypes

Results obtained by v. Holst (12) with Tupaias and from
studies in rats (2) it is obvious that animals of the
psychotype O show higher adrenocortical activities and corticosterone levels than the other psychotypes. Other endocrine
differences between A, B and O are apparent. this differences
may have negative effects on the resistence against infections. This hypothesis was investigated using male rats of
identical age and genotype (psychotype estimated according to
the previously mentioned behavioural examination). Mycoplasma
arthritidis was injected intravenously. The animals were caged
seperately or they persist living in groups of four per cage.
Following the infection, 20 different clinical characters were
continously recorded over a period of approx. 110 days (for
details see 13, 14). A remarkable reduction in body weight and
various signs of polyarthritis were important clinical
findings. Body weight decreased a few days post infection by
approx. 25 % and remained lowered until the end of the observation. Severity of weight reduction or polyarthritis was
calculated semiquantitatively for each animal. Values were

compared among groups of isogenic animals (14). A large variability was observed between the animals. Few animals showed
only a slight reduction in weight. Other animals lost 30 % or
more. The psychotypic influence on the severity of the reduction in body weight during the acute period of the infection
as well as during the total observation was estimated by
analysis of variance, modell II (Table 1). In both periods,
the variability is only to a small degree (approx. 10 - 30 %)
the result of the above mentioned psychotypic differences between the animals. This was apparent in only the genotypes AS
and DA. In rats of genotype LEW, the psychotypic influence on
the variance was insignificantly low.

TABLE 1. Analysis of variance (Modell II). The influence
of the psychotypes A, B, O on the variability of the body
weight reduction and polyarthritis in male rats infected
with M. arthritidis (in % of the total variability)

strain	N	weight reduction)1		polyarthritis	
		1-38	1-105	1-38	1-105
AS	24	26 % (*)	23 % (*)	10 % (*)	17 % (*)
AS (E)	10	30 % **	80 % **	70 % **	78 % **
LEW 5	24	31 % *	20 % (*)	1 %	1 %
LEW 4	20	12 %	1 %	1 %	1 %
LEW 3	24	5 %	7 %	13 %	1 %
LEW 2 (E)	16	1 %	1 %	13 %	1 %
LEW 1 (E)	10	1 %	1 %	1 %	1 %
DA	24	29 %	46 % (*)	31 % (*)	49 %
DA (E)	8	34 %	7 %	35 %	1 %
	160	∿20 %	∿20 %	∿13 %	∿16 %

)1 days under observation; (E) single caged
(*) p<0.1; * p<0.05; ** p<0.01

Repeatedly observed clinical alterations at the joints of
the legs were semiquantitativly recorded and their variance
calculated for the acute period and the total time of observation. Similar results as in body weight reduction were
obtained. There is a large variability of the severity of polyarthritis between animals. A few animals showed inflammation
at only one joint. The remainder showed 10 and more arthritic
joints. Analysis of variance shows that the variability of the
arthritic characters is caused to a small degree (10 - 30 %)
by the above mentioned psychotype. There is only a small interaction between the psychotypes and their susceptibility to an
artificial infection. Mechanisms of internal selection are to
a great extent independent from effects of external selection
as bacterial infections.

SUMMARY
Variability of the body weight remains in spite of high
standardization in inbred mice because it is mainly (∿60 %)
caused by an inborn determination mechanism creating
prospectively_a Gaußian distribution of such collectives of
animals.

Within groups of highly standardized inbred male rats sexual selection persists assigning the existence of three psychotypes among the animals different in their copulatory behaviour and performing centralized selection on the body weight distribution.

Susceptibility to M. arthritidis is only to a small degree (1 - 30 %) influenced by psychotypic characteristics of the male rats.

REFERENCES
1. Gärtner K, Meyer SD, Treiber A: Strain and sex specifity in the variability of the body weight in inbred mice. Z Versuchstierk, 21: 259-272, 1979
2. Gärtner K: Versuchstierkunde und "intangible variance" - eine dritte Komponente der kontinuierlichen Variabilität neben Erbgut und Umwelt. Verh Dtsch Zool Ges, 78: 61-75, 1985
3. Baunack E, Gärtner K, Schneider B: Is the "environmental" component of the phenotypic variability in inbred mice influenced by the cytoplasm of the egg? J Vet Med A, 33: 641-646, 1986
4. Gärtner K, Baunack E: Is the similarity of monozygotic twins due to genetic factors alone? Nature, 292: 646-647, 1981
5. Schwefel HP: Numerical optimization of computer models. Chichester: Wiley, 1981
6. Rechenberg I: Evolutionsstrategie. Optimierung technischer Systeme nach Prinzipien der biologischen Evolution. Stuttgart, Bad Cannstatt: Fromman-Holzboog, 1973
7. Gärtner K, Amelang D, Bredereck W, Tessmer Ch: Promiskuitive Kopulation bei Laboratoriumsratten. Fertilität, 2: 224-230, 1986
8. Huck UW, Lisk RD, Allison JC, van Dongen CG: Determinants of mating success in the golden hamster (Mesocricetus auratus): Social dominance and mating tactics under semi-natural conditions. Anim Behav, 34: 971-989, 1986
9. Kearsey MJ, Barnes BW: Variation for metrical characters in Drosophila populations. II. Natural selection. Heredity, 25: 11-21, 1970
10. Lucotte G: Selection stabilisatrice chez la caille domstique. Expérimentation animale, 7: 195-213
11. Singh, H, Nordskog AW: Breeding and genetics. Significance of body weight as a performance parameter. Poultry Science, 61: 1933-1938, 1982
12. v. Holst D, Fuchs E, Stöhr W: Physiological changes in male Tupaia belangeri under different types of social stress. Karger Biobehavioral Medicine Series, 2: 382-390, 1983
13. Velleuer R: Entwicklung eines Modelles zur Untersuchung genetischer und psychosozialer Einflüsse auf den Verlauf einer experimentellen Infektion mit Mycoplasma arthritidis bei der Ratte. Hannover, Tierärztl. Hochsch., Diss., 1986
14. Mensing KA: Verlauf und Schwere einer experimentellen Infektion mit Mycoplasma arthritidis bei der Ratte und Einfluß von Psychotypen. Hannover, Tierärztl. Hochsch., Diss., 1987

A.C. Beynen & H.A. Solleveld (Eds), New developments in biosciences: their implications for laboratory animal science, ISBN 978-94-010-7973-0
© 1988 Martinus Nijhoff Publishers, Dordrecht.

PROVOCATION OF LATENT INFECTIONS

N.-C. JUHR

Health status control is part of laboratory animal production and animal experiment and practiced as spf-status control, in grading schemes, or according to GLP-regulations.

Much work has been done to improve the control by using more sensitive and more specific methods.

Many years of experience show that control of production units and animals prior to the experiment is unsecured. Health problems occur during the experiment despite negative control results before starting the experiments.

If some experiments provoke latent infections, some type of challenge could be useful to detect latent infections (KARASEK 1970).

A set of provocation tests is proposed as part of health status control with the aim to reach secure results with fewer animals and less effort.

In many attempts a number of provocation tests proved useful on which - after a short communication (JUHR 1985) - will be reported.

Materials and methods

For health status control 20 animals each were selected from 12 charges to be controlled (animal-delivery, animal room, experiment).

10 animals were subjected to routine diagnostic measurements and 10 to one of the treatments for provocation.

Routine bacteriological control is practiced as culture of organ-material (spleen, liver, kidney, lung), blood-culture from heartpuncture in BHI-broth, lung and trachea culture in Mycoplasma media, and intestinal content selectively enriched in selenite broth for salmonella and cultured on Tergitol-agar.

A. Provocation of Latent Enteral Infections

As provocation tests to detect latent enteral infections we practice a medicamentous treatment of the animal with Cyclophosphamide, Tetrachlormethane, Chloralhydrate or Indomethacin in a dosage given in table 1.

TABLE 1. PROVOCATION OF ENTERAL INFECTIONS

Cyclophosphamide	80 mg/kg BW i.p.	SINGER et al. 1972
Tetrachlormethane	0.5 ml/kg BW s.c.	TAKENAKA and FUJIWARA 1975
Chloralhydrate	400 mg/kg BW i.p. (conc.:100 mg/ml)	FLEISCHMAN et al. 1977
Indomethacin	10 mg/kg BW oral on 3 days	BENONI et al. 1984

Cyclophosphamide has a special effect on the small gut and inhibits cell proliferation (ECKNAUER and LÖHRS 1976). This treatment results in a bacterial translocation as shown by BERG (1983).

Tetrachlormethane is liver toxic and has been reported to enhance B. piliformis infection (TAKENAKA and FUJIWARA 1975).

Chloralhydrate causes atony and distension of ileum and colon (FLEISCHMAN et al. 1977,. DAVIS et al. 1985) with stasis of intestinal content when given in concentrations higher than 125 mg/ml i.p.

Indomethacin leads to ulcerations in the distal jejunum and ileum and to enteritis (BENONI et al. 1984).

B. Provocation of Latent Respiratory Infections

Besides Cyclophosphamide for provocation of respiratory (SINGER et al. 1972) we use the catalase-blocking compound Aminotriazole (BRENNAN and FEINSTEIN 1969), or cause a local damage of the lung-epithelium through Hexamethylphosphoramide treatment (KIMBROUGH and GAINES 1966, OVERCASH et al. 1976), or produce a lung-edema by i.v. injection of a large volume of saline (BARROW 1968).

The dosage is given in table 2.

TABLE 2. PROVOCATION OF RESPIRATORY INFECTIONS

Cyclophosphamide	80 mg/kg BW i.p.	SINGER et al. 1972
3-Amino-1,2,4-triazole	2 mg/ g BW i.p.	BRENNAN and FEINSTEIN 1969
Hexamethyl-phosphoramide	1 g/l water	OVERCASH et al. 1976
physiol. NaCl-solution	10 ml/100 g BW i.v.	BARROW 1968

RESULTS

All bacteriological tests with our conventional methods revealed freedom of specified pathogens in the 12 charges investigated.

On the other hand our routine laboratory controls proved

that the system is sensitive and efficient when organisms were added to the sample in high dilutions from reference cultures of organisms searched for.

The results of health control measurements in animals previously subjected to a provocation treatment are tabulated for the two test groups in chronological order of the experiments.

A. Provocation of Latent Enteral Infections

TABLE 3. EFFECT OF TREATMENT ON DETECTION OF ENTERAL PATHOGENS

animals	treatment	number dead/treat. animals	result	
			pathol/histol.	culture/FITS *
Han : WIST conventional	Cyclophos-phamide 20 mg/250 g BW i.p.	1/10	necrotic foci in liver, ileum	*Bacillus piliformis
Han : NMRI conventional	Cyclophos-phamide 2,4 mg/30 g BW i.p.	3/10	spleeno-megaly	Streptobacillus moniliformis
DBA/N conventional	Tetrachlor-methane 1 mg/kg BW i.p.	3/10	necrotic foci in the liver	*Bacillus piliformis
Han : WIST conventional	Chloral-hydrate 400 mg/kg BW i.p.	2/10	atony of cecum and ileum, enteritis	Salmonella enteritidis
Han : WIST conventional	Indomethacin 10 mg/kg BW oral on 3 days	0/10	enteritis	Yersinia enterocolitica
C57BL conventional	Indomethacin 10 mg/kg BW oral on 3 days	0/10	ulcera in duodenum and ileum, enteritis	Citrobacter freundii

It was surprising, that morbidity - and in some groups mortality - occurred.

For laboratory work a diagnostic situation - with anamnesis, clinical and pathological picture - was evident.

Although each charge was only subjected to one treatment, in all cases diagnostic work resulted in a finding other than routine work without treatment.

A comparative evaluation of the efficiency of the different tests is not possible yet.

B. Provocation of Latent Respiratory Infections
TABLE 4. EFFECT OF TREATMENT ON MYCOPLASMA PULMONIS DETECTION

animals	treatment	number dead/treat. animals	lung-results histol. PLI*	culture	FITS
LEW/Crl conventional	Cyclophos- phamide 20 mg/250g BW i.p.	1/10	1,4	4/10	++
LEW/Crl conventional	Amino- triazole 500 mg/250 g BW i.p.	4/10	2,2	10/10	+++
Han : WIST SPF	HMPA 1 g/l water	1/10	1,6	1/10	-
Han : WIST conventional	HMPA 1 g/l water	6/10	2,0	6/10	+
Han : WIST SPF	physiol. NaCl-solut. 10 ml/100 g BW i.v.	0/10	1,4	1/10	-
Han : WIST conventional	physiol. NaCl-solut. 10 ml/100 g BW i.v.	1/10	2,0	6/10	++

*Peribronchial Lymphocytic Index (GIDDENS and WHITEHAIR 1969)

Based on these results we have changed our procedure for bacteriological spf-status control.
For routine control we now select 10 animals from each charge to be controlled and subject them individually or in pairs to one of the provocation-treatments.

REFERENCES

1. Barrow MV: Modified intravenous injection technique in rats. Lab.Anim.Care 18, 1968, 570 - 571.
2. Benoni G, Cuzzolin L, Raimondi MG, Velo GP: Indomethacin-induced intestinal lesions and fecal flora. Advances in Inflammation Research 6, 1984, 103 - 108.
3. Berg RD: Bacterial translocation from the gastrointestinal tracts of mice receiving immunosuppressive chemotherapeutic agents. Current Microbiol. 8, 1983, 285 - 292.
4. Brennan PC, Feinstein RN: Relationship of Hydrogen peroxide production by Mycoplasma pulmonis to virulence for catalase-deficient mice. J.Bact. 98, 1969, 1036 - 1040.
5. Davis H, Cox NR, Lindsey RJ: Diagnostic exercise: Distended abdomens in rats. Lab.Anim.Sci. 35, 1985, 392 - 394.

6. Ecknauer R, Löhrs U: The effect of a single dose of Cyclophosphamide on the jejunum of specified pathogenfree and germfree rats. Digestion 14, 1976, 269 - 280.
7. Fleischman RW, McCracken D, Forbes W: Adynamic ileus in the rat induced by chloral hydrate. Lab.Anim.Sci. 27, 1977, 238 - 243.
8. Giddens WE, Whitehair CK: The peribronchial lymphocytic tissue in germfree, defined-flora, conventional and chronic murine pneumonia affected rats. in: Germ-free biology ed. Mirand EA, Black N, New York: Plenum press, 1968, p. 75-84.
9. Juhr N-C: Gesundheitsüberwachung von Versuchstieren - Das Problem der latenten Infektion. Tierlaboratorium 10, 1985, 231 - 251.
10.Karasek E: Der Kortisontest zum Nachweis latenter Infektionen bei Versuchsmäusen. Z.Versuchstierkd. 12, 1970, 155-161.
11.Kimbrough R, Gaines TB: Toxicity of hexamethylphosphoramide in rats. Nature 211, 1966, 146 - 147.
12.Overcash RG, Lindsey JR, Cassell GH, Baker HJ: Enhancement of natural and experimental respiratory mycoplasmosis in rats by hexamethylphosphoramide. Amer.J.Pathol. 82, 1976, 171 - 190.
13.Singer SH, Ford M, Kirschstein RL: Respiratory diseases in cyclophosphamide-treated mice. I. Increased virulence of Mycoplasma pulmonis. Infect.Immun. 5, 1972, 953 - 956.
14.Takenaka S, Fujiwara K: Effect of carbon tetrachloride on experimental Tyzzer's disease of mice. Japan.J.Exper.Med. 45, 1975, 393 - 402.

A.C. Beynen & H.A. Solleveld (Eds), New developments in biosciences: their
implications for laboratory animal science, ISBN 978-94-010-7973-0
© 1988 Martinus Nijhoff Publishers, Dordrecht.

INTRODUCTION TO RECOMBINANT DNA TECHNOLOGY

J. VIJG, TNO Institute for Experimental Gerontology, P.O. Box 5815,
2280 HV Rijswijk, The Netherlands

PRINCIPLES
The characteristics of all living organisms, from simple viruses to
human beings are ultimately determined by the interplay of two types of
biomolecules: nucleic acids and proteins. All organisms are equipped with
one or more molecules (chromosomes) of deoxyribonucleic acid (DNA) as
ultimate template for their biological phenotype except for certain viruses
which have ribonucleic acid (RNA) as their primary source of information. As schematically depicted in Fig. 1, the information for the
synthesis of the many different proteins each cell relies on for its well
functioning can be retrieved from the cellular DNA via so-called
messenger RNA (mRNA). The processes of information retrieval, termed

EUKARYOTIC CELL

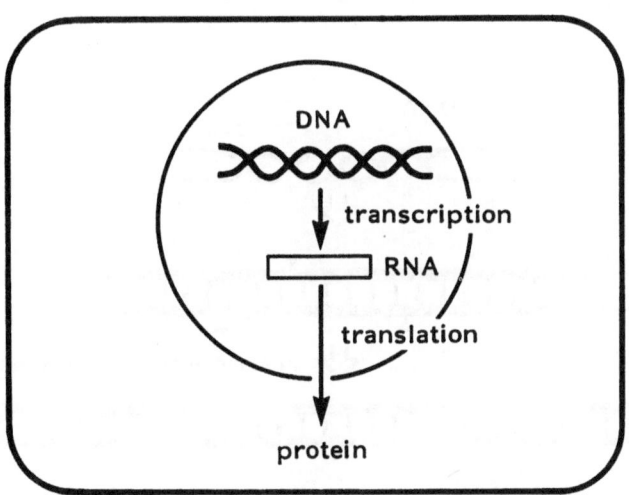

FIGURE 1. Information retrieval from the DNA takes place according to the
central dogma: DNA → RNA → protein.

transcription and translation, are themselves influenced by a great
number of endogenous and exogenous factors. Exogenous factors are, for
example, the environmental agents that can damage biomolecules (e.g.,
radiation and chemicals).

The principle that underlies recombinant DNA technology can be found in the nature of nucleic acid molecules. A typical (double-stranded) DNA molecule consists of two interwound polynucleotide chains. Each nucleotide in one chain - composed of a sugar, a phosphate group and a base - is specifically linked by hydrogen bonds between the bases to a nucleotide in the other chain. There are four different bases (and thus four different nucleotides), but no more than two types of pairing. Deoxyadenosine monophosphate (A) only binds to thymidine monophosphate (T) and deoxyguanosine monophosphate (G) only binds to deoxycytidine monophosphate (C). As a consequence of this specific base-pairing, certain pieces of single-stranded DNA recognize and bind only to their complementary counterparts and not to other DNA segments. However, it should be mentioned that a certain level of mis-pairing can be tolerated, that is, in addition to perfect-fitting DNA strands, also segments that are not completely complementary but only to some extent, are able to hybridize (bind) to each other. The situation for RNA is essentially the same, with the exception that the T (thymine) in the nucleotide thymidine monophosphate is replaced by a U (uracil), the sugar is now ribose instead of deoxyribose and RNA is single stranded. It can, however, very well basepair to a complementary DNA strand.

The complementation principle is applied by the cell to replicate its DNA and to transcribe DNA into RNA, and by the molecular geneticist to selectively isolate and propagate specific DNA sequences. This process of selective isolation, also called genetic cloning, is the basis of recombinant DNA technology. The sequences of interest can be derived from the total genomic DNA or the mRNA of the organism under study, for example, a virus, a bacterial pathogen, a parasite or a higher animal.

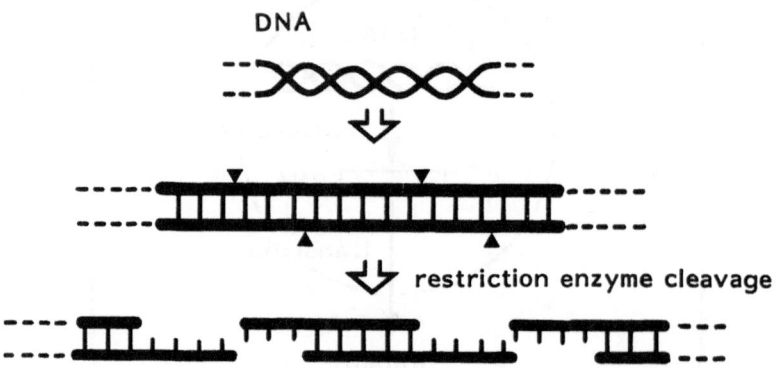

FIGURE 2. DNA can be cut into fragments by means of restriction enzymes. Restriction enzymes are isolated from bacteria and recognize specific sites in the DNA. The restriction enzyme EcoRI, for example, recognizes the sequence GAATTC in the double-stranded DNA and will cut the DNA at this site.

TECHNIQUES

Recombinant DNA technology, that is, the manipulation of genes and other DNA sequences, has been made possible by the discovery of restriction enzymes. Restriction enzymes are DNA-cutting enzymes derived from bacteria. The characteristic that makes them so important in recombinant DNA technology is their specificity. They do not cut DNA at random but only at certain sites, the sequence of which (four basepairs or larger) depends on the enzyme (Fig. 2). The occurrence of many different restriction enzyme-recognition sites in genomic DNA allows the investigator to excise virtually all DNA fragments desired and to insert them at other sites in the same or another genome, or to create totally new genomes at will. Resealing occurs with the help of enzymes called ligases. Thus, the key tools for recombinant DNA experiments are the large variety of restriction enzymes and the ligases. Another question is where to start when one wants to clone, that is, to specifically isolate and amplify, a certain DNA sequence, for instance a specific gene. For this purpose there are two strategies, based on either RNA or genomic DNA as the starting material.

Preparation of cDNA

It is not unusual to start with RNA as ultimate source of the DNA sequence to be cloned. This is the only possible way when utilizing an RNA virus. With eukaryotic cells, mRNA is an ideal source, especially when the gene in which one is interested is highly expressed. A large part of the cell message will then consist of the mRNA for the gene of interest. Some mammalian genes, the mRNA of which is present in small amounts are difficult to isolate and clone; this requires special techniques (1).

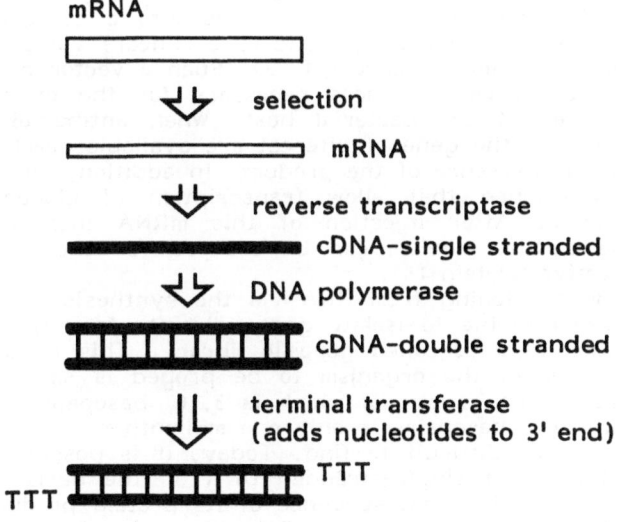

FIGURE 3. The preparation of double stranded cDNA. The final step, the synthesis of the T-tails, is necessary to allow insertion and ligation into a vector, which on its turn has been cut open with a restriction enzyme and provided with (complementary) A-tails.

It is necessary when starting from RNA to convert these molecules into DNA, since RNA itself cannot be cloned. This conversion process, shown in Fig. 3, is performed by a retrovirus enzyme, reverse transcriptase, which synthesizes complementary DNA (cDNA) from an RNA template and the four deoxyribonucleosides. The RNA portion of the RNA:DNA hybrid can then easily be removed by alkaline hydrolysis and a complementary second strand of DNA can be synthesized by means of a DNA polymerase. The now double-stranded DNA fragment (the cDNA) can be propagated – after its insertion and ligation into a suitable vector (a small circular or linear DNA fragment that is able to replicate in bacteria) – in a bacterial host. The bacteria are said to be transformed with the cDNAs.

Assuming that the starting material was total mRNA and that each mRNA species is cloned at least one time, the resulting vector-containing bacteria now represent a cDNA library, specific for the cell type or organism from which the mRNA was derived (see ref. 2 for recent methods to generate cDNA libraries). It is now necessary to identify the cDNA sequences that have been cloned. When the starting material is a mRNA population greatly enriched for a specific species, most of the clones will contain this specific cDNA. A mRNA population can be enriched for a specific mRNA species by separating the mRNAs on the basis of size, for example by gel electrophoresis. Subsequently, the different fractions are tested by in vitro translation for the protein products that they code for. In vitro translation can be performed by using an extract of reticulocytes, which contains the enzymatic machinery necessary for protein synthesis. However, when one wants to assess biological activity (for instance, when no other information of the protein product is available) it is necessary to transfer the mRNA to eukaryote cells, in which biologically active proteins can be produced.

Different individual clones of the cDNA library can be identified relatively easily when the library is made into a so-called expression vector. This can be a circular plasmid or a (linear) virus that is able to infect bacteria (see for instance ref. 3). Such a vector has been equipped with all the DNA sequences necessary for the expression of the cloned sequences in the bacterial host. When antibodies towards the protein product of the gene of interest are available, each clone can be analysed for the presence of the product. In addition, cloning strategies have been developed that allow transcription of cloned cDNAs into mRNAs in vitro. After injection of this mRNA into Xenopus laevis oocytes, the culture medium can be assayed for the presence of the biologically active protein (4).

In addition to cloning from RNA via the synthesis of complementary DNA it is also possible to isolate genes directly from genomic DNA via the preparation of a so-called genomic library. This is not so difficult when the genome of the organism to be probed is small. However, in case of humans, the genome of which is 3.10^9 basepairs long, a typical gene of about 3000 basepairs is only one millionth of the total amount of DNA and therefore difficult to find. Today, it is possible to chemically synthesize a piece of single-stranded DNA on the basis of information derived from the amino acid sequence of its protein product (1,5). One can screen a genomic library (or a cDNA library) using this oligonucleotide (typically 20 basepairs long) for clones homologous to the oligonucleotide.

Hybridization analysis

Specific DNA sequences that have been isolated and cloned can be used as DNA probes for the detection of sequences that are homologous or identical to themselves among a vast number of "foreign" sequences. For instance, viral sequences integrated in the DNA of mammals can be easily spotted by means of Southern hybridization analysis. This method, developed by E.M. Southern (6), is the most frequently used technique for genome analysis. Figure 4 is a schematic representation of an experiment in which the rat genome was analysed for sequences complementary to a human cDNA probe derived from a mRNA for an HLA class I antigen (1). As previously mentioned, there is often enough homology to allow hybridization under certain experimental conditions, even when perfect basepairing is precluded by a number of differences in the MHC genes of man and rat. The human probe in this experiment appeared to have enough homology to rat MHC genes to basepair to various MHC genes (MHC gene-containing DNA fragments). The basic steps in Southern analysis are restriction enzyme digestion of the genomic DNA (see above and Fig. 2), electrophoretic separation of the fragments according to their size, subsequent transfer of the pattern to nitrocellulose paper and hybridization of the size-separated DNA fragments on the filter paper with a labeled DNA probe (Fig. 4). DNA probes can be radioactively labeled by means of for instance nick translation. The non-radioactive nucleotides of the probe are replaced during this process by

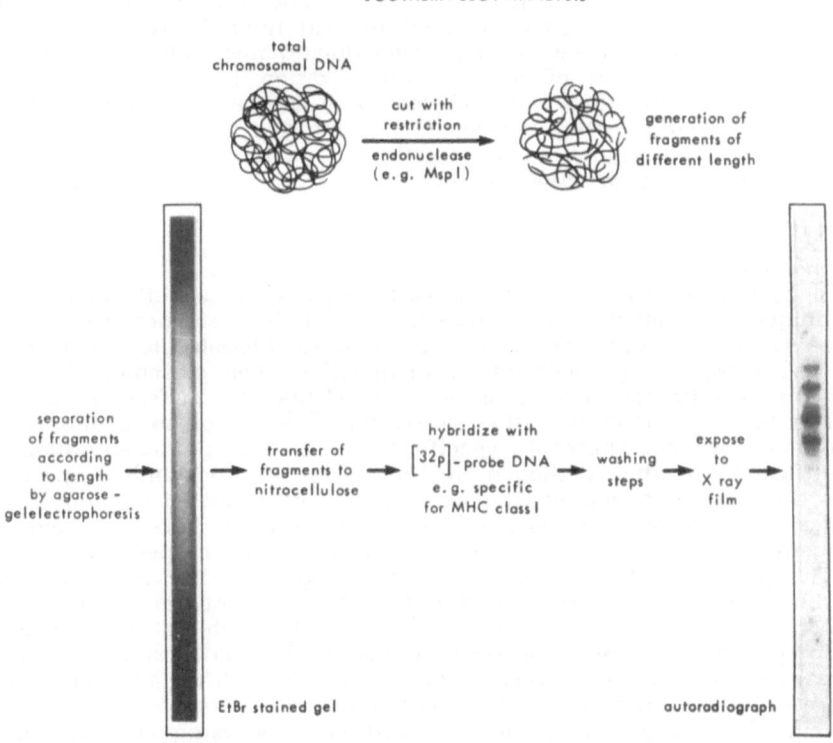

FIGURE 4. Schematic representation of the Southern blot hybridization protocol. For detailed explanation, see text.

^{32}p-labeled ones. After hybridization and the removal of unhybridized excess probe, the DNA fragments that contain sequences homologous to the DNA probe can be detected by autoradiography (Fig. 4). Alternatively, DNA probes can now be labeled by means of other non-radioactive detection systems (for a review, see ref. 7).

A comparable type of analysis using RNA is Northern hybridization. RNA molecules are size-separated and blotted to nitrocellulose paper in this technique. RNA species differing in size can be separately quantified in a certain tissue, cell sample or in body fluids or clinical isolates. A more rapid, though less specific, technique of quantifying mRNA species is the so-called quick blot method, developed by Bresser et al. (8). A sample of cells, a clinical isolate or body fluid is treated with proteolytic enzymes, solubilized in Na Iodide and passed through a nitrocellulose filter in this method. Depending on the exact conditions, either mRNA or DNA is selectively immobilized on the filter and can be detected and quantified by hybridization to a DNA (or RNA) probe. Oncogene expression in subsets of human T-lymphocytes has recently been accurately quantified using this method (9).

The previously described hybridization techniques are all based on the immobilization of RNA or DNA. Application of these techniques is only possible on whole cell populations or pieces of minced tissue. A different application of hybridization analysis is the identification of RNA or DNA species in single isolated cells. This approach has proved remarkably successful (for a review, see ref. 7) and more recently, Trask et al. demonstrated that this so-called in situ hybridization can even be performed with cell nuclei in suspension during flow cytometry (10).

A logical extension of in situ hybridization is hybridization histochemistry (11) in which the labeled DNA probe is incubated with a section of tissue. After appropriate washing and drying, the target sequences can be visualized in specific cell populations or tissue regions by autoradiography or fluorometry.

APPLICATIONS

Diagnostics

Since the specific character of each organism is a reflection of a set of unique biomolecules it is possible to distinguish between different organisms on the basis of differences in these biomolecules. This principle has important implications in diagnostics. For example, the set of proteins specific for the bacterium Escherichia coli differs greatly from that of humans. Generally, this means that the use of antibodies towards a protein (or part thereof), specific for a certain pathogen, allows the identification of that pathogen in clinical isolates or clinical specimens. This approach has allowed the development of rapid diagnostic tests in clinical microbiology. Monoclonal antibodies, which recognize only one antigen, have proved to be especially useful in this respect (12).

The principle of specific-protein recognition as the basis of distinguishing between one organism and the other or between their products is not limited to organisms that differ greatly such as mammals versus microorganisms. On the contrary, although different mammalian species have many proteins in common they nevertheless differ at certain sites. Therefore it is possible to raise antibodies that specifically react with proteins from only one species. In addition, by using antisera towards protein products of polymorphic gene systems, such as the major histocompatibility complex (13), it is possible to distinguish between individuals of one species.

It has become possible with the emergence of recombinant DNA technology to directly analyse the genes (or their mRNAs) that code for the various types of proteins characteristic for a cell or an organism. In addition to its great value in the production of vaccines or other poly-peptide products of cloned genes (12), recombinant DNA techniques are also useful in diagnostics. In this latter respect, they offer certain advantages as compared to immunodiagnostics. The presence of an infec-tive agent itself is directly indicated by its genome in DNA-based diag-nostics (14), whereas in immunodiagnostics, the pathogens themselves are usually not identified but rather one or more of their products. Furthermore, immunological detection of pathogens in body fluids, for instance serum, can be greatly disturbed by the presence in such fluids of antibodies towards the pathogen to be detected; this precludes accu-rate quantitation. Finally, it has become clear during the last decade that the cellular genotype is more amenable to rapid analysis than the cellular phenotype. This is most clearly illustrated by the advances that

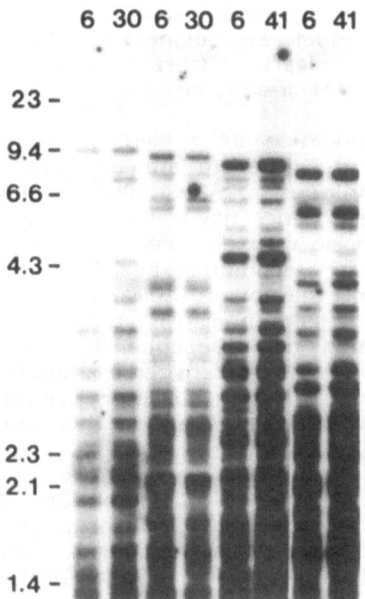

FIGURE 5. Southern blot hybridization analysis of DNA isolated from blood cells of a number of individual rats belonging to two different inbred strains, Wistar-derived WAG/Rij and Brown Norway BN/BiRij rats. For the principle of this method, see Fig. 4. The DNA probe used was an M13 virus which contains sequences that are homologous to the highly polymorphic minisatellites (18). The 8 lanes contain from left to right: 1 and 2, DNA from a 6- and a 30-month-old BN rat, digested with Hae III; 3 and 4, the same DNAs, digested with Sau3A; 5 and 6, DNA from a 6- and 41-month-old WAG/Rij rat, digested with Hae III; and 7 and 8, the same DNAs, digested with Sau3A.

has been made in the elucidation of the organization of complex gene systems such as the major histocompatibility system (13).

Another diagnostically interesting aspect of the possibility to analyse the genome itself instead of its products bears on sequences that cannot be cloned from mRNA. Protein-coding sequences are only a small part of the total genome (of both higher and lower organisms), which contains, besides introns and gene-flanking sequences, a number of so-called repetitive sequence families. Such a family typically consists of a consensus sequence which is present in a number of copies (sometimes a million), throughout the genome. Jeffreys et al. demonstrated that one such family, the so-called minisatellite repeats, are highly polymorphic (15). Indeed, they are so polymorphic that no two individuals, with the exception of monozygotic twins, can be found with the same arrangement of these sequences in their genomes. This characteristic makes probes for those sequences highly valuable as tools for identity testing, so-called DNA fingerprinting. This is important in forensic science (16) and for seeking the location of defective genes in inherited diseases (17).

These highly polymorphic probes can be used in laboratory animal science to determine the degree of genetic homogeneity in animal strains. This is illustrated by Fig. 5, which shows a series of Southern blot analyses of different individual rats belonging to two inbred strains. The minisatellite repeat pattern clearly differs between the two strains, but is identical among members of one strain.

Gene transfer

Another important application of cloned genes in laboratory animal science is the possibility to generate altered genomes. At the present time, a number of techniques are available to introduce genes into mammalian cells. Once these genes have entered the cell they are ligated and integrated in the genome with efficiencies as high as 10 % (19). The availability of gene transfer techniques has been considered a starting point for gene replacement therapy of humans (20). Another application that is already being practiced in many laboratories is to inject genes into fertilized eggs, which are subsequently reimplanted into foster mothers to give rise to so-called transgenic animals. The relevance of this aspect of recombinant DNA for laboratory animal science is beyond doubt and is discussed in detail elsewhere in this volume.

ACKNOWLEDGEMENTS

I would like to thank Drs. J.G.J. Bauman, L. Dorssers, S.K. Durham, J. Gossen, A. Kos and A.G. Uitterlinden for helpful discussions, Mr. E.J. van der Reijden and Mr. H.J. van Westbroek for the preparation of the figures and Mrs. G.A. Hofland for typing the manuscript.

REFERENCES
1. Sood, A.K. et al.: Proc. Natl. Acad. Sci. USA, 78, 616, 1981.
2. Dorssers, L. and Postmes, A.M.E.A.: Nucleic Acids Res., 15, 3629, 1987.
3. Young, R.A. and Davis, R.W.: Proc. Natl. Acad. Sci. USA, 80, 1194, 1983.
4. Noma, Y. et al.: Nature, 319, 640, 1986.
5. Edge, M.D. and Markham, A.F.: Biochim. Biophys. Acta, 695, 35, 1982.
6. Southern, E.M.: J. Mol. Biol., 98, 503, 1975.

7.   Bauman, J.G.J. et al.: in: J. Chayen and L. Bitensky (Eds.), Investigative Microtechniques in Medicine and Biology, Dekker, New York, pp. 41-87, 1984.
8.   Bresser, J. et al.: Proc. Natl. Acad. Sci. USA, 80, 6523, 1983.
9.   Gravekamp, C.: Nat. Immun. Cell Growth Regul., 6, 28, 1987.
10.   Trask, B. et al.: Science, 230, 1401, 1985.
11.   Coghlan, J.P. et al.: Anal. Biochem., 149, 1, 1985.
12.   Engleberg, N.C. et al.: New Engl. J. Med., 311, 892, 1984.
13.   Marx, J.L.: Science, 220, 937, 1983.
14.   Kos, A. et al.: Nature, 323, 558, 1986.
15.   Jeffreys, A.J. et al.: Nature, 314, 67, 1985.
16.   Lewin, R.: Science, 233, 521, 1986.
17.   Gusella, J.F.: Ann. Rev. Biochem., 55, 831, 1986.
18.   Vassart, G. et al.: Science, 235, 683, 1987.
19.   Capecchi, M.R.: Cell, 22, 479, 1980.
20.   Cline, M.J.: Mol. Cell. Biochem., 59, 3, 1984.

A.C. Beynen & H.A. Solleveld (Eds), New developments in biosciences: their implications for laboratory animal science, ISBN 978-94-010-7973-0

DNA HYBRIDIZATION AS A TOOL IN DIAGNOSING INFECTIOUS DISEASES. COMPARISON WITH OTHER METHODS.

A.D.M.E. OSTERHAUS

During the past decade several developments in the fields of immunology and molecular biology, like the hybridoma technology and gene cloning methods, have extensively broadened the diagnostic, preventive and therapeutic arsenal of human and veterinary medicine. Classical diagnostic procedures for infectious diseases mainly depended on the demonstration of specific agents or their antigens by (electron) microscopy or immunoassays - either directly in the clinical specimen or after an in vitro cultivation period - or on the demonstration of specific antibodies (table 1). The use of monoclonal antibodies greatly enhances the specificity and sensitivity of these methods.

TABLE 1. Diagnostic procedures for infectious diseases

Demonstration of
1 the agent → (electron) microscopy
2 specific antigens → immunoassays
3 1 + 2 after in vitro cultivation
4 specific enzymes → e.g. reverse transcriptase
5 specific antibodies → immunoassays
6 specific nucleic acids sequences → hybridization

The introduction of gene cloning technology has extended this diagnostic potential with the opportunity of demonstrating specific nucleic acid sequences in clinical specimens: Cloned DNA, transferred into e.g. bacterial plasmids cannot only be used to express certain gene products which may serve as antigens in diagnostic tests, but can also be used directly to detect corresponding nucleic acid sequences using nucleic acid hybridization methods. The ability of complementary strands of nucleic acid to form stable double-stranded molecules (hybrids) is the basis of this technology. Although hybridization methods of nucleic acids in solution have been described and allowed relatively fast reaction rates, the problem of separation of the formed double-stranded material on the one hand and of self-annealing of the labeled probe on the other hand, have initially prevented the application of hybridization methods for routine diagnostic purposes. Immunobilization of one of the two nucleic acid strands (present in the specimen) onto a solid matrix, made it possible to test large

numbers of samples simultaneously. Both DNA and RNA can be
bound to a number of solid supports of which nitrocellulose
due to the low cost and convenience of use is the most popular
for general applications. Especially DNA is extremely stable
and may be heated at high pH levels in order to free it from
proteins, RNA and membrane structures. The resulting DNA strand
can than be transferred and immobilized onto the solid matrix
by blotting or filtration. Only for fragments smaller than
about 200 nucleotides NC cannot be used and therefor alternative supports like diazotized papers are available. Also for
binding of RNA, usually alternative materials are used. Although the exact mechanism of the binding of DNA to NC is not
known, it is assumed to be non-covalent. The immobilized
nucleic acid immobilized onto the filter will not be rinsed
away even at high temperatures during washing procedures.
Such NC filters with DNA can be stored for several months. For
labeling of the probe, initially only radioactive labeling
methods were employed. The most common method is the procedure
of nick translation, in which ^{32}P-labeled nucleotides are
introduced into the specific DNA molecule,by nicking and re-
pairing the molecule using a DNAse and DNA polymerase respectively. The labeled DNA can than after hybridization to the
DNA from the clinical specimen be detected by autoradiography.
Since under the appropriate conditions the probe can be removed without washing away the DNA from the specimen, the same
membrane can be used to test with different probes. A disadvantage of this method is that prior to reaction with the immobilized DNA strands of the sample, the complementary labeled
strands may hybridize with each other, which may affect the
sensitivity of the method. Several methods of making single
stranded DNA probes, like the single stranded DNA M13 phage
primer extension or sandwich methods, and single stranded RNA
probes using the SP6 system result in increased specific activities and more sensitive procedures, especially when short
specific immobilized fragments are to be detected. Another
approach is the direct synthesis of oligonucleotides on the
basis of a known amino acid sequence which is then end-labeled.
In this way oligonucleotides of upto about 14 to 20 base pairs
may be constructed.
   Non-radioactive labeling techniques allowing the same sensitivity are urgently needed at present, because they will allow
a broader practical application of hybridization techniques
which is not confined to specialized laboratories. All methods
developed at this moment use an enzyme label which is either
biochemically or immunologically detected. In the first systems
biotin-labeled nucleotides are inserted into the nucleic acids
using methods similar to those used for radioactive labeling
procedures. The detection than depends on the specific interaction between biotin and streptavidin which is attached to
an enzyme molecule like alkaline phosphatase, which provides an
amplification step (fig.1). The immunological system depend on
the introduction of a hapten into the nucleic acid probe molecule, which may be recognised by a specific antiserum which is
also coupled to an enzyme (fig.2). The final recognition of hybrid
molecules results from the detection of coloured deposit which are

FIGURE 1. Biochemical detection of hybrids
          (v.d.Avoort, 1987)

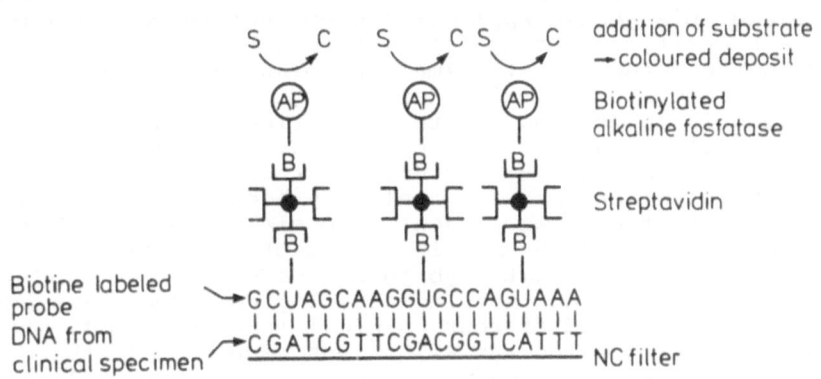

FIGURE 2. Immunological detection of hybrids
          (v.d.Avoort, 1987)

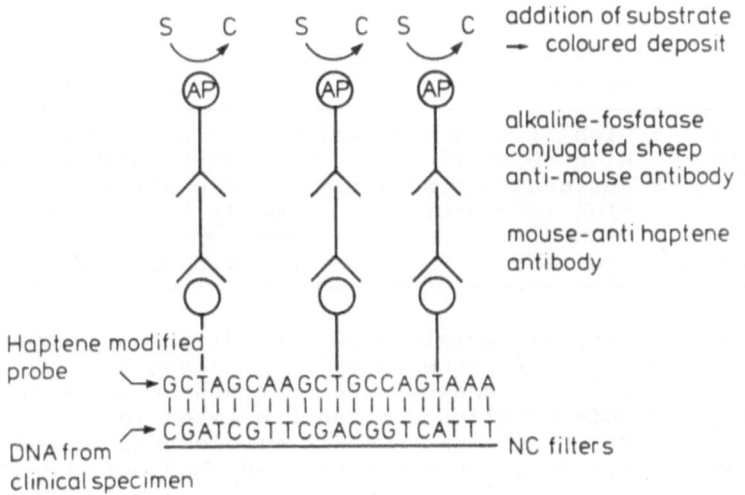

formed after addition of the suitable substrate. The whole field of generating non-radioactive probes is still in its infancy and it may be expected that the prospects of nucleic acid hybridization as a diagnostic tool will largely depend on developments in this field. Diagnostic applications of nucleic acid hybridization are apart from perinatal genetic screening mainly confined to the detection of animal and plant pathogens. Both for the fine identification of specific bacterial, viral and parasitic isolates, and for the detection of such pathogens in clinical or pathological specimens, the use of pathogen-specific probes generated by molecular cloning or other means in hybridization assays has already been introduced in a wide variety of fields at present. Especially the possibility to localize pathogen genomes in thin sections of tissue or in individual cells - in situ hybridization - has been used for the identification and showing aetiological involvement of a number of viruses in different clinical manifestations. The detection of measles virus and AIDS virus sequences in the brain from patients with subacute sclerosing panencephalitis and other brain disorders respectively and the presence of genomic sequences of various types of human papillomaviruses in different neoplastic genital lesions are probably the most striking recent examples.

Comparing the diagnostic possibilities and limitations of nucleic acid hybridization with those of the presently available enzyme immunoassay systems (table 2), it may first be stated that the specificity of the latter due to the greater specificity of the recognition site may always be higher.

TABLE 2.

	Immunoassays	NA hybridization assays
sensitivity	high (limits reached)	high(er) (labels!)
specificity	high (MoAbs!)	high(er) (probes)
detection	antigens/antibodies	nucleic acids
speed	high ($\leq$ 2 hrs)	low ($\geq$ 5 hrs - days)
simplicity	high	low
non-RA labels	available	being developed

Also the sensitivity of nucleic acid hybridization assays may, depending on the type of probe used, be at least as high as those reached for immunoassays, since antigen-antibody complexes only have dissociation constants between 10^{-9} and 10^{-12} mol/l. Furthermore infections of cells with certain viruses, in which no protein antigens are expressed, may only be detected by nucleic hybridization. However, the relative speed and handling simplicity of immunoassays and the well established use of non-radioactive labels in these assays, strongly favours their use at present. It may be expected that approvements in the development of sufficiently sensitive non-radioactive labeling methods, shortened and simplified hybridization procedures and the use of synthetic oligonucleotides will extend the field of applications of nucleic acid hybridization in routine diagnostic procedures in the near future.

A.C. Beynen & H.A. Solleveld (Eds), New developments in biosciences: their implications for laboratory animal science, ISBN 978-94-010-7973-0
©1988 Martinus Nijhoff Publishers, Dordrecht.

EMBRYO TRANSFER AND MANIPULATION

R.L. Gardner

INTRODUCTION

Mammalian experimental embryology is currently a rapidly expanding field in which new approaches and techniques are being devised and exploited at a growing pace. This means that it is now well beyond the scope of this presentation to deal comprehensively with the topics featured in its title. Consequently, no attempt has been made to focus on methodological details. Instead, the aim has been to make certain general observations on embryo transplantation, and to discuss the principal types of manipulation that have been performed on preimplantation stages of development. Furthermore, except where stated otherwise, the findings discussed in the following pages are based on work in the mouse, since this is the laboratory mammal on which most of the relevant studies have been done.

Embryo transfer
Substantial progress has been made in defining _in vitro_ culture conditions which enable normal maturation and fertilization of oocytes (Eppig & Schroeder, 1986; Hoppe, 1985), and which will support development of zygotes through to the blastocyst stage (Biggers, Whitten & Whittingham, 1971; Edwards, 1980). Further refinements will undoubtedly be introduced as more is learnt about the changing nutrient requirements of the early embryo (Leese & Barton, 1984). However, attempts to extend normal development _in vitro_ beyond the blastocyst stage have so far met with rather limited success, although early somite embryos can be obtained under certain conditions (Gonda & Hsu, 1980).

Recently, Beddington (1985) showed that, in a small proportion of cases, early postimplantation conceptuses could develop into morphologically normal advanced fetuses following orthotopic transfer to pregnant recipients. There is no reason to doubt that such fetuses would have gone to term, although no offspring have been obtained to date (R. Beddington, personal communication). It is to be hoped that the efficiency of this promising approach can be increased when more is known about materno-fetal relations during the early postimplantation phase of gestation in rodents. Meanwhile, the blastocyst remains the most advanced embryo known to be capable of completing development following replacement in its normal _in vivo_ environment.

Efficient techniques for transplanting blastocysts or earlier preimplantation embryos to the uteri or oviducts of suitably prepared recipient females have been developed for a variety of eutherian mammals. Recognition of the need to match donors and recipients both with respect to post-ovulatory age and site of transfer were important factors in their development (McLaren & Michie, 1956; Noyes et al., 1963). Thus, the best results are usually obtained if the transplanted embryos have developed for a longer time after coitus than the pseudo-pregnant uterus, especially if the embryos have resided in culture for any length of time (Bowman & McLaren, 1970). Also, oviductal stage embryos tend to fare poorly when transferred to the uterus (Noyes et al., 1963), although uterine stages can develop satisfactorily following transfer to the oviducts of early

147

post-coital recipients (Bronson & McLaren, 1970; Author's unpublished observations). However, such matching seems to be rather less critical in some species than in others. For example, introduction of 2- to 8-cell embryos directly into the uterus via the cervix in the human is standard practice at present in clinical _in vitro_ fertilization programmes (Edwards, 1985).

Non-surgical techniques for transplanting embryos to the uterus have been devised for a number of mammals (Kraemar, 1983). They are particularly useful in the larger species because they obviate the need for major surgery. This advantage is less obvious in the case of small laboratory mammals for which surgical procedures are simpler and immobilization of the conscious animal more difficult. Hence, although effective non-surgical transfer techniques were developed some yeare ago for the mouse (Marsk & Larsson, 1974; Moler, Donahue & Anderson, 1979), they have not been widely adopted. There would, however, seem to be a stronger case for further exploring the use of such techniques in species like the rabbit (Testart, 1969) which are very susceptible to adhesions following abdominal surgery that may jeopardise pregnancy (Author's unpublished observations).

There are several points to bear in mind in using embryo transfer as a tool in embryological and teratological research. One is the effect of transfer on the timing of embryonic development. It has become normal practice to transplant 4th day embryos to the uteri of 3rd day post-coitum (p.c.) pseudopregnant recipients in the mouse, thereby allowing for possible retarding effects of recovery, culture or manipulation on their development. In these circumstances, the transplanted embryos do not continue developing according either to their own schedule or to that of the recipient, but one that lies somewhere between the two (Aitken, Bowman & Gauld, 1977; Marsk, 1977). This can have important consequences when, for example, the response of transferred and native embryos to an agent having a narrow 'window' of teratological action during development is compared (e.g. Marsk, Larsson & Kjellberg, 1975).

It is useful in many types of investigation to be able to compare the development of treated and control embryos within the same recipient. This can be done in cases where they cannot be distinguished genotypically by transferring them to opposite oviducts or uterine horns. A factor that can confound the results of such bilateral transfers is the transmigration of embryos from one uterine horn to the other. The occurrence of this phenomenon was first documented in the mouse by McLaren and Michie (1954), who found that 4 out of 46 morulae or blastocysts transferred unilaterally to pregnant recipients implanted in the unoperated horn. It has been encountered very rarely by the present author in an extensive series of uni- or bilateral transfers in one random-bred strain and never found in another strain (unpublished data). Differences in anatomy of the cervix may account for the variation between strains in incidence of transmigration (J.H. Marston, personal communication). Another variable which may affect its occurrence is the amount of fluid that is normally injected into the uterine lumen with the embryos. Finally, McLaren (1963) and, more recently, Wiebold and Becker (1987) have provided evidence that the survival of both native and transferred embryos is greater in the right than the left uterine horn. For this reason, and also to guard against possible lateral differences in success of the transfer operation, it is important to randomise the allocation of treated and control embryos to the two sides of the reproductive tract.

A difficult problem is deciding whether to culture zygotes or early cleaving embryos to the blastocyst stage prior to transfer, or to return them directly to the oviduct following manipulation. Culture is obviously more convenient for monitoring the effects of intervention, but embryos develop more slowly in vitro than in vivo (Bowman & McLaren, 1970), and also tend both to arrest at the 2-cell stage (Whitten & Biggers, 1968; Goddard & Pratt, 1983) and become rather fragile (Harlow & Quinn, 1979). What one really wants to know is, of course, how the viability of in vitro grown blastocysts compares with that of their in vivo grown counterparts in terms of ability to complete normal development following transfer. Considering the increasing extent to which studies on various aspects of preimplantation development have come to depend almost entirely on in vitro culture, it is rather disconcerting to find a dearth of relevant data. Rates of normal development of blastocysts grown in culture that are cited in the literature range from 40-50% (e.g. Barton et al., 1985; Tsunoda et al., 1986), which is not dissimilar from that expected of those produced in vivo, to as low as 13% (Hoppe & Illmensee, 1977). This is clearly a topic that warrants more consideration than it has received in the past.

Culture to the morula or blastocyst is indicated when pre-morula stage embryos have to be divested of the zona pellucida. Whereas such denuded early embryos exhibit comparable rates of development to intact ones in vitro, they evidently do not survive at all in the oviduct (Bronson & McLaren, 1970; Modlinski, 1970). Since development of pre-morula stage embryos in the oviduct is impaired even by slitting the zona pellucida (Tsunoda et al., 1986; Gardner & Nichols, 1987), placing those from which the zona has had to be removed inside evacuated zonae is not an ideal solution. A most ingenious and rather more effective way of solving the problem is to embed the denuded embryos in tiny blocks of agar prior to transplantation (Willadsen, 1979; Tsunoda & McLaren, 1983). This is, however, a technically demanding procedure in species with small oviducts like the mouse so that it is probably worthwhile only if very poor rates of normal development are achieved with embryos grown in vitro.

Embryo manipulation
A wide variety of different manipulations have been carried out on the preimplantation mammalian embryo during the past 25 or so years. Most have entailed the addition or removal of cells or, alternatively, the transplantation of nuclei. A fourth category of manipulation involving introduction of exogenous DNA into the zygote or early embryo will not be discussed here as it is the subject of another presentation in this symposium.

1)   Addition of cells
The successful production of chimaeras by manipulating preimplantation embryos was first reported by Tarkowski (1961), although an inconclusive attempt to achieve this in the rat was reported earlier by Nicholas and Hall (1942). The essence of the approach, which has been widely adopted, is to aggregate pairs of whole zona-free cleaving embryos, usually at the 8-cell stage. Placing the embryos in contact at approximately 37°C facilitates their initial adherence (Mintz, 1962, 1971), as does prior exposure of one or both partners to phytohaemagglutinin (Mintz, Gearhart & Guymont, 1973), or to a polyclonal anti-mouse serum (Palmer & Dewey,

1983). It is, presumably, because of the remarkable capacity of the early embryo for regulating its size (Buehr & McLaren, 1974; Lewis & Rossant, 1982; Rands, 1986) that blastocysts formed by the aggregation of as many as 5-9 individual embryos can yield normal offspring (Petters & Mettus, 1984). So far, however, the greatest number of embryos in such multi-embryo aggregates that have been shown to contribute cells to resulting offspring is 4 (Petters & Markert, 1980).

One of the principal virtues of the aggregation technique is its simplicity. It is particularly useful for producing conceptuses that are chimaeric throughout both their embryonic and extraembryonic tissues (West et al., 1984). Furthermore, by making aggregates between embryos of different stages it is possible to impose a bias on the distribution of chimaerism (Spindle, 1982; Surani & Barton, 1984). The more advanced component in such heterochronic combinations tends to make a disproportionately large cellular contribution to the inner cell mass (ICM) and the less advanced one a disproportionately large contribution to the trophectoderm.

The aggregation technique is not, of course, restricted to the union of entire embryos. By partially or, indeed, fully dissociating cleavage stages, individual blastomeres or their daughter pairs can be combined with groups of 'carrier' blastomeres of a different genotype (Kelly, 1975). It is by means of this particular strategy that the developmental potential of blastomeres has been assessed most rigorously in the mouse (Kelly, 1975, 1977). Finally, the aggregation technique has also been used to incorporate cells other than blastomeres, notably embryonal carcinoma (EC) cells, into the preimplantation embryo (Stewart, 1982; Fujii & Martin, 1983). It has not, in general, proved a very effective means of encouraging such cells to participate in normal embryogenesis, although chimaeric offspring have been obtained thereby.

An alternative way of generating chimaerism during the preimplantation phase of development that has become increasingly popular in recent years is by injecting cells into the blastocyst (Gardner, 1968). It is more complicated than the aggregation technique because it depends on the use of carefully fashioned glass microinstruments held in micromanipulators (Gardner, 1978). However, various simplified versions of the original technique have been introduced that are entirely satisfactory providing dissociated cells rather than pieces of tissue are to be transplanted (Babinet, 1980; Bradley, 1987). Blastocyst injection offers greater scope for controlling both the level and distribution of chimaerism than the aggregation technique. It can readily be used for clonal analysis (Gardner, 1985; Gardner et al., 1985) and, depending on the type of cell that is transplanted, can yield cloning efficiencies exceeding 60% (Gardner, 1984).

The blastocyst injection technique was developed initially to enable investigation of both the normal fate of ICM and trophectoderm cells and the time of X-chromosome inactivation in the fetal lineage (Gardner, 1983; Gardner & Lyon, 1971; Gardner et al., 1985). It was adopted later for introducing EC cells into the early embryo (Brinster, 1974; Mintz, Illmensee & Gearhart, 1975; Papaioannou, et al., 1975). Transplantation of these malignant stem cells was undertaken for two reasons. The first was to assess their capacity for normal differentiation in an embryonic environment, and the second to explore their suitability for introducing new mutations that had been generated *in vitro* into mouse stocks via colonization of the germ-line. In practice, the frequency with which EC cells have been found to form functional gametes has proved too low to make their use for the latter purpose worthwhile (Papaioannou & Rossant, 1983). Fortunately, embryonic stem cells (called ES as opposed to EC cells) derived from blastocysts outgrown *in vitro* (Evans & Kaufman, 1981; Martin, 1981; Axelrod, 1984) have been found to possess all the useful attributes of EC cells without their principal drawbacks (Robertson, 1987). Thus, ES cells typically yield more extensive somatic chimaerism following injection into blastocysts than EC cells (Robertson & Bradley, 1986) and, more importantly, can produce satisfactory rates of germ-line colonization as well (Bradley *et al*., 1984).

Very recently, two groups have succeeded in producing mice that are deficient in hypoxanthine phosphoribosyl transferase (HPRT) activity by obtaining colonization of the germ-line of wild-type mice with mutant ES cells injected into blastocysts. One group used ES cells that had mutated spontaneously *in vitro* (Hooper *et al*., 1987), and the other cells in which mutation had been induced by insertion of a retrovirus into the gene (Kuehn *et al*., 1987). Although the hope that HPRT deficient mice will provide an animal counterpart of the human Lesch-Nyhan syndrome may not be realised, these studies clearly establish that this approach for introducing novel mutations into laboratory mammals is a viable one.

In considering addition of cells to the early embryo, discussion has been focussed so far on intra-specific chimaeras. However, similar manipulations have also been done using embryos belonging to different species of mammal. An early finding was that pairs of cleaving rat and mouse embryos would also aggregate to form morphologically normal integrated blastocysts in which cells of the two species could be identified in both the trophectoderm and ICM (Mulnard, 1973; Stern, 1973; Tachi & Tachi, 1980; Zeilmaker, 1973). Although the capacity for postimplantation development of the composite blastocysts was not tested in these studies, Mystkowska (1975) showed that chimaeras formed by sandwiching a cleaving bank vole embryo between two mouse embryos could progress to an early somite stage in the mouse uterus. More extensive postimplantation development of interspecific chimaeras was achieved by transferring mouse blastocysts which had been injected with rat ICM tissue to the mouse uterus (Gardner & Johnson, 1973, 1975). The fact that the trophectoderm of these embryos was composed entirely of cells of the same species as the uterine foster-mothers may have been significant since rat blastocysts transplanted to the mouse uterus fail to survive beyond implantation (Tarkowski, 1962; Tachi & Tachi, 1979). Nevertheless, the injection chimaeras experienced problems later in gestation and no individuals in which rat cells had persisted were detected among the viable offspring obtained in these experiments (Gardner & Johnson, 1975).

Viable interspecific chimaeras have been produced more recently between Mus musculus and M. caroli using individuals of the former species as uterine foster-mothers. Success was achieved in this case by means of embryo aggregation as well as by injecting caroli ICMs into musculus blastocysts or trophectoderm vesicles (Rossant & Frels, 1980; Rossant, Mauro & Croy, 1982; Rossant *et al*., 1983), despite the fact that caroli cells were present in trophectoderm derivatives in the majority of aggregation chimaeras analysed on the 10th day of gestation. Another interspecific combination that has yielded viable post-natal chimaeras by both techniques is that between the sheep and goat (Fehilly, Willadsen & Tucker, 1984; Meinecke-Tillman & Meinecke, 1984); these are species which, unlike M. musculus and M. caroli, do not form hybrids that are capable of completing gestation (West et al., 1977; McGovern, 1973). The M. musculus-M. caroli chimaeras have already provided important information on materno-fetal relations (Rossant et al., 1983), and also a very useful in vitro cell marker for lineage analysis (Rossant, 1985). Clearly the field of interspecific chimaeras is a most promising one, and is ripe for further exploitation.

Blastocyst reconstitution is another manipulation that is relevant here in connection with the production of chimaeras. It gives an entirely different pattern of chimaerism from that produced by embryo aggregation or blastocyst injection since it involves the combination of ICM and trophectoderm from different embryos (Gardner, 1971; Gardner, Papaioannou & Barton, 1973). Individual tissues of post-implantation conceptuses originating from reconstituted blastocysts are therefore composed of cells of one or other component genotype rather than a mixture of both. Initially, ICMs were isolated microsurgically and injected into vesicles produced by the short-term culture of mural trophectoderm that has been separated from other blastocysts with a micro-scalpel (Gardner, 1971; Gardner, et al., 1973). The procedure has been simplified by the introduction of immunosurgery for isolating ICMs (Solter & Knowles, 1975) and by making the ICM region of trophectoderm donor blastocysts herniate through a slit in the zona pellucida so that it can be removed after rather than before the reconstitution operation (Papaioannou, 1981).

Blastocyst reconstitution was first undertaken to establish the existence of an interaction between the ICM and overlying (polar) trophectoderm that seems to play an essential role in sustaining trophoblast proliferation (Gardner, et al., 1973). More recently, it has been used to investigate cell lineage relationships in the chorio-allantoic placenta (Rossant & Croy, 1985), the role of the trophoblast in maintenance of pregnancy (Rossant, et al., 1983), and the possibility that genomic imprinting shows lineage specificity (Barton, et al., 1985). However, because intact ICMs are used in the reconstitution operation, primitive endoderm and primitive ectoderm derivatives are necessarily of the same genotype in the resulting conceptuses.

Reconstituted blastocysts would be of even greater value if they could be assembled from trophectoderm, primitive endoderm and primitive ectoderm, each of which came from genotypically different blastocysts. The prospect of achieving this was encouraged by the discovery that micro-surgically separated primitive endoderm and ectoderm could recombine both rapidly and efficiently in vitro (Gardner, 1985b). Nevertheless, the need to use trophectoderm donor blastocysts that are significantly younger than the ICM tissue donors was a potential problem. This is unavoidable because endoderm cannot readily be separated from ectoderm in blastocysts recovered before the middle of the 5th day and, by this stage, mural trophectoderm has lost the capacity to substitute for deleted polar trophectoderm (Author's unpublished observations). However, the birth of 3 normal young in a pilot experiment in which 7 blastocysts were reconstituted by transplanting entire 5th day ICMs into 4th day trophectodermal vesicles showed that marked asynchrony between the two tissues was not incompatible with normal development. Hence, in this new reconstitution procedure, details of which will be published elsewhere, recombinant ICMs are injected into trophectoderm that is roughly 16 hours less advanced. In experiments undertaken so far, 16 out of 17 blastocysts reconstituted from genetically different tissues have implanted in vivo, 8 giving morphologically normal conceptuses on the 10th day of gestation. In 4 of the conceptuses each tissue analysed was composed entirely of cells of the same genotype as its blastocyst stage progenitor. Tissues of mixed composition were encountered in the remainder, indicating that the primitive endoderm or ectoderm or, in one case both, were contaminated with cells of the other component. Therefore, what might be termed triple tissue reconstitution of blastocysts is feasible and will, once the separation of endoderm and ectoderm is improved, be a useful addition to the repertoire of techniques for manipulating the preimplantation embryo.

2) Removal of cells

Initially, the aim of studying partial embryos produced either by destroying blastomeres or culturing them in isolation was to determine the extent to which the pre-implantation embryo could compensate for the loss of cells (Nicholas & Hall, 1942; Seidel, 1952; Tarkowski, 1959; Tarkowski & Wroblewska, 1967). Interest has shifted more recently towards using removal of cells as a means of typing embryos for sex or other genotypic characteristics (Gardner & Edwards, 1968; Epstein et al., 1978; McLaren, 1985), or producing genetically identical twins, triplets or quadruplets (Willadsen, 1981; Tsunoda & McLaren, 1983). As discussed by McLaren (1985), the success of such manipulations in different mammals depends mainly on the number of cells in the embryo at the time of blastulation. In species like the rabbit or sheep in which cell number is relatively high at this juncture, individual blastomeres isolated at the 4- or even the 8-cell stage can develop normally to term (Moore, Adams & Rowson, 1968; Willadsen, 1981). In those like the mouse in which it is substantially lower, complete development has been obtained only with blastomeres from 2-cell embryos (see Tsunoda & McLaren, 1983).

Cells have been removed from the preimplantation embryo at all stages, not just during early cleavage. For example, part of the mural trophectoderm has been excised from expanding rabbit blastocysts in order to type them and thereby control the sex ratio at term in this species (Gardner & Edwards, 1968). Conversely, specific removal of part of the ICM by suction was found to be compatible with normal development in the mouse (Lin, 1969). Bisecting the blastocyst perpendicular to the free surface of the ICM with a microscapel is a manipulation which ensures commensurate reduction in size of the trophectoderm and ICM (Gardner 1975). In practice, since the sharp-edged instrument used to produce 'half blastocysts' actually destroys cells in its path, embryos resulting from the operation will contain rather less than half the number of cells of intact ones. The majority of a series of such 'half-blastocysts' nevertheless implanted and approximately one quarter of those that did so went on to form normal mid-gestation fetuses in the mouse (Gardner, 1975). A similar operation can be performed on morulae without any loss of cells providing the embryos are first decompacted so that blastomeres are displaced rather than cut by the bisection needle (Kashiwazaki et al., 1987).

Cell removal has obvious applications in agriculture because it enables both sexing during the preimplantation phase of development and production of 2 or more embryos from one. It may eventually also play a role in clinical medicine by providing an alternative to amniocentesis or chorionic villus biopsy for antenatal diagnosis in certain circumstances (McLaren, 1985). However, the potential of such manipulations in the context of basic research should not be overlooked. For example, the production of identical twins could be very useful in studies on the effects of heredity versus environment in species in which inbred strains do not exist. Also, the value of chimaeras for studying sexual differentiation or characterizing developmental mutations would be greatly enhanced if the genotype of each component of composite embryos could be established before they were returned to the uterus. This is no longer just a remote possibility. Epstein et al., (1978) have shown that culture of one blastomere of a 2-cell mouse embryo can provide enough material to allow efficient karyotypic sexing of the twin embryo formed by the other one. Indeed, very recently Kashiwazaki et al., (1987) have made chimaeras by aggregating 4 blastomeres from each of two 8-cell embryos, using the remaining 4 blastomeres of both to determine their sex so that they could group aggregates of like composition for transfer.

3) Nuclear transplantation

The purpose of transplanting nuclei into enucleated eggs in Amphibia was to determine whether cellular diversification during embryogenesis is accompanied by stable alterations in the genome (Briggs & King, 1952; Gurdon, 1986). Attempts to apply this technique to the much smaller mammalian egg were prompted not only by this consideration but by the benefits that cloning might confer on animal husbandry. More recently, however, preoccupation with the problem of why diploid parthenogenetic embryos invariably fail to develop beyond mid-gestation has provided the main stimulus for undertaking this type of manipulation. Surprisingly, as will become clear later, these recent studies suggest that the scope for nuclear transplantation in mammals may be much more limited than was anticipated originally.

The earliest relevant experiments in mammals involved Sendai virus mediated fusion of embryonic or other cells with mouse eggs or early blastomeres (Graham, 1969; Baranska & Koprowski, 1970; Bernstein & Mukherjee, 1972; Lin, Florence & Oh, 1973). These were followed by attempts to transplant early embryonic nuclei microsurgically into unfertilized rabbit or fertilized mouse eggs which retained their own nuclei (Bromhall, 1975; Modlinski, 1978, 1981). Although the rabbit egg survived the actual operation better than the somewhat smaller mouse egg, it cleaved less well thereafter. Illmensee & Hoppe, (1981) subsequently devised a single microsurgical procedure for the introduction of a donor nucleus and removal of the host one which was compatible with good survival of mouse eggs, although the rate of cleavage was still low. The principal innovation was inclusion of cytochalasin B in the medium in which the eggs were kept both during manipulation and for an hour or so beforehand (Hoppe & Illmensee, 1977; Illmensee & Hoppe, 1981). Addition of colcemid or nocodazole, as well as cytochalasin B or D, to such media has become standard practice (McCrath & Solter, 1983; Barton, Surani & Norris, 1984), suggesting that the resistance of the egg to damage is enhanced by temporarily disrupting its cytoskeletal organization.

The introduction of a completely novel and less invasive technique for nuclear transplantation in the mouse by McGrath and Solter (1983) was unquestionably the most important step forward. Its key feature is that the plasmalemma of the egg is not actually penetrated by a micro-pipette at any stage in the operation. Suction is applied to the deformed surface of the egg so that one or both pronuclei can be pulled out of it still surrounded by part of the plasmalemma which seals as the pipette is withdrawn. Donor cells or karyoplasts are then injected inside the zona pellucida after being exposed to inactivated Sendai virus to promote their fusion with the egg. Very impressive rates of survival and normal development can be obtained when this technique is used to transplant pronuclei (McGrath & Solter, 1983; 1984a; Barton et al., 1984). A refinement introduced by Tsunoda et al., (1986) is to slit the zonae of a batch of eggs prior to removing their pronuclei. This enables enucleation to be done with a smooth-tipped pipette which is both easier to prepare and less likely to damage the eggs.

This new nuclear transplantation technique has already yielded two unexpected findings. One is that the presence of both a maternal and a paternal genome seems to be essential for normal development to term (Barton et al., 1984; McGrath & Solter, 1984a). The other is that, contrary to an earlier claim (Illmensee & Hoppe, 1981), embryonic nuclei appear to lose the capacity to support complete development of enucleated eggs even before they cease to be totipotent in the mouse (McGrath & Solter, 1984a).

The need for a genomic contribution from both parents was indicated by the inability of fertilised eggs to complete development following replacement of the maternal or paternal pronucleus with one of opposite parental origin (McGrath & Solter, 1984a; Barton et al., 1984; Surani et al., 1984). The diploid gynogenetic embryos closely resemble diploid parthenogenones in exhibiting fairly advanced fetal development accompanied by a marked deficiency of extra-embryonic tissues (Barton et al., 1984). The diploid androgenones, in contrast, show an earlier demise with failure of growth of the fetus rather than the membranes (Barton et al., 1984). In agreement with these findings, Mann and Lovell-Badge (1984) have povided evidence that the defective development of parthenogenones is due to their pronuclear rather than cytoplasmic constitution.

An obvious problem in evaluating these pronuclear transplantation experiments is that various egg components including part of the plasmalemma and cytoplasm are removed or introduced with each genome. Fortunately, compelling independent evidence that the results of such experiments reflect differential imprinting of the parental genomes during oogenesis and spermatogenesis is provided by certain genetic studies. These depend on the use of matings between mice that are heterozygous for reciprocal translocations or metacentric chromosomes to produce balanced zygotes through the union of gametes carrying complementary duplications and deficiencies. Such studies have revealed a number of regions of the genome which produce distinct phenotypes depending on whether both copies are contributed to the zygote by one parent or the other (reviewed in Cattanach, 1986). Genomic imprinting seems to be an enduring phenomenon since such divergent phenotypes may first become manifest relatively late in ontogeny (Cattanach & Kirk, 1985; Cattanach, 1986). This point has yet to be demonstrated conclusively, although non-equivalence of parental genomes has been shown recently to persist beyond the 2-cell stage when the embryonic genome is initially activated (Surani et al., 1986).

McGrath and Solter (1984a) have also used their technique to transplant nuclei from cleaving blastomeres to enucleated zygotes in the mouse. They found that the rate of development to the blastocyst stage was substantially lower with nuclei from 2-cell embryos than with pronuclear karyoplasts, and fell to only 5% using 4-cell embryos as donors. The success rate was even lower in a series of experiments in which they employed the wholly microsurgical technique of Illmensee and Hoppe (1981) rather than their own. A similar experience with cleavage stage nuclear donors has been reported by Surani et al., (1986). However, Robl et al., (1986) found enucleated 2-cell blastomeres to be better hosts than zygotes for nuclei from 8-cell embryos as judged by the ability of some to develop beyond implantation. Notwithstanding, all such conceptuses stopped developing by day 12. It therefore looks at present as though embryonic nuclei lose the capacity to substitute for the zygote nucleus well before the end of cleavage in the mouse. The fact that Willadsen (1986) has obtained positive results with nuclei from both 8- and 16-cell embryos in the sheep may reflect species differences in the timing of certain events such as activation of the embryonic genome. Thus, while the need for further work is clearly indicated, nuclear transplantation may well prove to be of rather limited value both for studying cellular differentiation and, at a more practical level, for cloning in mammals. Nevertheless, pronuclear transplantation has provided a powerful new approach for tackling certain hitherto intractable problems including the basis·of parental effect mutations (McGrath & Solter, 1984b; Renard & Babinet, 1986) and the time of activation of the embryonic genome with respect to expression of maternal as opposed to paternal alleles (Gilbert & Solter, 1985).

ACKNOWLEDGEMENTS

I thank Rosa Beddington, Chris Graham, Pam Hickey, Jennifer Nichols and Jo Williamson for help in preparing the manuscript and the Royal Society and Imperial Cancer Research Fund for support.

REFERENCES

Aitken, R.J., Bowman, P. & Gauld, I. (1977) The effect of synchronous and asynchronous egg transfer on foetal weight in mice selected for large and small body size. J. Embryol. exp. Morph. 37; 59-64.
Axelrod, H.R. (1984) Embryonic stem cell lines derived from blastocysts by a simplified technique. Dev. Biol. 101; 225-228.
Babinet, C. (1980) A simplified method for mouse blastocyst injection. Exp. Cell Res. 130; 15-19.
Baranska, W. & Koprowski, H. (1970) Fusion of unfertilized mouse eggs with somatic cells. J. exp. Zool. 174; 1-14.
Barton, S.C., Surani, M.A.H. & Norris, M.L. (1984) Role of paternal and maternal genomes in mouse development. Nature 311; 374-376.
Barton, S.C., Adams, C.A., Norris, M.L. & Surani, M.A.H. (1985) Development of gynogenetic and parthenogenetic inner cell mass and trophectoderm tissue in reconstituted blastocysts in the mouse. J. Embryol. exp. Morph. 90; 267-285.
Beddington, R.S.P. (1985) The development of 12th to 14th day foetuses following reimplantation of pre- and early primitive streak stage mouse embryos. J. Embryol. exp. Morph. 88; 281-291.
Bernstein, R.M. & Mukherjee, B.B. (1972) Control of nuclear RNA synthesis in 2-cell and 4-cell mouse embryos. Nature (Lond.) 238; 457-459.

Bowman, P. & McLaren, A. (1970) Cleavage rate of mouse embryos in vivo and in vitro. J. Embryol. exp. Morph. 24; 203-207.

Bradley, A. (1987) Production and analysis of chimaeric mice. In Teratocarcinomas and Embryonic Stem Cells: A Practical Approach (ed. E.J. Robertson), pp. 113-151. IRL Press, Oxford.

Bradley, A., Evans, M., Kaufman, M.H. & Robertson, E. (1984) Formation of germ-line chimaeras from embryo-derived teratocarcinoma cell lines. Nature 309; 255-256.

Briggs, R. & King, T.J. (1952) Transplantation of living nuclei from blastula cells into enucleated frogs' eggs. Proc. natn. Acad. Sci. USA; 38; 455-463.

Brinster, R.L. (1974) The effect of cells transferred into the mouse blastocyst on subsequent development. J. exp. Med. 140; 1049-1056.

Bromhall, J.D. (1975) Nuclear transplantation in the rabbit egg. Nature 258; 719-721.

Bronson, R.A. & McLaren, A. (1970) Transfer to the mouse oviduct of eggs with and without the zona pellucida. J. Reprod. Fert. 22; 129-137.

Buehr, M. & McLaren, A. (1974) Size regulation in chimaeric mouse embryos. J. Embryol. exp. Morph. 31; 229-234.

Cattanach, B.M. (1986) Parental origin effects in mice. J. Embryol. exp. Morph. 97 (Suppl); 137-150.

Cattanach, B.M. & Kirk, M. (1985) Differential activity of maternally and paternally derived chromosome regions in mice. Nature (Lond.) 315; 496-498.

Edwards, R.G. (1980) Conception in the Human Female. Academic Press, London.

Edwards, R.G. (1985) Current status of human conception in vitro. Proc. R. Soc. Lond. B 223; 417-448.

Eppig, J.J. & Schroeder, A.C. (1986) Culture systems for mammalian oocyte development: progress and prospects. Theriogenology 25; 97-106.

Epstein, C.J., Smith, S., Travis, B. & Tucker, G. (1978) Both X chromosomes function before visible X-chromosome inactivation in female mouse embryos. Nature 274; 500-503.

Evans, M.J. & Kaufman, M.H. (1981) Establishment in culture of pluripotent cells from mouse embryos. Nature (Lond). 192; 154-156.

Fehilly, C.B., Willadsen, S.M. & Tucker, E.M. (1984) Interspecific chimaerism between sheep and goat. Nature 307; 634-636.

Fujii, J.T. & Martin, G.R. (1983) Developmental potential of teratocarcinoma stem cells in utero following aggregation with cleavage-stage mouse embryos. J. Embryol. exp. Morph. 74; 79-96.

Gardner, R.L. (1968) Mouse chimaeras obtained by the injection of cells into the blastocyst. Nature (Lond). 220; 596-597.

Gardner, R.L. (1971) Manipulations on the blastocyst. In Advances in the Biosciences, No. 6: Schering Symposium on Intrinsic and Extrinsic Factors on Early Mammalian Development (ed. G. Raspe), pp. 279-296. Pergamon Press, Oxford.

Gardner, R.L. (1975) Analysis of determination and differentiation in the early mammalian embryo using intra- and interspecific chimaeras. In The Developmental Biology of Reproduction: 33rd Symposium of the Society for Developmental Biology (eds. C.L. Markert & J. Papaconstantinou), pp. 207-236. Academic Press, New York.

Gardner, R.L. (1978) Production of chimeras by injecting cells or tissue into the blastocyst. In Methods in Mammalian Reproduction (ed.

J.C. Daniel), pp. 137-165. Academic Press, New York.

Gardner, R.L. (1984) An in situ cell marker for clonal analysis of development of the extraembryonic endoderm in the mouse. J. Embryol. exp. Morph. 80; 251-288.

Gardner, R.L. (1985a) Clonal analysis of early mammalian development. Phil. Trans. R. Soc. Lond. B 312; 163-178.

Gardner, R.L. (1985b) Regeneration of endoderm from primitive ectoderm in the mouse embryo: fact or artifact? J. Embryol. exp. Morph. 88; 303-326.

Gardner, R.L. & Edwards, R.G. (1968) Control of the sex ratio at full term in the rabbit by transferring sexed blastocysts. Nature (Lond). 218; 346-348.

Gardner, R.L. & Johnson, M.H. (1973) Investigation of early mammalian development using interspecific chimaeras between rat and mouse. Nature New Biol. 246; 86-89.

Gardner, R.L. & Johnson, M.H. (1975) Investigation of cellular interaction and deployment in the early mammalian embryo using inter-specific chimaeras between the rat and the mouse. In Cell Patterning: Ciba Foundation Symposium 29, (new series), (eds. R. Porter & J. Rivers), pp. 183-196. Elsevier, Amsterdam.

Gardner, R.L. & Lyon, M.F. (1971) X-chromosome inactivation studied by injection of a single cell into the mouse blastoyst. Nature (Lond). 231; 385-386.

Gardner, R.L. & Nichols, J. (1987) Manuscript in preparation.

Gardner, R.L., Papaioannou, V.E. & Barton, S.C. (1973) Origin of the ectoplacental cone and secondary giant cells in mouse blastocysts reconstituted from isolated trophoblast and inner cell mass. J. Embryol. exp. Morph. 30; 561-572.

Gardner, R.L., Lyon, M.F., Evans, E.P. & Burtenshaw, M.D. (1985) Clonal analysis of X-chromosome inactivation and the origin of the germ line in the mouse embryo. J. Embryol. exp. Morph. 88; 349-363.

Gilbert, S.F. & Solter, D. (1985) Onset of paternal and maternal Gpi-1 expression in preimplantation mouse embryos. Dev. Biol. 109; 515-517.

Goddard, M.J. & Pratt, H.P.M. (1983) Control of events during early cleavage of the mouse embryo: an analysis of the 2-cell block. J. Embryol. exp. Morph. 73; 111-133.

Gonda, M.A. & Hsu, Y-C. (1980) Correlative scanning electron, transmission electron, and light microscopic studies of the in vitro development of mouse embryos on a plastic substrate at the implantation stage. J. Embryol. exp. Morph. 56; 23-39.

Graham, C.F. (1969) The fusion of cells with one- and two-cell mouse embryos. In Heterospecific Genome Interaction: The Wistar Institute Symposium Monograph 9 (ed. V. Defendi), pp. 19-33. The Wistar Institute Press, Philadelphia.

Gurdon, J.B. (1986) Nuclear transplantation in eggs and oocytes. J. Cell Sci. Suppl. 4; 287-318

Harlow, G.M. & Quinn, P. (1979) Isolation of inner cell masses from mouse blastocysts by immunosurgery or exposure to the calcium ionophore A23187. Aust. J. Biol. Sci. 32; 483-491.

Hooper, M., Hardy, K., Handyside, A., Hunter, S. & Monk, M. (1987) HPRT-deficient (Lesch-Nyhan) mouse embryos derived from germ-line colonization by cultured cells. Nature 326; 292-295.

Hoppe, P.C. (1985) Techniques of fertilization in vitro. In Reproductive Toxicology (ed. R.L. Dixon), pp. 191-199. Raven Press, New York.

Hoppe, P.C. & Illmensee, K. (1977) Microsurgically produced
    homozygous-diploid uniparental mice. Proc. natn. Acad. Sci. USA
    74; 5657-5661.
Illmensee, K. & Hoppe, P.C. (1981) Nuclear transplantation in Mus
    musculus: developmental potential of nuclei from preimplantation
    embryos. Cell 23; 9-18.
Kashiwazaki, N., Mizuno, J., Nagashima, H., Mizuno, A., Yamakawa, H.
    Yamanoi, J. & Ogawa, S. (1987) Sexing of half-embryos produced by
    microsurgical bisection of morulae and transfer, and production of
    two-sexed half-embryo chimaeras and homo-dizygotic twin chimaeras
    in mice. Exp. Anim. (in press).
Kelly, S.J. (1975) Studies of the potency of early cleavage blastomeres
    of the mouse. In The Early Development of Mammals, 2nd Symposium
    of the British Society for Developmental Biology (eds. M. Balls &
    E.A. Wild), pp. 97-105. Cambridge University Press, Cambridge.
Kelly, S.J. (1977) Studies of the developmental potential of 4- and
    8-cell stage mouse blastomeres. J. exp. Zool. 200; 365-376.
Kraemar, D.C. (1983) Intra- and interspecific embryo transfer. J. exp.
    Zool. 228; 363-371.
Kuehn, M.R., Bradley, A., Robertson, E.J. & Evans, M.J. (1987) A
    potential animal model for Lesch-Nyhan Syndrome through
    introduction of HPRT mutation into mice. Nature 326; 295-298.
Leese, H.J. & Barton, A.M. (1984) Pyruvate and glucose uptake by mouse
    ova and preimplantation embryos. J. Reprod. Fert. 72; 9-13.
Lewis, N.E. & Rossant, J. (1982) Mechanism of size regulation in mouse
    embryo aggregates. J. Embryol. exp. Morph. 72; 169-181.
Lin, T.P. (1969) Microsurgery of inner cell mass of mouse blastocyst.
    Nature 222; 480-481.
Lin, T.P., Florence, J. & Oh, J.O. (1973) Cell fusion induced by a virus
    within the zona pellucida of mouse eggs. Nature (Lond). 242;
    47-49.
Mann, J.R. & Lovell-Badge, R.H. (1984) Inviability of parthenogenones is
    determined by pronuclei not egg cytoplasm. Nature 310; 66-67.
Marsk, L. (1977) Developmental precocity after asynchronous egg transfer
    in mice. J. Embryol. exp. Morph. 39; 129-137.
Marsk, L. & Larsson, K.S. (1974) A simple method for non-surgical
    blastocyst transfer in mice. J. Reprod. Fert. 37; 393-398.
Marsk, L., Larsson, K.S. & Kjellberg, M. (1975) Developmental precocity
    in transferred embryos influencing the teratogen response to
    salicylate. J. Embryol. exp. Morph. 33; 907-913.
Martin, G.R. (1981) Isolation of a pluripotent cell line from early mouse
    embryos cultured in medium conditioned by teratocarcinoma stem
    cells. Proc. Natn. Acad. Sci. USA 78; 7634-7638.
McGrath, J. & Solter, D. (1983) Nuclear transplantation in the mouse
    embryo by microsurgery and cell fusion. Science 220; 1300.
McGrath, J. & Solter, D. (1984a) Inability of mouse blastomere nuclei
    transferred to enucleated zygotes to support development in vitro.
    Science 226; 1317-1319.
McGrath, J. & Solter, D. (1984b) Maternal T^{hp} lethality in the mouse is a
    nuclear, not cytoplasmic defect. Nature 308; 550-551.
McGovern, P.T. (1973) The effect of maternal immunity on the survival of
    goat x sheep hybrid embryos. J. Reprod. Fert. 34; 215-220.
McLaren, A. (1963) The distribution of eggs and embryos between sides in
    the mouse. J. Endocrin. 27; 157-181.
McLaren, A. (1985) Prenatel diagnosis before implantation: opportunities
    and problems. Prenatal Diagnosis 5; 85-90.

160

McLaren, A. & Michie, D. (1954) Transmigration of unborn mice. Nature (Lond). 174; 844.

McLaren, A. & Michie, D. (1956) Studies on the transfer of fertilized mouse eggs to uterine foster-mothers. 1. Factors affecting the implantation and survival of native and transferred eggs. J. exp. Biol. 33; 394-416.

Meinecke-Tillman, S. & Meinecke, B. (1984) Experimental chimaeras - removal of reproductive barrier between sheep and goat. Nature 307; 637-638.

Mintz, B. (1962) Formation of genotypically mosaic mouse embryos. Am. Zool. 2; 432 Abstr. 310.

Mintz, B. (1971) Allophenic mice of multi-embryo origin. In Methods in Mammalian Embryology (ed. J.C. Daniel), pp. 186-214.

Mintz, B., Gearhart, J.D. & Guymont, A.O. (1973) Phytohemagglutinin-mediated blastomere aggregation and development of allophenic mice. Dev. Biol. 31; 195-199.

Mintz, B., Illmensee, K. & Gearhart, J.D. (1975) Developmental and experimental potentialities of mouse teratocarcinoma cells from embryoid body cores. In Teratomas and Differentiation (eds. M.I. Sherman & D. Solter), pp. 59-82. Academic Press, New York.

Modlinski, J.A. (1970) The role of the zona pellucida in the development of mouse eggs in vivo. J. Embryol. exp. Morph. 23; 539-547.

Modlinski, J.A. (1978) Transfer of embryonic nuclei to fertilized mouse eggs and development of tetraploid mouse blastocysts. Nature 273; 466-467.

Modlinski, J.A. (1981) The fate of inner cell mass and trophectoderm nuclei transplanted to fertilized mouse eggs. Nature 292; 342-343.

Moler, T.L., Donahue, S.E. & Anderson, G.B. (1979) A simple technique for non-surgical embryo transfer in mice. Lab. Anim. Sci. 29: 353-356.

Moore, N.W., Adams, C.E. & Rowson, L.E.A. (1968) Developmental potential of single blastomeres of the rabbit egg. J. Reprod. Fert. 17; 527-531.

Mulnard, J. (1973) Formation de blastocystes chimeriques par fusion d'embryons de rat et de souris au stade VIII. C.R. Hebd. Seances Acad. Sci. Ser. D. Sci. Nat. (Paris) 276; 379-381.

Mystkowska, E.T. (1975) Development of mouse-bank vole interspecific chimaeric embryos. J. Embryol. exp. Morph. 33; 731-744.

Nicholas, J.S. & Hall, B.V. (1942) Experiments on developing rats II. The development of isolated blastomeres and fused eggs. J. exp. Zool. 90; 441-459.

Noyes, R.W., Dickmann, Z., Doyle, L.L. & Gates, A.H. (1963) Ovum transfers, synchronous and asynchronous, in the study of implantation. In Delayed Implantation (ed. A.C. Enders), pp. 197-209. University of Chicago Press, Chicago.

Palmer, J. & Dewey, M.J. (1983) Allophenic mice produced from embryos aggregated with antibody. Experientia 39; 196-198.

Papaioannou, V.E. (1981) Microsurgery and micromanipulation of early mouse embryos. Tech. Life Sci. Physiol P1(1) 116; 1-27.

Papaioannou, V.E. & Rossant, J. (1983) Effects of the embryonic environment on proliferation and differentiation of embryonal carcinoma cells. Cancer Surveys 2(1); 165-183.

Papaioannou, V.E., McBurney, M.W., Gardner, R.L. & Evans, M.J. (1975) Fate of teratocarcinoma cells injected into early mouse embryos. Nature (Lond). 258; 70-73.

Petters, R.M. & Mettus, R.V. (1984) Survival to term of chimeric morulae produced by aggregation of five to nine embryos in the mouse, Mus musculus. Theriogenology 22; 167-174.

Petters, R.M. & Markert, C.L. (1980) Production and reproductive performance of hexaparental and octaparental mice. J. Hered. 71; 70-74.

Rands, G.F. (1986) Size regulation in the mouse embryo I. The development of quadruple aggregates. J. Embryol. exp. Morph. 94; 139-148.

Renard, J-P. & Babinet, C. (1986) Identification of a paternal developmental effect on the cytoplasm of one-cell-stage mouse embryos. Proc. natn. Acad. Sci. USA 83; 6883-6886.

Robertson, E.J. (1987) Embryo-derived stem cell lines. In Teratocarcinomas and Embryonic Stem Cells: A Practical Approach (ed. E.J. Robertson), pp. 71-112. IRL Press, Oxford.

Robertson, E.J. & Bradley, A. (1986) Production of permanent cell lines from early embryos and their use in studying developmental problems. In Experimental Approaches to Mammalian Embryonic Development (eds. J. Rossant & R.A. Pedersen), pp. 475-508. Cambridge University Press, Cambridge.

Robl, J.M., Gilligan, B., Critser, E.S. & First, N.L. (1986) Nuclear transplantation in mouse embryos: assessment of recipient cell stage. Biol. Reprod. 34; 733-739.

Rossant, J. (1985) Interspecific cell markers and lineage in mammals. Phil. Trans. R. Soc. Lond. B 312; 91-100.

Rossant, J. & Croy, B.A. (1985) Genetic identification of tissue of origin of cellular populations within the mouse placenta. J. Embryol. exp. Morph. 86; 177-189.

Rossant, J. & Frels, W.I. (1980) Interspecific chimaeras in mammals: successful production of live chimaeras between Mus musculus and Mus caroli. Science 208; 419-421.

Rossant, J., Mauro, V.M. & Croy, B.A. (1982) Importance of trophoblast genotype for survival of interspecific murine chimaeras. J. Embryol. exp. Morph. 69; 141-149.

Rossant, J., Croy, B.A., Clark, D.A. & Chapman, V.M. (1983) Interspecific hybrids and chimeras in mice. J. exp. Zool. 228; 223-233.

Seidel, F. (1952) Die Entwicklungspotenzen feiner isolierten Blastomeren des Zweizellenstadiums in Saugetieren. Naturwissenschaften 39; 355.

Solter, D. & Knowles, B.B. (1975) Immunosurgery of mouse blastocysts. Proc. Natn. Acad. Sci. USA 72; 5099-5102.

Spindle. A. (1982) Cell allocation in preimplantation mouse chimaeras. J. exp. Zool. 219; 361-367.

Stern, M.S. (1973) Chimaeras obtained by aggregation of mouse eggs with rat eggs. Nature (Lond). 243; 472-473.

Stewart, C.L. (1982) Formation of viable chimaeras by aggregation between teratocarcinomas and preimplantation mouse embryos. J. Embryol. exp. Morph. 67; 167-179.

Surani, M.A.H. & Barton, S.C. (1984) Spatial distribution of blastomeres is dependent on cell division order and interactions in mouse morulae. Dev. Biol. 102; 335-343.

Surani, M.A.H., Barton, S.C. & Norris, M.L. (1984) Development of reconstituted mouse eggs suggests imprinting of the genome during gametogenesis. Nature 308; 548-550.

Surani, M.A.H., Barton, S.C. & Norris, M.L. (1986) Nuclear transplantation in the mouse: heritable differences between parental genomes after activation of the embryonic genome. Cell

45; 127-136.

Tachi, S. & Tachi, C. (1979) Ultrastructural studies on maternal embryonic cell interaction during experimentally induced implantation of rat blastocysts to the endometrium of the mouse. Dev. Biol. 68; 203-223.

Tachi, S. & Tachi, C. (1980) Electron microscopic studies of chimeric blastocysts experimentally produced by aggregating blastomeres of rat and mouse embryos. Dev. Biol. 80; 18-27.

Tarkowski, A.K. (1959) Experiments on the development of isolated blastomeres of mouse eggs. Nature 184; 1286-1287.

Tarkowski, A.K. (1961) Mouse chimaeras developed from fused eggs. Nature (Lond). 190; 857-860.

Tarkowski, A.K. (1962) Interspecific transfers of eggs between rat and mouse. J. Embryol. exp. Morph. 10; 476-495.

Tarkowski, A.K. & Wroblewska, J. (1967) Development of blastomeres of mouse eggs isolated at the four- and eight-cell stage. J. Embryol. exp. Morph. 18; 155-180.

Testart, J. (1969) Comparison de differentes techniques de transplantation des blastocysts chez la lapine. Ann. Biol. Anim. Biochem. Biophys. 9; 351-360.

Tsunoda, Y. & McLaren, A. (1983) Effects of various procedures on the viability of mouse embryos containing half the normal number of blastomeres. J. Reprod. Fert. 69; 315-322.

Tsunoda, Y., Yasui, T., Makamura, K. Uchida, T. & Sugie, T. (1986) Effect of cutting the zona pellucida on the pronuclear transplantation in mice. J. exp. Zool. 240; 119-125.

West, J.D., Bucher, T., Linke, I.M. & Dunnwald, M. (1984) Investigation of variability among mouse aggregation chimaeras and X-chromosome inactivation mosaics. J. Embryol. exp. Morph. 84; 309-329.

West, J.D., Frels, W.I., Papaioannou, V.E., Karr, J.R. & Chapman, V.M. (1977) Development of interspecific hybrids of Mus. J. Embryol. exp. Morph. 41; 233-243.

Whitten, W.K. & Biggers, J.D. (1968) Complete development in vitro of the preimplantation stages of the mouse in a simple chemically defined medium. J. Reprod. Fert. 17; 399-401.

Wiebold, J.L. & Becker, W.C. (1987) Inequality in function of the right and left ovaries and uterine horns of the mouse. J. Reprod. Fert. 79; 125-134.

Willadsen, S.M. (1979) A method for culture of micromanipulated sheep embryos and its use to produce monozygotic twins. Nature 277; 298-300.

Willadsen, S.M. (1981) The developmental capacity of blastomeres from 4-cell and 8-cell sheep embryos. J. Embryol. exp. Morph. 65; 165-172.

Willadsen, S.M. (1986) Nuclear transplantation in sheep embryos. Nature (Lond). 320; 63-65.

Zeilmaker, G.H. (1973) Fusion of rat and mouse morulae and formation of chimaeric blastocysts. Nature (Lond). 242; 115-116.

A.C. Beynen & H.A. Solleveld (Eds), New developments in biosciences: their implications for laboratory animal science, ISBN 978-94-010-7973-0
© 1988 Martinus Nijhoff Publishers, Dordrecht.

STRAIN PRESERVATION OF RODENT EMBRYOS. POSSIBILITIES AND LIMITATIONS

H.J. HEDRICH and I.C. REETZ

1. INTRODUCTION

More than a decade ago Whittingham et al. (50) were the first to report on mammals that developed from embryos frozen to subzero temperatures. This successful recovery of frozen-thawed mouse embryos started a new era based on Mazur's theoretical analysis of cell freezing (23). Cryobiology since has become a promising branch of biomedical science. Since these early reports a number of papers have been published on variations or different freezing methods as well as fundamental mechanisms acting during freezing and thawing. Still, however, a number of questions remain unsolved. Irrespective of this fact, frozen mouse embryos can be successfully recovered. With respect to rat embryo freezing information is rather scarce.

The purpose of this presentation is to outline the objectives, the technical and biological aspects, as well as the accomplishments and limitations of freeze-preservation of laboratory rodents.

2. OBJECTIVES OF CRYOPRESERVATION OF STRAINS

Modern biomedical research requires the availability of various special strains of mice and rats to achieve the most valuable results in the particular field of interest. To serve these purposes a large number of strains have been developed throughout the world such as standard inbred, congenic and segregating inbred, and recombinant inbred strains as well as strains with numerical and structural rearrangements of chromosomes. Table 1 gives an estimate on the number of strains of laboratory rodents available. It totals more than 1200 strains of mice. The number of rat strains is estimated as somewhere in-between 400 and 500. The cost of just maintaining a breeding nucleus of one strain under specified patho-

TABLE 1   Number of Strains and Stocks of Mice and Rats Available For Biomedical Research*

	Mice	Rats
Outbred stocks	120	70
Inbred strains	250	130
Congenic and segregating inbred strains	300	140
Recombinant inbred strains [No. of sets]	300 [25]	60 [4]
Strains with numerical and structural rearrangements of chromosomes	120	
Mutant carrying strains and stocks	110	50
Total	1200	450

* All numbers are approximative values based on Festing (1980), Green (1981), Butcher and Howard (1986), and Hedrich (unpublished).

A number of strains and stocks is being maintained by several holders (>20) and may not be identical despite the same designation.

gen free conditions is rather high (up to 11,000.- DM per annum, this is equivalent to about 5500.- US$). The number of animals per strain needed on an annual basis is varying and ranges from constant to frequent to occasional to very rare demand. Thus, the effective costs per animal used in research may be enormously high. Institutes involved in large scale breeding and/or in maintaining breeding nuclei of laboratory animals have therefore to decide upon which strains ought to be kept and for which a sufficient quality control program can be run. Consequently, it is not possible to keep available as many strains as necessary to fit the actual but varying demand of the research community. Thus quite often the specific animals needed for a particular type of research have to be ordered from other, sometimes overseas, sources. In this case embryo banking could help to overcome this limitation and moreover reduce the actual costs by maintaining rarely used breeding nuclei only in liquid nitrogen. Indeed, a few embryo banks have been established throughout the world, however with different objectives and using different techniques.

The successful freezing of murine embryos and the recovery of live offspring may now solve an urgent problem in laboratory animal science, which is a lack of space in the animal houses and the need to preserve the gene pool of laboratory rodents, i.e. preserving the multiplicity of the various strains and to prevent genetic drift in the populations maintained.

All breeding colonies may be subject to extinction by accidental loss. Such a loss can be caused by a disease outbreak, decline in reproductive performance, genetic contamination or uncontrolled environmental events. A frozen embryo repository will allow for an almost immediate replacement of the colony. In this context one has, however, to bear in mind that many inbred strains are refractory to high rates of revitalization.

Sublines of strains are likely to differ at various gene loci. Factors that alter the genetic constitution of an inbred strain are outcrossing, differential fixation of originally heterozygous loci and mutation. The effects of genetic contamination and the separation of strains at an early stage of inbreeding are obvious. Spontaneous mutations, however, alter the genetic constitution of a strain at a constant but almost unnoticed rate. The average mutation frequency per gene per generation has been generally accepted as $1x10^{-5}$ and assuming $3x10^4$ structural gene loci in the genome anyone gamete carries 0.3 mutations per generation. Bailey (2) has provided a formula that calculates the probability of encountering at least one mutant when comparing individuals from two sublines which have been separated for a specified number of generations since subline dichotomy. Introducing the above mutation rate into the formula, the chance that two sublines differ at least at one locus after 10 generations of separation is 99.92%. One means of controlling this subline divergence would be a regular restocking of breeding colonies (at set intervals) from an embryo-bank (15).

3. TECHNICAL AND BIOLOGICAL ASPECTS

The methodological part will mainly concentrate on the freezing of mouse embryos, as most research has been done on this species and indeed the embryo-repositories established routinely freeze-preserve only mouse embryos. Features concerning collection, preservation, and transfer of rat embryos will be discussed below.

3.1. Donor Preparation and Embryo Collection

3.1.1. Superovulation. In order to obtain a sufficiently large number of ova per pedigreed parental cage the females are treated with gonadotrophins. The aim of hormone treatment with PMS and HCG is to synchronize mating and ovulation and to increase the yield of eggs. The results of superovulation are varying and depend on several factors (13), i.e. hormone dose, interval between PMS and HCG injection, age of females treated, seasonal effects, and hormone batch (variable choriogonadotropin content in PMS preparation).

For superovulation a fixed schedule is applied, e.g. PMS is injected at 5:00 pm followed by HCG 46-48 hours later. Females are mated individually at the time of HCG injection. On the next morning they are checked for plugs. The day of plug is registered as day 1 of fertilization. Hormone doses for PMS and HCG are identical and range from 4 to 10 international units.

Not all inbred strains respond alike. Whereas CD2F1 or other hybrids have an average ova yield of 30 to 40 with 70-80% of all females plugged, inbred strains shed only about 15 to 20 eggs with 45 to 50% of the females having plugs. The hormone doses applied have to be adjusted for each strain and vary between different batches and brands of PMS (36). In general more than twice the usual number of oocytes may be ovulated and fertilized making a sufficient supply of embryos available for embryo freezing.

3.1.2. Embryo Collection. For freezing different stages of embryos (two-cell embryos to early blastocysts) may be used. We use eight-cell embryos because they seem to be best suited for this purpose.

Eight-cell embryos are collected on day 3 of pregnancy (which is about 60-66 hours apart from mating). Under a dissecting microscope the excised oviducts are flushed from the infundibular opening by inserting a 30-33 gauge Hamilton needle attached to a syringe filled with a modified Dulbecco's phosphate buffered saline solution (PB1, 47). After three successive washings in PB1 embryos with unimpaired morphology are transferred to the freezing containers by hand-drawn micropipettes. Various containers are in use: 2 ml polycarbonate-tubes with screw caps, glass or polypropylene ampoules, or plastic straws (Ø 3 mm, 63.5 mm length to be sealed by metal and plastic bulbs, e.g. Minitüb®; or mini-paillettes, so called 'French straws', Ø 2 mm, 133 mm length to be sealed by various methods such as specifically designed plugs or by heat). The former straws appear to be best suited for freezing, as they do not cause problems because of leakage and sterility. Furthermore, they are easier to handle and upon recovery nearly no embryos get lost. With respect to sterilizing heat-labile embryo-containers with ethylene oxide one has to consider the cytotoxic effect of the absorbed-retained gas and must not use such gas sterilized plastics until a sufficient post-sterilization aeration (~ 3 weeks) is provided (39). There are however straws available that withstand heat-sterilization, e.g. Minitüb®.

3.2. Freezing Procedures

The original technique as described by Whittingham et al. (50) and Wilmut (55) requires a controlled slow freezing and slow thawing procedure with dimethylsulfoxide (Me$_2$SO) or glycerol as the cryoprotectant. The effect of the cryoprotectants is thought to colligatively reduce the concentrations of both the extracellular and intracellular electrolytes, and the extent of cell shrinkage at a given subzero temperature (40). Regardless of the mechanisms of injury proposed to act on mammalian cells being frozen, embryos can be successfully recovered after freezing.

Which of the following procedures is to be used depends on various factors mostly defined by laboratory equipment and cryobiological expertise. Reliable storage and revitalization is affected by many factors such as cooling and thawing rate, type of medium and cryoprotective agent, storage temperature and duration, developmental stage of the embryo, type of recipient used for embryo transfer, and the transfer technique. The decision for a specific method should be based on a consistantly high rate of revitalization. This is an important factor to the economics of an embryo-bank. Once a feasible technique has been established this should possibly not be altered.

3.2.1. Slow Freeze-Slow Thaw Proceedure. Freezing of embryos at a low speed (Fig.1A) of 0.3-0.8°C/min. to -80°C permits the embryos to undergo progressive dehydration, thus preventing intracellular crystallization. Thawing has then to be slow (4-20°C/min., Ø 8°C/min) in order to allow the cells to rehydrate without deleterious side-effects (49). The original protocol as developed by Whittingham et al. (50), is a manually operated slow-freeze technique using dewar flasks: Equilibration of embryos with Me$_2$SO as the cryoprotectant (final molality 1.5 to 2.0) at 0°C for 10 min. Transfer to a cooling bath maintained at -5°C. After 5 min seeding (induction of extracellular ice formation at a temperature at which the latent heat released during crystallization will not cause extreme temperature changes) by touching the tubes at the air-medium-interface with a metal rod precooled in liquid nitrogen. After a further equilibration period of 5 min the tubes are transferred to a freezer which has been precooled to -5°C. The temperature is then lowered at a constant rate of 0.5°C/min to -80°C. Then the tubes are removed and immediately placed into liquid nitrogen.

3.2.2. <u>Rapid Cooling Procedures.</u> In contrast to the original view that embryos survive storage at -196°C only if slowly frozen to -80°C it could be shown that mouse and rat embryos survive even rapid cooling to subzero temperatures provided that thawing is fast at about 300-500°C/min.

Embryos frozen by the various 'fast' techniques are thought to be in a metastable state. According to Leibo (20) very slight alterations during warming and cryoprotectant removal might produce disastrous consequences. It also has to be mentioned that embryos requiring fast thawing must be held at -196°C during storage (and transport). On the contrary slowly fro-

FIGURE 1   Temperature - diagram of three freeze-thaw procedures:

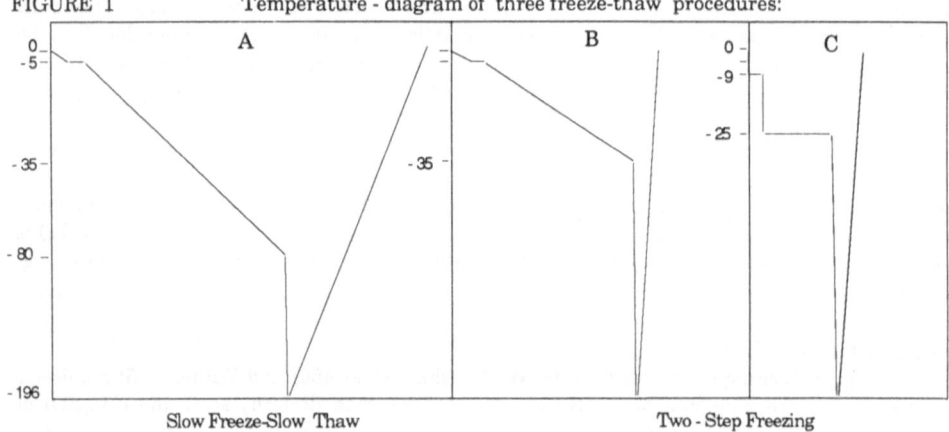

Slow Freeze-Slow Thaw   Two - Step Freezing

zen embryos are less susceptible to damage at higher subzero temperatures (33,35) and therefore can be transported on dry ice (49).

3.2.2.1. <u>Two-Step Freezing</u>. The procedure of freezing mouse embryos at a low rate (0.3°C/min) to a temperature in-between -30°C and-40°C following the conventional protocol with subsequent immersion in liquid nitrogen (53; Fig.1B) has certain advantages (52). For instance it is less time consuming and secondly, less expensive and more practical types of apparatus are available or may be designed. However, a differential susceptibility of embryos at various stages of development with different molar concentrations of various cryoprotectants has been observed (28).

A straightforward two-step freezing technique (Fig.1C) has been described by Wood and Farrant (56). The first step involves seeding of the samples after 2 min in an seeding bath maintained at -9°C. Immediately thereafter the tubes are transferred to a constant temperature ethanol bath at -20°C and hold between 15 and 30 min before being rapidly cooled in liquid nitrogen.

We have obtained the most constant results by applying a modification of the former technique. For freezing plastic straws (Minitübs®) are used instead of other containers because they can be better marked and almost no embryos get lost during manipulation. The straws are loaded as depicted in Fig. 2.

FIGURE 2

a   b   c   d   b   e

a, metal bulb; b, air space; c, freezing medium with embryos; d, seeding point; e, plastic bulb

In a programmable automatic ethanol cooling bath the embryos are equilibrated at 0°C in freezing medium (PB1) with 1.5M 1,2-propanediol as the cryoprotectant (26, 27, 38) for 5 min, then cooled to -6°C at a rate of 1°C/min, seeded and then slowly cooled to -32°C at a rate of 0.3°C/min, held for about 5 min at -32°C and transferred directly to liquid nitrogen. The straws are always kept in horizontal position and only lifted up for seeding.

Thawing at a rate of ~300°C/min is achieved by warming the straws in a water bath (20°C) or at room temperature for 40 sec.

3.2.2.2. Vitrification (Quick freeze-fast thaw procedure). Recently several research groups have reported that mouse and even rat embryos would survive freezing after rapid cooling by directly plunging into liquid nitrogen (5, 19, 34, 42, 54). This quick freeze-fast thaw procedure requires the use of a highly concentrated aqueous solution of cryoprotectants. At sufficiently low temperatures, these solutions become so viscous that they turn into an amorphous state without any formation of ice. This process has been termed vitrification. Most groups use glycerol (3.0M-4.0M) as a permeable and succrose (0.25M-1.0M) as a non-permeable cryoprotectant. Before freezing the embryos need to be dehydrated. At temperatures below 4°C embryos can tolerate exposure to a concentrated solution of cryoprotectants and the associated osmotic dehydration (34). This dehydration which should not exceed a certain time (~15 min) has the advantage of allowing the cells to tolerate rapid dilution of the cryoprotectants after thawing.

Though this method does not require an elaborate bio-freezer and appears to be rather simple it has not yet replaced the more conventional techniques. Still the conditions for predehydration and cryoprotectant removal require further optimizing (54). Post-thaw survival is variable for the different developmental stages of the embryos and may depend on the type and concentration of cryoprotective solution used for vitrification. In contrast to these requirements dehydration and removal of cryoprotectant is less critical for the survival of embryos frozen by the two-step method as used in our laboratory.

3.3. Thawing

The requirements for thawing embryos are defined by the freezing procedure and the cryoprotectants used as mentioned above. The manipulation depends on the cryo-containers used. In tubes and ampoules regularly a few embryos get lost because they stick to the wall, whereas with straws nearly all embryos are recovered. Upon defined thawing and after the cryoprotectant has stepwise been diluted out, the embryos are put through several (~3) washes in PB1. Only those embryos that appear to be morphologically unaffected by microscopic inspection are selected for further processing, i.e. culture for 12-24 hours in embryo culture medium (46) or immediate transfer to pseudopregnant recipients.

3.4. Transfer to Pseudopregnant Surrogate Mothers

Pseudopregnancy is induced by mating female mice with a vasectomized male of proven sterility. Selecting proestrus females will increase the yield of recipients. This procedure is timed such that a synchronous (i.e. same chronological stage) or asynchronous (i.e. recipient stage minus 1 day of embryo development) transfer can be performed. A close synchrony between embryo development and endometrial differentiation enables a large proportion of transferred embryos to develop into live offspring. Asynchrony is advantageous if embryos have suffered a developmental delay during manipulation in vitro.

Transfer of embryos to surrogate mothers in a fully allogeneic combination has been shown to evoke maternal immune reactions that differ from those obtained from semi-allogeneic transfers or natural pregnancies caused by allogeneic males (8). In transplanted offspring a hypo-reactivity of spleen cells towards maternal cells was seen, suggesting some form of immunological tolerance caused either by passage of maternal cells into the fetal compartment and/or the activation of suppressor memory by a cell-free antigenic substance crossing the placenta. Such effects are confined to the transplanted animals and are not heritable

(24) in contrast to Uphoff's hypothesis (44).

Most groups are using F1-hybrids as surrogate recipients because of better results as with pure inbreds. As the revitalized offspring will not be used immediately for experiments but to establish a breeding nucleus the reactivity found will not be of importance as it is not passed on to their offspring.

A serious problem is raised by the finding of McCullagh (24) who demonstrated that in rat embryo transfer experiments in which embryo and surrogate mothers were fully allogeneic with respect to each other a significantly impared rate of survival was obtained. These results need further investigation as to the genetic basis (major or minor histocompatibility loci). One consequence might be to transfer embryos only to closely related surrogate mothers or to hybrids carrying at least the MHC haplotype of the embryo donor strain.

Morulae (day 3.5) can be transferred with good results into the uteri of day 2.5 recipients and blastocysts into day 3-3.5 recipients (32). Earlier preimplantation stages fail to develop into live offspring upon uterine transfer (11). However, all preimplantation stages can be transferred to the oviduct of day 1 pseudopregnant recipients with good success (31,43). The best results are achieved with 4- to 8-cell embryos. With this technique the embryos need to have an intact zona pelucida, which is not a requirement in uterine transfers (6). Several authors have reported that preimplantation stages of mouse and rat embryos developed to term after non-surgical embryo-transfer (5 ,19, 34, 42, 54). Except for Moler et al. none of the other groups obtained results compatible with those of the surgical techniques. Since the number of pedigreed embryos of one frozen batch is limited and highly valuable it is advisable to use a surgical method.

On the basis of several considerations we have decided for the oviduct transfer:(i) Four- to eight-cell mouse and rat embryos survive the freeze-thaw protocol as applied in our laboratory with good results. Nearly 90% appear morphologically normal. (ii) Intermediate culture and production of advanced pseudopregnant females (day 2.5 to 4.0) require a sophisticated timing especially if transfers shall be performed on gnotobiotic recipients. It is not a rare occasion that pedigreed embryos are thawed and cultured but at the time of transfer no properly timed recipients are available. (iii) With immediate transfer to the oviduct an embryo-batch is not thawed until proper pseudopregnant recipients are available. The proportion of weaned offspring from transferred embryos at present is approximately 65% for hybrid and ~52% for inbred donors. Considering the above 90% recovery rate of morphologically normal embryos and a pregnancy rate upon transfer of $\geq$ 50% for inbred strains an overall success of ~25% is obtained.

3.5. Freezing of Rat Embryos

Successful freezing of rat embryos obtained from outbred or F1-hybrid donors has been reported ((9, 17, 25, 26, 48, 57). All of the above mentioned types of freezing have been employed. Though the results obtained are far behind those from mice the freezing of rat embryos causes the minor problems in establishing an embryo bank. A requirement for successful freezing is an abundant supply of embryos which in mice usually is achieved by superovulation. This has not proven very feasible for this species even by using immature rats (48, 57). Culture of rat embryos is also considered suboptimal as contrasted to results obtained in mice. Therefore transfer of 4- to 8-cell stage embryos into the oviduct of pseudopregnant recipients appears to be the most suitable method. For handling and freezing a tissue culture medium such as TCM 199 supplemented with 10% inactivated rat serum is preferred rather than a modified Dulbecco's phosphate buffered saline solution. Finally it is difficult to produce timed pseudopregnant animals. Neither cervical stimulation with a vibrator (10) nor the application of a vaginal tampon (58) has been very effective. The use of two vasectomized males and the use of an otoscope as a vaginal speculum to check for plugs has given more promising results (36).

3.6. Set-up of an Embryo Bank

The capability of successfully revitalizing frozen-thawed embryos is the first and decisive step in setting-up an embryo bank which one can imagine as a circuit which starts and ends with the live breeding nuclei of strains included and which covers the collection, freezing, storage, and revitalization of embryos. All methods then have to be established and managed on a routine basis.

In respect to inbred strains only embryos from single brother-sister matings of the foundation colony should be frozen together in one cryotube (termed 'embryo-batch') to secure the pure inbred status. In respect to mutants as well as congenic strains in which the background strain is conventionally maintained or preserved it is not obligatory to adhere to a strict inbreeding system. A few generations of backcrossing will establish again a clearly defined system. Each embryo-batch is to be regarded as a potential breeding nucleus. One batch should contain a minimum of 10 embryos providing at least 2 to 3 offspring. To achieve a 99.9% probability of one fertile breeding pair 8 to 10 such batches are required. On these grounds we have decided that 20 embryo-batches per strain will provide a save back-up.

To yield sufficient numbers of pedigreed embryo donors the foundation colonies have to be expanded to a sufficient size. This size depends on the number of embryos that can be collected from one normally mated or superovulated donor. If a strain is refractory to superovulation, like e.g. CBA, closely related offspring are mated. This could be either matings of the male or of one male offspring from a BxS breeding cage with all female offspring from successive litters. This will allow pooling of embryos from several cryo-tubes until an adequate sample for embryo-transfer has been obtained without interrupting the inbred status severely.

Each freeze run needs to be controlled not only by a temperature recording but also by a vital test, which means that an additional embryo-batch not to be used for the gene pool is included and shortly thereafter revitalized. F1-hybrids are best suited as embryo donors for such a control-batch in each freeze run or in tests to evaluate various methods, e.g. (BALB/c x DBA/2)F1.

Records set up have to include conditions of the freeze run, a strain description, as well as identification and storage location. The physical conditions of the freezing procedure including results of viability tests obtained from the particular control batch are required as a documentation of a correct freezing technique and to further provide information on the thawing procedure to be applied. For each stock in the repository a description of the strain or mutant with particular information about phenotype, reproductive performance, and strain history should be kept on a file. Informations concerning identification require a complete pedigree information, like parentage, genotype, generation, code number of the embryo-batch, number and developmental stage of embryos frozen, and a precise storage location.

To detect changes due to genetic drift in the original population or to detect changes brought about by manipulating the embryos, a set of various marker systems defining a genetic profile of the strains has to be available. A genetic monitoring program (16) is always required in order to definitely exclude any mix-up that might have happened at any stage from the collection of embryos til transfer.

4. ACHIEVEMENTS AND POSSIBILITIES

To date only in a few laboratories embryo-banks have been set up or are being developed, e.g.the Jackson Laboratory, Bar Harbor, Maine and the Oak Ridge National Laboratory, Tennessee, U.S.A., the MRC Radiobiology Unit, Harwell, and the MRC Embryology and Experimental Toxicology Unit, Carshalton, England, and the Central Institute for Laboratory Animal Breeding, Hannover, Germany. The mouse embryo repository at the Jackson Laboratory by far is the most advanced. It is currently in progress (since 1978) to preserve embryos from 500 different strains (29). There are promising indications that other laboratories will also start with the freeze preservation of mouse embryos.

The two major objectives for preserving strains by means of cryopreservation of embryos are economical and genetic considerations. Though the latter point appears to be the most important to scientists by protecting invaluable strains and animal models of human diseases against loss we are always faced with the financial cut offs threatening experimental animal colonies. It is therefore important to know the extent of savings embryo freezing could provide. A cost-benefit analysis has been performed by Lyon (21) expressing the amount of work and expense involved in freezing and storing embryos in terms of the expense involved in keeping a strain traditionally. The costs of freezing a sample of 500 embryos has been considered equivalent to 3 months normal maintenance. To revitalize a strain, surrogate mothers have to be bred, by which the thawed embryos can be gestated and nursed. For this procedure another 3 months have been estimated. Freezing and thawing thus would be equivalent to half a year of normal maintenance. If a strain is stored frozen in liquid nitrogen for one year a 50% and for a five year period a 90% saving would be achieved.

We have calculated the costs of traditionally maintaining an SPF breeding nucleus, as well as the costs of freezing 20 embryo-batches, storing in liquid nitrogen and revitalization of a strain, based on actual expenses for personnel, investments, and current expenses at our institute. In terms of current costs for savely freeze-preserving a strain by 20 batches and for the maintenance of the embryo bank calculated over a five years period an average annual saving of ~ 60% is achieved. With every frozen strain added to the embryo bank and with each additional year of maintenance in liquid nitrogen this saving is noticeably increased.

Whittingham et al. (51) have studied the effects of increased levels of γ-radiation at -196°C. Whilst the freezing technique exerted a significant effect on survival, even at high radiation doses (100x background) survival was only slightly impaired. Upon revitalization all progeny bred normal, showing no chromosomal anomalies or increase in mutation rate. It has also been shown that the cryoprotectants themselves have radioprotective properties (1). These data confirm the thesis that cryopreservation of embryos is a means to prevent genetic drift.

If embryo collection, cryopreservation, culture and transfer to specified pathogen free surrogate mothers is carried out aseptically the raised offspring are free from parasites, most pathogenic bacteria and murine viruses provided that the zona pelucida is intact (37, 41). Denuded embryos as well as hatched blastocysts are susceptible to cytotoxic effects of virus, e.g. mouse hepatitis virus, that is present in the medium (18). Embryos from contaminated colonies can be frozen at the breeder site, whereas revitalization may be performed in association with a gnotobiotechnical unit when germfree or SPF surrogate mothers are available.

5. LIMITATIONS

Despite the advantages that are evident for freeze preserving strains of mice and rats one cannot neglect that there are also certain limitations. Although there is a general notion that freeze preservation of strains would cease any further genetic diversification, as there is no further replication and thus mutations cannot be propagated, a possible damage to the DNA by background ionizing radiation cannot be overcome by repair mechanisms. Furthermore, there is one information by Bailey and Mobraaten (3) that in freeze-preserved and revitalized mice skin graft tests revealed an increased frequency of histocompatibility mutations as compared to normal control mice. These data, however, need further verification.

Freezing of rat embryos still has not advanced as much as freezing of mouse embryos. According to the published results the rate of rat embryos developing into live fetuses, respectively offspring is less than one third of the mouse results. Rat embryo banking still is at its infancy and most laboratories freeze only mouse embryos.

The progress of embryo banking is slow if striktly pedigreed embryos ought to be frozen because breeding nuclei (foundation colonies) are usally small sized and not all required kinship matings (brother-sister or father-daughter) will yield enough fertile matings at anyone trial. If only mutant genes or in the case of congenic strains the amount of backcrossing need to be preserved the time required to obtain a sufficient number of freezeable batches is much shorter as kinship can be neglected. Although there have been quite optimistic calculations on how many strains may be frozen annually according to our experience two skilled technicians will be able to freeze 450 to 500 batches per year, which is equivalent to about 25 strains (3 technicians can freeze approximately twice as many). In view of the vast number of strains available for biomedical research (Table 1) it is impossible for one group to perform all freezing by itself.

Though an exchange of frozen mouse embryos among the laboratories involved in freeze preservation of rodents would tremendously aid laboratory animal medicine and biomedical research we are far away from such an exchange program. Each laboratory was set up with different objectives and therefore different logistics. Most of these are build for intramural research only. None of the freezing techniques employed is alike. Each laboratory uses its special modification of one of the above mentioned techniques. This would require that all groups engaged in a frozen embryo exchange programme will have to be able to manage the different methods.

6. CONCLUSIONS

With the different freezing techniques reported a save freeze preservation of genetic stocks of mice is possible. The results obtained in this species even when freezing inbred strains are sufficient and have been shown to be a highly economic method. Further improvement of the technique, primarily concerning the yield of pedigreed embryos and the transfer technique would substancially reduce the costs and speed up the operation. Freezing of rat embryos still is at a developmental stage and unfortunately has not gained the attention required.

In view of the many strains of laboratory rodents available it is impossible for one institution to preserve all. Regarding the overall economic situation this task calls for an international joint-adventure.

REFERENCES

1. Ashwood-Smith MJ: Current concepts concerning radioprotective and cryoprotective properties of dimethyl sulfoxide in cellular systems. Ann. N.Y. Acad. Sci., 243:246-256, 1975.
2. Bailey DW: Sources of subline divergence and their relative importance for sublines of six major inbred strains of mice. In: Morse HC (ed): Origins of Inbred Mice.New York: Academic Press,1978: 197-215.
3. Bailey DW and Mobraaten LE: Changes in Mice Derived from Frozen-Thawed Embryos. The Jackson Laboratory Scientific Report 1980-81: 36.
4. Beatty RA: Transplantation of mouse eggs. Nature, 168:995,1951.
5. Biery KA, Seidel GE, and Elsden RP: Cryopreservation of mouse embryos by direct plunging into liquid nitrogen. Theriogenology, 25:140,1986.
6. Bronson RA and McLaren A: Transfer to the mouse oviduct of eggs with and without the zona pellucida. J. Reprod. Fert., 22:129-137, 1970.
7. Butcher GW and Howard JC: The MHC of the laboratory rat, *Rattus norvegicus*. In: Weir DM(ed) Genetics and Molecular Immunology. Oxford: Blackwell,1986: 101.1-101.18.
8. Cibotti R, Kinsky RG, and Voisin GA: Immunological relationships between murine surrogate mothers and transplanted embryos, as studied by local GVH reactivity. J. Reprod. Immunol., 9: 225-236, 1986.
9. Chupin D and DeReviers MM: Quick freezing of rat embryos. Theriogenol., 26: 157-166, 1986.
10. DeFeo VJ: Vaginal-cervical vibration: A simple and effective method for the induction of pseudopregnancy in the rat. Endocrinol., 79: 440-442, 1966.

11. Dickmann Z and Noyes RW: The fate of ova transferred into the uterus of the rat. J. Reprod. Fertil., 1: 197-212, 1960.
12. Festing MFW: International Index of Laboratory Animals. Fourth Edition, MRC, 1980.
13. Gates AH: Maximizing yield and developmental uniformity of eggs. In: Daniel Jr. JC (ed): Methods in Mammalian Embryology. San Francisco: W. H. Freeman and Co., 1971: 64-75.
14. Green MC (ed): Genetic Variants and Strains of the Laboratory Mouse. Stuttgart: Gustav Fischer Verlag, 1981.
15. Hedrich HJ: Aiming at genetic constancy of inbred strains via genetic monitoring and cryopreservation. Proceedings of the 7th ICLAS Symp., Utrecht 1979. Stuttgart: G. Fischer Verlag, 1980: 329-331.
16. Hedrich HJ: Overview of the State of the Art in Genetic Monitoring. In: Melby EC and Balk MW (ed): Biomedical Research: The Importance of Laboratory Animal Genetics, Health and Environment. Proceedings of the Fifth Charles River International Symposium on Laboratory Animals. New York: Academic Press, 1983: 75-99.
17. Kiehm DJ, Rouleau A, Vanderhyden B, and Armstrong DT: In vitro fertilization of rat oocytes following cryopreservation. Theriogenol., 27: 242, 1987.
18. Kraft V, Reetz IC, and Hedrich HJ: unpublished results.
19. Krag KT, Koehler IM, and Wright RW: A method for freezing early murine embryos by plunging directly into liquid nitrogen. Theriogenology, 23:199, 1985.
20. Leibo SP: Introduction to Embryo Freezing. In: Zeilmaker GH (ed): Frozen Storage of Laboratory Animals. Stuttgart:G. Fischer Verlag, 1981: 1-19.
21. Lyon MF: Implications of Freezing for the Preservation of Genetic Stocks. In: Mühlbock O (ed): Basic Aspects of Feeze Preservation of Mouse Strains. Stuttgart: Gustav Fischer, 1976 57-65.
22. Marsk L and Larsson KS: A Simple Method for Non-Surgical Blastocyst Transfer in Mice. J. Reprod. Fertil., 37: 393-398, 1974.
23. Mazur P: Kinetics of water loss from cells at subzero temperatures and the likelihood of intracellular freezing. J. Gen. Physiol., 47: 347-369, 1963.
24. McCullagh P: The immunological consequences allogeneic rat embryo transfer. Aust. J. Biol. Med. Sci., 63: 645-650, 1985.
25. Miyamoto H and Ishibashi T: Survival of frozen-thawed mouse and rat embryos in the presence of ethylene glycol. J. Reprod. Fert., 50: 373-375, 1977.
26. Miyamoto H and Ishibashi T: The protective action of glycols against freezing damage of mouse and rat embryos. J. Reprod. Fert.,54: 427-432, 1978.
27. Miyamoto H and Ishibashi T: Survival of Mouse Embryos Frozen-Thawed Slowly or Rapidly in the Presence of Various Cryoprotectants. J. Exp. Zool., 226: 123-127, 1983.
28. Miyamoto H and Ishibashi T: Liquid nitrogen vapour freezing of mouse embryos. J. Reprod. Fert., 78: 471-478, 1986.
29. Mobraaten LE, Davisson MT, and Taylor BA: The Jackson Laboratory Scientific Report 1985-86: Genetic Resources, 1986:101.
30. Moler TL, Donahue SE, and Anderson GB: A Simple Technique for Nonsurgical Embryo Transfer in Mice. Lab. Anim. Sci. 29: 353-356, 1979.
31. Noyes RW and Dickmann Z: Survival of ova transferred into the oviduct of the rat. Fertil. Steril., 12: 67-79, 1961.
32. Rafferty Jr. KA : Methods in experimental embryology of the mouse. Baltimore: Johns Hopkins Press, 1970.
33. Rall WF: The Role of Intracellular Ice in the Slow Warming Injury of Mouse Embryos. In: Zeilmaker GH (ed): Frozen Storage of Laboratory Animals. Stuttgart: Gustav Fischer Verlag, 1981: 33-44.
34. Rall WF and Fahy GM: Ice-free cryopreservation of mouse embryos at -196° C by

vitrification. Nature 313:573-575, 1985.

35. Rall WF, Reid DS, and Farrant J: Innocuous biological freezing during warming. Nature, 286: 511-514, 1980.

36. Reetz IC and Hedrich HJ: unpublished results.

37. Reetz IC, Wullenweber-Schmidt M, Kraft V, Hedrich HJ, Meyer B, Sickel E: Beitrag zur hygienischen Sanierung von Mäuseinzuchtstämmen mittels Embryotransfer. Dtsch. tierärztl. Wschr, 94: 39, 1987.

38. Renard JP and Babinet C: High survival of mouse embryos after rapid freezing and thawing inside plastic straws with 1,2- propanediol as cryoprotectant. J. Exp. Zool., 230: 443-448, 1984.

39. Schiewe MC, Schmidt PM, Bush M, and Wildt DE: Toxicity potential of absorbed-retained ethylene oxide residues in culture dishes on embryo development in vitro. J. Anim. Sci., 60: 1610-1618, 1985.

40. Schneider U and Mazur P: Relative Influence of Unfrozen Fraction and Salt Concentration on the Survival of Slowly Frozen Eight-Cell Mouse Embryos. Cryobiol. 24: 17-41, 1987.

41. Singh EL: The disease control potential of embryos. Theriogenol., 27: 9-20, 1987.

42. Takeda T, Elsden RP, and Seidel GE: Cryopreservation of mouse embryos by direct plunging into liquid nitrogen. Theriogenol., 21: 266, 1984.

43. Tarkowski AK: Experiments on the Transplantation of Ova in Mice. Acta Theriol., 2: 251-267, 1959.

44. Uphoff DE: Maternal Influences: Their Immunologic Aspects. In: Mühlbock O (ed): Basic Aspects of Feeze Preservation of Mouse Strains. Stuttgart: G. Fischer Verlag, 1976: 85-102.

45. Vickery BH, Erickson GI, and Bennett JP: Nonsurgical Transfer of Eggs through the Cervix in Rats. Endocrinol.,85: 1202-1203, 1969.

46. Whittingham DG: Culture of mouse ova. J. Reprod. Fert. Suppl., 14: 7-21, 1971.

47. Whittingham DG: Embryo banks in the future of developmental genetics. Genetics, 78: 395-402, 1974.

48. Whittingham DG: Survival of rat embryos after freezing and thawing. J. Reprod. Fert., 43: 575-578, 1975.

49. Whittingham DG: Sensitivity of Mouse Embryos to the rate of Thawing. In: Zeilmaker GH (ed): Frozen Storage of Laboratory Animals. Stuttgart: Gustav Fischer Verlag, 1981: 21-32.

50. Whittingham DG, Leibo SP, and Mazur P:Survival of mouse embryos frozen to -196° C and -269° C. Science, 78: 411-414, 1972.

51. Whittingham DG, Lyon MF, and Glenister PH: Re-establishment of breeding stocks of mutant and inbred strains of mice from embyos stored at -196° C for prolonged periods. Genet. Res. Camb., 30: 287-299, 1977.

52. Whittingham DG, Wood M, Farrant J, Lee H, and Halsey JA: Survival of frozen mouse embryos after rapid thawing from -196° C. J. Reprod. Fert., 56: 11-21, 1979.

53. Willadsen SM: Factors affecting the survival of sheep embryos during deep-freezing and thawing: In: Elliot K and Whelan J (ed): The Freezing of Mammalian Embryos. Amsterdam: Elsevier, 1977:175-189.

54. Williams TJ and Johnson SE: A method for one-step freezing of mouse embryos. Theriogenol., 26: 125-133, 1986.

55. Wilmut I: The effect of cooling rate, embryos during freezing and thawing. Life Sci., 11: 1071-1079, 1972.

56. Wood MJ and Farrant J: Preservation of Mouse Embryos by Two-Step Freezing. Cryobiol., 17: 178-180, 1980.

57. Wood MJ and Whittingham DG: Low Temperature Storage of Rat Embryos. In: Zeilmaker GH (ed): Frozen Storage of Laboratory Animals. Stuttgart:G. Fischer Verlag, 1981:119-128.

58. Yang WH: Induction of Pseudopregnancy in Rats by Vaginal Tampon. Endocrinol., 82: 423-425, 1968.

A.C. Beynen & H.A. Solleveld (Eds), New developments in biosciences: their implications for laboratory animal science, ISBN 978-94-010-7973-0

THE GENERATION OF TRANSGENIC ANIMALS AND THEIR USE IN FUNDAMENTAL RESEARCH

A. Berns

Divison of Molecular Genetics of the Netherlands Cancer Institute and the Department of Biochemistry of the University of Amsterdam, Plesmanlaan 121, 1066 CX Amsterdam, The Netherlands.

Introduction

With the advent of recombinant DNA techniques any gene can be in principal isolated and analyzed physically. In order to study the function of gene products and regulation of gene expression, one of the most powerful and straightforward methods is transfer of isolated genes into living cells, and the subsequent analysis of the effects. The value of this approach might be best illustrated by the functional identification of a number of genes, directly involved in tumor induction, thereby increasing enormously our knowledge of fundamental processes of growth and differentiation. However, gene transfer studies in cell culture have also their limitations: Only a restricted number of cell types can be propagated in culture, while retaining their tissue specific characteristics. Furthermore, the effects of gene products on complex interactive systems as present in complete organisms, are difficult to assess in vitro. The solution to this problem is both elegant and simple: The introduction of genes, which can be defined up to the nucleotide sequence level, into all cells of a mammalian organism. This can be achieved by gene transfer into pre-implantation embryos, from where the introduced gene is transferred to all cells of the organism as integral part of the host genome. The combination of the long standing experience of embryologists to isolate and grow embryos with the technology of the molecular biologist makes this possible. This not only allows the testing of gene functions in their most natural environment, it is also instrumental in studying the interaction of defective genes, whether present as inherited disorders or present as the consequence of somatic mutations as observed in many cancers.
In the subsequent paragraphs the various aspects of this approach will be discussed in more detail:

i.   Why transgenesis? What parameters can be investigated by transgenesis that are not accessible for study by more convential techniques?
ii.   What methods are available to generate transgenic animals? Which is the preferred method for what purpose?
iii. What have we learned from studies with transgenic mice? For example in cancer research? What can be expected from this technology in the near future?
iv.   What are the applications outside the area of fundamental research? Can it be used to enhance animal production systems? Can it be used for producing new pharmaceuticals?
v.   Are any risks involved?

I. WHY TRANSGENESIS ?

Multicellular organisms have evolved from simple cell aggregates, which in the course of evolution have progressed from "sharing" tasks to "dividing" tasks. This has resulted in an increasing sophistication of cell specializations. Dividing tasks in multicellular organisms requires extensive communication between cells in order to coordinate the many different functions exerted by the specialized cell lineages. Although the genetic content between the different somatic cells of a mammalian organism is (at least for 99.99%) identical, the biochemical constitution, physiological appearance, and functional commitment can greatly differ. These special features are easily lost if cells are explanted and put into culture, in which specific instructions for maintaining the specialized differentiation level are no longer present. It is quite possible that for a number of cell types "in vitro" culture will never sufficiently be able to mimic the "in vivo" situation to retain even the most prominent features of the differentiated cell, simply because "in vivo" the structural organization of the different cells composing a tissue form an essential element of the specific instruction itself. Therefore, in order to study gene function in a number of specialized cell types, one has to rely on analyses in vivo. Transgenesis allows such analyses.
However, transgenic animals not only can serve as substitute for our inability to develop a suitable cell culture system. Complex interactive processes, intricately associated with complex organisms, can often only be studied in a system in which a series of different interacting cell lineages are present: the intact organism. This is most apparent for processes like mammalian development, immunosurveillance, development of cancer, and aging. In the case of cancer the subsequent steps which finally lead to a highly malignant tumor are hard to discern. Transgenesis allows analysis of genes which play an instrumental role in these processes.
The mouse is the preferred model system for these studies. It combines characteristics like low cost, easy handling, and short generation time with an excellent genetic characterization, coupled to the availability of a large number of inbred strains. Presently, studies using transgenic mice concentrate predominantly on the following aspects:

1. Analysis of the effects of (mutant) gene products. In these studies the biological consequences of overexpression or ectopic expression of a particular (modified) gene is primarily analyzed.
In large areas of developmental biology, immunology, and cancer research, fields which have extensive overlaps, transgenic mice offer unique opportunities to dissect these complex processes.
2. Analysis of gene regulatory functions with the aim to define the nucleotide sequences in or near structural genes, which are responsible for the tissue- and developmental-specific expression.
3. Generation of mutants. Genetic well-defined mutants constitute a powerful tool in studying gene function. This is especially relevant for the study of mammalian development and genetic diseases. The approach to be taken is the targeted inactivation of defined endogenous mouse genes via homologous recombination or retroviral insertion. This aspect will be in focus for the years to come and will, if attainable, more or less routinely, constitute a major breakthrough in mammalian genetics (in other systems, from E. coli to Drosophila, the availability of large numbers of mutants has been a decisive factor for our present knowledge).

4. As a model system for the application of gene transfer in larger domestic animals. These studies concentrate on aspects like: production of pharmaceuticals; quality of existing animal products; energy efficiency of the animal production system; and disease resistance of livestock.

II METHODS TO GENERATE TRANSGENIC ANIMALS

All methods to introduce new genetic information into the germ line of animals are based on the principle that once the new genes are inserted into the chromosomal DNA of the early embryo, they will automatically become part of all cell-types of the organism, including germ cells. This has been proven a successful approach. Preferred target cells for manipulation are: The fertilized zygote, the preimplantation embryo, and pluripotential embryonic cell lines. Depending on the target cell used, gene transfer into these cells has been achieved by microinjection, DNA transfection, and infection using retroviral vector systems. Three major routes can be discerned (see also figure 1):

1. Gene transfer into the zygote. This has so far been the most efficient procedure to generate transgenic animals. A high efficiency of stable gene transfer (up to approximately 30% of the animals born) can be obtained if DNA is directly microinjected into one of the pronuclei (preferentially the male pronucleus) (3,21). Efficiency of transfer is highest with linear DNA molecules. Typically, approximately 100 molecules are injected in a volume of 1 picoliter. Injection into mouse pronuclei is relatively easy, as murine ova are translucent.

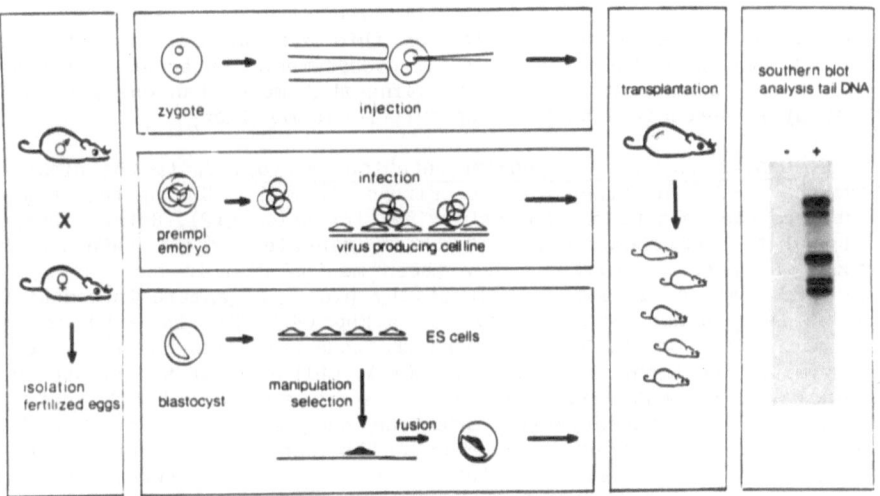

Figure 1: Methods of producing transgenic mice based on the manipulation of embryonal target cells at three different developmental stages.

Injection is a particularly straightforward procedure if interference contrast optics (Nomarski) are being used. The injected DNA is integrated in the chromosomal DNA of the host predominantly in concatemeric form (in head to tail arrays). Molecules up to 50 kbp long have been efficiently incorporated in this way. Integration into the chromosomal DNA occurs early in the preimplantation phase, as approximately 70-80% of the transgenic animals are heterozygous for the transgene, and transmit the trait to their offspring according to Mendelian expectations. The insertion of large DNA fragments results in general in a better controlled expression of the gene, probably because of a more "natural micro-environment". Many genes have been shown to be expressed in a developmental- and tissue-specific fashion. These include genes like elastase, insulin, ∝-fetoprotein, albumin, myosin, crystallin, immunoglobulin, MHC class I, T-cell receptor, collagen, beta-and alpha-globin, vimentin, and various others (for overview see ref. 14,21; unpublished results). For the expression of some of these genes, a small region upstream of the coding region is sufficient to confer the tissue-specific expression behaviour, while other genes require intronic sequences or sequences located downstream of the structural gene in order to be properly expressed. This already indicates that the expression behaviour of fusion genes (regulatory region of one gene fused to the structural genetic information of another gene) is not always predictable. Fusion genes might even show a complete new pattern of expression due to the combination of (different) regulatory elements, as has indeed been observed (25).

The number of DNA copies incorporated can vary from 1 to several hundred. No relation has been observed between the number of copies and the level of expression. Expression levels up to those of the cognate endogenous gene have been reported frequently. In a few instances significant higher levels of expression were found (21). The site of integration of the transgene is probably responsible for this effect. Evidence has been provided that more than 1 copy of a concatemeric integrated transgenic array can be transcribed (19), indicating that more than one gene within the array is accessible to the transcriptional machinery.

2. <u>Gene transfer into pre-implantation embryos</u>. Up to the blastocyst stage the developing embryo can be cultured <u>in vitro</u>. During this time the blastomers can be a target for retroviral infection (12). Retroviruses are equipped to efficiently insert their genetic content into the host chromosomal DNA. Retroviruses have been manipulated in such a way that they can serve as carrier for almost any piece of genetic information as long as it does not exceed a size of 8 kbp (4,7,30). In order to allow efficient infection of the blastomers the zona pellucida has to be removed by enzymatic treatment or acid tyrode (11). Culturing of the blastomers on a monolayer of virus-producing cells is an easy and highly efficient method (13,26). Alternatively, infection can be performed in a later stage: Virus or virus producing cells can be injected into the blastocoele. Virus released into the blastocyst cavity will very frequently infect the Inner Cell Mass (ICM) cells that will form the embryo. Most of the founders will be mosaic for the transgene, as incorporation generally occurs only in a subset of the cells which form the embryo. In addition, the founder might contain various retroviral insertions at different chromosomal sites, which in general will segregate in the offspring.

With a low efficiency it is also possible to introduce genes into the germ

line by intra-uterine retroviral infection of the midgestation embryo
(23).
Although germ-line insertions are easily obtained by retroviral infection
of preimplantation embryos, this technique has so far met with limited
success as to the proper expression of the transferred genes. However, the
field is moving fast, and new vectors have been designed which are likely
to perform better (30). This might turn out to be a suitable alternative
for those species, that produce zygotes which are difficult to
microinject.

   3. Gene transfer into embryonal stem cells (ES). It has become
possible to grow cells derived from preimplantation embryos, for many
generations, while preserving their capacity to contribute to all tissues
of the mouse after fusion with preimplantation embryos (2). This now
allows the genetic manipulation of a large numbers of cells in vitro and
selection for a particular genetic trait before the ES cell is fused to
preimplantation embryos (8). This approach is especially promising for the
generation of defined mutant mouse strains by site-specific insertional
inactivation. The principles of this approach have been worked out, and
mouse strains inactivated at the HPRT locus have been recently reported
(16). The next step will be the inactivation of an autosomal gene, for
which no selection can be exerted in the ES cell culture phase.

WHAT HAS BEEN LEARNED FROM TRANSGENIC EXPERIMENTS

Here follow a few general observations which are not surprizing to most
molecular geneticists any more, but which are nevertheless remarkable:

   1. If a transgene (including its authentic regulatory region(s)) is
expressed at all, it is generally expressed in the correct developmental-
and tissue-specific fashion. Very frequently expression levels are close
to the levels of expression of the endogenous cognate gene. This indicates
that the site of integration is not pivotal in the regulation of
expression, and the organization of the chromosomal DNA is apparently
flexible enough to incorporate and allow correct expression of large
pieces of exogenous DNA.
   2. Even if the transgene has been derived from another species, the
developmental- and tissue-specific expression is maintained. This
indicates that the expression is regulated by transacting factors, which
have been sufficiently conserved in evolution in order to recognize and
act upon regulatory sequences present in DNA from different species.

In a number of research areas transgenesis has become an important genetic
tool. Three examples will be described briefly.
   a. Transgenesis to reveal the position and nature of regulatory
sequences required for the tissue-specific expression of a gene. The
liver-specific alpha-fetoprotein gene is a good example. The regulatory
region has been mapped in great detail via transgenesis with gene
constructs, harboring different amounts of flanking DNA sequences. This
analysis shows that for a proper regulation various DNA domains can be
required which cooperate to confer the correct level and specificity of
expression (10).
   b. Transgenesis to reveal the molecular mechanism underlying
immunosurveillance. The studies in which rearranged immunoglobulin genes
have been introduced into the germ line of mice have been most rewarding.

These studies have shown that backfeed mechanisms are operational that inhibit rearrangement of the endogenous genes (27). Furthermore, it became apparent that regions in the Immunoglobin transgenes are subjected to continuous somatic mutation, allowing further modulation of the antibodies (19), and formation of idiotypic networks (28). Similar studies are currently being performed with T-cell receptor genes.

    c. Transgenesis to reveal the mechanisms of oncogenic transformation. In transgenic mice the direct involvement of distinct genes in the tumorigenic process can be easily assessed by testing the effects of a high level of expression of that gene in a particular tissue using fusion gene constructs. In this respect I would like to highlight the observations made in transgenic mice bearing c-myc oncogene constructs. Transgenic mice carrying the c-myc gene coupled to the lymphoid-specific IgH (immunoglobulin heavy chain enhancer) regulatory element almost invariably develop B-cell malignancies (1). However, the data indicate that high myc expression by itself is not sufficient to induce the fully malignant phenotype, as the outgrown tumors are of monoclonal origin (18). This shows that malignant transformation requires additional unknown events within the pre-neoplastic cell population, that result from the high expression of the c-myc transgene. Similar conclusions have been drawn from mammary adenocarcinomas arising in transgenic mice carrying a myc gene fused to the regulatory region of the murine mammary tumor virus (24). Here the hormonal condition seemed a prerequisite for the occurrence of additional events, since tumors were only observed in forced bred female mice. Recently, it has been shown that also the presence of the transgenic BPV-1 genome in the mouse genome predisposes these mice for fibropapillomas of the skin (17). In all the transgenic mouse strains mentioned here, tumor development required additional events. Tumor induction in experimental systems provides the explanation for this observation: To obtain the fully malignant phenotype, the successive activation of different oncogenes is required. Therefore, transgenic mouse strains harboring each a different activated oncogene can be used to assay directly the contribution of each of these genes to tumor development by crossing experiments. Most of these studies have just started, and the answers to the most relevant questions only begin to emerge. No doubt a wealth of information will become available in the next years.

IV APPLICATION IN DOMESTIC ANIMALS

Obviously the technological advancements in molecular biology and embryology have attracted the attention of those using animal production systems. The transgene technology could be employed for a number of purposes:

    1. Production of new pharmaceuticals. Especially those products which otherwise can only be produced in mammalian cell culture might be good candidates for production in large domestic animals. Examples are the ξ-1 antitrypsin, plasminogen activator, clotting factors. The most appropriate production site seems the mammary gland, which allows excretion via the milk. Expression of these proteins could be directed to the mammary gland by using the promoter and regulatory regions of genes normally expressed in the lactating gland like caseins and whey proteins. Although many aspects have to be solved (proper processing; stability, and purification from the milk) crude calculations show that a limited number of animals would be sufficient to supply the world's need for these important proteins (6).

2. Alteration of the quality of products presently produced. One could think of "humanizing" the milk by expressing a transgene in the lactating gland which encodes the enzyme beta-galatosidase, so as to break down the lactose. Alternatively, in some cases growth-curves might be influenced by transgenic systhesis of higher levels of the growth-releasing factor or growth hormone (9).

3. Alteration of the resistance of livestock against disease. Especially if sensitivity for a particular disease is mediated by the Major Histocompatibility Complex (MHC) genes, as has been shown in man, transgenesis with a different MHC class I gene could increase resistance of a particular valuable strain of livestock. We have shown that in mice a heterologous MHC class I antigen can function properly (15; our unpublished results). Therefore, even the use of MHC genes from other species might be contemplated.

Many laboratories have started to invest in germline gene transfer in domestic animals. Successful transfers have now also been reported for rabbits, pigs, goats, sheep, cattle, and chicken. Although the efficiency is in general much lower than observed with mice (5), the technology is now open for further improvements, and it can be other reasonable efficiencies will be reached for other species as well.

V. RISKS

None of the gene transfer methods available today can target the transgene to a specific chromosomal locus. Only by selecting for a rare homologous recombination event, as is in principle possible with ES cells which can be propogated in vitro, site-specific alterations can be evoked. As this selection is impossible with preimplantation embryos, the integration of transgenes cannot be controlled (as yet) in the direct manipulation of embryos. Two aspects need consideration in this respect:

a. The site of integration can affect the proper functioning of the transgene, e.g. by integration within heterochromatin the transgene might become silent, or, alternatively, by integration near a strong cis-acting element, it might be expressed aberrantly.

b. The integration might affect the proper functioning of the locus. The transgene could disrupt a host gene or interfere with its regulation. The integration might also cause deletions and more complex rearrangements as has been reported in a few instances (21). In case micro-injection is used as a gene transfer method, one can envisage that chromosomal damage can ensue from the physical intrusion by the injection needle. In that case the defect is not marked by the transgene. The analysis of a large number of transgenic mice shows that "insertional mutagenesis" occurs with a frequency of approximately 10%. The actual number might even be higher, as some of the mutants might have been unnoticed, and as dominant developmentally lethal mutations are not scored. Mutations which have been observed are: A $\xi(I)$ collagen defect (an integrated provirus interrupts the first intron of the mouse collagen gene, resulting in recessive embryonic lethals (23)); abnormal limb formation (by DNA insertion in morphogenetic loci controlling limb formation (20,29)), and sterility (by DNA insertions in regions probably of importance for spermatogenesis (22)).

So if germline transfer is considered to generate new herds of livestock, special attention is required to assertain the "genetic fitness" of the transgenic founder (I will not discuss the need to maintain sufficient genetic variation in the various species, which should also be a "point to consider" in classical breeding programs). Another aspect which should be contemplated is the increasing concern by the general public with respect to genetic manipulation. Clear guidelines are required to settle both practical as well as ethical issues with respect to germline manipulation of livestock.

References

1. Adams, J.M.,A.W. Harris,C.A. Pinkert,L.M. Corcoran,W.S. Alexander,S. Cory,R.D. Palmiter, and R.L. Brinster 1985. The c-myc oncogene driven by immunoglobulin enhancers induces lymphoid malignancy in transgenic mice. Nature 318:533-538.
2. Bradley, A.M.,M.J. Evans,M.H. Kaufman, and E. Robertson 1984. Formation of functional germ line chimeras from embryo-derived teratocarcinoma cells. Nature 309:255-256.
3. Brinster, R.L.,H.Y. Chen,M.E. Trumbauer,M.K. Yagle, and R.D. Palmiter 1985. Factors affecting the efficiency of introducing foreign DNA into mice by microinjecting eggs. Proc. Natl. Acad. Sci. USA 82:4438-4442.
4. Cepko, C.L.,B.E. Roberts, and R.C. Mulligan 1984. Construction and applications of a highly transmissible murine retrovirus shuttle vector. Cell 37:1053-1062.
5. Church, R.B. 1987. Embryo manipulation and gene transfer in domestic animals. TIBTECH 5:13-19.
6. Clark, A.J.,P. Simons,I. Wilmut, and R. Lathe 1987. Pharmaceuticals from transgenic livestock. TIBTECH 5:20-24.
7. Cone, R.D., and R.C. Mulligan 1984. High-efficiency gene transfer into mammalian cells: Generation of helper-free recombinant retrovirus with broad mammalian host range. Proc. Natl. Acad. Sci. USA 81:6349-6353.
8. Gossler, A.,T. Doetschman,R. Korn,E. Serfling, and R. Kemler 1986. Transgenesis by means of blastocyst-derived embryonic stem cell lines. Proc. Natl. Acad. Sci. USA 83:9065-9069.
9. Hammer, R.E.,R.L. Brinster,M.G. Rosenfeld,R.M. Evans, and K.E. Mayo 1985. Expression of human growth hormone-releasing factor in transgenic mice results in increased somatic growth. Nature 315:413-416.
10. Hammer, R.E.,R. Krumlauf,S.A. Camper,R.L. Brinster, and S.M. Tilghman 1987. Diversity of Alpha-Fetoprotein Gene Expression in Mice Is Generated by a Combination of Separate Enhancer Elements. Science 235:53-58.
11. Hogan, B.L.M.,F. Constantini, and E. Lacy 1986. Manipulation of the mouse embryo: a laboratory manual. Cold Spring Harbor Laboratory, Cold Sprong Harbor.
12. Jaenisch, R.,H. Fan, and B. Croker 1975. Infection of preimplantation mouse embryos and of newborn mice leukemia virus: Tissue distribution of viral DNA and RNA and leukemogenesis in the adult animal. Proc. Natl. Acad. Sci. USA 72:4008-4012.
13. Jahner, D.,K. Haase,R. Mulligan, and R. Jaenisch 1985. Insertion of the bacterial gpt gene into the germ line of mice by retroviral infection. Proc. Natl. Acad. Sci. USA 82:6927-6931.
14. Krimpenfort, P., and A. Berns 1987. Gene transfer in mammalian embryos. Human Reproduction :in press.

15. Krimpenfort, P.,G. Rudenko,F. Hochstenbach,D. Gussow,A. Berns, and H. Ploegh 1987. Crosses of two independently derived transgenic mice demonstrate functional complementation of the genes encoding heavy (HLA-B27) and light (beta2-microglobulin) chains of HLA class I antigens. EMBO J.: in press.

16. Kuehn, M.R.,A. Bradley,E.J. Robertson, and M.J. Evans 1987. A potential animal model for Lesch-Nyhan syndrome through introduction of HPRT mutations into mice. Nature 326:295-298.

17. Lacey, M.,S. Alpert, and D. Hanahan 1986. Bovine papillomavirus genome elicits skin tumours in transgenic mice. Nature 322:609-612.

18. Langdon, W.Y.,A.W. Harris,S. Cory, and J.M. Adams 1986. The c-myc oncogene perturbs B lymphocyte development in Eu-myc transgenic mice. 47:11-18.

19. O'Brien, R.L.,R.L. Brinster, and U. Storb 1987. Somatic hypermutation of an immunoglobulin transgene in transgenic mice. Nature 326:405-409.

20. Overbeek, P.A.,S.P. Lai,K.R. van Quill, and H. Westphal 1986. Tissue-specific expression in transgenic mice of a fused gene containing RSV terminal sequences. 231:1574-1577.

21. Palmiter, R.D., and R.L. Brinster 1986. Germline transformation of mice. Ann. Rev. Genet. 20:465-499.

22. Palmiter, R.D.,T.M. Wilkie,H.Y. Chen, and R.L. Brinster 1984. Transmission distortion and mosaicism in an unusual transgenic mouse pedigree. Cell 36:869-877.

23. Schnieke, A.,K. Harbers, and R. Jaenisch 1983. Embryonic lethal mutation in mice induced by retrovirus insertion into the alpha1(I) collagen gene. Nature 304:315-320.

24. Stewart, T.A.,P.K. Pattengale, and P. Leder 1984. Spontaneous mammary adenocarcinomas in transgenic mice that carry and express MTV/myc fusion genes. Cell 38:627-637.

25. Townes, T.M.,H.Y. Chen,J.B. Lingrel,R.D. Palmiter, and R.L. Brinster 1985. Expression of human beta-globin genes in transgenic mice: effects of a flanking metallothionein→human growth hormone fusion gene. Mol. Cell. Biol. 5:1977-1983.

26. van der Putten, H.,F.M. Botteri,A.D. Miller,M.G. Rosenfeld,H. Fan,R.M. Evans, and I.M. Verma 1985. Efficient insertion of genes into the mouse germ line via retroviral vectors. Proc. Natl. Acad. Sci. USA 82: 6148-6152.

27. Weaver, D.,F. Costantini,T. Imanishi-Kari, and D. Baltimore 1985. A transgenic immunoglobulin Mu gene prevents rearrangement of endogenous genes. Cell 42:117-127.

28. Weaver, D.,M.H. Reis,C. Albanese,F. Costantini,D. Baltimore, and T. Isamichi-Kari 1986. Altered repertoire of endogenous immunoglobulin gene expression in transgenic mice containing a rearranged mu heavy chain gene. Cell 45:247-259.

29. Woychik, R.P.,T.A. Stewart,L.G. Davis,P. D'Eustachio, and P. leder 1985. An inherited limb deformity created by insertional mutagenesis in transgenic mouse. Nature 318:36.

30. Yu, S.F.,T. Von Ruden,P.W. Kantoff,C. Garber,M. Seiberg,U. Ruther,W.F., Wagner, E.F., Anderson, and E. Gilboa 1986. Self-inactivating retroviral vectors designed for transfer of whole genes into mammalian cells Proc. Natl. Acad. Sci. USA 83:3194-3198.

A.C. Beynen & H.A. Solleveld (Eds), New developments in biosciences: their implications for laboratory animal science, ISBN 978-94-010-7973-0
© 1988 Martinus Nijhoff Publishers, Dordrecht.

ON THE INHERITANCE OF BLOOD CHARACTERS IN MICE

R. KLUGE, K.G. RAPP, and K. BUROW

1. INTRODUCTION

The phenotypic variation of quantitative blood characters in rats and mice is known to be caused to a remarkable part by genetic influences (Burow et al., 1983).
But up to the present day little is known about the allelic background of this genetic variance. The analysis of single marked chromosomes can help to get more information about the way of inheritance of quantitative characters.
Monogenic inherited protein variants can be used as chromosome markers. The transfer of such marked chromosomes from parents to their offspring can be traced by suitable experiments, e.g., the analysis of F_2-populations or backcrosses derived from inbred strains. In the evaluation it can be tested whether a defined chromosome or chromosome combination has an influence on quantitative traits (Geldermann, 1975). The principle of such an experimental design is shown in Fig. 1 by a backcross system.

In the present study, inbred mice were used to analyse the genetics of blood characters because these animals are defined for a great number of monogenetic traits (Green, 1981) and differ with respect to several quantitative blood values as could be shown by own preliminary tests.

2. MATERIALS AND METHODS

2.1. Animals

A backcross was set up with mice of the inbred strains AKR/NHan and C57BL/6JHan which were bred and maintained under SPF conditions at the Central Institute for Laboratory Animal Breeding in Hannover, W. Germany.

AKR males were crossed with C57BL females to produce a F_1-generation, from which females were backcrossed to C57BL males. A total of 2317 animals (1165 males; 1152 females) were obtained in the backcross generation and included in the analysis.
The animals were fed a diet with 0.2 % of total fat (type 1006, Altromin, Lage, W. Germany). Food and water were given ad libitum.
The litters were standardised to 10 pups immediately after birth.
The animals were sacrificed at an age of exactly 11 weeks.

2.2. Analysis of blood

Blood was collected by heart puncture under deep ether anesthesia in monovets treated with EDTA to avoid clotting. The blood was subsequently measured in a modified Coulter Counter Sr (Coulter Electrics, Krefeld, W. Germany).

186

Fig.1. THE ANALYSIS OF QUANTITATIVE TRAITS USING MARKER GENES

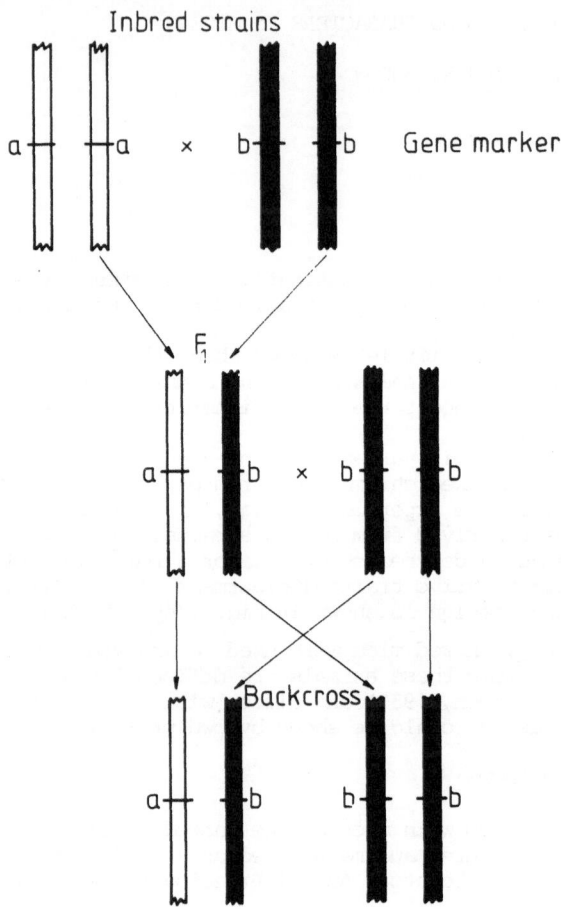

Seven characters were determined from each sample: white blood cell count (WBC), red blood cell count (RBC), hemoglobin content (HGB), hematocrit (HCT), red cell volume (MCV), mean corpuscular hemoglobin content (MCH), and hemoglobin concentration per cell (MCHC).
The last value was not included in the statistical evaluation because it was calculated by the counter and not directly measured.

2.3. Electrophoresis

Four enzymes were used as chromosome markers:
NADP-isocitrate dehydrogenase (locus: Idh-1 on chromosome 1),
glucosephosphate dehydrogenase (locus: Gpd-1 on chromosome 4),
glucosephosphate isomerase (locus: Gpi-1s on chromosome 7), and
esterase-1 (locus: Es-1 on chromosome 8).

For these enzymes the AKR and C57BL/6 strains express different alleles resulting in different phenotypes after electrophoresis. The typing was done in starch gel. The special methods and conditions are given in detail by Kluge (1980).

2.4 Statistics

The quantitative blood values were compared between the homozygous and the heterozygous marker genotype groups of the backcross generation in an analysis of variance. The differences were then tested for significance by F- and t-tests. Each marker locus was considered separately in the analysis.

3. RESULTS

The main results are summarized in Tab. 1. It shows the chromosomes 1 and 7 to be strongly involved in the development of the characters of the red blood picture.
Between 10 and 25 % of the phenotypic standard deviation of these traits could be explained by the different genotypes. For nearly all traits similar effects were determined in both sexes. In addition, chromosome 8 expressed a high influence on MCV in females and similar effects in both sexes concerning MCH. Nearly no significant genotype differences could be calculated for Gpd-1 indicating that genes of the segment of chromosome 4 do not play an important role in the determination of the blood characters under investigation. WBC was affected by none of the four marked chromosome sections.

TABLE 1. Effects of marked chromosome sections on quantitative blood characters (significant differences between homozygous and heterozygous genotype groups are given as percentage of the respective standard deviation)

| Blood traits | Sex | Chromosome markers | | | |
		Idh-1 Chrom.1	Gpd-1 Chrom.4	Gpi-1s Chrom.7	Es-1 Chrom.8
WBC	♂♂				
	♀♀				
RBC	♂♂	22,6			
	♀♀	23,3	13,3		
HGB	♂♂			12,7	
	♀♀	14,6		14,5	
HCT	♂♂	7,1			
	♀♀	13,3		13,3	
MCV	♂♂	16,0		19,0	
	♀♀	17,2		18,3	19,3
MCH	♂♂	15,4		25,6	12,8
	♀♀	17,1		17,1	13,3

4. DISCUSSION

The theoretical background of this way of analysis is the assumption that genes which are responsible for a certain quantitative character are not distributed randomly in the genome but are located to a major extent on a few chromosomes being inherited more or less as a genetic unit. There are several wellknown examples of gene complexes from which the associated genes determine proteins of related metabolic functions, e.g., the major histocompatibility complex of the mouse (Benacerraf and Germain, 1978), the esterase cluster of chromosome 8 in mice (Peters, 1982), and the genes determinating the scutellum bristles of Drosophila (Rendel, 1979). The results found in our investigation indicate the same tendency for the genes of at least some of the quantitative blood characters. Only some chromosomes seem to be responsible for the major part of the genetic variance of the traits under investigation.

The results of this study confirm those of Kluge and Geldermann (1982) for characters of body composition and body size of the same backcross mice as described here.
Interestingly, chromosome 1 marked by Idh-1 was found to have no effects on body weight, fat or body length while it has the greatest influence on blood characters.
In contrast, chromosome 4 (Gpd-1) and especially chromosome 8 (Es-1) are particularly involved in traits of body fattening but only to a minor extent in the development of blood cell characteristics.

Comparing both character groups with respect to sex it becomes obvious that most of the genotype differences of blood traits are similar for males and females. This may give a hint on the identity of genes in both sexes which are responsible for the development of these characters.
In contrast, most of the effects found particularly for body fat content and related traits were only significant in females indicating that body fat is at least partly regulated genetically differently in males and females.

However, it is important to notice that the relations found between blood characters and single chromosomes first of all are true in detail under the conditions of this particular experiment. They have to be confirmed by crosses of other inbred lines.

In this connection it has also to be considered that in the F_1-females one meiosis occured during the development of the gametes for the backcross animals. Thus there was the chance of recombination resulting in linkage brakes. Because of this fact we cannot assume the whole chromosome to be marked by a single locus but a segment which may differ in size among the individuals.
The calculated effects between marker and quantitative trait can be expected to be in most cases of indirect manner. That is, the marker alleles are not involved themselves in the regulation of the metric trait.
Therefore the calculated relations are to be expected the smaller the greater the distance on the chromosome between marker and the genes which determine the quantitative character.

Thus it would be of great advantage to consider two or more markers on a certain chromosome to recognize the recombinant animals and to map the questionable "quantitative" genes more precisely on a particular segment of the chromosome.

In addition to recombination heterosis has to be noticed as a reason of masking or misjudging the effects of single chromosome sections. The size of heterosis has to be calculated by reciprocal backcrosses.

A first evidence that sections of the chromosomes 1 and 7 of the mouse are really concerned with the regulation of blood characters we got from a study with well defined and genetically controlled outbred Han:NMRI mice (Rapp et al., 1985). Only small effects could be expected because of the great number of recombinations which could have occured since founding the population.
However, nearly fixed relations of about the same size as in the backcross could be found between chromosome 1 and chromosome 7 and characters of the red blood picture. This findings confirm the assumption that both chromo-
somes carry genes which influence number and size of red blood cells.
In the NMRI study we used two markers for chromosome 7: Gpi-1s and Hbb. The relations between Hbb and the metric traits were found to be very close but not significant for Gpi-1s. This may indicate the "blood" genes to be located only at a small map distance from the Hbb locus.
However, in the backcross resulting from the inbred lines the effects could also be assured for Gpi as marker because of the only one possible recombination. Unfortunately we could not type Hbb variants in the back-cross. Of course its effects could be expected to be even closer than in the NMRI population.

Summarizing the results of our studies it seems to be possible to assign at least some of the "quantitative blood genes" to certain chromosomes of the mouse by the use of marker genes.

REFERENCES

1. Benacerraf B, Germain RN: The immune response genes of the major histo-compatibility complex. Immunological Review 38, 70 - 119, 1978.
2. Burow K, Rapp KG, Kluge R: Strain differences in inbred rats: influence of strain, sex and age on heamatological traits. 2nd FELASA Symposium, Malmö, Sweden, 1983.
3. Geldermann H: Investigations on inheritance of quantitative characters in animals by gene markers. Theoretical and Applied Genetics 46, 319 - 330, 1975.
4. Green MC: Genetic variants and strains of the laboratory mouse. Gustav Fischer Verlag, Stuttgart, New York, 1981.
5. Kluge R: Beziehungen zwischen markierten Chromosomenabschnitten und quantitativen Merkmalen bei der Labormaus. Dissertation, Göttingen, 1980.
6. Kluge R, Geldermann H: Effects of marked chromosome sections on quanti-tative traits in the mouse. Theoretical and Applied Genetics 62, 1 - 4, 1982.
7. Peters J: Nonspecific esterases of Mus musculus. Biochemical Genetics 20 (5/6), 585 - 606, 1982.
8. Rapp KG, Kluge R, Burow K: Studies on the genetics of quantitative traits in Han:NMRI mice using monogenic markers. 23rd GV-SOLAS Meeting, Veldhoven, Netherlands, 1985.
9. Rendel JM: Canalisation and selection. In: Quantitative Genetic Varia-tion (ed Tompson JN, Thoday JM), Academic Press, New York, San Francis-con, London, 1979.

A.C. Beynen & H.A. Solleveld (Eds), New developments in biosciences: their
implications for laboratory animal science, ISBN 978-94-010-7973-0
© 1988 Martinus Nijhoff Publishers, Dordrecht.

INVESTIGATING GENETIC VARIABILITY BETWEEN THE MHS HYPERTENSIVE STRAIN OF
RATS AND ITS NORMOTENSIVE CONTROL, MNS

BARBER BR, TORIELLI L, FERRANDI M, FERRARI P, SALARDI S, PARENTI P, DUZZI L.
Farmitalia Carlo Erba - Istituto Ricerche Nerviano (Milano) - Italy

High blood pressure (HBP) in man is considered a major risk factor in
cardiovascular heart disease; the most frequent cause of death in modern
civilizations. Most states of HBP are classifed as "essential hypertension"
as distinct from "secondary hypertension" which is due to some known organ
dysfunction or other cause.

The causes of essential hypertension are unknown but may not always be
the same in different patients, and the many demiological investigations
and family studies so far carried out on human populations to detect genetic markers - useful to clinical prediction of the disease in young persons-
or associated physiological anomalies - which may give insight into the
causes - are often confounded by the multifactorial aetiology and ethical
limits to its physiological investigation. From such studies it has come to
be accepted by most that HBP in man is an inheritable trait controlled by
genes at several loci the effects of which are strongly influenced by
environment (1).

Powerful statistical methods have been applied to population genetics
but inbred strains of laboratory animals may offer some advantages as, once
a gene effect has been characterized, appropriate assortative matings and
crosses can be designed to evaluate its expression on various genetic back-
grounds under rigidly controlled environmental conditions. There now exist
at least eight strains of rats, plus relative controls, selected as models
for essential hypertension and each possesses different characteristics for
HBP and different associated traits (2). Among other inbred strains, which
have not been subjected to any selection for blood pressure (BP), this
trait shows a considerable polymorphism (3).(Fig. 1).

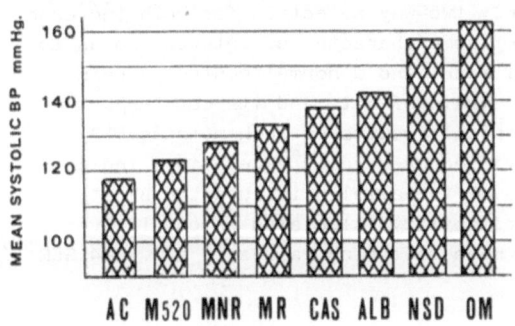

Fig. 1 - Mean systolic blood
pressures in some inbred rat
strains.

Drawn from data by C.T.Hansen,
1972.

If homology of gene action is accepted between the two species, single spontaneously hypertensive rat strains could be considered as equivalents to separate subgroups of HBP patients and collectively, together with others characterized for mean systolic blood pressure and haematological traits, as a global model for the human population.

CHARACTERIZATION

The scope of this paper is to discuss practical applications of standard breeding methods to investigations into the causes of HBP in the Milan hypertensive strain (MHS) of rats, the results of which have been fully reviewed elsewhere (4,5,6). In summary, parameters of characters with possible connections to pressure control which distinguish MHS from its counterpart the Milan normotensive strain (MNS), include: lower heart rate; lower relative kidney weight, plasma vasopressin, renin and plasma urine osmolalities; higher left ventricular weight; faster excretion of a saline load, faster Na^+/K^+ cotransport across red blood cell (RBC) membranes and, in the prehypertensive stage of life; higher water intake, urine excretion, glomerular filtration rate and a faster sodium uptake across renal brush border membranes, (7). Blood pressure in adult MHS rats is 169 ± 0.4 and in adult MNS 130 ± 0.5 (mmHg, mean$\pm$SEM). Body weights and growth rates are similar.

Cross transplantation experiments have shown that, HBP is induced in MNS recipients of MHS kidneys while in the reciprocal transplant, MHS blood pressure is lowered (8) also the differences in RBC parameters between the two strains are maintained in F1 hybrid recipients of their respective bone marrow cells (9). This latter experiment indicates that the cell defect responsible for RBC abnormalities in MHS is primary and genetically determined in the hematopoietic stem cells.

BREEDING STRATEGIES

The problems involved in investigating the causes of any genetically determined disease, the phenotypes of which show quantitative inheritance, depend on distinguishing the effects of all the genes related to other functions from those expressing the defect. Mendel's 2nd law, of independent assortment, warns us that association between any pair of traits controlled by genes at different loci does not necessarily imply either linkage, cause or effect, and a convincing biological reason for the association as well as serious research is needed to support any such hypothesis.

In the development of strains of animal models for human hereditary illnesses it is normal practice to apply two-way selection for both increase and decrease in the measurable value of some character of interest so as to create, simultaneously, a pathological model and a normal control strain. In the case of the Milan hypertensive strain of rats and its counterpart MNS the responses to selection were typical results of such methods (10). The manner in which the characters deviated over early generations indicate a progressive recombination and fixing of the alleles influencing BP at different loci. The coefficient of heritability calculated on cumulative selection differential of parents and mean offspring values is $h^2=0.46\pm4$ SEM (11).

However, the inbreeding process to which strains are subjected also rapidly fixes miriads of other heritable characters in unknown combinations. It might be useful to remember that, the consequences of mating two pairs of offspring from every family in subsequent generations derived from a single initial mating, will create as many strains with different characteristics as desired. If, to each half of such a pedigree selection were to be applied for only one of two opposing aspects of a given variable, almost every line produced would eventually possess a distinct virtually homozygous background which might, or might not, influence its expression. The results of standard two-way selection will represent only one line from each half of the pedigree, but which one?

The full success of the initial selection process may also be compromised by fixation of "wrong" alleles or residual heterozygosis due to opposing selection pressures e.g; for fertility or litter size. If we look at the possible combinations of only 2 alleles at each of 3 loci with additive effects influencing the same character in a F2 segregating population (Fig. 2), we will see that 50% of genes are in heterozygosis but triple homozygotes are not only found in the two extreme of the 7 phenotypic classes formed; they also exist in 3 subjects in each of the 2nd and 4th classes, although not all their fixed genes are of the same sign.

Such tables are based on the simplifying assumption that each allele increases or decreases the character by the same amount, but in nature it is more likely that each will have a different degree of influence in one direction or the other. If the influence is small genotypes could easily be confused during selection. A similar situation may be present in MNS which does not have as low a BP as some inbred strains reported above and, in cross-immunization reactions involving certain RBC membrane proteins (12), which differ between MHS and MNS, shows inconsistences consonant with heterozygosis at the loci determining them. However this might also be explained by various immunological factors. Matings designed to confirm this hypothesis are in course which, besides the basic research interest, will form an important genetic quality control.

Analysis of data from classic genetic crosses (13)- which consist of samples from two parental inbred strains, their F1 hybrids and three segregating populations derived by intercrossing the F1 offspring and reciprocally backcrossing them to the original strains - form a useful approach to many genetic problems. The results of such a study (11) have confirmed that HBP in MHS rats fits a model for additive, polygenic inheritance controlled by at least 2-3 loci with some dominance for normotension. The degree of dominance is 58% and estimates of heritability range from 45 to 64% according to the sample and methods used. In the F2 generation some correlations between characters of interest have been estimated which indicate genetic associations between RBC Na^+-K^+ cotransport and RBC volume (r= -0.45); RBC Na^+-K^+ cotransport and mean BP (r= 0.52); RBC volume and relative kidney weight (r= 0.46) (9).

There was no negative correlation between HR and BP, such as might be expected from the values of the parental strains, although the respective

194

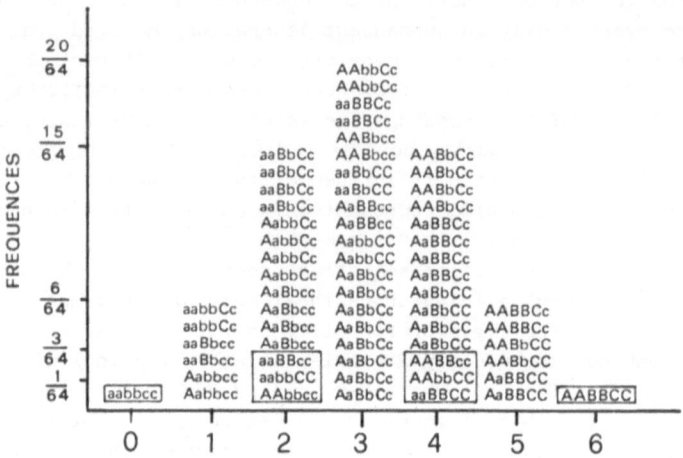

NUMBER OF ALLELES WHICH INCREASE THE CHARACTER

Fig. 2 - Independent assortment of alleles at 3 loci in F2 hybrids.

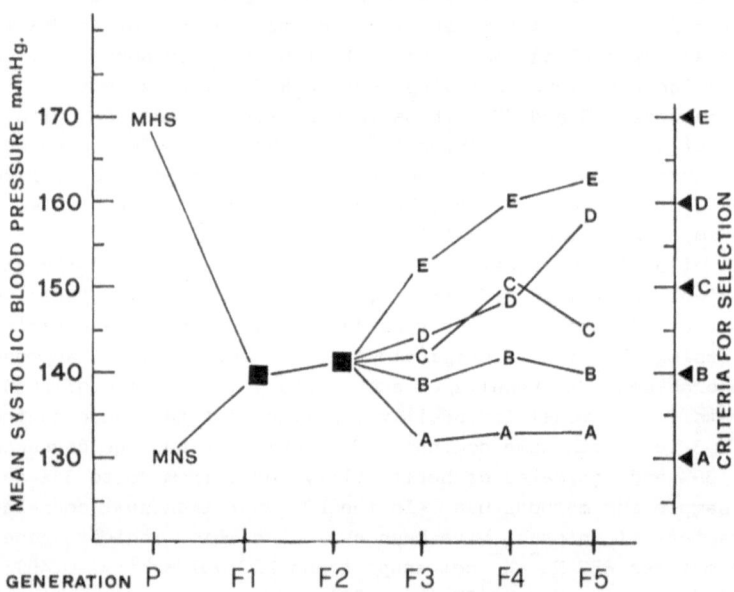

Fig. 3 - Response to selection in MHS/MNS recombinant lines.

distribution frequencies of these parameters in the six populations studied indicate similar modes of inheritance. Selective mating of extreme pheno-types have, however, established dissociation in 2 recombinant lines (unpublished). The low BP line of these has a mean HR identical to MHS and mean BP equal to MNS. By applying modifications of the cross-backcross-intercross methods described by Green (14) to this line, a congenic low BP control strain to MHS is now being established. The"coefficient of relation ship"(Falconer's "coancestry f" (15):the probability that two animals will inherit the same gene by descent) between the congenic line and MHS is at present f= 96.4% after 5 cycles of such mating.

Much controversy has ensued as to whether BP value frequencies in human populations are determined by many polygenes and best represented by a continuous distribution or comprise several underlying distributions, each determined by a discrete genetic mechanism (16). The inheritance of metric traits can, however, be satisfactorily explained also, by a polyfac-torial model based on simple transmission of genes according to Mendelian segregation (17).

We saw above that, the genotypes in a segregating F2 generation com-prise 50% heterozygotes, consequently it is expected that the mean value of families in F3 will be similar to the mean of the F2 parents which produced them. By positive assortative mating of phenotypes from an F2 population it should, therefore, be possible to fix single pairs of alleles in various combinations within only a few generations. The problem is to decide which value level to select for and how·many.

The aim of this selection would be to stabilize the mean BP at dif-ferent levels in each line of descent according to whether 1,2,3 or more genes for HBP had been fixed. If, however, a selected value were to be intermediate to that determined by a given set of homozygous genes, the result would be forced heterozygosis at one or more loci and should be revealed by alternation from one generation to the next, of the mean BP, around the level actually expressed by the homozygous state. Eventually the established lines could be investigated by a diallel cross analysis (13) of all the characters considered to be associated with HBP.

An initial approach along these lines has produced some results (un-published) with MHS/MNS recombinants. Despite the limits to numbers of animals it is possible to breed and measure for BP when many lines are maintained, the means have followed closely the expected values (Fig.3)for 3 generations and the family standard deviations of these means are half that of the foundation colonies, which suggests that the environmental in-fluences might have less effect on the expression of single genes than when they act in concert; in other words the signal may be getting clearer!

REFERENCES

1. Sing CF, Skolnick M (eds): Genetic analysis of common diseases: Appli-cations to predictive factors in coronary disease. Progress in clini-cal and biological research Vol. 32 N.Y. Alan R Liss Inc. 1979

2. Lovenberg W, Horan M (eds): Genetic rat models for hypertension: Guidelines for breeding, care and use. Hypertension 9 n.1 (suppl.) Jan. 1987

3. Hansen CT: A genetic analysis of hypertension in the rat. In, Spontaneous hypertension - Its pathogenesis and complications. Tokyo, Igaku Shoin Ltd 1972

4. Bianchi G, Ferrari P, Barber BR: The Milan hypertensive strain. Handbook of hypertension Vol.4: Experimental and genetic models of hypertension. W. de Jong (ed) Elsevier Science Publications BU. 1984

5. Ferrari P, Cusi D, Barber BR: Erythrocyte membrane and renal function in relationship to hypertension in rats of the Milan hypertensive strain. Clin. Sci. 63: 61s-64s, 1982

6. Bianchi G, Cusi D, Ferrari P, Barlassina C, Barber BR, Vezzoli G, Lupi P, Polli E: Renal mechanisms in the pathogenesis of essential hypertension. Frontiers in Cardiology for the eighties. Donato L. and L'Abbate A. (eds) Academic Press 13-16, 1984

7. Parenti P, Hanozet GM, Bianchi G: Sodium and glucose transport across renal brush border membranes of Milan hypertensive rats. Hypertension vol.8 n. 10 Oct. 1986

8. Fox U, Bianchi G: The primary role of the kidney in causing the blood pressure differences between the Milan hypertensive strain (MHS) and normotensive rats. Clin.Exp.Pharmacol.Physiol. Suppl.3, 71-74, 1976

9. Bianchi G, Ferrari P, Trizio D, Ferrandi M, Torielli L, Barber BR, Polli E: Red blood cell abnormalities and spontaneous hypertension in the rat: A genetically determined link. Hypertension vol.7 n.3, 319, 1985

10. Bianchi G, Barber BR, Ferrari P, Duzzi L: The Milan hypertensive strain of rats and its control normotensive strain. Hypertension 9 suppl.I I-30/I-33, 1987

11. Schlager G, Barber BR, Bianchi G: Genetic analysis of blood pressure in the Milan hypertensive strain of rat. Can. J. Genetic Cytol.28, 1986

12. Ferrandi M, Salardi S, Modica R, Bianchi G: Antigenic properties of erythrocyte membrane from an hypertensive strain of rats. Annals of New York Academy of Sciences (in press) 1987

13. Festing MW: Notes on genetic analysis. In, Inbred strains in research London. Mac Millan Press Ltd 1979

14. Green EL (ed): Chap.2 Breeding systems. In, Biology of the laboratory mouse 2nd ed. New York McGraw-Hill 1966

15. Falconer DS: In Chapt 5 III. Pedigreed populations and close inbreeding. Introduction to quantitative genetics. Edinborough, Oliver and Boyd 1961

16. McManus J: Bimodality of blood pressure levels. Statistics in Med. vol.2 253-258, 1983

17. Cavalli-Sforza LL, Bodmer WF: The genetics of human populations. Chap.9, San Francisco, W.H. Freeman and Co. 1971

A.C. Beynen & H.A. Solleveld (Eds), New developments in biosciences: their implications for laboratory animal science, ISBN 978-94-010-7973-0

GENE MAPPING AND LINKAGE HOMOLOGY

L.F.M. VAN ZUTPHEN AND M.G.C.W. DEN BIEMAN
Department of Laboratory Animal Science, Veterinairy Faculty, University
of Utrecht, P.O.Box 80.166, 3508 TD Utrecht, The Netherlands.

Information on the linkage map of laboratory animals is of value for
the genetic quality control of inbred strains. A reliable method for
genetic monitoring is provided by the characterization of inbred strains
for a given set of monogenic markers and the regular control of this
genetic profile. The marker genes used for this system should be
sufficiently distributed among the genome.

Gene mapping is also of value for validation of genetic models. The
number of genetic disorders with a fully established homologous condition
in animals and man is still limited. Phenotypic resemblance does not
prove similarity of the causative factor. Validation of a genetic
disorder in animals as a model for human disease depends on evidence for
homology between the causative genes. Genes are assumed to be homologous
if DNA sequence homology exsists, or, if biochemical homology between the
gene products in two different species can be demonstrated. Homology can
also be assumed when the causative genes are mapped within homologous
linkage groups. Detailed criteria for gene homologies are given by Lalley
and McKusick (1985).

Several methods are being used for mapping genes in laboratory
animals. Conventional methods depend on family studies. Animals differing
for the genes under study are crossed and back-crossed and linkage is
studied in the progeny by calculating recombination frequencies.

The development of the concept of recombinant inbred strains (Bailey,
1971) and, more recently, of recombinant congenic strains (Demant and
Hart, 1986) substantially increased the possibilities of gene mapping in
laboratory animals. Also *in vitro* techniques like somatic cell hybridization in combination with recombinant DNA techniques and *in situ*
hybridization, provide new powerful techniques for the assignment of
genes to chromosomes. These methods contribute considerably to the rapid
increase of our knowledge of the linkage maps. This increased knowledge
enables the comparison of maps of different species to one another in
order to establish linkage homologies, due to evolutionary conservation
of chromosome segments.

Most striking linkage homology is seen when the X-chromosomes of
different species are compared. The dosage compensation mechanism by
random inactivation of one of the X-chromosomes in each cell of the early
female embryo necessitates the conservation of the total X-chromosome.
Although this mechanism does not apply to autosomes, there is also
increasing evidence for the existence of linkage homology for large parts
of the autosomal genome (Lalley & McKusick, 1985; Van Zutphen, 1986).
Linkage homology not only exists between closely related species, but
also between mouse and man. This is exemplified in Table 1 for mouse
chromosome 8.

TABLE 1. MOUSE-HUMAN LINKAGE HOMOLOGY (mouse chromosome 8)

Mouse chrom.	Mouse marker symbol	Marker name	Human marker symbol	Human chrom.
8	Aprt	Adenine phosphoribosyl transferase	APRT	16q
8	Got-2	Glutamate oxaloacetate transaminase-2	GOT2	16q
8	Hp	Haptoglobin	HP	16q
8	Mt-1,2	Metallothionein	MT1,2	16q
8	Tat	Tyrosine aminotransferase	TAT	16q
8	Ctrb	Chymotrypsinogen	CTRB	16
8	Es-6	Caboxylesterase	ESB3	16
8	Gr-1	Gluthione reductase-1	GSR	8p
8	Prp	Proline rich proteins	PRP	12

Nine genes presumed to be homologous with genes of mouse chromosome 8 have been mapped in man. Seven of these are linked on human chromosome 16, whereas one gene (GSR) is on human chromosome 8 and another (PRP) on human chromosome 12. Thus, Table 1 also demonstrates that linkage homology is not a rule without exceptions.

Mouse chromosome 8 contains Es-6, a gene that is part of a complex system of carboxylesterase genes. The carboxylesterases (EC 3.1.1.1) form a group of non-specific enzymes with a wide distribution among animal species. The physiological function of these esterases is unknown but the enzymes are of special interest for the laboratory animal scientist because of their high degree of polymorphism and because of their pharmacogenetic and toxicogenetic aspects. The genetic background of these esterases has been studied extensively in different species of laboratory animals. The concept of homology can be illustrated more in detail by the results of these studies.

Two distantly linked clusters of highly polymorphic esterase genes have been found in the rabbit (Van Zutphen et al., 1977; Fox and Van Zutphen, 1979; Van Zutphen et al., 1987), the mouse (Womack and Sharp, 1976; Peters, 1982; Nash and Von Deimling, 1982; Von Deimling, 1984) and the rat (Hedrich and Von Deimling, 1987). In addition, a number of esterase genes have been mapped outside these clusters.

Figure 1 summarizes our present knowledge on linkage of esterase genes in laboratory animals and man. Presumed homologies are indicated by dotted lines. The presumed homology is not based on linkage relationships only, but also on information about biochemical proporties of the gene products, such as substrate specificity, sensivity for inhibitors, molecular weight and isoelectric point (see references indicated in Fig. 1). It must be stressed, however, that the available biochemical data are still fragmentary. The proposed homologies are tentative and need to be substantiated by a systematic approach.

Solid information on homology of esterase genes in laboratory animals improves the rational choice of the experimental animal for testing estertype products. For exemple, it has been demonstrated that Es-10 esterase of the rat can hydrolyse a number of pharmacologically unrelated esters, like the hypocholesterolaemic drug clofibrate, the local anaesthetics procaine and butanilicaine, the monoamine oxidase inhibitor

Fig. 1

PRESUMPTIVE HOMOLOGOUS ESTERASE GENES

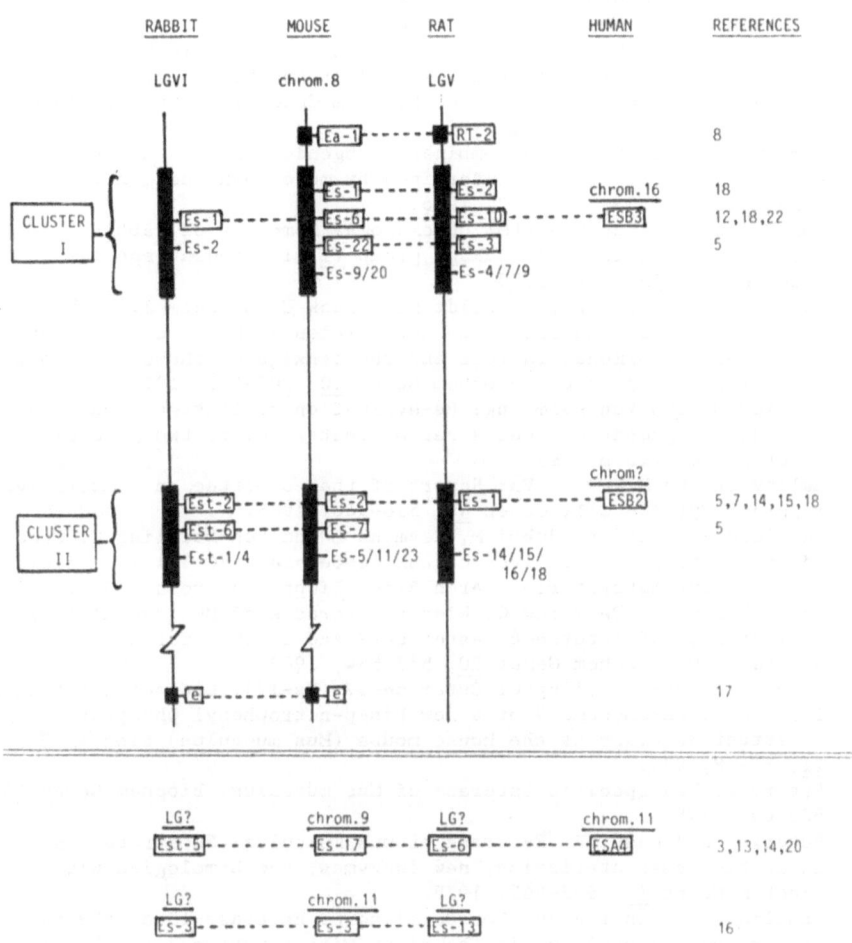

isocarboxazide and the insecticid malathion (Mentlein et al., 1987).The proposed homology of rat *Es-10* with rabbit *Es-1* and mouse *Es-6* would imply that the esterases produced by these genes have similar catalytic properties and can also hydrolyze these estertype products. This has recently been demonstrated for clofibrate (Van Zutphen and Den Bieman, 1987). We have started a comparative study on these esterases in laboratory animals and man in order to investigate some of the pharmacogenetic and toxicogenetic aspects more in detail.

REFERENCES

1. Bailey DW: Recombinant-inbred strains. An aid to identify linkage and function of histocompatibility and other genes. Transplantation 11, 325-327, 1971.
2. Bender K, Nagel M and Günther E: Es-6, a further polymorphic esterase in the rat. Biochem Genet 20, 221-228, 1982.
3. Deimling O von: MNL 68, 65-66, 1983.
4. Deimling O von: Esterase-23 (Es-23): Characterization of a new carboxylesterase isozyme (EC 3.1.1.1) of the house mouse, genetically linked to Es-2 on chromosome 8. Biochem Genet 22, 769-782, 1984.
5. Deimling O von; pers comm 1987.
6. Demant P and Hart AAM: Recombinant congenic strains. A new tool for analyzing genetic traits determined by more than one gene. Immunogenetics 24, 416-422, 1986.
7. Fox RR and Van Zutphen LFM: Chromosomal homology of rabbit (Oryctolagus cuniculus) linkage group VI with rodent species. Genetics 93, 183-189, 1979.
8. Gasser DL, Silvers WK, Reynolds HM, Black G and Palm J: Serum esterase genetics in rats: two new alleles at Es-2, a new esterase regulated by hormonal factors and the linkage of these loci to the Ag-C blood group locus. Biochem Genet 10, 207-217, 1973.
9. Hedrich HJ and Von Deimling: Re-evaluation of linkage group 4 of the rat and assignment of twelve carboxylesterases to two gene clusters. J Hered 1987 (in press).
10. Lalley PA and McKusick VA: Report of the committee en comparative mapping. Cytogen Cell Genet 40, 536-566, 1985.
11. Mentlein R, Ronai A, Robbi M, Heymann E and Von Deimling O: Genetic identification of rat liver carboxylesterases (EC 3.1.1.1) isolated in different laboratories. Arch Bioch Bioph (in press) 1987.
12. Nash HR and Von Deimling O: Kidney esterases of Mus muscules: Further polymorphism of esterse-6, esterase-9 and a new esterase, esterase-20. Biochem Genet 20, 537-554, 1982.
13. Otto J and Von Deimling O: Esterase-17 (Es-17): Characterization and linkage to chromosome 9 of a new bis-p-nitrophenyl phosphate resistant esterase of the house mouse (Mus musculus) Biochem Genet 21, 37-48,1983.
14. Peters J: Non-specific esterase of Mus musculus. Biochem Genet 20, 585-606, 1982
15. Peters J and Nash HR: Esterase of Mus musculus: Substrate and inhibition characteristics, new isozymes, and homologies with man. Biochem Genet 16, 553-563, 1978.
16. Scholze M, Günther E and Von Deimling O: Esterase-13 of the rat (Rattus norvegicus) and its homology with esterase-3 of the house mouse (Mus musculus). Comp Biochem Physiol 78B, 713-718, 1984.
17 Searle AG: Clues to homologous regions in mammalian autosomes. Cytogenet Cell Genet 16, 430-435, 1976.
18. Womack JE and Sharp M: Comparative autosomal linkage in mammels: Genetics of esterases in Mus musculus and Rattus norvegicus. Genetics 82, 665-675, 1976.
19. Zutphen LFM van, Den Bieman MGCW and Bouw J: Serum esterase genetics in rabbits. IV The prealbumin and β-globulin systems. Biochem Genet 15, 989-1000, 1977.

20. Zutphen LFM van, Fox RR and Den Bieman MGCW: Genetics of two tissue esterase polymorphims (Est-4 and Est-5) in the rabbit. Biochem Genet 21, 773-780, 1983.
21. Zutphen LFM van: Genetics of laboratory Animals. In: Laboratory Animals. World Animal Science C2. EJ Ruitenberg and PJW Peters (eds). Elsevier Science Publishers, Amsterdam 1986, pp 47-84.
22. Zutphen LFM van and Den Bieman MGCW: Unpubl. results, 1987.
23. Zutphen LFM van, Den Bieman MGCW and Koster A: Esterase genetics and clofibrate hydrolysis in rat and rabbit. Z Versuchst.k 1987 (in press)
24. Zutphen LFM van, Den Bieman MGCW, Von Deimling O and Fox RR: Genetics of a tissue esterase polymorphism (Est-6) in the rabbit (Oryctolagus cuniculus). Biochem Genet 1987 (in press).

A.C. Beynen & H.A. Solleveld (Eds), New developments in biosciences: their implications for laboratory animal science, ISBN 978-94-010-7973-0

A NEW METHOD TO PRODUCE ARTIFICIAL MONOZYGOTIC TWINS IN MICE [+]

P. AGRAWALA[*], E. BAUNACK[o], J. HAHN[*]

[*]Abt. f. exp. Fortpflanzungsbiologie der Klinik f. Gynäkologie des Rindes der Tierärztlichen Hochschule Hannover, [o] Zentrales Tierlaboratorium der Medizinischen Hochschule Hannover

1. INTRODUCTION
   The low viability of artificially produced mouse twins described by these authors (Gärtner and Baunack 1981) and others (for references see Tsunoda and McLaren 1983) was the reason to test the validity of the EDTA-method described in this paper.
Monozygotic twins are a very useful instrument for many biological methods in laboratory animal science. Their advantage is the high degree of uniformity due to identical genotypes because of their origin from a common fertilized ovum (Gärtner and Baunack 1981). Due to the logistical difficulties involved in assembling sets of naturally conceived twins and the uncertainty in the diagnosis of zygosity, there have been efforts to induce monozygotic twinning in mice with various methods. Kaufman and O'Shea (1978) injected pregnant females with vincristine sulfate. Naruse et al. (1983) induced artificial fission of the inner cell mass in mouse blastocysts by exposure to vinblastine sulfate after removal of the zona pellucida. Both procedures result in many fetuses with gross malformations and only few monozygotic twin pairs (Baunack, unpublished). Techniques for splitting mouse preimplantation embryos to produce genetically identical individuals have been described by several authors (Markert and Seidel 1981, Baunack and Gärtner 1981, Dyban and Sekirina 1984, Nagashima et al. 1984) but unfortunately success rate was rather low. In the present study, mouse monozygotic twins were produced by splitting and reaggregating preimplantation stages using a simple method. Viability of half embryos was estimated after in vitro culture and transfer to pseudopregnant recipients.

2. PROCEDURE
2.1. Materials and methods
   Series 1 was carried out on mice of the inbred strains C57BL/6J and CBA/NHan and on the mutants C57BL/6JA^{w-J} and C57BL/6Ja^{t} (Green, 1981). All animals were barrier maintained in polycarbonate cages with wooden granula bedding in groups of 15 at controlled environmental conditions (room temperature 22 $\pm$ 2 ^{o}C, relative humidity 55 $\pm$ 5 %, light from 7 a.m. to 7 p.m.). A commercial pelleted diet (Altromin 1320) and tap

[+] supported by Deutsche Forschungsgemeinschaft, SFB 146, B 9

water from Makrolon bottles were available ad libitum.
Embryos were collected in Medium 16 (Whittingham 1971), by
flushing the oviducts and uteri, as early 4- to 8-cell stages
in the morning or as compacted stages (8- to 16-cell) in the
evening of day 3 of pregnancy (day of vaginal plug = day 1)
from naturally ovulated and singly (C57BL) or 1 o : 3 $\varphi\varphi$
(C57BL mutants and CBA) paired females. The zona pellucida was
removed with 0.5 % pronase (Bowman and McLaren 1970) and the
embryos were washed twice in Medium 16. Each zona-free embryo
was then placed in 1 ml of 0.5 % EDTA (Ethylenediamine-Tetra-
acetic Acid, Sigma^o) in Ca^{++}- and Mg^{++}-free PBS (Dulbecco's
PBS without Ca^{++} and MG^{++}, Serva^o) in tissue culture
multidishes (Nunclon^o) for 1 - 7 min at room temperature in
order to achieve desintegration and decompaction,
respectively. Then the blastomeres were dissociated by gently
pipetting the embryos several times. After a short washing
interval in Medium 16, groups of 2 - 8 blastomeres were
agglutinated with an extended, closed capillary tube in small
drops of a solution containing 0.01 ml of the PHA
(Phytohemagglutinin, Sigma^o) stock solution (25 mg PHA in 5 ml
PBS without Ca^{++} and Mg^{++}) and 0.09 ml Medium 16 (Mintz 1973).
After agglutination, embryos remained in the PHA solution for
another 1 - 2 minutes and were then cultured pairwise in
Medium 16 at 37 ^{o}C and 5 % CO_2 in air for 17 - 20 hours and 40
- 44 hours. After these intervals, embryos were checked
morphologically for development (Figure I).

Series II was carried out with hybrid $CB6F_1$/Han mice. These
animals were kept conventionally at 22 ^{o}C, 55 % relative
humidity and light from 6 a.m. to 6 p.m.; they had free access
to pelleted food (Altromin 1324) and tap water. Females were
superovulated with 5 I.U. PMSG followed by 5 I.U. HCG 48 hours
later and paired singly with isogenic males. Control for
vaginal plugs followed 12 hours after HCG injection. 77 hours
after HCG injection, embryos were flushed from the uterine
horns with culture medium (PBS + 0.4 % BSA, Serva^o) and
pooled. Only morphologically intact compacted morulae were
used for the experiments. Zona pellucida was thinned with
prewarmed 0.5 % pronase (Sigma^o) for 4 - 5 min and then
completely removed by pipetting the embryos with a glass
capillary tube. Desintegration of blastomeres was carried out
as in series I. Agglutination was done in 0.05 % PHA in PBS
+ 0.4 % BSA as above.
Reaggregated groups were singly cultivated in microdrops of
the culture medium under silicon oil for about 16 hours at
37.2 ^{o}C, 5 % CO_2 in air and 100 % relative humidity.
Subsequently the embryos were morphologically evaluated; late
morulae and early blastocysts were transferred to the uteri of
day 3 pseudopregnant recipients. Recipients were sacrificed on
day 14 of pseudopregnancy; number of live fetuses and late
resorptions were counted.

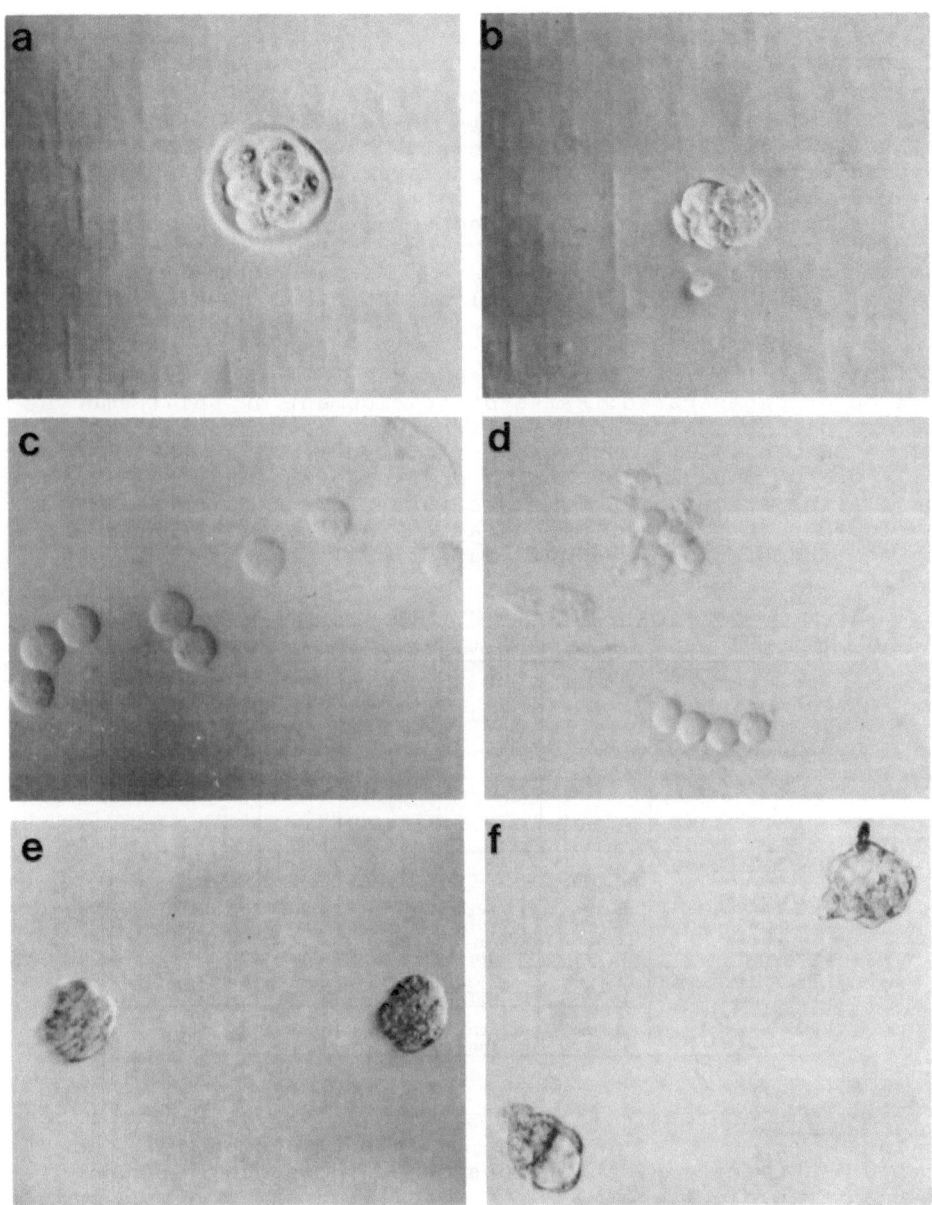

FIGURE I. An 8-cell mouse embryo dissociated with 0.5 % EDTA, the halves agglutinated with 0.0 % PHA and cultured in-vitro for 17 - 20 hours and 40 - 44 hours.
a) 8-cell stage b) denuded 8-cell stage c) 8 dissociated blastomeres d) monozygotic pair of 4 blastomeres each e) monozygotic pair after 17 - 20 hours in-vitro culture developed to compacted morulae f) monozygotic pair after 40 - 44 hours in-vitro culture developed to blastocysts

3. RESULTS

Series I. Table 1 summarizes the development of 'half'
embryos of the 4 investigated mouse strains C57BL/6JHan,
C57BL/6Ja^{t}Han, C57BL/6JA^{w-J}Han and CBA/JHan. A total of 186
intact preimplantation embryos (4-16-cell stages) was
collected from day 3 pregnant females. These embryos were
denuded with 0.5 % pronase, dissociated with 0.5 % EDTA and
the blastomeres agglutinated to form two 'halves' with 0.05 %
PHA. Of the 364 'halves' cultured in vitro, 10 (3 %) developed
to 6-16-cell stages after 17 - 20 hours, 271 (74 %) to
compacted morulae and 7 (2 %) to early blastocysts. 38
'halves' reaggregated in the culture dish and then developed
to compacted morulae, which makes a total developmental rate
for twin embryos of 90 %.
36 (10 %) of the in vitro cultured 'halves' either failed to
develop or degenerated. Rates of development of split embryos
did not differ statistically between the investigated strains.
Of the 364 'halves', 158 were cultivated in vitro for another
day (40 - 44 hours). 108 (68 %) developed to early blastocysts
and blastocysts, 50 (32 %) did either show no further
development or signs of degeneration. Again, differences
between the strains were not statistically significant.

TABLE 1
THE VIABILITY OF MOUSE TWIN EMBRYOS PRODUCED WITH E D T A AFTER IN VITRO CULTURE 3.2.87 - 15.5.87

strains	C57BL/6JHan	C57BL/6Ja^{t}Han	C57BL/6JA^{w-J}Han	CBA/JHan	total
stages collected:	79	51	32	24	186
4 - 8	55	49	17	22	143
9 - 16	24	2	15	2	43
n blastomeres in dissociated and agglutinated embryos:	total 'halves': 158	total 'halves': 95	total 'halves': 63	total 'halves': 48	total 'halves': 364
2	15	15	1	12	43
3	38	26	13	12	89
4	68	50	23	20	161
5	17	4	8	-	29
6	12	-	4	1	17
7 - 10	8	-	14	1	23
after 17 - 20 h in vitro:					
total development	142 (90%)	87 (92%)	55 (87%)	42 (92%)	326 (90%)
n 6-16-cell stages	4 (3%)	6 (4%)	—	—	10 (3%)
n compacted morulae	125 (79%)	61 (64%)	47 (75%)	38 (79%)	271 (74%)
n early blastocysts	3 (2%)	—	4 (6%)	—	7 (2%)
n no development	16 (10%)	8 (8%)	8 (13%)	4 (8%)	36 (10%)
n reaggregated	10 (6%)	20 (22%)	4 (6%)	4 (8%)	38 (10%)
after 40 - 44 h in vitro:	total n = 40	total n = 50	total n = 44	total n = 24	total n = 158
total development	27 (68%)	31 (60%)	34 (77%)	16 (67%)	108 (68%)
n compacted morulae	2 (5%)	—	—	1 (4%)	3 (2%)
n blastocysts	25 (63%)	31 (62%)	34 (77%)	15 (63%)	105 (66%)
n no development	13 (33%)	19 (38%)	10 (23%)	8 (33%)	50 (32%)

Series II. Of 35 morulae 12 (39 %) were halved, the halves
cultured in vitro for 12 hours, underwent recompaction and
blastulation and were then transferred to isogenic pseudopregnant recipients.
Table 2 shows the preliminary results of transfer. A total of
24 'halves' was transferred to 3 recipients. Recipients 1 and
2 both received 5 'halves' on each uterine horn, recipient 3
received 5 undissociated, denuded embryos on the right
uterine horn and 4 'halves' on the left. On day 14 of
pregnancy, the recipients were sacrificed; number of live
young was estimated as 7 (29 %) and number of late resorptions
as 3 (12.50 %). Of the 5 control embryos, 4 (80.0 %) developed
into live fetuses.

TABLE 2

Preliminary results of the viability of CB6F$_1$/Han twin embryos produced from compacted embryos with the help of 0.5 % EDTA and cultured in vitro for 12 hours, after transfer to the uteri of day 3 pseudopregnant recipients

recipient no.	number of twin embryos transferred		number of live young on day 14		number of late resorptions	
	right	left	right	left	right	left
1	5	5	2	1	-	1
2	5	5	1	1	1	1
3	(5 control embryos)	4	(4)	2	(-)	-
Total	10	14	3 (without controls)	4	1	2
			7 (29.17 %)		3 (12.50 %)	

4. DISCUSSION

For many biological investigations identical twins provide advantageous conditions because of their complete identity. They are also a way to decrease the number of animals necessary for the experiment. The dissociation of mouse preimplantation embryos with 0.5 % EDTA in Ca^{++}- and Mg^{++}-free Medium 16 takes 1 - 7 minutes. This difference is due to the fact that in the intermitotic period adhesion of sister blastomeres decreases considerably and dissociation is facilitated.

The cleavage rate of halved embryos was not to any extent inferior in comparison with whole embryos, but naturally resulted in morulae half size of normal embryos. Regular establishment of ICM takes place irrespective of cell number, if there is at least one blastomeres shielded by the others, which means that there should be twelve blastomeres at time of blastulation. As secretion of blastocoelic fluid seems to represent a cytoplasmic activity which embryos undertake after a certain definite period of time and irrespective of the number of blastomeres (Tarkowski and Wroblewska 1967), also the blastocysts developed from halved embryos are half normal size. For postimplantation development a minimum of 3 ICM cells is necessary (Rossant 1976); so the transfers were only done with morulae and blastocysts from 4- to 8-cell stages, because transfers of blastocysts from 2-cell halves would have probably shown poor postimplantation developmental success. Even if only 3 transfers with a total of 24 half embryos were performed as yet, it seems obvious, also from the in-vitro culture results, that the EDTA method for twinning mouse embryos presents a reasonably simple and effective method.

SUMMARY
1. Monozygotic mouse twin embryos established by desintegration with 0.5 % EDTA and agglutination with 0.05 % PHA
   after denudation with 0.5 % pronase show good in-vitro
   development after 17 - 20 hours (90 %) and 40 - 44 hours
   (68 %).
2. Transfer results of monozygotic mouse twin embryos
   established with the EDTA method are satisfactory with 29 %
   live young and 12.5 % late resorptions on day 14 of
   pregnancy.

REFERENCES
1. Baunack E, Gärtner K: Influence of micromanipulation of
   ova on the postnatal development of mice (AKR). In:
   Culture Techniques, D Neubert and H-J Merker (eds.).
   Berlin, New York: Walter de Gruyter & Co., 1981
2. Bowman P, McLaren A: Viability and growth of mouse embryos
   after in-vitro culture and fusion. J Embryol exp Morph,
   23: 693-704, 1970
3. Dyban A, Sekirina G: Study of preimplantation development
   of monozygotic twins. Experiments on mouse embryos. Soviet
   J Developmental Biol, 12:130-139, 1982
4. Gärtner K, Baunack E: Is the similarity of monozygotic
   twins due to genetic factors alone? Nature, 292:446-447,
   1981
5. Green M: Genetic variance and strains of the laboratory
   mouse. Stuttgart, New York: Gustav-Fischer-Verlag, 1981
6. Kaufman, MH, O'Shea KS: Induction of monozygotic twinning
   in the mouse. Nature, 276:707-708, 1978
7. Markert C, Seidel G (eds.): Parthognesis, identical twins
   and cloning in mammals. In: New Technologies in Animal
   Breeding. New York, Acad. Press, 1981
8. Mintz B, Gearhart J, Guymont A: Phythohemagglutinin-
   Mediated Blastomere Aggregation and Development of
   Allophenic Mice. Developmental Biol., 31: 195-199, 1973
9. Nagashima M, Matsui K, Sawasaki K, Kano Y: Production of
   monozygotic mouse twins from mircrosurgically bisected
   morulae. J Reprod Fert, 70: 357-362, 1984
10. Naruse I, Kano S, Shoji R: Experimental Induction of Two
    Inner Cell Masses in Mouse Embryos by Vinblastine
    Treatment In Vitro. Teratology, 28: 215-218, 1983
11. Rossant J: Postimplantation development of blastomeres
    isolated from 4- and 8-cell mouse eggs. J Embryol exp
    Morph, 36(2):283290, 1976
12. Tarkowski A, Wroblewska J: Development of blastomeres of
    mouse eggs isolated at the 4- and 8-cell stage. J Embryol
    exp Morph, 18(1): 155-180, 1967
13. Tsunoda Y, McLaren A: Effect of various procedures on the
    viability of mouse embryos containing half the number of
    blastomeres. J Reprod Fert, 69: 315-322, 1983
14. Whittingham DG: Culture of mouse ova. J Reprod Fert Suppl,
    14: 7-21, 1971

A.C. Beynen & H.A. Solleveld (Eds), New developments in biosciences: their
implications for laboratory animal science, ISBN 978-94-010-7973-0
© 1988 Martinus Nijhoff Publishers, Dordrecht.

GENETIC ANALYSIS OF MULTIGENIC TRAITS USING THE RECOMBINANT CONGENIC
STRAINS

P. DEMANT, A.A.M. HART, and L.F.M. VAN ZUTPHEN
The Netherlands Cancer Institute, Amsterdam, and Department of Animal Science,
University of Utrecht, Utrecht, The Netherlands.

When presenting a new genetic tool several questions have to be
answered: What is its aim? Why has it been necessary to develop a new
genetic tool? What are its advantages and its disadvantages? How does it
compare with other research possibilities and approaches? and finally, how
should it be optimally used?

The recombinant congenic strains have been designed to analyze the
quantitative genetic traits, which are controlled by several genes. The
multigenetically determined traits are usually much more difficult to
analyze than the traits which are controlled by a single gene. When
comparing two strains which differ at a trait controlled by several genes,
we do not know which of the genes is responsible for the difference. In
fact we do not even know how many genes are involved. On the other hand the
difference caused by a single gene, for example an enzyme polymorphism, is
readily analyzed by genetic methods.

In the past the genetic differences between strains have been analyzed
using backcrosses and F_2 hybrids. A qualitatively new range of
possibilities has been opened by invention of the recombinant inbred
strains (RIS) by Bailey (1; 2).

A series of RIS is produced by incrossing of F_2 hybrids of two inbred
strains. Each RIS contains an approximately even proportion of genes of
each parental strain. For each RIS, however, the set of genes received from
one parental strain is different and unique. The invention of RIS
revolutionized mouse genetics for two main reasons. First, as their
genotypes become typed and known, a series of RIS serves as a permanent and
universal equivalent (except for lack of heterozygosity) of a fully
genotyped F_2 population. This has an enormous economic advantage for
linkage work, because any polymorphic trait may be analyzed with the same
set of RIS. Second, individuals of a RIS can be regarded (again, except for
their homozygosity) as multiple identical replicas of a F_2 hybrid mouse.
Thus, a quantitative phenotype, as opposed to a single F_2 individual, can
be established with reasonable certainty in animals of a single RIS, and
its linkage relationship tested.

The use of recombinant inbred strains led to discovery of a great number
of linkages. Most of these new linkages, however, concern genetic traits
controlled by a single gene. The application of recombinant inbred strains
to studies of quantitative traits controlled by several genes did not lead
often to the expected results. In a great number of situations the values
observed in recombinant inbred strains created between two strains which
differ in a quantitative character, represent a continuous range between
the two parental values. The lack of distinct classes of phenotypes in this
situation is the cause of the failure to assign any meaningful linkage. As
a result, the expectation that the analyses of interstrain differences in
susceptibility to a variety of tumors will lead to closer understanding of

209

the nature of the tumorigenic processes has been fulfilled only partially.

Looking back into the history of the mouse genetics, which has been connected from the very beginning with the interest in the resistance against tumors, it is obvious that with the exception of some genes related to the viral oncogenesis and to chemical processing of carcinogens, not a single gene has been identified as a tumor-resistancegene proper. It is true that the analysis of the genes which were discovered in another way, like histocompatibility genes or the mutations obese, hairless and others, has revealed that they are also involved in tumor resistance (3; 4). This has led to a number of interesting findings. However, we do not know whether these genes are really the most important genes. Most likely, other genes are more important for tumor resistance. For example, although the extreme difference in susceptibility to lung tumors between strains A and C57BL was described half a century ago (5), the responsible genes cannot be identified, even using the recombinant inbred strains (6).

The advantages and limitations of the use of the recombinant inbred strains have been discussed expertly by Bailey (7). The nature of the difficulties encountered when analyzing quantitative multigenetically determined traits has been further discussed by Briles et al. (8). We have shown that the nature of the segregation of the genes in recombinant inbred strains intrinsically distorts or obscures the correlation between phenotype and genotype and thus prevents us reaching conlusions about the identity of genes involved. It can even lead to assignment of genetic control to the genes which actually are not involved at all (9).

In order to overcome these difficulties, we have proposed a new genetic tool, the recombinant congenic strains (9). With the use of limited backcrossing and subsequent brother-sister mating, the genetic material of one inbred strain (the donor strain) can be dispersed on a background of another inbred strain (the background strain), so that a manageable number of recombinant congenic strains (20-25) is created, each strain carrying only a small proportion of genes (10-15%) of the donor strain. In this way non-linked genes controlling the same trait, e.g. susceptibility to a certain type of tumor, become separated in different recombinant congenic strains and can be studied individually (Table 1).

The main advantage of recombinant congenic strains is that they separate the individual genes - components of the multigenic system - into separate recombinant congenic strains. In this way two new possibilities are open. First, because it is known which recombinant congenic strain carries these genes, we can now proceed with the genetic analysis of each of these genes separately and to establish their linkage. Second, we can study the function of each of these genes separately, using the pertinent recombinant congenic strains in comparison with the background strain. Finally, we can study the interaction of the relevant genes by combining them, using appropriate hybrids between the recombinant congenic strains. Obviously, as the recombinant congenic strains have been produced between strains which differ in a number of quantitative traits, involving not only tumor resistance, but also the differences in hormonal metabolism or in immunological factors a.o., they can be used with advantage to many different purposes. The main disadvantage of the recombinant congenic strains, which is the same as the main disadvantage of the recombinant inbred strains, is that they can be used only to analyze those traits or genes which differ between the two inbred strains originally used to produce the series of recombinant strains.

TABLE 1.   A QUANTITATIVE TRAIT STUDIED WITH R.I.S. AND R.C.S. (A MODEL)

PARENTAL STRAINS: PHENOTYPE: X - 80%   U - 30%

THREE NON-LINKED GENES:	LOCUS 1	X = 40%	U = 0%
	LOCUS 2	X = 0%	U = -50%
	LOCUS 3	X = 40%	U = 80%
IRRELEVANT GENE:	LOCUS N	X = 0%	U = 0%

A. R.I.S.

	A	B	C	D	E	F	G	H	I	J	K	L	M	N	O	P
LOCUS 1	X	X	X	X	X	X	U	U	X	X	U	U	U	U	U	U
LOCUS 2	X	X	X	X	U	U	X	X	U	U	X	X	U	U	U	U
LOCUS 3	X	X	U	U	X	X	X	X	U	U	U	U	X	X	U	U
LOCUS N	X	X	X	X	U	U	U	U	X	X	X	X	U	U	U	U
PHENOTYPE	80	80	100	100	30	30	40	40	70	70	80	80	0	0	30	30 %

B. R.C.S.

	A	B	C	D	E	F	G	H	I	J	K	L	M	N	O	P
LOCUS 1	X	U	U	U	U	U	U	U	U	U	U	U	U	U	U	U
LOCUS 2	U	U	X	U	U	U	U	U	U	U	U	U	X	U	U	U
LOCUS 3	U	U	U	U	X	U	U	U	U	U	U	U	X	U	U	U
LOCUS N	U	U	U	X	U	U	X	U	U	U	U	U	U	U	U	U
PHENOTYPE	70	30	80	30	0	30	30	30	30	30	30	30	40	30	30	30 %

   The second disadvantage of recombinant congenic strains, relative to the recombinant inbred strains, is the greater variance of linkage determination (9). This is greater than that of linkage determined using recombinant inbred strains. As a consequence, any suspected linkage determinant using recombinant congenic strains must be confirmed in subsequent backcross or F_2 hybrid tests. However, the same procedure has been advocated by Bailey (7) for any linkage which emerges from the tests of recombinant inbred strains. This necessity is even more underlined by the consideration of Silver and Buckler (10) who have shown, using Bayesian statistics, that the a priori chance of finding the linkage using recombinant inbred strains is much lower than previously assumed. Hart (in preparation) has applied the equations of Silver and Buckler (10) to calculate the probability of finding a linkage for any gene, using a series of recombinant inbred strains which have been typed for a number of markers. The results of these calculations for series of 7, 14 or 25 recombinant inbred strains, which were typed for 10, 20, 60, 100, 160, or 200 markers, are shown in Table 2.
   The results show that the probability of finding linkages for any given gene using recombinant inbred strains is rather small. Even when using a series of 25 recombinant inbred strains typed for 200 markers, the chance of finding a linkage with 95% confidence for any new gene is only 0.504. This probability is higher when we are satisfied with a lower confidence limit (80%), but then the use of subsequent backcross or F_2 tests to confirm the linkage is an absolute imperative and the chance of a negative result is considerably higher.

Table 2.   PROBABILITY OF DETECTING LINKAGE (WITH 95% CONFIDENCE) WITH
          a̲ MARKERS AND b̲ RECOMBINANT INBRED STRAINS

a	b		
	7	14	25
10	0	0.013	0.034
20	0	0.026	0.069
60	0	0.077	0.190
100	0	0.135	0.296
160	0	0.192	0.429
200	0	0.234	0.504

A potential disadvantage of recombinant congenic strains is that it may not be possible to produce a series of recombinant congenic strains between any desired pair of inbred strains. The main condition of applicability of recombinant congenic strains is that each individual recombinant congenic strain in the series receives a small proportion of the genes from the donor strain, approximately 12%, and that in the whole series of recombinant congenic strains the genome of the donor strain is represented almost totally. The proposed breeding scheme achieves this aim if there is no conscious or unconscious selection for certain genotypes. However, this cannot be guaranteed a priori and it is conceivable that combinations of the genes in some pairs of the inbred strains may interact in such a way that the resulting recombinant congenic strains contain too many genes of the donor strain and therefore do not have sufficient resolution power, or that they contain too few genes of the donor strain and therefore the series as such does not cover sufficiently the genome of the donor strain. It appears therefore necessary to control at certain intermediate stages of breeding whether the segregation of genes of the donor strain and the background strain is compatible with the expectation. The preliminary tests of the four unlinked enzyme genes show that the segregation of the genes of the STS strain on the BALB/c background in the recombinant congenic strains tested conform with the expectation (data not shown). We plan to perform this test in each series which we are producing to ascertain that the actual segregation is compatible with the expected use of the strains.

Table 3 shows how the recombinant congenic strains fit into the spectrum of genetic tools, which are available for analysis of strain differences. For the traits, which are controlled by a single gene, the recombinant inbred strains are the best available possibility to establish the linkage of the responsible gene. Recombinant inbred strains mix the genomes of the two parental strains in equal proportions in each recombinant inbred strain and therefore the variance of the linkage estimation is rather low.

The second tool, the recombinant congenic strains, can be used for analyses of traits controlled by two to five or six genes. The genes of one parental strain are mixed with the genes of the other parental strain in proportion 1:8, therefore the variance of linkage is higher and the finding of linkage for the relevant genes is more difficult than with the recombinant inbred strains.

Finally, for traits which are controlled by more than six genes, congenic strains have to be produced for each gene separately by backcrossing, using selection for the studied trait. An example of this approach are the congenic strains for histocompatibility genes produced by Snell (11).

Table 3.   APPROPRIATE SYSTEMS FOR GENETIC ANALYSIS OF INTERSTRAIN DIFFERENCES

Number of genes involved	Genetic test system	Proportion of donor strain genome	Detection of linkage
1	Recombinant inbred strains	0.50	Possible
2-5	Recombinant congenic strains	0.125	Possible but more difficult
6 or more	Congenic strains	<0.01	Not likely

Presently, the following series are in different stages of preparation: (a) OC.B genes of the B10.O20/Dem (C57BL/10-H-2^{pz}) spread over the O20/A (H-2^{pz}) background. This series is relevant to the study of complement genes, androgen regulation, and mammary and lung tumorigenesis. (b) HC.B and BC.H-involving genes of C3H/Sn and C57BL/10 strains. These are two of the most well-characterized inbred strains with many quantitative and qualitative polymorphisms, including tumor-resistant genes, and genes controlling resistance to prolactin-induced mammary tumors. (c) CC.ST-spreading genes of the STS/A strain over the BALB/cHe background. This provides another combination of relatively alien genetic material and is relevant also to the study of hormonal induction of mammary tumors.
   We hope that, when finished, they will be useful in a number of applications.

REFERENCES

1. Bailey DW: A search for genetic background influences on survival time of skin grafts from mice bearing Y-linked histoincompatibility. Transplantation 3: 531-534, 1965
2. Bailey DW: Recombinant-inbred strains. An aid to identify linkage and function of histocompatibility and other genes. Transplantation 11: 325-327, 1971
3. Heston WE: Genetics of neoplasia. In: Burdette WJ(ed): Methodology in Mammalian Genetics, pp. 247-268, Holden-Day, San Francisco, 1963
4. Demant P: Histocompatibility and genetics of tumor resistance. J. Immunogenet. 13: 61-67, 1986
5. Bittner JJ, and Little CC: The transmission of breast and lung cancer in mice. J. Heredity 28: 117-121, 1937

6. Malkinson AM, Nesbitt M, and Skamene E: Susceptibility to urethan-
   induced pulmonary adenomas between A/J and C57BL/J mice: Use of AxB and
   BxA recombinant inbred lines indicating a three-locus genetic model. J.
   Natl. Cancer Inst. 75: 971-974, 1985
7. Bailey DW: Recombinant inbred strains and bilineal congenic strains.
   In: Foster HL, Small JD, and Fox JG(eds): The Mouse in Biomedical
   Research, pp. 223-239, Academic Press, New York, 1981
8. Briles DE, Benjamin Jr. WH, Huster WJ, and Posey B: Genetic approaches
   to the study of disease resistance with special emphasis on the use of
   recombinant inbred mice. Curr. Top. Microbiol. Immunol. 124: 21-35,
   1986
9. Demant P, and Hart AAM: Recombinant congenic strains - a new tool for
   analyzing genetic traits determined by more than one gene.
   Immunogenetics 24: 416-422, 1986
10. Silver J, and Buckler CE: Statistical considerations in linkage
    analysis using recombinant inbred strains and backcrosses. Proc. Natl.
    Acad. Sci. USA 83: 1423-1427, 1986
11. Snell GD: Histocompatibility genes of the mouse. II. Production and
    analysis of isogenic resistant lines. J. Natl. Cancer Inst. 21: 843-
    877, 1958

A.C. Beynen & H.A. Solleveld (Eds), New developments in biosciences: their implications for laboratory animal science, ISBN 978-94-010-7973-0
© 1988 Martinus Nijhoff Publishers, Dordrecht.

ACUTE DERMAL TOXICITY: MORPHOLOGICAL RESPONSE OF THE HAIRLESS MICE SKIN ORGAN CULTURE

W. PITTERMANN, F. BARTNIK AND K. KÜNSTLER

Henkel KGaA, Henkelstraße 67, 4000 Düsseldorf 1

1. INTRODUCTION
The non-immunological irritancy potential of a chemical substance on skin or mucous membranes is one of the informations required for product development or marketing procedures. The risks of skin or mucous membrane irritancy have to be known for official registration as well as for occupational aspects, transportation or consumer use (1).

The animal model worldwide in use for the assessment of this criteria is the skin or mucous membrane of the rabbit. Type and degree of the potential irritation are assessed visually following a modified Draize scheme (2,3). Eventually the test substance is graded. The increasing knowledge of the biochemical mechanisms of skin irritancy and public and professional discussion on animal welfare stimulated the development of more or less alternative methods and the use of different animal species too (4).

In contrast to the worldwide accepted rabbit model these alternative methods are rather laboratory-specific and are not completely comparable. Their validation does not depend only on methodic progress but just as well on systematically performed comparative studies. Without a further validation a national or supranational recognition cannot be achieved.

Therefore "acute dermal toxicity" was studied on three different models in a granted project (5).

Treated fibroblast cell cultures were not suitable for a routine assessment from a pathological or morphological point of view. Histological sections taken from the chorionallantonic membrane of fertilized and incubated hen eggs treated with the compound appeared to be rather uniform. The findings concentrated on changes of the blood circulation system. However, these alterations are much more recognizable in a macroscopic rating (6).

The histological examinations of metabolically and structurally intact samples of a skin organ culture offer quite a number of interesting aspects to the morphologist.

The result can be compared with other animal studies that concern local effects on skin or mucous membranes. The transferability of the morphological data to human conditions rather improves the possibilities of an evaluation of risks. By the preparation of histological slides this IN VIVO/IN VITRO assay is documented at the same time. Beyond that, the local effects are microscopically assessed and they can be related to different mechanism of irritancy.

2. PROCEDURE

The procedure for the skin culture assay detailed in the literature (6,7,8) has been modified quite a number of times. Applied irritating substances affect an enzyme release within the skin explant. The higher the skin irritancy, the higher also is the enzyme activity in the culture medium, inasmuch as no enzyme inhibition prevails. The evaluated activities bring about the rating of the test substance.

In the course of our specific modifications, hairless mice were introduced together with a diminution of the skin explants in use. Pre-trials evidenced that in female hairless mice higher activity of enzymes could be released via irritating substances than in males. The age of the animals up to 264 days affects no relevant influence on the tests.
The diminution of the skin area to a mere 50 mm2 allows a significant reduction of animals per dose and also a better handling of the samples. After sacrificing the animal and preparation of the ventral and dorsal skin 20 explants per animal can be obtained. 7 specimens are used per dose and substance.

Out of the actual postulates of, 'reduction, refinement and replacement of animal studies' a reduction of the animal number and refinement by escaping pain is achieved.

After the total incubation of the explant over 40 hours the culture medium is evaluated for its enzyme activity and replaced by formaldehyde. The fixed samples are embedded, cut and stained routinely with hematoxylin and eosin.

3. RESULTS AND DISCUSSION

An accurate performance requires as low autolysis as possible. Viability of the skin culture can be determined in the epidermis.

Mitotic processes that exist in the Str. basale diminish in frequency during the incubation period. If fixing substances in high concentration are added after the pre-incubation period of 16 hours mitosis is stopped and fixed. The Str. basale retains the character of a matrix cell storage and the epidermis remains thinner than without stopping mitosis. If the explant is autolytic, lowered staining of epidermis and dermis as well as increased desintegration are found. Furthermore, mitotic activity of the epidermis is inhibited.

The results that are morphologically feasable mostly concentrate on the epidermis. The variation af alterations generally is greater than in comparable IN VIVO studies. The dynamics of inflammation, vascular permeability, edema and infiltration by polymorphnuclear leukocytes are completely absent under culture conditions. Therefore, the very image of the changes is only determined by the concentration, physico-chemical properties of test substances, time of effects and histological preparation.

For example, the cells of the Str. spinosum are very enlarged by 50 % sodium hydroxide whereas the nuclei became pyknotic and the Str. granulosum vanished. 50 % sulfuric acid causes a slight swelling of the epidermal cells and a desintegration. At the same time, focally arranged cells with eosinophilic cytoplasma and pyknotic nuclei can be observed in the epidermis.

Pure morpholine results in extensive shrinking of chromatin, whereas bromo acetic acid effects a lysis of the upper epidermal layers.

Such alterations are completely masked in IN VIVO studies due to immediate exsudation and penetration of blood components. Additional specific alteration can be found in the upper keratinization layers of the Str. corneum and Str. granulosum. The administration of sulfuric acid causes a accelerated differentiation resulting in a focal para - or even hyper-keratosis.

After administration of 5 % sodium hydroxide the Str. granulosum remains unchanged, whereas the cells of the Str. spinosum are altered in the way described above.

The alterations within these upper epidermic layers also are sometimes the only changes induced by the test substances that are evident in low dose ranges.

The administration of silver nitrate evidences by its brown-dark precipitations the barrier effect of the Str. corneum. In low dosing, these precipitates are only found on the Str. corneum. With increasing concentration, the precipitation effects additional layers of the epidermis and dermis coating them with confluent brown black material.

On the edge of the epidermal areas that are changed by the applied substances i. e. silver nitrate, reactive zones within the Str. basale occur. Densely packed elongated cells lie in the intact area of the Str. basale, comparable to the epidermal reaction of wound healing.

Severe swelling or shrinking processes that affect the complete explant, occur in high dose ranges of sulfuric acid, silver nitrate, sodium hydroxide and morpholine. Specific alterations of the hair follicles or sebaceous glands that have been directly caused by the substance tested have not yet been found.

In accordance to the results obtained by Helman et al it can
be stated that the cultured skin has retained the properties
of a morphologically differentiated reaction on the toxicological effects of different chemical substances.

A high correlation between enzyme activity and morphological
findings was observed by Helman et al (7) in the comparable
skin culture assay. Similar results were obtained in our own
studies.

In this survey, the enzyme activities of the test group are
brought into relation with the activities of control group
treated with the solvent only and graded (Table 1). Deviations from the concentration of the test substance dependance
are possible.

Physico-chemical properties of a substance such as high or
low pH may inhibit enzyme activities in situ by denaturation.

TABLE 1
Results of Skin Culture Assay with Different Compounds

COMPOUND	CONCENTRATION ($ug\ cm^{-2}$)	EFFECTS in vitro	Morphology	in vivo skin irritation
Sodium hydroxide	5000	moderate	positive	corrosive
	500		positive	
	50		negative	
Sulfuric acid	5000	moderate	positive	corrosive
	500		positive	
	50		negative	
Silver nitrate	5000	minor to	positive	corrosive (?)
	500	moderate	positive	
	50		(positive)	
Tween 20	10000	none	negative	no irritant
Morpholine	10000	severe	positive	irritating
	1000		positive	
	100		negative	
Bromo acetic acid	1000	severe	positive	corrosive
	100		(positive)	
	10		negative	
Formaldehyde*	10000	moderate	positive	irritating
	1000		positive	
	100		negative	

*: active Substance

After application of sulfuric acid the activity is partially inhibited whereas after silver nitrate a complete inhibition results. That ist valid for enzymes LDH and GOT, but not for the consumption of the substrate glucose which is a good criterium for viability assessment.

The comparison of morphological data with IN VIVO data is limited by the absence of specific inflammatory reactions of the dermis in skin organ culture. This means that factors such as vascular permeability, cellular infiltration and activity of mediators are quite absent.

So a clear distinction between the grade "corrosive" and "irritant" is not yet possible. The criteria for a differentiation may be found in provable changes of the Str. basale and the dermis.
Further studies and experience are necessary with the morphological results of skin organ culture as a model for acute dermal toxicity.

4. CONCLUSIONS
Morphology of the hr/hr mice skin organ culture assay

- complements the biochemical results
- improves the transferability of the results
- differentiates epidermal lesions
- enables a documentation of the assay
- can be used by routine

R E F E R E N C E S

1. German Chemical Law 16. Sept. 1980
2. OECD Guidelines for Testing of Chemicals No. 404
   Acute Dermal Irritation/Corrosion
3. OECD Guidelines for Testing of Chemicals No. 405
   Acuty Eye Irritation/Corrosion
4. E. Patrick, H. I. Maibach and A. Burkhalter: Mechanisms of chemically Induced Skin Irritation. I. Studies of Time Course, Dose Response and Components of Inflammation in the Laboratory Mouse. Toxicol. Appl. Pharmacol. 81, 476 - 490 (1985)
5. Bartnik, F. G., K. Künstler, N.-P. Lüpke, W. Sterzel and S. Wallat: Possible Replacement of Animal Experiments in local Tolerance Testing on Skin and Mucous Membranes. Symposium 'Alternatives to Animal Experiments in Risk Assessment' Berlin 1987
6. Pittermann, W. Unpublished Results (1986)
7. Kao, J.; Hall, J; and J. M. Holland: Quantification of Cutaneous Toxicity: An in Vitro Approach Using Skin Organ Culture. Toxicol. Appl. Pharmacol. 68, 206 - 217 (1983)

8. Helman, R. G.; Hall, J. W.; and J. Y. Kao: Acute Dermal Toxicity; In Vivo and In Vitro Comparisons in Mice. Fundamental and Applied Toxicol. 7, 94 - 100 (1986)
9. Gibson, W. T.; M. R. Teall: Interactions of C_{12} Surfactants with the Skin. Changes in Enzymes and Visible and Histological Features of Rat Skin Treated with Sodium Lauryl Sulphate. Fd. Chem. Toxic. 21, 587 - 594 (1983)

Part of the work presented was sponsored by the Bundesministerium für Forschung und Technologie FRG.

A.C. Beynen & H.A. Solleveld (Eds), New developments in biosciences: their implications for laboratory animal science, ISBN 978-94-010-7973-0
© 1988 Martinus Nijhoff Publishers, Dordrecht.

THE DUTCH RABBIT IN TOXICITY TESTING
CHEMICAL - INDUCED CREATINE KINASE RELEASE;
A SPECIES- AND STRAIN-SPECIFIC RESPONSE ?

T.D. YIH, A.J.M. DEGEN, J. KAMERMAN AND H.F.P. JOOSTEN
Organon Int. B.V., Drug Safety R and D Labs.,
P.O. Box 20, 5340 BH Oss, The Netherlands

INTRODUCTION
Within the framework of preclinical safety testing a compound having cardiovascular activity (CC), was studied in rabbits.
For the study of its teratogenic potential, CC was administered orally to Dutch rabbits from day 6 to 18 of pregnancy at dose levels of 25, 50 and 100 mg/kg body weight.
Food consumption decreased in a dose-related manner. At 100 mg/kg no live foetuses could be recovered on day 29 of pregnancy; at 50 mg/kg only half of the number of dams had live young, whereas at 25 mg/kg no embryotoxicity was observed.
Inhibition of food consumption was found in association with overfilling of the stomach, contraction of its pyloric part, empty intestines and high plasma levels of creatine kinase (CK). These effects were not present in oral toxicity studies in Wistar rats and Beagle dogs.
New studies were designed to investigate species- and strain-specificity and the underlying mechanism of toxicity.

MATERIALS AND METHODS
Rabbit strains
The following strains were used:
- Dutch rabbits (spf) supplied by the Broekman Inst., Someren, The Netherlands
- New Zealand White (NZW) rabbits from the same breeder
- Chinchilla rabbits from HSD/Cpb, Zeist, The Netherlands
- Himalayan rabbits (HY/CR) from Charles River, France
If not stated otherwise adult male animals were used.
Husbandry
Animals were housed individually in galvanized iron cages measuring 50x50x40 cm (lxbxh) under standard conditions of temperature (21 +/- 3 C), humidity (60 +/- 20%) and artificial illumination (12 hrs off, 12 hrs on).
They had free access to tap water from bottles and to pelleted food (LKK 20, Hope Farms, Woerden, The Netherlands).
The study with Himalayan rabbits was carried out at RL-CERM, Riom, France.
Dosing
For oral and intraperitoneal administration CC was suspended in an aqueous medium of 0.5% gelatin containing 5% mannitol.
The suspension was administered intragastrically through a polyethylene catheter (Pharmaplas Disposable Catheters).

For intravenous administration CC was dissolved in 30 % polyethylene-glycol (PEG300) in saline to the limit of solubility of 1.7 mg/ml. The drug solution was infused into the marginal ear vein at a pump rate of 0.5 ml/min.
Clinical observations, ECG recordings and autopsy
Animals were daily observed for behavioural changes and clinical signs.

Electrocardiographic recordings were obtained by bipolar leads I, II, III and unipolar leads aVR, aVL and aVF (Cardiscript IV, Schwartzer, W-Germany).
For autopsies animals were killed by destruction of the brain stem (Supercash pistol, Accles and Shelvoke Ltd., No. 78130) followed by rupture of the diaphragm.

Laboratory investigations

Plasma from heparinized blood, collected by puncture of the marginal ear vein, was assayed for creatine kinase (CK) at 30 C according to the recommendations of the GSCC (1) and in some studies for lactate dehydrogenase (LDH) also at 30 C (2).

Analysis of results

Pre-dosing values of food intake and plasma CK showed large inter-individual variation. To reduce variation each animal served as its own control by expressing the post-dosing values as a percentage of the pre-dosing value. A decrease in food intake was considered significant for values lower than 75 % and an increase of CK for values higher than 200 %.

RESULTS

Intragastric route of administration

Following a single i.g. dose of 100 mg/kg food intake decreased abruptly. The number of faecal droppings was sharply decreased during a subsequent period of 7 hrs. Food intake was restored to the pre-dosing value within 2 or 3 days.
Plasma CK levels in samples collected 24 hrs post-dosing were increased as compared with the pre-dosing value.
From a study with i.g. dose levels of 10, 20 and 30 mg/kg it appeared that the minimum dose level giving a 2-fold increase in CK activity was 30 mg/kg. Food intake at that dose level was depressed to about 40 % of the pre-dosing value.
The time course of the CK rise was studied in a female rabbit using an i.g. dose level of 100 mg/kg. Plasma CK activity increased sharply between 16 and 24 hrs post-dosing. The values before dosing and at 16, 20 and 24 hrs after dosing were 277, 646, 30000 and 57400 U/L, respectively.
Following multiple dosing the suppression of food intake lasted the entire treatment period and returned to normal thereafter. The CK rise was first observed 24 hrs after the 1st dose, whereas subsequent doses caused a low or undetectable CK rise.

Intravenous route of administration

Following intravenous administration of CC at dose levels of 6.9 or 10.7 mg/kg a slight reduction in food intake was observed. In contrast to the oral route the peak level of CK activity was observed 7 hrs post-dosing.
The values were 19300 and 13700 U/L at 7 and 24 hrs post-dosing, respectively. A significant rise occurred already at 2 hrs post-dosing (1896 U/L).

In the rabbit which received 10,7 mg/kg plasma CK and LDH levels were determined. The peak level for LDH was also found at 7 hrs post-dosing.

Intraperitoneal route

A high plasma CK level was also observed 24 hrs following an intraperitoneal injection of CC (100 mg/kg), despite the fact that at autopsy a significant amount of test compound was found as a precipitate in the abdominal cavity. At 24 hrs post-dosing the plasma CK level was 31600 U/L (more than 80-fold the pre-dosing value).

Other rabbit strains

Intragastric administration of CC (100 mg/kg) to NZW, Chinchilla

and Himalayan rabbits did not give uniform results.
Inhibition of food intake was observed in all strains, although
not to the extent as observed in the Dutch rabbit (1-36 %). For
the NZW and Chinchilla rabbits food intake was about 50 % of the
pre-dosing value. The post-dosing values in the Himalayan
rabbit ranged from 8 - 72 %.
With respect to the CK activity a clear rise was measured in the
Himalayan rabbit (308 - 1618 %). Three out of five Chinchilla
rabbits showed a doubling in plasma CK activity, whereas no
clear CK rise was observed in the NZW rabbit.

Table 1

| Strain | No. | Relative values at 24 hrs post-dosing | | | | | |
| | | CK | | | Food intake | | |
		Mean	SD	Range	Mean	SD	Range
Dutch	5F	6681	8485	736 - 20722	9	6	4 - 20
Dutch	14M	398	275	196 - 1094	8	9	0 - 36
Himalayan	4M	779	589	442 - 1618	34	32	8 - 76*
Chinchilla	5M	190	63	122 - 265	58	2	56 - 59**
NZW	4M	171	53	104 - 217	51	9	43 - 65

* received a fixed ration of 150 g/day;** measured in only 2
animals

DISCUSSION
Guidelines for embryotoxicity testing of drugs require two test
species, a rodent and a non-rodent species. The rat and the
rabbit are the selected species for routine use.
Dose selection is of critical importance. In a well designed
study the high dose should cause some maternal toxicity, whereas
the intermediate and low dose group should be free from maternal
toxicity. Toxicity in the high dose group, however, should not
be too severe. It is well known that a strong disturbance of
the physiological balance of the pregnant female may affect
intra-uterine development and may result in foetal loss or
abnormal embryogenesis or foetogenesis.
Maternal toxicity thus imposes limitations to dose levels.
Because of the strong inhibiting effect of CC on food intake in
the Dutch rabbit, teratogenicity testing in this rabbit strain
was questionable. Moreover, the anorectic effect and the plasma
CK rise had to be evaluated.
The strong inhibition of food intake, the observation of
overfilling of the stomach with contraction of its pyloric part
and the rise of plasma CK levels are at present considered as
specific to the rabbit as these effects were absent in rats
(Wistar and Sprague Dawley strain) and Beagle dogs.
The data suggest that these effects are caused by a primary
action of the test compound on the stomach wall involving
contraction of the pyloric part and prevention of further food
intake as a logical consequence. Pyloric contraction might lead
to CK release from the muscular tissue. Subsequent doses would

224

either cause additional release of CK or no release. This might depend on the degree of contraction of the pyloric muscle.

Initially, the possibility that direct local contact between the inner stomach wall and the test compound is conditional for the anorectic and CK response was considered. This has been ruled out, however, by the data from the i.p. and especially from the i.v. studies.

High plasma CK levels are indicative for muscle damage (3). If the heart muscle is involved, especially the MB CK-isoenzyme would be increased. In case of damage of skeletal muscle or the musculature of the intestinal tract the MM CK-isoenzyme fraction would be elevated. Preliminary studies indicated that the CK rise is due to the MM-type isoenzyme fraction. This would exclude the heart as the source of CK release. This is in agreement with the normal ECG pattern and the absence of clinical signs related to heart damage.

It is interesting in this context, that a rise in CK, especially of the MM CK-isoenzyme fraction has been observed in a Japanese study following i.v., p.o. and s.c. administration of methamphetamine, a psychotropic compound (4).

All data obtained so far point to the pyloric muscle as the target tissue of toxicity. There is, however, no clear relationship between the degree of food intake inhibition and CK rise. The very high CK levels observed in the i.v. study at 10.7 mg/kg for instance, were accompanied by a rather moderate decrease of food intake to 72 % of the pre-dosing value.

The studies in the different rabbit strains indicate that the Dutch rabbit, although the most sensitive, is not unique with respect to the observed toxicity. The lower response in the Chinchilla and NZW rabbits could be due to either lower sensitivity of the pyloric muscle or to a difference in bioavailability of the compound at the target tissue.

An indication for the possible role of the latter is the difference in time course of the CK release following intravenous and intragastric administration.

A definite answer can be obtained from data on plasma levels of the test compound in the different rabbit strains.

Summarizing, the release of CK in the Dutch rabbit is species-, but not strain-specific. The NZW strain is the least sensitive strain.

Future studies aim at the identification of the target tissue releasing CK and the explanation of the peculiar position of the NZW rabbit.

Acknowledgements
The participation of Dr. Lecoin (Riom, France) and the assistance of Mr. Felet, v. Keulen, Kouwenberg and Willemse is gratefully appreciated as are the members of the Clinical Chemistry group for determinations of enzyme activities.

REFERENCES
1. Z. Klin. Chem. Klin. Biochem. 15, 249-51, 1977
2. Z. Klin. Chem. Klin. Biochem. 10, 281-91, 1972
3. Pearce, J. Br. Med. J. 2, 167, 1965
4. Aoki, T.,Namiki,H. and Tokudome, S. Jap. J. Alcohol Stud. Drug Depend. 17, 226-36, 1982.

A.C. Beynen & H.A. Solleveld (Eds), New developments in biosciences: their
implications for laboratory animal science, ISBN 978-94-010-7973-0
© 1988 Martinus Nijhoff Publishers, Dordrecht.

INTERTRIAL - INTERVAL IN THE "WATER ESCAPE TEST" IN MICE

H. Heinecke

1. INTRODUCTION

In 1961 ESSMAN and JARVIK published a "water escape test"
in mice as motivating factor in animal learning. The
biological basis of such tests is that mice like other
rodents are able to swim and in consequence they are also
able to ashore. This is essential to survive for the animals,
and under this aspect it is an unconditioned reflex. "The
present technique has been shown to be effective for demonstrating the rapid acquisition of orientation response in
mice" (ESSMANN and JARVIK 1961). It is a useful test in
behavioural research, as a very easy and quick learning test
in mice. It will be possible to get results without a long
training period after three or four trials in one hour. In
the standard procedure the Intertrial-Interval (ITI) is
15 min (ESSMAN and JARVIK 1961, HEINECKE 1987) or 24 hours
(FESTING 1973).
FESTING 1973 described the influence of ether narcosis on
the form of the learning curves. His result showed that the
learning effect (the reduction of the time needed to escape
in the following trials) is longer between the first and
the second and the second to the third trial in anaesthetized
mice than in the control animals.
We used this test in mice to study the Intertrial-Interval.

2. MATERIAL and METHODS

2.1. Animals

For the study we used male and female inbred and F1-hybrid
mice from our own breeding colony: AB/Jena (albino),
C57BL/6Jena (black) and ABD2F1-hybrid (agouti); 5 to 6 weeks
old with a body weight from 19 up to 23 g dependent on the

strain or hybrid. Microbial status SPF (bacteriologically
and parasitologically MRC category xxxx, and free of the
studied viruses: adenovirus, ectromelia, EDIM/LIVM, PVM,
Reo-3 and Sendai and also free of Mycoplasma, Toxoplasma
and Leptospira (five serogroups)).

Five mice/Makrolon cage typ II were kept under partial
barrier conditions: autoclaved wood shaving once a week,
tap water acidified with HCl pro analysis on a pH of 2.7 to
3.0 and pasteurized pellet food (IMET-H 3, MADRY 1978) ad
libitum; temperature 20 to $22^{o}C$ relative humidity from about
50 to 70 % not regulated. Natural and fluorescent light
c. 100 to 250 lx from 05.45 to 19.45 CET. The experiments
were performed in an experimental room under the same
conditions between 08.00 to 10.00 CET during different
seasons of the years 1985/1986.

2.2. Apparatus and procedure

The apparatus consisted of a Makrolon cage type III
(45 x 28 x 16 cm) with an escape platform at 32° angle in
one corner. The cage was filled with water (8 cm deep,
temperature $27^{o}C$). The mouse was held by the tail in the
corner opposite the platform and was dropped into the water.
The time was recorded from the dropping into the apparatus
until the mouse climbed with all four legs the escape ramp.
The given Intertrial-Intervals (ITI) were 1.5 up to 2 seconds,
1, 2, 3 and 5 min respectively for two consecutive trials.

2.3. Statistical analysis

All mice were tested only three consecutive trials. When we
used short ITI mice from several populations got tired if we
took four consecutive trials. The time required to escape was
recorded in seconds. A limit of 60 seconds was set. Animals
that failed to escape after 60 sec. were removed and given a
score of 60 sec. Animals that reached the platform in one or
two seconds were excluded, based on the opinien that during
such a short time a correct search strategy is impossible.

A log transformation of the escape time, however, gave a distribution that was approximately normal (FESTING 1973). The statistical differences between the log 10 transformed data were evaluated with the multiple t-test and the exact Fisher test. The level of significance was $p \leq 0.05$.

3. RESULTS and DISCUSSION

The table 1 gives the mean response latency of female mice of two inbred strains and one F1-hybrid population for reaching and climbing the escape ramp. The results shows, that an extremely short ITI of 1.5 up to 2 seconds had a negative influence only on learning speed of the female AB/Jena mice. This effect was not observed in several experiments with ABD2F1-hybrid and C57BL/6Jena female mice. The same negative results were found in male mice (data not given) of the two inbred strains and F1-hybrids (Fig. 1). These studies showed

- that the water escape test according to ESSMAN and JARVIK can also be used for studies of the Intertrial-Interval;

- that the extremely short ITI of 1.5 up to 2 seconds influenced only the female AB/Jena mice. FESTING (1973) found after post-trial etherization no strain differences in short memory. It shows, that the mechanisms are different.

I am indebted to Dr. U. Lundberg for helpful discussion and Mrs. R. Brückner and K. Reum for technical assistance.

REFERENCES

1. Essman WB, Jarvik ME: Psychol. Rep. 8, 58 (1961)
2. Heinecke H: Z. Versuchstierkd. 30, No 1/2 (1987)
3. Mackintosh NJ: The Psychology of Animal Learning. New-York - London - San Francisco. Acad. Press 1974
4. Madry M: Versuchstierernährung. In: Köhler D, Madry M, Heinecke H: Einführung in die Versuchstierkunde Band II: Angewandte Versuchstierkunde. Jena. VEB Gustav Fischer Verlag 1978

TABLE 1: Intertrial-Interval (ITI) in female mice
(log 10 water escape time $\bar{x} \pm s_{\bar{x}}$ in sec
+ = significant, in parenthesis = number of
animals failed the limit of 60 sec)

AB/Jena

ITI	N	1.	2.	3. Trial
2 sec	9 (1)	$1.1^{\pm}0.06$	$1.0^{\pm}0.07$	$0.8^{\pm}0.14$
	9	$1.1^{\pm}0.13$	$1.0^{\pm}0.16$	$0.7^{\pm}0.16$
	8	$1.3^{\pm}0.13$	$1.1^{\pm}0.21$	$1.1^{\pm}0.18$
	26	$1.2^{\pm}0.06$	$1.0^{\pm}0.15^{-}$	$0.9^{\pm}0.16^{+}$
1 min	10	$0.9^{\pm}0.08$	$0.6^{\pm}0.12$	$0.5^{\pm}0.12$
	10	$1.0^{\pm}0.08$	$0.6^{\pm}0.14$	$0.7^{\pm}0.11$
	9 (1)	$1.1^{\pm}0.12$	$0.7^{\pm}0.16$	$0.5^{\pm}0.10$
	29 (1)	$1.0^{\pm}0.05$	$0.6^{\pm}0.14^{+}$	$0.6^{\pm}0.11^{+}$
2 min	9 (1)	$1.2^{\pm}0.12$	$0.8^{\pm}0.13$	$0.4^{\pm}0.08$
	10 (2)	$1.2^{\pm}0.09$	$0.6^{\pm}0.10$	$0.7^{\pm}0.13$
	9 (1)	$1.3^{\pm}0.11$	$1.1^{\pm}0.15$	$1.0^{\pm}0.17$
	28 (2)	$1.2^{\pm}0.06$	$0.8^{\pm}0.13^{+}$	$0.7^{\pm}0.13^{+}$
3 min	9	$1.2^{\pm}0.11$	$0.8^{\pm}0.09$	$0.7^{\pm}0.17$
	10	$1.1^{\pm}0.05$	$0.7^{\pm}0.10$	$0.7^{\pm}0.13$
	10	$1.0^{\pm}0.07$	$0.7^{\pm}0.12$	$0.7^{\pm}0.12$
	29	$1.1^{\pm}0.04$	$0.7^{\pm}0.10^{+}$	$0.7^{\pm}0.14^{+}$
5 min	10 (1)	$1.2^{\pm}0.10$	$0.7^{\pm}0.08$	$0.8^{\pm}0.11$
	10	$1.0^{\pm}0.07$	$0.6^{\pm}0.06$	$0.5^{\pm}0.07$
	10	$1.1^{\pm}0.11$	$0.9^{\pm}0.15$	$0.6^{\pm}0.09$
	30	$1.1^{\pm}0.06$	$0.7^{\pm}0.10^{+}$	$0.6^{\pm}0.09^{+}$

ABD2F1-hybrids

ITI	N	1.	2.	3. Trial
	10	0.8 ± 0.06	0.5 ± 0.09	0.6 ± 0.10
2 sec	10	1.0 ± 0.13	0.4 ± 0.10	0.4 ± 0.10
	10	1.0 ± 0.13	0.6 ± 0.09	0.3 ± 0.07
	30	0.9 ± 0.09	$0.5 \pm 0.09^+$	$0.4 \pm 0.09^+$
	8	0.8 ± 0.09	0.6 ± 0.12	0.4 ± 0.15
1 min	8	0.9 ± 0.15	0.6 ± 0.11	0.4 ± 0.06
	9	1.1 ± 0.11	0.5 ± 0.09	0.4 ± 0.13
	25	0.9 ± 0.11	$0.5 \pm 0.11^+$	$0.4 \pm 0.15^+$
	10	1.0 ± 0.11	0.8 ± 0.11	0.7 ± 0.11
2 min	10	0.9 ± 0.07	0.7 ± 0.11	0.5 ± 0.09
	10	1.0 ± 0.08	0.6 ± 0.06	0.5 ± 0.14
	30	1.0 ± 0.09	$0.7 \pm 0.09^+$	$0.6 \pm 0.11^+$
	9	0.9 ± 0.10	0.6 ± 0.10	0.5 ± 0.13
3 min	10	1.2 ± 0.11	0.7 ± 0.11	0.6 ± 0.13
	9	1.2 ± 0.08	0.9 ± 0.09	0.7 ± 0.13
	28	1.1 ± 0.10	$0.6 \pm 0.09^+$	$0.5 \pm 0.14^+$
	9	0.8 ± 0.10	0.7 ± 0.10	0.5 ± 0.09
5 min	10	1.0 ± 0.05	0.7 ± 0.10	0.7 ± 0.14
	10	1.2 ± 0.08	0.7 ± 0.11	0.7 ± 0.05
	29	1.0 ± 0.09	$0.7 \pm 1.0^+$	$0.7 \pm 0.10^+$

C57BL/6 Jena

ITI	N	1.	2.	3. Trial
2 sec	10	(1)1.0 ± 0.08	$0.6 \pm 0.09^+$	$0.4 \pm 0.07^+$
1 min	10	1.1 ± 0.07	$0.6 \pm 0.11^+$	$0.4 \pm 0.13^+$
2 min	9	(1)1.1 ± 0.08	$0.6 \pm 0.10^+$	$0.4 \pm 0.10^+$
3 min	11	1.0 ± 0.07	$0.5 \pm 0.08^+$	$0.5 \pm 0.12^+$

FIGURE 1: Water escape learning curves in female and
male AB/Jena mice on different Intertrial-
Interval (ITI in log 10 sec)

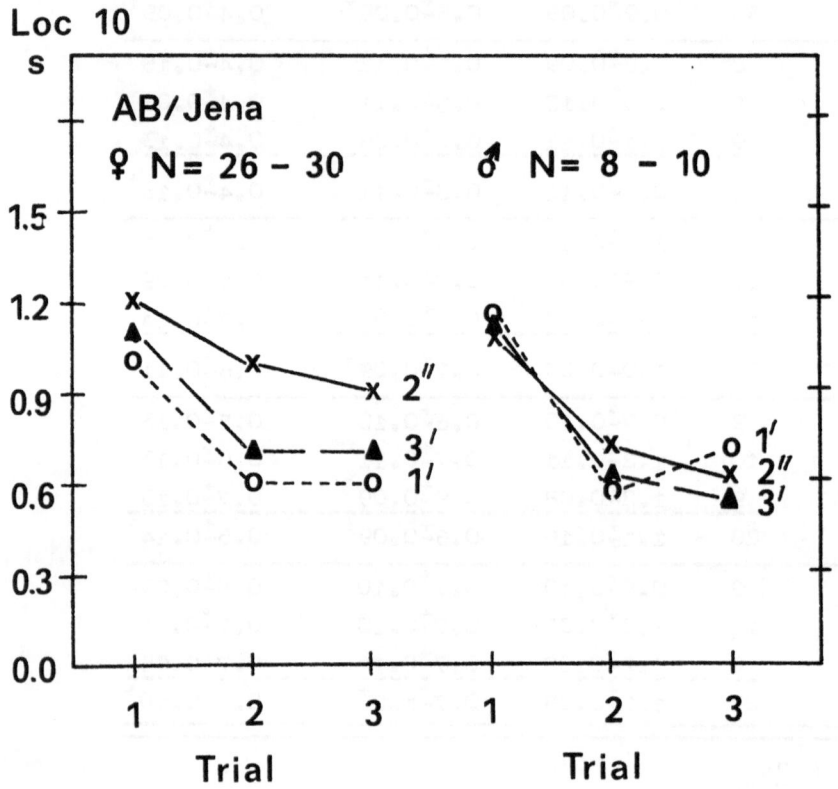

Author's address: Prof. Dr. H. Heinecke, Central Institute
for Microbiology and Experimental Therapy of the Academy of
Sciences of the GDR, Beutenbergstraße 11 J e n a, DDR 6900

A.C. Beynen & H.A. Solleveld (Eds), New developments in biosciences: their implications for laboratory animal science, ISBN 978-94-010-7973-0

BEHAVIOUR, HOUSING AND WELFARE OF NON-HUMAN PRIMATES

TREVOR B POOLE

1. INTRODUCTION

Unlike the vast majority of laboratory animals, primates are not domesticated and may even be wild caught, so that the laboratory must aim to provide suitable conditions for keeping a wild animal in captivity. Even zoos, which have a long experience of keeping these animals and where spatial considerations are less critical, have difficulty in providing a sufficiently complex and stimulating environment for non-human primates. Markowitz (1) drew attention to the behavioural impoverishment and abnormal activities displayed by primates and other higher mammals in zoological gardens.

All non-human primates are adapted to living in a situation where there is a very varied input of sensory information. They are relatively long-lived and build up a lifetime of experience to cope with environmental changes, using learning and intelligence to solve everyday problems of survival. They carry out a complex daily routine which includes foraging, exploration, avoiding predators, and elaborate social behaviour. Highly social species such as baboons, macaques and chimpanzees often carry out complex social manipulations, which involve nepotism, reciprocal altruism and rivalry for resources and status. Most species are hostile to strangers and some defend territories, patrol them and advertise their presence to rival groups. Even when inactive, a primate requires a secure nesting place and chimpanzees and orangutans construct nests out of tree branches.

Food requirements are complex, because most primates are omnivores utilising a wide range of food sources (2). Not only do many of them have a good geographical knowledge of their home range but they may even remember the fruiting times of different trees; they also learn from older individuals which foods are edible and where they are to be found.

It is clear from the foregoing that the wild environment cannot be simulated in the laboratory and that a captive primate is forced to live in a situation in which it is unable to do most of the things for which its behaviour is adapted. It is therefore essential to examine the advantages and disadvantages experienced by a primate, when captive, in a laboratory cage.

2. THE CAPTIVE ENVIRONMENT

2.1 Attributes

Theoretically, captivity offers many advantages for the non-human primate. There is no shortage of food, it is protected from predators, competition from conspecifics, diseases and parasites, while medical treatment is available should it fall sick.

On the other hand, captive primates have restricted space, may be under-stimulated by living in an over-predictable situation and have a very limited ability to control or manipulate their environment in a beneficial way. In addition, they experience a lack of appropriate stimuli to elicit whole ranges of behaviour and frequently are isolated or placed in inappropriate social situations. Should there be any fear-inducing stimuli, captive primates are unable to avoid them.

2.2 Guidelines for keeping captive primates

A variety of guidelines have been drawn up with the aim of ensuring that primates in captivity are kept under humane conditions. Examples are those from the United States (3) and Europe (4). The weakness of such attempts is that the end result is a table with exact figures, usually relating a particular cage to the weight of an animal, such figures are not based on scientifically established environmental and behavioural needs but largely on economics and current practice. The unsatisfactory nature of such guidelines becomes apparent when it is realised that one of the commonest primates in captivity (Macaca fascicularis) is unable to sit in the European recommended cage in a normal posture with its tail freely suspended. Because of the lack of appropriate data, on the behaviour of primates in cages of different dimensions, published guidelines from other countries also differ (5). UFAW is currently carrying out research on the housing of two species of old world monkey (Macaca fascicularis and Papio).

The Council of Europe guidelines provide a single table of dimensions for housing, which classifies primates solely on the basis of weight, even though rodents are divided into several species! One of the most unsatisfactory aspects of these guidelines is that they fail to take into account the social structure of the species. Marmosets, squirrel monkeys and macaques all show different social organisations and these determine how compatible social groups may be created in captivity, even the Hamadryas baboon has a quite different sociobiology from its close relatives in the cynocephalus group. Managers of primate units should be familiar with these facts and take them into account when planning primate housing.

3. WELFARE OF CAPTIVE PRIMATES
3.1 Assessment

Behavioural measures provide the best way of assessing the welfare of captive primates, assuming that the animals are in good physical health and provided with an adequate and balanced diet. Physiological measures of stress are less sensitive because the animal may use coping techniques to normalise its endocrine profile, these efforts at homeo-stasis have their own costs, which will not be apparent. The most obvious indicators of poor welfare are the presence of various types of abnormal behaviour.

3.2 Abnormal behaviour

Abnormal behaviour may be defined as inappropriate behaviour which is maladaptive and does not promote the success and survival of the individual or its offspring. Such behaviour is not goal-orientated, but it must be appreciated that there are normal behaviours where the goal may not be obvious, for example, social play. Erwin & Deni (6) divided abnormal behaviour into two types, firstly, qualitative abnormal behaviour, which is outside the animal's normal behavioural repertoire, for example, self mutilation or repetitive stereotypies, and, secondly, quantitative abnormal behaviour where the behavioural elements are in the natural repertoire, but there are errors in context, sequencing, frequency or duration. These two categories are not completely mutually exclusive, but provide useful working definitions.

There are a variety of causes of abnormal behaviour in captive non-human primates. The individual may, during its infant or juvenile development, lack particular experiences which are necessary for it to develop a normal adult ethogram. The captive primate may be attempting to cope with an environment which is outside its adaptive range, so that it suffers from a lack of appropriate behaviour in its repertoire. The environment may be so artificial that the animal misreads the environmental signals and thus behaves in a maladaptive manner. Finally, the low level of significant environmental stimulation may lead to the animal building up an abnormal repertoire of behaviour as an alternative to inactivity.

It is important that the manager of the primate unit should be familiar with the commoner types of abnormal behaviour and a brief listing follows.

Stereotyped behaviours: functionless activities which are repeated regularly. Such behaviour is not observed in wild and the animal appears to be developing an alternative ethogram. Cronin (7) found that naloxone temporarily eliminates behavioural stereotypies in pigs and suggested that their performance many induce endorphin production. This can be regarded as a form of coping behaviour and is an indicator of poor welfare, reducing the animal to a caricature of its natural self. Stereotyped behaviour may include elaborate sequences which are built up over time.

There are two common types of stereotypy, firstly, inactive ones where the animal appears to be stuck in a fixed position, for example, holding the face or ear, poking the face or eye, holding a leg, tail or portion of fur, and fur plucking. Secondly, active stereotypies include repeated horizontal or vertical circling of the cage, charging the cage, swaying the body from side to side repeatedly, head flicking, weaving from side to side, repeated somersaulting or rocking with arms clasped to the body.

Autoaggression: the monkey attacks parts of its body such as its arm or tail. Such attacks may result in injury, but these lesions may be accidental, as a result of the repetitive nature of the activity.

Learned helplessness: the primate becomes passive and dependent on its keeper, it may behave as a permanent juvenile and cannot cope with any complexity or make decisions.

Abnormal level of activity: the individual may be relatively inactive, being unresponsive and slow moving, it does not utilise all the captive environment. It may even fail to interact with conspecifics. There is a general lack of curiosity and a failure to react to novel objects. In contrast, hyperactivity may develop where the animal carries out pointless activities such as pacing, fidgeting, or repeatedly running round the cage (see stereotypies above). The animal may be excessively alert and react to the slightest stimulus.

Feeding: captive primates may overeat or overdrink and often become obese. Another vice of captivity is coprophagy; wild primates have no contact with their faeces, which usually just fall to the ground. Thus, they have not evolved toilet behaviour and faeces may be investigated and accidentally or deliberately ingested; a similar situation may develop with regard to eating vomitus. Anorexia may also be a problem in some captive primates.

Abnormal social behaviour: there are a very large range of social behaviours which may be adversely affected by the captive environment,
these will not be described in detail. They include hyperaggressiveness,
often resulting from the inability of the loser to escape from an
aggressor, juvenility even as an adult, abnormalities of sexual and
reproductive behaviour, abnormal mother-infant interactions which include
neglect of the young and sometimes, infanticide by a mother. Often
abnormalities of social behaviour are developmental, resulting from
an unnatural early social environment (8).
Restricted behavioural repertoire: in captivity elements of the natural
repertoire may be missing or rarely observed. This may result from
the restrictions imposed by the environment or may be symptomatic of
a generally low level of activity.

4. IMPROVING THE CAPTIVE ENVIRONMENT
4.1 Methods
A number of strategies have been utilised to improve the environment
of captive primates so that they can optimise their behavioural reper-
toires and eliminate abnormal behaviours.
Naturalistic manipulations aim to simulate the wild situation, by the
provision of natural materials. This approach has its limitations
in the laboratory, where space is always restricted, however, the animal
should be provided with a home cage large enough to carry out normal
locomotory behaviour, such as leaping, running. climbing and jumping
from perch to perch, if this is not practicable then the primate should
have regular access to a large exercise cage (9). The use of natural
materials for cage inclusions, such as wooden perches or a woodchip
substrate, are also beneficial (10).
Developmental abnormalities can be avoided by providing the young animal
with as natural a social environment as possible (8). Maternal deprivation is particularly serious and early weaning should be avoided.
Young baboons spend the first 4 months of their life dependent on their
mothers and remain semi-dependent until the age of 18 months (11) so
that early separation from the mother should be avoided to ensure that
normal young are produced. Marmosets and tamarins need to remain in
the parental group to acquire experience of carrying and caring for
infants; this ensures that they make adequate parents themselves.
Unless it is not possible, for specific reasons, non-human primates
should be kept with compatible individuals of their own species and
experiments designed to avoid social isolation. This is both advantageous to the wellbeing of the animal and also results in better science.
The advantage of a social companion lies in the fact that it provides
a complex and relatively unpredictable set of stimuli with which the
individual can interact, it also allows the animal to employ the normal
tactile behaviour towards a conspecific (eg grooming, sitting in contact,
playful behaviour).
While it is important to be aware that a compatible social grouping
in captivity may differ considerably from that usually found in natural
conditions, it is important to be familiar with the natural sociobiology
of the species. Two examples of the application of this knowledge
will be given. Firstly Erwin (12) found that harems of pig-tailed
macaques given two rooms showed more aggression and bullying than those
in a single room. The reason appears to be that the dominant male
can only control aggression among females if they are in view. All-female
groups showed more fighting than those with a dominant male. The second
example is the Hamadryas baboon at Emmen and Rhenen zoos. This species

has an unusual social structure for a baboon or macaque, with male/female bonded harems. In both zoos males with a group of females were found to steal and abuse infants. This arose because the male had been placed with a group of strange females instead of selecting them individually and building up a bond with each one. As a result he had no control over the females which refused to be herded and as a consequence he behaved abnormally to infants.

The primate's physical environment can also be enriched by the addition of devices which directly increase the animal's opportunity to carry out more complex or time consuming behaviour.

The usual method of placing food in the cage at a readily accessible site eliminates the need for an animal to carry out foraging behaviour which, in the wild ranges from 11-55% of the time during daylight hours (13). Devices which increase the time captive primates spend foraging have proved highly effective. Artificial termite mounds for great apes extend the animals' behavioural repertoire and occupy their time (14). Using a woodchip substrate to the cage and scattering small food items in it increase foraging time and reduce aggression (10). Other methods of increasing foraging time are the provision of live insect prey or artificial gum trees for common marmosets (15) and puzzle boxes which deliver pellets if the animal manipulates them correctly.

Objects or devices which can be manipulated provide novelty and stimulation (1), but are most effective when linked to a food reward. Such devices include reaction timers, where the animal is rewarded for a faster reaction to a human player or computer, and reward-delivering machines of the Skinner box type, even noughts and crosses (tic tac toe) has been used with great apes. Quite elaborate learned tasks can readily be incorporated into the animal's behavioural repertoire, great apes have been trained to obtain tokens from a machine with which they can 'buy' food from a slot machine. These arrangements help to keep the animals active and interested even though they bear little resemblance to the wild situation. One note of warning must be sounded, namely that an animal may become obsessed with a reward-producing device and spend a great deal of time using it repetitively. It is therefore important that the device should only be 'on' for restricted periods of time and that there should be a signal which tells the animal when the device is activated.

Contact with the caretaker is extremely important and should include handling and training primates to co-operate in procedures, such as weighing, injections and being moved from place to place. Where the experiment necessitates social isolation, the caretaker can play a vital role as a substitute companion.

It is generally easier to keep the smaller, more primitive monkeys in enriched environments. Common marmosets are a good example and they behave normally when kept in a small cage 1m^3 with five other members of a family group, even though, in the wild, they range over a hectare. Some species are particularly unsuitable in the laboratory because of their needs for space and complexity, for example, brachiating species such as, gibbons. The close genetic relationship between ourselves and the pongidae makes these animals unsuitable for laboratory experiments on ethical grounds. According to Benveniste and Todaro (16) human and chimpanzee have 99.8% of their genes in common, which makes us as closely related to a chimpanzee as a horse is to an ass.

4.2 Assessing the effectiveness of environmental manipulation

The best way to assess the effectiveness of environmental improvements is by recording the animal's behaviour before, after a change and several months later. If possible the animal should be observed on days when the improvement is accessible and days when it is not. Any or all of the following behavioural changes can be taken to be indicative of improvement:

An increase in the animal's behavioural repertoire (ie a larger number of different types of behaviour). Clearly this must not include aversive and fearful behaviour.

A reduction in the frequencies of abnormal and stereotyped behaviour (although the latter may be difficult to eliminate, if they have been long-established).

Changes in activity pattern. Where an individual has been apathetic and inactive, greater activity should be shown, on the other hand hyper-activity (such as pacing and restlessness) should be reduced, and overall activity may actually decrease. Levels of activity may be compared with those recorded from the field and diurnal wild primates spend 23-70% of the time during daylight hours in resting (13).

Better utilisation of the environment, such as greater use of available cage space.

5. CONCLUSIONS

If Primates are to be kept in laboratories, they present unique problems. They are highly active, arboreal animals and because of their high level of intelligence they suffer if they do not have the opportunity to carry out a wide range of adaptive behaviours. They actively seek stimulation and, if this is missing, are prone to develop abnormal behaviour patterns.

Every effort should be made to enrich the environment of non-human primates in a way which utilises their behavioural capabilities and enables them to interact and have greater control over their situation. The effectiveness of such improvements should be assessed scientifically and care must be taken to ensure that the ensuing behaviour is not abnormal.

All those who keep non-human primates in the laboratory should consider the biology and behavioural repertoire of the species and, on the basis of this knowledge, seek to optimise their social environment. Where it is essential to use a primate model in biomedical research, preference should be given to the small, monogamous and more productive species such as the common marmoset because it is relatively easy to provide these animals with an adequate environment social and physical in the laboratory.

REFERENCES

1. Markowitz H: Behavioral Enrichment in the Zoo, Van Nostrand Reinhold, New York, 1982.
2. Richard AF: Primates in Nature, WH Freeman & Co, New York, 1985.
3. ILAR: Guide for the care and use of laboratory animals, National Institutes of Health Publication, No 80-23, Animal Resources Programme, Bethesda, Maryland, 1985.
4. European Convention for the protection of Vertebrate Animals used for Experimental and other Scientific Purposes, Appendix A: Guidelines on Accommodation and Care, Report No 123, HMSO, London, 1985.
5. Royal Society/UFAW: Guidelines on the care of laboratory animals and their use for scientific purposes, Pt 1 Guidelines on housing and care of laboratory animals, preprint July 1987, UFAW, Potters Bar, 1987.
6. Erwin J, Deni R: Strangers in a strange land : abnormal behaviours or abnormal environment? In: Captivity and behaviour (eds) J Erwin, TL Maple, G Mitchell, Van Nostrand Reinhold, New York, 1979.
7. Cronin GM: The development and significance of abnormal stereotyped behaviours in tethered sows, University of Wageningen, Wageningen, 1985.
8. Mason WA: Early socialization. In: Primates : the road to self-sustaining populations, (ed) K Benirschke pp 321-329, Springer-Verlag, New York, 1986.
9. Hearn JP: Marmosets and tamarins. In: The UFAW Handbook on the care and management of laboratory animals. (ed) TB Poole, Longman Scientific and Technical, Harlow, 1987.
10. Anderson JR, Chamove AS: Allowing captive primates to forage. In: Standards in laboratory animal management, UFAW, Potters Bar pp 253-256, 1984.
11. Altmann J: Baboon mothers and infants. Harvard University Press, Cambridge, USA, 1980.
12. Erwin J: Environments for captive propagation of primates : Interaction of social and physical factors. In: Primates : The road to self-sustaining populations. (ed) K Benirschke pp 297-305 Springer-Verlag, New York, 1986.
13. Herbers JM: Time resources and laziness in animals, Oekologia 49, 252-262, 1981.
14. McEwen P: Environmental enrichment : an artificial termite mound for orang utans, UFAW, Potters Bar, 1986.
15. McGrew WC, Brennan JA, Russell J: An artificial "gum tree" for marmosets (Callithrix j. jacchus), Zoo Biology 5 : 45-50, 1986.
16. Benveniste RE, Todaro GJ: Evolution of type C viral genes : evidence for an Asian origin of man, Nature 261, 101-108, 1976.

A.C. Beynen & H.A. Solleveld (Eds), New developments in biosciences: their implications for laboratory animal science, ISBN 978-94-010-7973-0

HOUSING AND WELFARE OF LABORATORY RODENTS

G. CLOUGH

1. INTRODUCTION
The aim of this paper is to review the published recommendations for the environmental requirements of laboratory rodents. An attempt is made to assess whether or not these recommendations are likely to give rise to discomfort or lack of well-being.

2. REVIEW OF ENVIRONMENTAL FACTORS
In view of the tendency towards the standardisation of housing conditions for laboratory rodents it is interesting to see that different countries' Guidelines for the levels at which the environmental factors concerned should be set still vary considerably. Table 1 gives a summary of the current recommendations for rats and mice as published in America, Canada, Germany, Sweden, the United Kingdom, and the twenty or so other European countries represented by the Council of Europe (1,2,3).

TABLE 1. Summary of the range of published environmental recommendations for rat and mouse rooms from America, Canada, Council of Europe (23 countries) Germany, Sweden, U.K., UFAW(UK).

FACTOR	SPECIES	RANGE	TOLERANCE WITHIN THE RANGE
TEMPERATURE (°C)	Mouse	18–26	±1.5 to ±4.0
	Rat	18–26	±2.0 to ±4.0
RELATIVE	Mouse	40–70	±5.0 to ±15.0
HUMIDITY (%)	Rat	40–70	±2.5 to ±15.0
LIGHT Intensity(lx)	Both	<60 to 400	Uniform to ±50
L:D (Hours)	Both	12:12 to 14:10	Not specified
SOUND	Both	ca.50dBA &' ≯NR45 to <85dB	Not specified
VENTILATION	Mouse	8–20	Not specified
RATE (ac/hr)	Rat	10–20	Not specified

Although the data imply that mice are less tolerant to variations in temperature and more tolerant to variations in H than are rats, and that the minimum ventilation rate for mice is lower than that for rats, there is no evidence that this is the case. It is suggested that these apparent differences have arisen through historical 'accident' related to the ambient conditions prevailing in the countries of origin of the various recommendations.
    Closer examination of the publications referred to will show that most of the authorites do warn that with the room light intensities given, albino

strains housed in the upper cages in a room are likely to suffer retinal damage. The majority also mention that the ventilation system should be 'draught free' – though only Sweden support this by stating that linear air speed in the region of the animal cages should not exceed 0.1m/s. There is little agreement on sound levels and considerable confusion over what these should be. This confusion is demonstrated by the higher recommendation for permitted sound levels quoted in Table 1 as ">85dB" which is the figure given in both the American and Canadian guidelines; though neither authority give the source of their information, it almost certainly derives from a paper by Anthony in 1963 (4) who recommended that noise levels should not exceed 85dBA.

2.1. Differences between the macro- and micro-environments. All the published guidelines give recommendations for the environmental conditions that it is thought should be maintained in animal rooms. It is quite clear, however, that the welfare of rodents living in cages is more closely related to the cage environment than to that of the room itself. There is now considerable evidence (1,2) that even in rooms with apparently good control systems, the differences between the macro- and micro-environments can be very great. This is shown by the data in Table 2 (1,5)

TABLE 2. Summary of measured differences between room and cage conditions.

FACTOR	ROOM LEVEL	CAGE LEVEL	% VARIATION(Room=100)
TEMPERATURE (°C)	21	22 to 27	+2 to +29
LIGHT Intensity(lx)	350	2 to 250	-30 to -99
VENTILATION RATE(ac/hr)	26	14 to 24	-8 to -53
AMMONIA LEVEL (ppm)	<10	0 to >250	0 to +>2000
5μ PARTICLE COUNT (/1 air)	3.5	2058 to 18282	+588 to +>500000

From this it can be seen that whereas temperature, ammonia levels and particle counts are often higher in cages than in the room, light intensities and ventilation rates therein are likely to be lower. It is worth emphasising that the differences reported were all found in cages and rooms being maintained according to current recommendations and in which the recorded room conditions appeared to be quite stable.

2.2. Relationship between room conditions and the welfare of the animals. Knowing that large differences such as those shown above can occur, let us now consider whether such conditions are likely to interfere with the animals' welfare.

2.2.1. Temperature. Detailed studies in Japan have shown for both rats (6) and mice (7) that a wide range of reproductive, haematological and other physiological parameters do not differ significantly over two generations if the room temperature remains within the range 16-26°C. From those results it is reasonable to conclude that none of the current temperature recommendations are likely to result in poor welfare for those species.

2.2.2. Relative humidity. Because laboratory rodents do not sweat significantly, ambient RH assumes increasing importance in relation to their ability to lose heat as the dry bulb temperature rises. Reference to Table 1 shows that some current recommendations do, at the upper extreme, allow an RH of 70% at 26°C. In a recent pilot study of RH in relation to mouse breeding performance at 21°C, it was found that the pre-weaning mortality of BALB/c mice was 45% at around 40%RH and 21% at around 70%RH (8). Clearly neither of these figures is acceptable and it seems likely that the situation would be worse if the temperature was raised to 26°C. Other than this preliminary evidence, however, there is no new information since this matter was last reviewed (1). The conclusion at that time was that current guidelines seem reasonable and are unlikely to cause great problems to the animals. However, as high RH encourages ammonia production which in turn increases the animals' susceptibility to respiratory infection (9), it is certainly not good for the welfare of the animals. It is perhaps partly for this reason that several authorities give warning to 'avoid 40% and 70% RH for extended periods'. The RH of air is, of course, closely related to its ability to take up water and, as shown in Table 3, this subject is very relevant to the ventilation rates required in a room.

2.2.3. <u>Ventilation rates</u>. Animals living in cages have no choice but to obtain all the air they breathe from their immediate surroundings. Using a currently recommended cage size, mouse stocking densitiy and environmental specifications for a fully stocked mouse room, the data in Table 3 show just how quickly the water content of the cage air can rise to the permissable upper limit of RH; also shown is the corresponding air change rate that is required to ensure that the RH does not rise above 70%.

TABLE 3. Ventilation rates required to maintain cage air within the extremes of RH shown in Table 1 (ie 40-70%)

SITUATION: 3 X 25g mice in 2.4 1 air (Type I cage) @ 18°C/40%RH
Time mice take to produce enough water to raise RH to 70% @ 26°C:
          Respiratory water only - 10mins (= 6 ac/hr)
          Total water production - 1.88mins (= 31 ac/hr)
SITUATION: 3 X 25g mice in 2.4 1 air (Type I cage) @ 19°C/40%RH
Time mice take to produce enough water to raise RH to 70% @ 23°C:
          Respiratory water only - 7mins (= 8.5 ac/hr)
          Total water production - 1.33mins (= 45 ac/hr)

This does show that the recommended ventilation rates for rodent rooms (8-20 air changes per hour) are certainly not excessive. This is particularly true when it is known that many commonly used air distribution systems have efficiencies as low as 30-60% (2,3). It should be emphasised that the figures used to calculate the data for Table 3 only take into account the water contributed to the air by the animals; with the addition of water from other sources such as personnel, floor washing etc., even higher air change rates would be required. It is for this reason it is worth repeating that specifying room air change rates probably has little relevance to the ventilation rate achieved within the cages. From a practical point of view, whilst specifying cage ventilation rates is probably unrealistic, it is certainly not advisable to specify a room ventilation rate without simultaneously giving some indication of the efficiency that the air distribution system must achieve. The latest UK guidelines that are to be implemented under the new legislation go some way towards this by stipulating that the required air change rates must be "... distributed throughout the room".

2.2.4. <u>Lighting</u>. Most of the recommendations for lighting intensities do now take into account the sensitivity of the eyes of albino rodents to light levels above 60 lux. This is reflected in the general lowering of the levels indicated in guidelines current in 1984 (2) and those published more recently (1,3), though care will still be needed to protect animals in the upper cages in a room, particularly when transparent cages are in use. There is no evidence that the recommendations for the control of photoperiod are likely to affect rodents adversely - though it is important to note that even very short flashes of light (1 second or less) occurring during the dark phase of the L:D cycle should be avoided as they can result in significant physiological changes in photoperiodically sensitive species such as the hamster (10). Similarly, though little work has been done to determine the possible effects of different wavelengths of light on animals, what evidence there is supports the view that neither fluorescent nor tungsten lights are likely to cause significant adverse effects.

2.2.5. Sound. The relevance of sound to laboratory animals has been reviewed (11) as has its importance in relation to animal welfare and experimental results (9); the following are a few examples of known effects on rodents: − physical damage to the cochlea
                    − hypertension
                    − changes in body weight
                    − changes in immune response and tumour resistance
                    − effects on reproductive function
                    − cannibalism
                    − audiogenic seizures
                    − audioconditioning related to atypical drug responses
                    − changes in blood chemistry and cellular distribution
For present purposes it is necessary to note that although laboratory rodents cannot hear low frequency sounds, they can hear (and indeed produce and communicate by) high frequency sounds in the range 10-70kHz; remember that man regards any sounds of around 18-20kHz and above as 'ultra-sound' as they are of too high a pitch to be heard. In practice this means that situations that may appear silent to a person, can be very noisy and stressful to small rodents. The importance of this fact is emphasised in a recent publication entitled 'Sources of ultrasound in a laboratory' and Table 4 gives a few examples of those sources which the authors regard as of probable significance.

TABLE 4. Examples of sources of ultrasound of probable significance. (Items marked '*' Produce no sound audible to man)

SOURCE	INTENSITY(dB)	FREQUENCY RANGE(kHz)
Stopper rotated in glassware	127	4 − 160
Squeaky swivel chair	103	2 − 160
Vacuum hose	102	3 − 140
Tap running into sink	98	2 − 250
Computer monitor (Colour)*	70	16 − 160
Computer monitor (Monochrome)*	61	16 − 94

They stress the potential hazards associated with devices such as computer monitors which appear silent to man because they do give rise to significant levels of ultrasound within the range 30-70kHz to which rodent species are particularly sensitive (13). They also stress the potentially hazardous nature of the ultrasonic components of the other sounds listed as well as dripping taps, telephone bells and squeaky door closers; squeaking wheels on rodent cage racks have also been found to give rise to sounds as high as 100dBA which included frequencies in the range 40-60kHz (1). From these reports it is clear that ambient ultrasound is probably not uncommon in laboratories and animal houses and that it occurs at frequencies and levels which give cause for concern.

The most specific recommendation concerning noise is that in the draft Guidelines issued by the Royal Society and UFAW in the UK (1). This states not only that in the empty room the background sound level should be 'about

50dBA' (as previously), but also that it must not exceed Noise Rating Curve (NRC) 45. The aim of this extra condition is to exclude sounds with a strong component at one particular frequency. This may seem like a step in the right direction, but only insofar as it indicates some awareness of the problem of defining sounds in a meaningful way. In practical terms the highest octave band used in specifying NR curves has a mid frequency of only 8kHz (1) to which small rodents are relatively insensitive. In other words, like the dBA scale, the Noise Rating system is based entirely on the auditory responses of the human ear and so is equally irrelevant to rodents. As the sound frequencies to which rodents are most sensitive are ultrasonic it follows that high frequency sounds are likely to be more 'damaging' to them than sounds in the human audible range; indeed it is known that some mice suffer audiogenic seizures in response to purely ultrasonic sounds (9). Until further information about animals' responses to sounds becomes available, it is recommended that all housing guidelines should include a warning of the sensitivity of laboratory animals to high frequencies, with the added advice that the known sources of such sounds should be used with due consideration and awareness of the potential problems they may cause.

2.2.6. <u>Conclusions</u>. Table 5 summarises the present situation. From that it can be seen that the currently recommended levels of room temperature and RH are probably satisfactory; those for ventilation rates and lighting conditions, however, need to be applied with care because of the large differences which can so easily be created between the micro- and macro-environments. Recommendations for sound levels, based as they are on the human ear, are largely irrelevant to animals and it is this factor which is most likely to unknowingly interfere with their behaviour and welfare.

TABLE 5. Relationship between published environmental guidelines for and the welfare of laboratory rats and mice

FACTOR	RANGE	RELEVANCE TO WELFARE
TEMPERATURE	22±4°C	Unlikely to have significant adverse effects
RELATIVE HUMIDITY	55±15%	As above if achieved in cages; 55±10% (already adopted by some) allows greater margin for removal of water from cages.
VENTILATION RATE	8-20 ac/hr	<15 in fully stocked rooms probably acceptable provided it is associated with efficient air distribution system.
LIGHT		
Intensity	60-400lx	350-400 satisfactory for staff working; care needed to avoid damage to eyes of albinos in upper cages
Photoperiod (L:D)	12:12 to 16:08	Unlikely to have significant adverse effects.
Wavelength		Lack of information on wavelength effects but no evidence that daylight-type fluorescent or tungsten lights have adverse effects.
SOUND	ca.50dBA & NRC 45 to <85dB	Current guidelines all related to human ear function and are irrelevant to animal hearing. This factor is the one most likely to give rise to discomfort and lack of well-being in rodents

REFERENCES

1. Clough G: Are the published recommendations for the environmental requirements of common laboratory animals compatible with their comfort and well-being? (Paper presented at the ICLAS-CEMIB-FESBE Regional Scientific Meeting, Brazil, November 1986) In press.
2. Clough G: Environmental factors in relation to the comfort and well-being of laboratory rats and mice. Proc Symp LASA/UFAW: 'Standards in Laboratory Animal Management'; 7-24. UFAW, 8 Hamilton Close, South Mimms, Potters Bar, Hertfordshire, EN6 3QD, UK. ISBN 0 900767 36 7, 1984
3. Clough G: The Animal House: Design, equipment and environmental control. Ch.8 in "The UFAW Handbook on the Care and Management of Laboratory Animals", 6th ed. Longman Group (UK) Ltd, 1987.
4. Anthony A: Criteria for acoustics in animal housing. Lab Anim Care, 13, 340-350, 1963
5. Raynor TH Steinhagen WH & Hamm TE Jnr: Differences in the microenvironment of a polycarbonate caging system: bedding vs raised wire floors. Laboratory Animals 17; 85-89, 1983
6. Yamauchi C, Fujita S, Obara T, Ueda T: Effects of room temperature on reproduction, body and organ weights, food and water intakes and haematology in rats. Lab Anim Sci, 31; 251-258, 1981
7. Yamauchi C, Fujita S, Obara T, Ueda T: Effects of room temperature on reproduction, body and organ weights, food and water intakes and haematology in mice. Lab Anim Sci, 32; 1-11, 1983
8. Donnelly HT: Personal communication, 1987
9. Clough G: Environmental Effects on Animals used in Biomedical Research. Biol Rev 57; 487-523, 1982
10. Ellis DH & Follett BK: Gonadotrophin secretion and testicular function in Golden hamsters exposed to skeleton photoperiod with ultra-short light pulses. Biol Reprod, 29; 805-818, 1983.
11. Gamble MR: Sound and its significance for laboratory animals. Biol Rev 57; 395-421, 1982
12. Sales GD & Wilson KJ: Sources of ultrasound in a laboratory. Report of a UFAW Vacation Scholarship. UFAW, 8 Hamilton Close, South Mimms, Potters Bar, Hertfordshire, EN6 3QD, UK.
13. Brown AM: High levels of responsiveness from the inferior colliculus of rodents at ultrasonic frequencies. J Comp Physiol Biochem, 6; 1-73, 1973

A.C. Beynen & H.A. Solleveld (Eds), New developments in biosciences: their implications for laboratory animal science, ISBN 978-94-010-7973-0

HOUSING CONDITIONS AND EXPERIMENTAL RESULTS

WOLF H. WEIHE

INTRODUCTION
The relationship between experimental results and the housing of animals has several aspects. I shall deal with three aspects and present a concept for housing directed not only to the researcher but also to those who decide on the licences for research.
The first aspect concerns the health of the animals. According to the definition of the World Health Organization human health is not only the absence of disease but also a state of total physical and mental wellbeing. Laboratory animal science focussed for a long time on the eradication of diseases in animals, while only a small group of scientists was concerned with welfare. Recently welfare and well-being of the animal have gained wide attention as part of the animal protection movement. Well-being now has a higher priority than the absence of disease (8).
The second aspect concerns the fact that "The requirement for the housing of experimental animals depends largely upon the nature of the experiments being performed" (7). There is an incongruence between the national and international guidelines for the care of animals and the requirements for their use in experiments. Guidelines for housing laboratory animals are chiefly based on practical experiences in a few countries where many animal laboratories exist. They are generalizations without specified consideration of the physiology of the various animal species and requirements for particular experiments (16). The less understanding there is of the nature of a particular experiment the more emphasis is put on generalized principles. There is a tendency among officials of the licencing bureaus and committees on animal protection and bioethics to refer to a general denominator for housing of animals in experiments often referred to a standardization.
The third aspect concerns the variability of biological parameters in experiments. Ten years ago Gärtner et al. (12) stated that the reduction of variability of biological parameters in experimental animals is the principal aim of laboratory animal science. This led to a "method-oriented-research" concentrating on methodological sophistication and losing track of the question being researched. It has now been accepted that even under highly standardized conditions of housing an "intangible variation" of morphological and physiological parameters persists (11). The problem is to identify the permissible range of variation of parameters and the influence of housing on the intangible variation.

THE CATEGORIES OF VARIABLES
An experiment on animals is based on four categories of variables:
1. Housing, 2. Experimental animal, 3. Experimental measure, and
4. Result, i.e. response.

The variables of category 2 are intrinsic, the variables of categories 1 and 3 are extrinsic. The results of an experiment are always expressions of the regulation responses of the animal coping with the intrinsic and extrinsic stimuli. The multiple regulating mechanisms to maintain a balanced homeostasis were often neglected in the past by seeing the animals as a "measuring instrument" that could be adjusted by keeping it under restricted conditions. Behaviorism has supported this way of thinking because "the strict behaviorist, such as Skinner, sees no need to speculate about the inner workings of the organism" (13). The argument was that the variance of the results is affected by the variance of the variables within the categories housing, experimental animal and experimental measure. Consequently to control the variance of the result a control of the variables affecting it was demanded. The simple conclusion was that in practice the smaller the number of elements or factors and their variance within the categories housing and experimental animal the narrower the variance of the responses to the measure of the experiment would be. In the early period of laboratory animal science reports on variation of results with change of housing conditions was automatically interpreted as proving a causal relationship. There is no reason to infer that a causation follows from an established relation or correlation as long as this is not experimentally verified. For general application a list of agreed on conditions was developed known as standardization which provided a practical basis for the national guidelines for the care of animals.

The variability of responses of the animal are determined by its physical condition and the genom. It was thought that the intangible variance within the genom could be reduced and controlled through breeding for homozygosis. Using inbred mice for testing pentobarbital McLaren and Michie (21) observed an increased variance of the sleeping time of mice instead of the expected reduction of variance compared to outbred strains. Today the genom of mice can be characterized by genetic monitoring with much precision which has helped to trace the contribution of the genom to the total variance of responses.

THE CATEGORIES OF EXPERIMENTS
There are two categories of experiments: 1. basic investigations, and 2. biological testing.

Guidelines are directed towards the requirements for consistent results in biological testing but do not primarily consider the requirements of basic research or the well-being of the animals. Results of basic investigations are expected to be gained under natural realistic conditions. They should be conclusive, reliable, significant, and reproducible. The participating intrinsic and extrinsic influences on a result should be investigated in additional experiments. Quantitative precision with a small variance is not requested in every instance. On the contrary a wide variance can be of heuristic value and may lead to unforeseen discoveries about the interaction of operative factors and regulation mechanisms. Emphasis is on a realistic experimental set up while the amount of variance of results is of secondary importance. A narrow uniform variance of results is requested only for biological testing which is limited to selected well known regulation mechanisms or established animal models. Emphasis is on restricted or even unrealistic model situations selected for the purpose of susceptibility to the stimulus. The latter is mandatory and has therefore a higher priority than the well-being of the animal.

The problem in the design of an experiment is whether emphasis should be on the search for and characterization of responses to known and unknown stimuli, as it is in basic research, or on the precision of known responses, as in biological testing. The art in the design of an experiment is to stabilize the regulation pathways in such a way that they are as unaffected as possible by stimuli from the environment apart from those selected as the measure or procedure as the sole variable. In planning an experiment the researcher should ask himself whether his experiment belongs to the category of basic research or to biological testing or whether it is between the two. Unless the category of the experiment is clearly understood at the onset the stipulations for the housing conditions of the animals for the particular experiment may be inappropriate.

HOUSING AS ECOSYSTEM

Housing may be defined as an artificial separation of a space to provide attennuated conditions with elimination of unwanted factors from the natural environment. Housing includes husbandry.

The variables of the category housing have attracted the attention of researchers in a very inconsistent fashion: some such as temperature have been overemphasized because they can be measured easily, others have been ignored because monitoring is complicated or expensive or they have not been considered important. Lang and Vesell (17) noticed that in eight leading biological journals only the species was mentioned in all articles gender in 45.5%, animal density in 8.0%, and photoperiod in 6.9%. When Clough (6) eight years later did a comparable screening he noticed that now gender was mentioned in 95%, animal density in 32% and photoperiod in 42%.

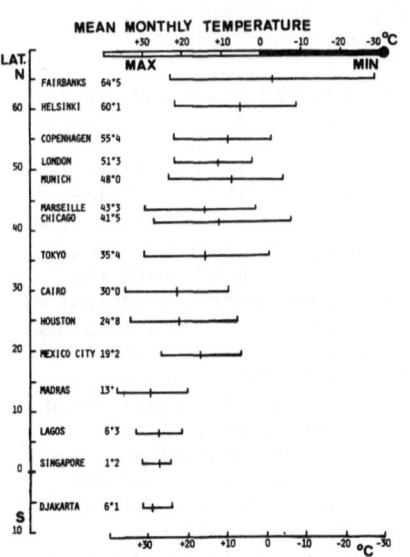

Fig. 1. Mean maximum and minimum monthly temperature range and mean annual temperature (vertical line) of university cities on a global scale.

The variables of housing can be separated according to the four sections of the housing ecosystem:
1. physical variables, 2. chemical variables, 3. biological variables, 4. psychosocial variables.
Compared to the natural ecosystem the sum total of variables of the housing ecosystem is restricted and partly controllable. The housing ecosystem is by no means a truly closed system within the natural ecosystem because there are energy exchanges, leaks, and carriers such as man which maintain links between the outside and inside. In the biological sector of the ecosystem it has been possible to exclude factors such as the pathogenic or all microorganisms, but in the other sectors a comparable separation has not been so easy to achieve. Examples for physical variables are air temperature, humidity, motion and pressure, radiation, light, noise and the living space given by the cage. Chemical variables are food, water, air and pollutants, including ammonia and disinfectants. Biological variables are viruses, bacteria, parasites and other living organisms. Psychosocial variables are gender, groupe size, species mixing, handling and attendants including the researcher. These elementary variables are diversified in a great number of factors. Temperature for

example can be defined as mean daily temperature, daily maximum or minimum, daily range, length of defined deviation from the mean and others. On a global scale there is a wide range of mean monthly temperature and mean annual temperature depending on latitude as illustrated for some university cities in Fig. 1. This has not been considered in the present guidelines which advocate room temperatures that are unrealistic for hot climates.

EFFECTS OF HOUSING

Housing thus appears as a sum of operative factors each of which has a characteristic variation with a specific controllability. Any elimination of variables or restriction to a fixed value of the natural variation of a factor leads to a modification of the animal through adaptation. The principal question is how much variation of the housing factors is permissible without interferring with the result and how much variation is necessary to maintain the health and well-being of the animal.

The responses of the animal to the variables in the ecosystem may be classified in five types according to their degree of strength (Table 1).

Table 1. Types of response according to strength of stimulus.

Type	Feature of Regulation Mechanism
Neutralization	Suppression, receptor down regulation, deprivation
Modulation	Transitory shifts, sensitization, desensitization
Modification	Continual shift, adaptation, habituation,
Interference	Alteration of system function
Destruction	Damage of function and structure

Examples for neutralization are neutral temperature, reduction and deprivation of sustaining factors such as light intensity, length of photoperiod, anaerobic gut flora in gnotobiotic animals and individual isolation. Neutral conditions for one or more regulation mechanisms lead to their suppression which can be seen as an aberation from health with a resulting overshooting response to stimulation. Modulation of regulation follows the impact of zeitgeber such as light and noise, feeding, handling and cleaning. It involves the activity of the neuroendocrine system of which the pineal gland with the rhythmic production of melatonin under the control of the sympathetic system has an important role in immune and stress responses (1,19,24,26,29) and the regulation of adrenergic receptors (9). Recent research has revealed the great importance of the enkephalins and endorphins of the endogenous opioid system as modulators (23,25). Modification is the process and the result of adaptation to a changed condition to relieve homeostasis with establishment of efficiency of regulation to every kind of variable. Best known examples are heat or cold, light, handling scheme or nutrients. Modulation and modification are reversible with different time constants noticeable by a change of the functional range. Interference is the failing of a regulation mechanism due to an infection, a toxin, malnutrition, mutilation or deprivation of a vital factor such as a vitamin. The strongest effect is expressed as destruction when units of regulation systems are damaged leading to their collapse with subsequent failing of dependent and related regulation mechanisms ultimately ending in the breakdown of regulation and death. Examples are intolerable conditions of extreme heat or cold, thirst, starvation, neglect, toxic concentrations of pollutants or devastating diseases such as infections and cancer.

As housing permits the elimination of environmental factors many efforts were made to reduce their number in the assumption that with each factor eliminated the variation of responses would decrease. The environment was emptied of as many factors as appeared dispensable leaving only the bare conditions required for survival. This was particularly favored in exper-imental psychology by introducing the Skinner box (13,18). For some time the aspects of housing were determined by economic considerations including working convenience for the caretakers: e.g. keeping rats on wire mesh to facilitate cleaning and reducing cage size to place more animals per room. The increase of ammonia concentration in the air was given much attention though it may simply serve as an indicator of the need for cleaning some cages more often than others (15). For the essential life sustaining fac-tors the aim was and still is to control them with a minimum of variation assuming that any variation of these factors might cause an increase of the variaton of the dependent responses.

THREE AND FOUR-VARIABLE EXPERIMENTS

The responses to environmental factors are investigated in three-variable experiments consisting of the variables housing, animal and result. In three-variable experiments commonly the animal as variable is defined and standardized and the factors of housing varied. The results of such experi-ments in basic and environmental physiology and psychology provide the principle knowledge in anatomy, physiology and regulation of environment-body interraction which is used in setting up four-variable model exper-iments. The crucial point is that regulation is always modulated and modi-fied by the housing conditions and thus the particular housing condition influences the result in any case. For many four-variable experiments in basic research it is unknown to what extent these influences are important and interfere with the response to the measure. To arrive at a full under-standing of the variation of a response to the variation of a measure the effects of the other two categories of variables have to be investigated consecutively. Working under standardized conditons of housing and animal it must be concluded that the result of a four-variable experiment will be valid only under these defined conditions. They may well be highly artifi-cial for the animal and certainly not a reliable reflection of the whole range of conditions to which an animal is exposed in the natural ecosystem. Two examples may be given here to illustrate the limitations of standard-ization in basic research. In Fig.2 a combination of two types of a four-variable experiment is presented. The toxicity of chlorpromazine was tested in mice at various ambient temperatures (3). The dosage of the pharmacon and the ambient temperature were varied. The toxicity between 13 and 38°C ranged from 11 mg/kg at 13°C to 350 mg/kg at 28°C. The variance of each toxicity test was small but the variance of toxicity due to the change of housing conditions exceeded 300 times. A similar experiment with pro-caine over an even wider temperature range resulted in a small difference of toxicity (27). For biological testing it would be reasonable to demand only one temperature for

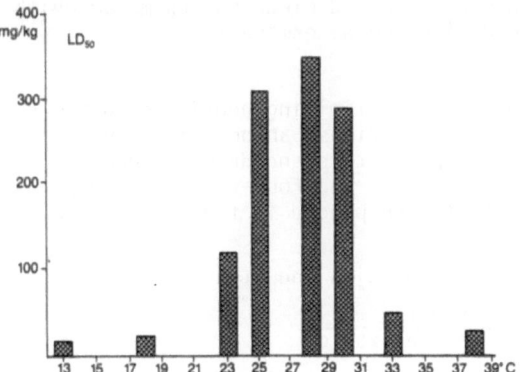

Fig. 2. Relationship between ambient temperature and toxicity of chlorpromazine in mice (3).

housing to facilitate comparability with other pharmaca. For basic research concerned with the effects of the pharmacon the change of toxicity with the variance of housing temperature initiated further research to investigate the mechanism of the decline of regulation capacity above and below $28^{\circ}C$ as expressed by the increasing toxicity (Fig.2). The mechanism of action of chlorpromazine with interferring and destructive side effects in the CNS and the periphery was slowly revealed while the drug was used to investigate the mechanisms of temperature regulation (4).

Nerem et al (22) produced atherosclerosis in two groups of rabbits by feeding a diet enriched with 2% cholesterol. In one group the rabbits were managed with the usual basic care in feeding and cage cleaning, in the other group each rabbit was "handled, stroked and talked to" and visited several times daily to establish a psychosocial relationship with the investigator. Sudanophilia was 60% less in the handled group than in the unattended group. The authors wondered "if no attention was given to social environment in the experimental protocol, what was the real effect observed?" Clearly for a four-variable experiment in which a variation of both measure and housing is applied, the decision on the housing conditions cannot be made in advance. One non-stressful condition can be provided and then research proceeds by systematically investigating the effects of factors with a possible influence in a multi-dimensional fashion changing the strength and duration of the factors.

CONDITIONS OF HOUSING

From the qualification of the five types of effects of housing (Table 1) some rules can be derived which form the basis for the choice of conditions for stimulated but unstrained homeostasis favoring well-being. The aim is to work out a balance of conditions which provide enough stress to stimulate regulation but not too much to cause strain. Physiological regulation including behavior should not be restricted but allowed to operate freely so that regulation can be expressed naturally. Variance of results need not be the determining yardstick. The main rule is to provide conditions for housing that eliminate interferring and destructive influences such as infections, pollutants, toxins, dangers, accidents, novelty, uncertainty, aggressors or extreme physical factors and reduce but not eliminate the intensity of modulating and modifying vital conditions such as climatic factors, nutrition, handling, group and cage size and cleaning scheme on which the animals depend. How can these conditions be determined?

REGULATION MAPPING

After the concept of reducing variance through limiting housing conditions had failed new ways had to be tried. The conditions can be worked out by mapping responses of measured biological parameters and drawing zones separated for the five types of effects from three- and four-variable experiments (Table 2). The five types of effects are placed in five regulation zones.

Table 2. Relationship between extent of regulation and effect.

Regulation Zone	Effect
Minimum regulation	Neutralization
Stimulated regulation	
Desirable condition	Modification
Permissible condition	Adaptation
Undesirable condition	Interference
Failing regulation	Destruction

The aim is to determine for each factor the range of desirable and permissible conditions within the zone of stimulated regulation by mapping the responses to varying the conditions of individual housing factors.
The approach is demonstrated in a simplified model of temperature regulation. In Fig. 3 metabolic rate of the rat is correlated with ambient temperature (14). There is a minimum metabolic rate of the fasting rat from 28

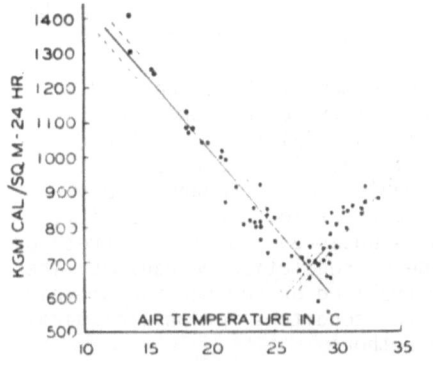

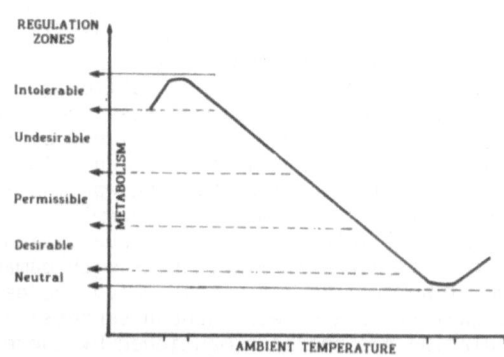

Fig. 3. Relationship between
ambient temperatures and
metabolism in rats (14).

Fig. 4. Model for regulation zones
in the temperature-metabolism
diagram of rodents.

to 30°C known as the thermoneutral zone. With decreasing ambient temperature the animal responds with increasing metabolic rate up to a minimum temperature when regulation fails, known as the lower tolerance temperature. The same is observed at the upper tolerance temperature. The zones of Table 2 are then inserted for the range between the neutral temperature and the lower and upper tolerance temperatures (Fig.4).
According to the model in Fig.4 neutral temperature in the zone of minimum regulation is not desirable because it forces restriction on the regulation mechanisms. The desirable zone is determined by a range of ambient temperature considered to be a favorable stimulus to maintain balanced regulation during feeding and change of day and night (5). Outside this zone as under natural climatic conditions an effort for adjustment is required and adaptation established as long as the conditions do not exceed the adaptive capacity. When the strength of the stimulus is valued by the effort required to cope with it the word stress is commonly used. Any housing factor can be a potential stress by straining homeostasis. In quantitative terms this can be seen in the example of Fig.4. The strength of a factor on the abscissa is expressed by the change of the distance between the minimum and maximum response on the ordinate. The stimulus approaches the strength of a stress causing strain with increasing distance from the range of the zone of desirable conditons. The regulating capacity of the animal must be taken into account which is high in the young adult and declines with advancing age (20) or under the influence of experimental measures.
If for a given factor a wide range on the abscissa was tested but little difference of response observed on the ordinate it may be concluded that this is not a stress factor (2). This may well be because the chosen range of variation of the investigated factor was not wide enough.
High light intensity for night-active animal species such as rats would be comparable with unwanted neutrality conditions because it would suppress

motor activity and related regulation mechanisms while low light intensity would increase them. The change from desirable to undesirable light intensities remains to be demonstrated experimentally. Every housing factor is a potential stress because, as Paracelsus has observed, it is the quantity which makes the poison. A special example of regulation mapping is the bacterial contamination of animals. Germ-free animals are living in the neutralization zone because they are deprived of important factors such as anaerobic bacteria for enteral metabolism, SPF animals would belong to the zone of desirable conditions and animals infested with pathogens to the zone of undesirable conditions.

Regulation mapping is the application of environmental physiology and psychology for practical purposes. The rule of introducing zones for factors in housing has been mentioned by Yamauchi (28) for laboratory animals and Ewbank (10) for farm animals. Yamauchi listed four zones: optimum, goal, recommendable and worst allowable limits. Ewbank speaks of physiological stress meaning a harmless stimulus of low level intensity, medium level intensity at which adaptation is observed and of overstress or distress at high level intensity with damage of regulation mechanisms. The zone of neutralization within which the neutralized or reduced factor gains the quality of a stress through supressing the conditons for physiological regulation has not been mentioned by these authors.

CONCLUSION

The mapping of regulation responses to varying housing factors may be evaluated separately for the two categories of experiments. The conditions for the care of animals in basic research should be such as to warrant a truly healthy animal with respect to absence of disease and well-being. Such housing conditions are collectively within the desirable range of the zone of stimulated regulation. For biological testing neutral conditions should be provided for the selected regulation mechanism so that the maximum range of reaction capacity between the zone of minimum regulation and the zone of failing regulation can be utilized (Fig.4) Such conditions of housing which are specifically determined by the selected regulation mechanism are not necessarily identical with those for well-being. Health of the animal is desirable but not mandatory as long as the sensitivity of response is high and the variance small.

For basic research housing conditions should be designed to meet the requirements of the experiment within the desirable or permissible range of stimulated regulation for the welfare of the animal. The aims for suitable conditions for the welfare of the animal and for the experiment in basic research are concurrent, not divergent. Consideration of the welfare of the animal ensures that a realistic not a distorted animal model is used. The presently available guidelines for the housing of experimental animals need to be reviewed under this aspect because they have been worked out at a time when requirements for biological testing were given more weight than those for basic research.

REFERENCES

1. Andrews RV and Belknap RW 1985: Metabolic and thermoregulatory effects of photoperiod and melatonin on Peromyscus maniculatus acclimatization. Comp.Biochem.Physiol.A 82: 725-729.

2. Baumans V, Stafleu FR and Bouw J 1987: Testing housing system for mice - the value of a preference test. Z.Versuchstierk. 29: 9-14.

3.Berti C and Cima L 1954: Einfluss der Temperatur auf die pharmakologische Wirkung des Chlorpromazins. Arzneimittel.Forsch. 5: 73-74.

4.Borbely AA and Loeppe-Hinkkanen M 1979: Phenothiazines. In: Body Temperature, Regulation, Drug Effects and Therapeutic Implications. P Lomax and E Schönbaum (ed), Dekker, New York, 403-426.

5.Büttner D 1987: Influence of strain specific patterns of locomotor activity on the daily pattern, minimal, mean and maximal oxygen consumption in the rat. Z.Versuchstierk. 20: 121-128.

6.Clough G 1982: Environmental effects on animals used in biomedical research. Biol.Rev. 57: 487-523.

7.Clough G 1987: The animal house: design, equipment and environmental control. In: The UFAW Handbook on the Care and Management of Laboratory Animals. T Poole (ed), Longman Group, Harlow, p.112.

8.Curtis SE 1985: What constitutes animal well-being? In: Animal Stress. GP Moberg (ed), Amer.Physiol.Soc., Bethesda, Maryland, 1-14.

9.Davies AO and Lefkowitz RJ 1981: Regulation of adrenergic receptors. In: Receptor Regulation. RJ Lefkowitz (ed), Chapman-Hall, London, 84-121.

10.Ewbank R 1985: Behavioral responses to stress in farm animals. In: Animal Stress. GP Moberg (ed), Amer.Physiol.Soc., Bethesda, 71-80.

11.Gärtner K 1985: Versuchstierkunde und "intangible variance" - eine dritte Komponente der kontinuierlichen Variabilität neben Erbgut und Umwelt. Verh.Dtsch.Zool.Ges. 78: 61-75.

12.Gärtner K, Bube P, Flamme A, Peters K and Pfaff J 1976: Komponenten biologischer Variabilität und die Grenzen ihrer Manipulierbarkeit. Z.Versuchstierk. 18: 146-158.

13.Giannelli MA 1986: Three blind mice, see how they run: a critique of behavioral research with animals. In: Advances in Animal Welfare Science 109-164.

14.Herrington LP 1940: The heat regulation of small laboratory animals at various environmental temperature Amer.J.Physiol. 129: 123-139.

15.Hirsjärvi PA and Väliaho TU 1987: Microclimate in two types of rat cages. Lab.Animals 21: 95-98.

16.Irving III GW 1985: Regulations and guidelines for animal care: problems and future concerns. In: Animal Stress. GP Moberg (ed), Amer.Physiol. Soc., Bethesda, Maryland, 281-296.

17.Lang CM and Vesell ES 1976: Environmental and genetic factors affecting laboratory animals: impact on biomedical research. Fed.Proc. 35: 1123-4.

18.Lockwood R 1986: Anthropomorphism is not a four-letter word . In: Advances in Animal Welfare Science 1985. MW Fox and LD Mickley (eds). Martinus Nijhoff Publ., Dordrecht, 185-199.

19.Lynch HJ and Deng MH 1986: Pineal responses to stress. J.Neural.Transm. (Suppl) 21: 461-473.

20.McCarty R 1986: Age-related alterations in sympathetic-adrenal medullary responses to stress. Gerontology 32: 172-183.

21.McLaren A and Michie D 1954: Are inbred strains suitable for bio-assays? Nature (Lond), 173: 686-687.

22. Nerem RM, Levesque MJ and Cornhill JF 1980: Social environment as a factor in diet-induced atherosclerosis. Science, 208: 1475-1476.

23. Plotnikoff NP, Faith RE, Murgo AJ and Good RA (eds) 1986: Enkephalins and Endorphins. Stress and the Immune System. Plenum Press, New York.

24. Reiter RJ 1986: Normal patterns of melatonin levels in the pineal gland and body fluids of humans and experimental animals. J.Neural.Transm. Suppl) 21: 35-54.

25. Rodgers RJ and Hendrie CA 1984: On the role of endogenous opoid mechanisms in offence, defense and nociception. Prog.Clin.Biol.Res. 167: 27-41.

26. Seggie J, Campbell L, Brown GM and Grota LJ 1985: Melatonin and N-acetylserotonin stress responses: effects of type of stimulation and housing conditions. J.Pineal Res. 2: 39-49.

27. Sievers RF and McIntyre AR 1937: The effect of temperature upon their toxicity of procaine for white mice. J.Pharmacol. 59: 90-92.

28. Yamauchi C 1985: Environmental Control of Laboratory Animals. Japanese with English insertions. Tokyo.

29. Yocca FD and Friedman E 1984: Effect of immobilization stress on rat pineal beta-adrenergic receptor-mediated function. J.Neurochem. 42: 1427-1432.

A.C. Beynen & H.A. Solleveld (Eds), New developments in biosciences: their implications for laboratory animal science, ISBN 978-94-010-7973-0
© 1988 Martinus Nijhoff Publishers, Dordrecht.

BASIS OF THE EUROPEAN GUIDELINES: FACTS OR INTUITION?

Dr. P. de Greeve

INTRODUCTION

In its endeavor to harmonize European legislation since 1961 the Council of
Europe has drawn up a number of international conventions in the field of
animal welfare.
"The Convention for the Protection of Vertebrate Animals used for experimental and other scientific purposes" - Convention number 123 - was opend
for signature on the 18th March 1986 (1). At this moment eight Member
States and the European Communities have signed it. So, 16 countries have
obliged themselves to give effect to the provisions of the Convention by
translating them into national legislation.
The Convention for the Protection of Vertebrate Animals used for experimental and other scientific purposes is designed primarely to restrict both
the number of experiments and the number of animals used (a) by imposing
strict conditions on procedures, (b) by exploiting better the results of
procedures and (c) by encouraging the development of alternative methods.
The second objective is to benefit the welfare of animals as far as it is
compatible with the purpose of the procedure. This second objective has
been set out in article 5 with a reference to Appendix A. This Appendix
contains a set of guidelines which are aimed to be of help in the implementation of the provisions of the Convention.
On 24 november 1986 an EC Directive was adopted by the Council of Environmental Ministers of the European Communities. This directive includes the
same guidelines.
This paper consists of the following parts:
After a short outline of the origin, the content of the guidelines and
their status, it will deal with the question whether the guidelines will be
able to meet the requirements. Then, information on the implementation of
the guidelines in The Netherlands will be given. Finally, some recommendations will be made.

THE EUROPEAN GUIDELINES

Following Recommendation 621 dated 21 January 1971 the Assembly recommended
that the Committee of Ministers establish a committee. Its terms of reference should be to draft international legislation including the conditions
under which and the scientific grounds on which experiments with live animals could be authorised.
This committee, since 1977 known as the Ad Hoc Committee of Experts for the
Protection of Animals (CAHPA) consists of senior civil servants, often as-
sisted by specialists. The CAHPA meetings were attended by observers from
some non-member States and from organizations, both governmental (EC) and
non-governmental (World Society for the Protection of Animals, Federation
of Veterinarians in Europe and other organizations).

It deals with all issues on animal welfare, such as the slaughter of animals, international transport of animals, pet animals.
CAHPA started its work on the elaboration of the draft convention at its third meeting (January 1978).
During 10 plenary meetings the draft convention was prepared.

Looking back I will refer to two issues: a. the origin and the content and b. the status of the guidelines.
a. The origin and content.
CAHPA's negotiations have led to the conclusion that for his own well-being, man sometimes has to make use of animals, but that he has the moral obligation to ensure that the animal's health and well-being is protected as much as possible.
At its fourth meeting (october 1978) CAHPA agreed to take a preliminary draft presented by Mr. George Vallier, the French representative and then chairman of CAHPA, as a basis for its further work which had among others the following wording:
  "Every animal to be used in an experiment shall in accordance with its
  physiological and ethological needs, be provided with accommodation, an
  environment, which permits freedom of movement and food and care
  appropiate to its health and well-being".
The draft was considerably improved on proposals of the second chairman Mr. G.P. Pratt (U.K.) in the meeting of a working group in January 1981 at Paris.
At the eighth meeting it was proposed to adopt provisions for the accommodation and care of animals. These provisions should be similar to the provisions in the European Convention for the Protection of Animals kept for Farming purposes. The Committee, however, felt it better to supply the convention with detailed recommended standards in an Appendix, which had to be elaborated. It was agreed that a German "Gutachten" as well as the U.S. guidelines for the care and use of laboratory animals prepared by the Committee on care and use of Laboratory Animals would constitute a useful basis for the elaboration of European standards. The Norwegian delegation was ready to study this question in more detail. On the fourth of May 1981 our distinguished collegue Dr. Stian Erichsen submitted a draft for Appendix A called: "Recommended Standards for the Accommodation and Care". During two consecutive meetings the draft was elaborated by a working group. The experts were unanimous in recognising that the text prepared by Dr. Erichsen, who in the meantime was elected as the third president of CAHPA, was in accordance with their wishes, especially with regard to the proposed structure. The experts thought, however, that the document would benefit by being shortened; indeed this document would very likely be used by persons having sufficient expertise in the field of accommodation and care of animals. In addition, it was the intention to mention a bibliographical documentation in the Appendix, for further detailed information. In CAHPA drafting Document 81/24 seven different Guides are mentioned, all containing information on the design of animal facilities and the daily practice in the animal house. I would like to draw your attention to two of these Guides.
The first one is the publication of the Society for Laboratory Animal Science titled "Recommendations concerning the Planning, Structure and Construction of Animal Facilities for Institutes Performing Animal Experiments", edited in 1980. These recommendations were translated from the revised third German edition of the so-called "Gutachten" published in 1980 (2). The revisions were the result of both discussion stimulated by earlier

editions and experience gained since 1971.

In the closing remarks of the first edition of the German "Gutachten" of 1977 the need for a scientific evaluation of empirically established data had already been stressed. Sixteen German experts in the field of labora- tory animal science had emphasized in this Gutachten the necessity to draw up morphological and physiological parameters to evaluate housing condi- tions. This work still has to be done.

The second Guide to be mentioned is the Guide to the care and use of ex- perimental animals of the Canadian Council on Animal Care (CCAC)(3). This guide consists of two volumes forming a continuing basis of the CCAC as- sessment program. Volume 1 edited in 1980 and reprinted in 1984 is made up of ten chapters dealing with all important issues of animal care, such as: the responsibility for the care of the experimental animal, laboratory ani- mal care, standards for experimental procedures. In Appendix XIX of this Guide general reference works, handbooks and journals dealing with the care and use of experimental animals are listed. The second volume, edited in 1984 deals with the care and use of representative classes of animals.

The reader will be impressed by the extent of the information. Indeed, a wonderful and useful Guide.

Unfortunately, the definitive text of the Appendix A of the Convention does not include any additional bibliographical information. In my opinion with this additional information the guidelines have gained in strength and in appreciation.

The guidelines contain tables and figures.

The TABLES indicate minimum requirements for the keeping of animals. They originally were proposed by Dr. Erichsen. It must be emphasized that they are based on current standards from various sources such as the Gesell- schaft für Versuchstierkunde and the Institute of Animal Resources (ILAR). It must also be underlined that such minimum standards are mainly the re- sult of long practical experience and not of systematic scientific re- search.

So, the 9th CAHPA meeting recognized that it was difficult to fix figures for dimensions and sizes of accommodations due to the large variety in body size, age, etc.

The FIGURES give the minimal required space for one animal as well as for groups of animals.

In a meeting in November 1981 the Working group considered proposals from Mr. P. Svendsen from Denmark. This method presents some real advantages in particular from an economic point of view, because one can continue to utilize current cages by reducing the number of animals per cage.

Following the proposals of Mr. Svendsen it was necessary of course to es- tablish the optimal standards of animals per cage and species.

New proposals for the tables and figures were established in november 1981. On 29 April 1983 the draft Convention, including the Appendix A was sub- mitted to the Committee of Ministers. In the course of the examination of the draft a number of amendments were proposed, including some minor cor- rections to certain tables and figures relating to Appendix A.

As decided by the Committee of Ministers a meeting of technical experts chaired by Dr. Erichsen took place on 7 March to prepare the final examin- ation of the Convention by the Committee of Ministers.

On 31 May 1985 the Committee of Ministers adopted the Convention as sub- mitted by CAHPA.

b. The status of the Guidelines of Appendix A has been frequently discussed.
At the seventh meeting (April 1980) it was argued that standards for the
accommodation of animals should be elaborated but it would be not practical
to fix mandatory rules, which should be part of the convention itself. It
was therefore decided that such standards should be included in an Appendix
to the Convention, which might be detailed. The obligatory minimum remained
fixed as follows:
"Recommended standards on accommodation and care are given in Appendix A".
In Januari 1981 it was again concluded that it was understood that the
standards should have no mandatory character and the Drafting Group put the
last sentence of this article between brackets:
("Recommended standards on accommodation and care are given in Appendix").
In the next meeting the CAHPA replaced the title of the Appendix A "Recom-
mendated Standards for the Accommodation and Care" by the following wording: "Guidelines on Accommodation and Care"; it was agreed that the words
Recommended Standards should be replaced by the word Guidelines throughout
the whole text of the Appendix, including the tables. The experts felt that
the word "Guidelines" would sound less mandatory than the words "Recommended Standards" and would offer a greater flexibility and thus permit a
more efficient application.
Animal protection organizations were tenacious in their pursuit to
strengthen the guidelines, because in their opinion guidelines without a
binding nature render the Convention totally useless.
During the meeting of March 1985 the CAHPA discussed this matter again.
To emphasize the objective of the Appendix A and to better justify its position in the draft the experts reworded the last sentence of article 5:
   "In the implementation of this provision, regard should be paid to the
   guidelines for the accommodation and care set out in Appendix A to
   this Convention".

CAHPA found it necessary to make clear the status of these guidelines with
an introductory note in the final draft:
   " Unlike the provisions of the Convention itself, they are not mandatory: they are recommendations to be used with descretion, designed as
   a guidance to the practices and standards which all concerned should
   conscientiously strive to achieve. It is for that reason that the term
   "should" has had to be used throughout the text even where "must"
   might seem to be more the appropriate word. For example it is self-
   evident that food and water must be provided."

DISCUSSION

Referring to the title of this paper it is clear that the guidelines are
based on knowledge as the result of long experience and secondly that they
are not based on systematic scientific research starting from the needs of
the animal.
With respect to the general care and accommodation mentioned in article 5
the guidelines deal with the basic requirements of laboratory animals: the
physical facilities, the environment in the holding room and its control
are well described. The guidelines will be of great help in designing appropriate animal quarters.
However, only few critical and objective data on space requirements for
animals are available. Comparision of guidelines reveal remarkable diffe-

rences leading Fox to the conclusion that "the process whereby cage size requirements are determined is illogical and unscientific" (4). Nevertheless, caging systems based on successful experience and professional judgement must be utilized untill results of a scientific evaluation of husbandry systems are available.

Laboratory scientists are bound to pay utmost attention to the care of laboratory animals for two reasons:
a. for humane reasons: man has ethical responsibilities towards animals used in experiments and
b. for scientific reasons: research with bad, inappropriate animals i.e. non standardized or stressed animals is poor research!
The influences of housing conditions on the animal as well as on the experiment are well documented but scattered in the literature. In the previous paper a very interesting survey has been given. Nevertheless, I will draw your attention to a publication of Riley (5).
Discussing the influence of environmental stress on tumour development Riley prod·iced arguments in support of a low stress environment for animal housing. He mentioned nine factors most essential for low-stress animal housing:
- no recirculation of noxious air that has been in previous contact with animals;
- partial sound proving of the animal storage shelves;
- elimination of animal room vibrations and high pitched sounds of centrifuges, vacuum cleaner ventilation fans and other noisy laboratory equipment.
- eliminations of draught, air turbulence, and wind-tunneleffects;
- precise light control to establish circadian rythms and to regulate light intensity exposure;
- segregation of males and females with respect to transmissible odours, pheromones, and other stress inducing signals;
- segregation of experimental animals that are experiencing stress from normal or control animals;
- introduction of special minimum stress animal handling techniques and cage cleaning procedures and
- avoidance of draughty, uncomfortable and stressful wire bottomed cages.
Riley made a strong plea for a systematic approach of low-stress animal housing.
In addition I want to draw your attention to the publication of Dr. Michael Fox: "Laboratory Animal Husbandry: Ethology, Welfare and Experimental Variables" (4). This book should be added to the list of handbooks that students and furture scientists are obliged to read, because it contains a lot of information on aspects of housing conditions and animal care in experimental procedures.

A critical survey of housing systems currently in use will clearly indicate that by designing restricted environments secondary consideration is given to the needs of laboratory animals because they conflict with economic and other factors.
According to Wallace (6) in the design of cages and other restricted environments a list of known and suspected animal needs should be compared with a list of human requirements, i.e for husbandry and experimentation.
Designing a new mouse cage - presently well-known as the Cambridge cage - she listed mouse needs and human requirements (table 1).

MOUSE NEEDS	HUMAN REQUIREMENTS

+ activity
+ eating
+ drinking
+ sleeping
+ defecating
+ urinating
+ nesting
	= in relation to animal
	  + confinement
	  + productivity
	  + health
	= in relation to the cage
	  + hygiene
	  + cost
	  + comfort for the handler
	  + design should be adaptable

Table 1: Mouse needs and human requirements

Starting point in determining the mouse's needs was: "the mouse's wishes and convenience as deduced from behaviour studies".

Each of the needs and requirements she described shortly, for instance:

"Activity: a living space permitting exploration, exercise, grooming and social interaction where territory can be marked, containing material providing sensory stimulation and adaptable for sleeping and nesting. Dry ventilated and cooler than animal's body temperature"

and one of the most important human requirements:

"Cost: materials and their manufacturing must be cheap. The design must be easy to mass-produce with a minimum of hand labor. The parts must be durable in use washing, storing, assembly and handling".

Comparing these two listings Wallace designed the Cambridge cage, which now has been in use for over twenty years. Preference tests clearly demonstrated the preference of mice for the Cambrigde cage in relation to other cages presently used (7).

According to Wallace this sort of thinking should underlie the design of all restricting environments for animals in the 80's. However, in 1987 still exist restricted confinements in which the animal cannot make normal postural adjustments such as stretching out his legs in a natural resting position. According to the spirit of the Convention laboratory animal scientists are bound to eliminate these housing systems.

As a result of the growing concern for the welfare of animals the behavioural and psychological well-being of laboratory animals has begun to receive increasing attention during the last decades. The study of the welfare of animals has grown into a new science: the animal welfare science. Papers dealing with scientific and ethical aspects of animal welfare are well published in the annual series "Advances in Animal Welfare Science". Untill now three volumes have been edited. They give an interesting survey of topics and areas of investigation of this young, interdisciplinary science.

In an elaborate contribution in the 1984 edition Wemelsfelder (8) deals with the question whether it is possible to study animal well-being scientifically. Discussing basic needs of animals she describes categories of behavioural activities being considered to correspond with the most basic needs of animals:

+ eating and drinking behaviour   + social behaviour
+ explorative behaviour and locomotion   + sleeping behaviour
+ play behaviour

Wemelsfelder concludes that an environment with too little stimuli and a (fundamental) lack of the possibility of self-expression deprives human beings and animals. Environmental deprivation, physical and social, result in deprived, bored animals. The environment can be improved by providing toys such as sticks and balls for cats, dogs, rabbits and other laboratory animals. For non human primates anti-boredom devices have been developed such as the grape board and the honey-fishing board which have reached a considerable degree of sophistication (9). Provisions like ropes, bars and perches are also appropriate for climbing. Meeting need of search behaviour in zoo's behavioural engineering is used as an aid in the maintanance of healthy animals: the animal has to work for its food instead of being fed on regular times (10). Feeding primates a mixture of tasty corn (grain) on the bottom of their cage will give the animals the opportunity to collect their food. The well-being of the animals concerned has improved considerably.

Most laboratory animals such as mice, rats, dogs and monkeys are social animals. For such animals a social environment is of crucial importance. Some of the possible kinds of social deprivation are early weaning, prevention of the development of a parent offspring bond, the keeping of animals in single sex groups or in isolation.

Some effects of social deprivation are higher mortality, increased aggression, displacement behaviour, distorted behaviour.

The keeping of animals in isolation needs special consideration not only for ethical reasons but also for scientific reasons. This is because being caged separately could inadvertently introduce a number of variables, which could question the validity of any research conclusions derived from such animals. Isolation stress appears to evoke heightened metabolisme and adrenocortical hyper-activity.

This year in the serie World Animal Science a book has been published under the title "Laboratory Animals". It is meant as a comprehensive volume on a number of aspects of laboratory animal science (11). The editors, Ruitenberg and Peters, have expressed the hope that this would inspire more interest in laboratory animal science and particularly assist in creating the proper attitude towards the use of animals. In chapter 2.3.8 discussing the aspects of caging, however, it is put forward that "larger primates - i.e. larger than squirrel monkeys - that are to be used in experimental procedures should be single caged".

With respect to the known distressing effects of social isolation on monkeys this statement needs further explanation.

Militzer concludes that a judgement of the suitablity of husbandry systems for laboratory animals should be based on an evaluation of the behavioural activities of the animals concerned (12). For instance the frequency of play and resting behaviour is a degree for the appropriateness of a husbandry system. Knowledge of the repertoire of behaviour the animal displays is a first requisite.

Brummer (13) made an attemp to determine the status of well-being and non well-being of rabbits. Present knowledge of the behaviour of wild rabbits and observations obtained from domestic rabbits served as standards. Two interesting remarks of Brummer with respect to guidelines of the Council of Europe are:

1. The biological behaviour of the single-kept domestic rabbit best approximates that of the wild rabbit and
2. With respect to cage size the minimal recommended cage floor for breeding animals heavier than 2,5 kg is 6400 cm^2; this is approximately 2,5 times the European standard and over four times the current German standard.

The findings of Brummer stress the urge for scientifically evaluation of the guidelines!

DISCUSSION AND IMPLEMENTATION OF THE GUIDELINES IN THE NETHERLANDS

In The Netherlands the Convention has received a critical acceptance.
At this moment its provisions are being implemented into daily practice. The guidelines of Appendix A are not to be adopted straight away. The Dutch Society of Laboratory Animal Science has set up four working groups to elaborate guidelines for the housing of small rodents, guinea pigs, rabbits, dogs and cats. Their reports will be sent to the government and are aimed to help in the implementation of the guidelines.
The following examples indicate the discussion which is going on.
1. MICE
   - There is a discussion about the optimal frequency of cage cleaning: once a fortnight or two times a week. The difference in opinion stems from the question what - from an ethological point of view - stresses mice more: the handling by the caretaker or a 'dirty' micro-environment. The influence of contact bedding such as wood shavings in controlling ammonia build-up has been recently outlined (14).
   - In the University of Groningen research is performed on mice originally obtained from the wild. While visiting the animal facility the Inspectorate noticed that the number of mice per cage was not in accordance with the guidelines. On instruction of the Inspectorate the number of animals was reduced. However, as a result of the increased area the mice started fighting (territorial behaviour). After consultation with the Inspectorate the former situation was re-established and the fighting stopped.
2. RATS
   - According to the Standard Operation Procedure of a farmacological experiment groups of three rats weighting approximately 200 grams are housed in standard cages type III with floor area of 750 cm^2. During the experimental period the rats gain weight to 400 till 500 gram. According to figure 9 of the Appendix A the minimum floor area required for these rats is 950 cm^2. However, such a cage is not available. According to the Standard Operation Procedure the rats have to remain together as an experimental group. Using a cage type IV (1815 cm^2) would cost more and require more space. As suggested by Schlingmann, a Dutch animal technician, the floor area for these rats has been enlarged by uniting two cages type III connected by a perspex tube giving the rats the opportunity to use two cages (15). To activate the rats food and water can be supplied on different racks. The system has proven to be useful. On request technical details can be obtained from Mr. Schlingmann.
3. DOGS
   - Dogs are social animals adapted for running; they have a strong sense of territory. These qualities are probably among the most important basic factors to be considered in providing satisfactory housing for dogs. The most important need for dogs is contact with other dogs or with humans. Deprivation of social contact may be even more inhumane than deprivation of space and exercise.
   Dogs should not be caged unless required by the procedure. Caged dogs should be exercised at least once a day. This can be achieved by walking the dog on a leash or releasing the dog from its cage into the corridor of the animal room. Under circumstances which allow for close

personal attention to dogs it is easier to obtain employees who are fond of animals. Such persons rarely accept or keep a job where the dogs are in constant stress through unnecessarily close confinement.
- In a report of the Dutch Inspectorate "The housing of laboratory dogs" a survey is given of the present literature. Licence holders are obliged by the Inspectorate to accommodate their dogs according the guidelines of the Convention.
- As in other countries such as Switzerland there is no concensus in the Netherlands between experts on the use of grid floor for the keeping of dogs. A working group consisting of persons experienced in the keeping of dogs are discussing the pros and cons of grid floor versus solid floor. For sanitation and cleaning reasons grid floors are prefered in many laboratory facilities. According to table 8 of the guidelines for housing dogs in pens grid floors should not be used unless the procedure requires it. The opinion is growing that conditions should be made upon the quality of grid and that only approved grids should be used. The use of purpose bred dogs accustomed to the accommodation on grid floor will strongly be underlined.

4. PRIMATES
- On the initiative of the Veterinary Inspectorate a group of experts discussed the various aspects of housing Macaca's. The recommendations of the working group have been adopted by the Inspectorate. Fortunately, all institutes have agreed upon adjustment of inadequate confinements before 1 January 1990.
- Based on behavioural studies in the wild as well in zoo's and following the example set by several institutes the Primate Centre in Rijswijk has set up a housing system for their breeding colonies of Macaca's in the open air. The environment consists of natural ground cover, an elevated climbing structure and a heated sleeping shelter. Starting this new method of housing the ethologist in charge had to convince several animal caretakers, that Macaca's do not suffer from the Dutch climate. Indeed, they do not suffer even in the winter period, for such climatic conditions are also found in their natural habitat.
At this moment the condition of the Macaca's in terms of well-being is considered to be good.
In this way ethology findings are applied to the care of laboratory animals.
- Boot et al. (16) investigated the influence of housing conditions on the pregnancy outcome in Macaca fascicularis. Female monkeys were kept under 3 different housing conditions: individually in type A cage (45 x 45 x 60cm), individually in type B cages (70 x 70 x 100cm) and as couples in type B cages. More succesful pregnancies (90%) were recorded for females housed individually in type B cages than for females housed individually in type A cages. This and other findings lead the author to the conclusion that housing conditions can have a profound influence on reproductive success in cynomolgus monkeys.

CONCLUSION

1. The Convention is a step forward to the protection of laboratory animals in Europe. The guidelines stimulate the discussion on the welfare of the laboratory animal. The Convention is just a starting point...!

2. It is clear that in many aspects the recommendations mentioned in Appen-

dix A are minimum requirements. The guidelines of the Convention will have relatively little effect in those countries having stricter national regulations. Member states have to be aware that the establishment of national guidelines looser than those mentioned in the Appendix A, cannot be justified.

3. There is a need for more information on the well-being of laboratory animals on the basis of behavioural studies.

RECOMMENDATIONS

1. Housing systems as well as handling procedures are to be evaluated to determine the response of the animal to different conditions of care using the behaviour as an indicator of the well-being of the animals.

2. It is the task of laboratory animal scientists to demand the application of ethology into the care of the laboratory animal.

3. Despite legislation and regulations the motivation for improvement of the welfare of the laboratory animal must primarily come from within the individual scientist and all persons in charge of the daily care. Therefore education of future scientists in respect to the attitude towards animals is utmost important. Students should be taught the ethical side of using animals in experiments early in their training. They have to be taught to recognize signs of discomfort, such as distress or pain.
To improve the practice of the daily animal care the Canadian Council on Animal Care formulated eleven principles of animal care in 1980 to be fixed at the wall on several places in the animal laboratory. This stimulating example was followed in 1984 by the Swiss Academy of Medical Sciences and the Swiss Academy of Sciences editing a likewise leaflet. In the Netherlands similar recommendations are in preparation.

4. A FELASA-Workinggroup on the evaluation of housing systems should be established to update the guidelines in order to broaden and improve the protection of animals, that are to be used in scientific research considered essential for human interest.

ACKNOWLEDGEMENT

Mr. E. Ausems of the general secretary of the Council of Europe in Strassbourg informed me about the history of article 5 and the development of Appendix A, for which I am very grateful.

REFERENCES

1. European Convention No. 123: European convention for the protection of vertebrate animals used for experimental and other scientific purposes, 1986. Council of Europe, Section des Publications, Strassbourg.
2. Recommendations concerning the Planning, Structure and Construction of Animal Facilities for institutes performing animal experiments, 1980, Society of Laboratory Animal Science (Gesellschaft für Versuchstierkunde), Basle.
3. Guide to the care and use of experimental animals, 1980. Canadian Council on Animal Care (CCAC).
4. Fox, M.W., 1986. Laboratory Animal Husbandry, State University of New

York Press.

5. Riley, V. 1981. Psychoneuroendocrine influences on immunocompetence and neoplasia. Science 212, 1100-1109.

6. Wallace, M.W., 1982. Some thoughts on the laboratory cage design process. Int. J. Stud. Anim. Prob. 3: 234-42

7. Baumans, V., Stafleu, F.R. and Bouw, J., 1987. Testing housing system for mice - the value of a preference test. Z. Versuchstierkd. 29, 9-14.

8. Wemelsfelder, F., 1984. Boredom in animals. In: Advances in Animal welfare Science, eds. Fox, M.W. and Mickley, L. 115-154. The Netherlands: Martinus Nijhoff.

9. Goodall, J., 1979. Anti-boredom Devices for Primates. In: Comfortable Quarters for Laboratory animals, Animal Welfare Institute, Washington.

10. Schmidt, M.J. and Markowitz, H. 1979. Behavioral engineering, In: Comfortable Quarters for Laboratory animals, Animal Welfare Institute, Washington.

11. Solleveld, H.A., McAnulty, P., Ford, J., Peters, P.W.J. and Tesh, J. (1986). Breeding, housing and care of laboratory animals: p. 6. In: Laboratory Animals: Laboratory animal models for domestic production, World Animal Science C2, eds. Ruitenberg, E.J. and Peters, P.W.J.. Elsevier Science Publishers B.V. Amsterdam, Oxford, New York, Tokyo.

12. Militzer, K., 1986. Beurteilung der Tiergerechtheit von Haltungssystemen für Labortiere. In: Wege zur Beurteilung tiergerechtiger Haltung bei Labor-, Zoo- und Haustieren. Schriftenreihe Versuchstierkunde 12. Verlag Paul Parey, Berlin und Hamburg.

13. Brummer, H., 1986. Symptome des Wohlbefindes und des Unwohlbefindes beim Kaninchen unter besonderer Berücksichtigung der Ethopathien. In: Wege zur Beurteilung tiergerechtiger Haltung bei Labor-, Zoo- und Haustieren. Schriftenreihe Versuchstierkunde 12. Verlag Paul Parey, Berlin und Hamburg.

14. Hisjärvi, P.A. and Väliaho, T.U. 1987. Microclimate in two types of rat cages, Lab. Animals (1987) 21, 95-98.

15. Schlingmann, F., 1986. Modificatie bestaande dierkooien teneinde te voldoen aan de door de Raad van Europa gestelde richtlijnen, Biotechniek 25, 65-67.

16. Boot, R., Leussink, A.B. and Vlug, R.F., 1985. Influence of housing conditions on pregnancy outcome in cynemolgus monkeys (Macaca fascicularis) Lab. Anim. 19, 42-47.

A.C. Beynen & H.A. Solleveld (Eds), New developments in biosciences: their implications for laboratory animal science, ISBN 978-94-010-7973-0
© 1988 Martinus Nijhoff Publishers, Dordrecht.

GNOTOBIOTIC ANIMALS IN NUTRITION RESEARCH

B. RATCLIFFE
The Polytechnic of North London, Holloway Road, London N7 8DB, UK

1. INTRODUCTION

At the beginning of this century, Metchnikoff (1907) speculated that certain microorganisms of the intestinal tract and their 'putrefactive' activities were potentially deleterious. He believed that their harmful effects could be interrupted or suppressed by the consumption of foods such as yoghurt. His observations of the longevity of peasants in Eastern Europe who regularly consumed large amounts of yoghurt, added weight to this.

Although his ideas seem somewhat misguided now, Metchnikoff was amongst the first to realise that intestinal microorganisms might be of importance to health and that their activity depended to some extent on dietary intake Since that time, there has been an increasing awareness of the involvement of the microflora in the functioning of the alimentary canal. In fact, current dietary recommendations are the antithesis of Metchnikoff's and now dietary fibre is consumed in quantity to promote the 'putrefactive' activities of bacteria in the large intestine.

Investigations of the interactions between diet and the gut microflora of man are limited by ethical considerations and the inaccessibility of the gastro-intestinal tract. Generally, studies have been confined to the examination of faecal material or to the use of patients with partially resected intestines (ie. colostomy, ileostomy).

The use of animal models permits access to all parts of the gut, either *post mortem* or by surgical cannulation. Gnotobiotic animals, in particular, may be used to facilitate studies of the intestinal tract under germ-free (GF) conditions or when associated with one or more microbial species.

In 1885, Pasteur formulated ideas for a fundamental study of the relationships between diet and intestinal microorganisms using GF animals. By the turn of the century, GF animals had been successfully derived and maintained for short periods of time (Nuttal & Thierfelder, 1896; Schottelius, 1902). These early gnotobiotic studies were hampered by a lack of knowledge of the nutrient requirements of the experimental animals. Much of the early work concentrated on making up this deficit and on methods of providing sterile, 'complete' diets for these animals. By 1946, the technological expertise was such that Reyniers and his colleagues in the USA were able to rear and maintain albino rats over several generations in GF conditions. Similar successes were obtained in Europe shortly afterwards (Gustafsson, 1948).

Having developed the techniques for obtaining GF animals, it was then possible to purposely put microorganisms into their intestines either as single species or as various combinations of species up to, and including, the provision of a complete 'normal' microflora.

The objectives of this review are to indicate the importance of gnotobiotic animals to nutrition research and to show something of the scope and variety of use of such animals. Research centres world-wide, including the

UK, USA, USSR, Europe and Japan, are currently involved in nutritional experiments with gnotobiotic animals.

2. RODENTS

It is implicit in the choice of experimental model, that it has some similarity to man, at least in the specified area of research. Although the use of rodents has been criticized frequently, they have been used in a wide variety of studies. Rats have probably been chosen most frequently but mice and guinea pigs have often been used. A variety of factors, including availability of animals, suitability of facilities and the particular subject of research, determines the final choice of species.

Rats and mice, which are used in gnotobiotic research, are almost invariably obtained from breeding colonies of GF animals maintained in isolators. The difficulties of hand-rearing new-born, hysterectomy-derived rats and mice dictates that this is usually reserved for the establishment of such breeding colonies *de novo*. Once a breeding colony of GF rats has been obtained, hysterectomy and subsequent fostering of the pups to GF dams can be used to introduce new genetic strains or to produce fresh breeding stock.

Guinea pigs are more precocious than rats or mice at birth and can take solid food almost immediately and so they can more reasonably be obtained by hysterectomy or hysterotomy. Techniques for deriving and raising small mammals for gnotobiotic work have been reviewed recently (Saito & Nomura, 1984).

2.1. Research using rodents

GF rodents have been shown to require dietary supplements of vitamins B and K in excess of the requirements of their conventional (CV) counterparts (Gustafsson, 1948, 1959; Mickelsen, 1956; Daft *et al*, 1963). It was suggested that bacterial synthesis of these vitamins normally helped to fulfill nutritional requirements. It was not clear, however, whether vitamins from bacterial synthesis were available to rats through absorption from the gut or via coprophagy. If it was the latter, microbial synthesis would have no significance for the supply of vitamins to man. Recent evidence from human studies has indicated, however, that absorption from the gut of vitamin B_{12} from bacterial synthesis could be important under some circumstances (Albert *et al*, 1980).

There has been considerable research activity concerning dietary fibre and its potential effects on gastro-intestinal physiology and health (Spiller & Kay, 1980; Stasse-Wolthuis, 1981). Regular consumption of fibre at a reasonable level in the diet appears to reduce the risk of a number of disorders of the digestive system including colon cancer. Furthermore, it appears to affect the cycling of bile acids in the gut and certain types of fibre may have a hypocholesterolaemic action. It is not known how such effects are mediated but they appear to involve interactions between dietary fibre and the gastro-intestinal microflora. So several workers have chosen gnotobiotic animals as experimental models. Sacquet and his colleagues in France have used GF and CV rats to examine the effects of various fibre sources including pectin and wheat bran on cholesterol and bile acid metabolism (Sacquet *et al*, 1982a, 1982b, 1983, 1985). Their results indicated that the microflora was an important mediator in the responses to fibre but that the type of fibre was also a significant factor. For instance, they found that pectin had a hypocholesterolaemic effect under some circumstances but in CV rats wheat bran increased plasma cholesterol.

Recent work in the U.K. has examined the effects of fibre on gut epithelial cell renewal in rats (Goodlad *et al*, 1987). Supplementation of the diet with fermentable fibre appeared to increase crypt cell production rate. Studies with GF rats showed that fibre *per se* could not elicit the responses seen in the presence of a gut microflora (Goodlad and Ratcliffe, unpublished). Similar research with mice has led to the suggestion that the variable effects of different types of dietary fibre are related to their suitability as substrates for bacterial fermentation (Komai *et al*, 1982).

The organic acids produced in the gastro-intestinal tract by microbial activity may have wide ranging effects including directly trophic effects for the mucosal epithelia, lowering of pH in the large intestine and a contribution to the energy intake of the host after absorption from the gut. Certainly, organic acids appear to be taken up from the large intestine of man (McNeil *et al*, 1978; Ruppin *et al*, 1980; Høverstad *et al*, 1982) and may provide up to 10% of daily energy needs (McNeil, 1984). Work with gnotobiotic mice, monoassociated with *Clostridium difficile* or **Escherichia coli**, examined the production of short chain fatty acids by these microorganisms *in vivo* and it was concluded that variations in the composition of the gut microflora might be more important in determining the type and quantities of short chain fatty acids in the intestines than variations in the type of substrate available for fermentation (Høverstad *et al*, 1985).

Another research area, where the production of volatile fatty acids may be important and where gnotobiotic animals are proving to be useful, is the study of 'novel' or 'unavailable' carbohydrates. Considerable interest centres around the potential use of such carbohydrates as low energy bulking agents for the food industry. Substances such as maltitol (α-D-glucopyranosyl-(1-4)-D-sorbitol), lactitol (β-D-galactopyranosyl-(1-4)-D-sorbitol) and lactulose (β-D-galactopyranosyl-(1-4)-D-fructofuranose) have been thought to be poorly hydrolysed by mammalian enzymes and to have low cariogenicity. When maltitol was given to GF rats, however, there was a significant uptake and hydrolysis of maltitol indicating that gut bacteria are not essential for the metabolism of this compound (Gärtner *et al*, 1979; Lian Loh *et al*, 1982). Metabolizable energy values for lactulose and lactitol have been obtained in CV animals (Bird *et al*, 1985) but so far gnotobiotic studies with rats have been confined to the effects of lactulose on nitrogen metabolism (Bird, 1986).

Other areas of nutrition research where gnotobiotic rodents are used, to mention but a few, include studies of the effects of antigen-free diets on nitrogen metabolism (Umehara *et al*, 1984), the effects of the gut microflora on the metabolism of amino acids and proteins (Tsuda *et al*, 1982), the effects of diet on blood proteins (Bolotskikh *et al*, 1981), interactions between Maillard reaction products and mineral absorption (Andrieux & Sacquet, 1984), intestinal enzyme activity (Juhr, 1980) and the role of bile in the absorption of fat (Demaine *et al*, 1982).

Research in nutrition is inextricably linked with the assessment of food safety or food toxicology. In addition to the potentially beneficial products of bacterial metabolism such as vitamins and short chain fatty acids, endogenous and exogenous compounds in the digesta may be converted to toxic, mutagenic or carcinogenic products as they pass along the gut. Bile acids and cholesterol have been implicated as substrates for conversion to carcinogens by bacterial activity (Hill, 1983) and so gnotobiotic models have been used to examine the effects of diet and the gut microflora on the metabolism of these substances (Gustafsson *et al*, 1981; Weidema *et al*, 1985; Borgström *et al*, 1986; Saxerholt & Midtvedt, 1986). The potential protective effects of dietary fibre have been referred to earlier. Other studies in this area have concentrated on the activities of the so-called 'biotransformation' enzymes β-glucosidase, β-glucuronidase, nitrate reductase and nitro-reductase which are involved in the production of potentially harmful substances (Rød & Midtvedt, 1977; Gadelle *et al*, 1985; Ward *et al*, 1986).

2.2. Rodents associated with a human microflora

Criticism can be levelled at the use of CV rats because they have a species-specific microflora which has intrinsically different character-istics to the human gastro-intestinal microflora and so might be expected to respond quite differently under some circumstances. Attempting to over-come these limitations various workers have developed the use of gnoto-biotic rodents which are associated with a human faecal microflora ('Hu'). Amongst the earliest work in this area was a study of the aetiology of parasitic infection with *Blastocystis hominis* using guinea pigs associated with a human microflora (Phillips & Zierdt, 1976). Prior to this Vossen and Van der Waaij (1973) experimented with the introduction of a human flora to mice. Several workers have since been involved in the validation of this system using mice (Raibaud *et al*, 1980; Hazenberg *et al*, 1981) and rats (Sacquet *et al*, 1984; Mallett *et al*, 1987).

Fuller and his colleagues in the UK have developed the use of 'Hu' rats for a number of nutritional studies (Cole & Fuller, 1987; Mallett *et al*, 1987). The 'Hu' rats were obtained by dosing GF rat pups at 1 day old with a suspension of a faecal flora collected from a healthy adult human. Subsequently, the microflora was passed on by dosing GF rats with a sus-pension of colon contents from a 'Hu' rat or by normal contact of rat pups with their 'Hu' dams (Cole & Fuller, 1987). Alternatively, GF rats were dosed with human faecal material at 4 weeks of age (Mallett *et al*, 1987). Although it is generally held that the gut microflora is indigenous and specific to a particular host species, 'Hu' rats were maintained in isolators and the human microflora persisted in the intestines in relative-ly stable populations. The human microflora may not function in an exactly similar way in rat intestines but the system serves as a living 'culture vessel'. Fuller's group has examined the intestinal profiles for 'bio-transformation' enzymes in 'Hu' rats and has found them to be broadly similar to the profiles found in human faeces (Mallett *et al*, 1987). When CV and 'Hu' rats and human subjects were given dietary supplements of pectin (5g/100g diet for rat, 18g/day for man), CV rats responded with significant increases in the activity of nitrate reductase. Such changes were not observed in the caecal contents of 'Hu' rats or in human faecal material. This model has also been used to examine the effects of butter, beef fat and yoghurt on two 'biotransformation' enzymes (Cole & Fuller, 1987).

Sacquet *et al* (1984) found no advantage to the use of 'Hu' rats for the study of bile acid metabolism. In GF rats the predominant bile acids are cholic and β-muricholic acids. Faecal material from gnotobiotic rats which had been associated with human faecal microflora still contained β-muri-cholic acid but this bile acid is not usually found in human faeces. CV rats, however, have very much reduced levels of this acid and significant levels of hyodeoxycholic acid appear.

Ducluzeau *et al* (1984) used human volunteers and 'Hu' mice to examine some of the effects of the consumption of bran.

2.3. Advantages and disadvantages

Rats and mice can be successfully bred and maintained over many generations in GF conditions obviating the need for surgery and labour-intensive hand-rearing. In addition, they have short generation times and produce large numbers of offspring per litter. They are easy to handle and relatively docile. Their small size means that transfers from one isolator to another are simple and that large numbers can be maintained, either as breeding stock or as experimental subjects, in a relatively small amount of isolator space. Furthermore, rats are omnivorous and will readily adapt to a wide range of experimental diets. Rodents have been used as laboratory animals over many years and an enormous wealth of knowledge of their 'biology' is available in published literature.

Nevertheless, there may be considerable differences between the physiology of man and rodents and extrapolation of results from experimental studies to the human situation is fraught with pitfalls. The use of gnotobiotic rodents associated with one or two species of bacteria can be criticized because these bacteria may function in an entirely different manner once the constraints of the 'microecosystem' of the conventional gut, including competition from other microbial species, are lifted. Alternatively, this does allow the study of microbial species in a living system which may be preferable to studies *in vitro*.

The disadvantages of using a rodent microflora as a model system have already been mentioned and these are surmounted to some extent by the use of 'Hu' rodents. A major problem with GF rodents is that they invariably develop enormously enlarged caeca when compared with CV animals (Coates, 1975) and they display anomalies in the functioning of the large intestines(Gordon & Bruckner, 1984). This can lead to problems in interpreting the findings of GF and CV comparative studies.

Coprophagy is a major problem with rodents, particularly rats. It is possible that responses to changes in diet and/or the gut microflora which are monitored via faeces or digesta may reflect changes occurring after second passage through the system. Where it is appropriate and possible, steps should be taken to avoid or minimize coprophagy by experimental animals.

Man exhibits a unique pattern of growth and development which could not be simulated in rodents and nutritional requirements may differ considerably.

3. PIGS
3.1. Advantages and disadvantages
Pigs have many anatomical and physiological similarities to man (Glauser 1966; Pond & Houpt, 1978; Dodds, 1982) and they have been used for a wide range of biomedical studies (Bustad & McClellan, 1966; Mount & Ingram, 1971 Panepinto & Phillips, 1986; Hughes, 1986; White *et al*, 1986; Fettman, 1986) Pigs are increasingly favoured as models in studies of nutrition and gastro-intestinal physiology. They are omnivorous and will consume food in daily meals whereas rodents tend to have nocturnal eating habits. Pigs have a protracted suckling period (up to 2-3 months) and so can be of use as models for infant nutrition. The digestibility of protein in mixed diets is similar for pigs and man (Forsum *et al*, 1981) and gut hormones show a similar pattern following a meal (Adrian & Bloom, 1981).

Pigs do not practise coprophagy to the extent seen in rodents and it is easier to take steps to prevent it by the appropriate use of metabolic cages. GF pigs have less marked abnormalities of the large intestine than rodents and caeca show only marginal increases in size.

Pigs are disadvantaged for gnotobiotic studies by their large mature size and rapid growth rate. This demands large isolators and facilities for their maintenance or they may only be kept for short periods. In addition, they consume large amounts of feed and water and produce corresponding amounts of faeces and urine. These disadvantages may be overcome to some extent by the use of miniature pigs which have been kept for over 5 months in relatively small Gustafsson-type isolators (Ratcliffe & Fordham, 1987). Göttingen miniature pigs tend to be quite variable in response for some parameters, however, and this may be related to their genetic heterogeneity. The large domestic breeds are less variable because they are usually pure-bred lines. The use of a breed of miniature pig such as the Yucatan, which is derived from one primary gene pool (Panepinto & Phillips, 1986), might overcome this problem.

In common with rodents, much is known about the nutritional requirements and physiology of pigs. Although they have a long gestation period (mean 114 days), a regular supply of litters can be obtained from commercial herds. The isolator space for two small pigs, however, would easily support twenty rats. It follows that obtaining sufficient numbers of replicates in a nutritional experiment with gnotobiotic pigs places considerable demand on isolator space.

It is not feasible to breed pigs under GF conditions and for most gnotobiotic studies pigs are obtained *ex utero* and hand-reared. GF pigs have been obtained by hysterectomy (Young *et al*, 1955) or hysterotomy (Landy & Ledbetter, 1966) and, more recently, without surgical techniques (Ducluzeau *et al*, 1976; Ratcliffe & Fordham, 1987). The methods used by Ratcliffe and Fordham (1987) make the procurement of GF pigs very much simpler and easier with the added advantages that pigs are obtained after normal parturition, so the risk of prematurity is avoided, and piglets which are not required for GF work can be returned to their dam for normal suckling.

Once established in a GF isolator, baby pigs readily consume diets based on sterilized bovine milk. Miniats (1984) reviewed the techniques for producing and rearing GF farm animals.

3.2. Research using pigs

Gnotobiotic pigs are widely used for studies of disease pathogenicity and immunology. They are less extensively used for nutrition research than rats but a suprisingly wide subject area is covered.

Recently, gnotobiotic pigs have been used in studies of food antigenicity (Kovaru *et al*, 1982; Ratcliffe *et al*, 1987). Other work has examined the effects of dietary fibre and the gut microflora on dietary nitrogen digestibility (Ratcliffe & Low, 1985). Also, gnotobiotic pigs have been used to extend the studies of 'novel' carbohydrates (see 2.1.) and it has been shown that lactitol can be hydrolysed to some extent in GF conditions (Ratcliffe, unpublished). Associating GF pigs with a human faecal microflora, that is 'Hu' pigs, could extend the usefulness of these animals in nutrition research.

4. OTHER ANIMALS

Chickens and other birds have a long tradition of use in gnotobiotic research (Coates, 1975). They are relatively easy to obtain and maintain but their phylogenetic remoteness from man must limit their use as an animal model.

Lagomorpha have been used in a variety of gnotobiotic studies and their susceptibility to dietary-induced changes in cholesterol metabolism has led to their use as a model in a number of CV studies (Jadidi *et al*, 1987). But the validity of such use is questionable since Lagomorpha are normally coprophagous and herbivorous.

Pre-ruminant lambs and calves have been used in isolators to study both fundamental and applied aspects of nutrition (Peyraud *et al*, 1981). They seem more subject to microbial contamination than pigs and it is unlikely that their use would give any advantage over pigs.

5. ALTERNATIVES TO GNOTOBIOTIC ANIMALS

While gnotobiotic animals are relatively easy to obtain, they nevertheless require some specialized equipment, the use of aseptic techniques and sterile food and water. Various workers have attempted to study interactions between the diet and the gut microflora by using antibiotics or sulphur-containing drugs to suppress the activities of intestinal microorga..isms (Mason & Just, 1976; Sandaradura & Bender, 1986). The level of dose of the antimicrobial drug has to be kept low to avoid confounding effects from the drug action *per se* but high enough to produce significant depression of bacterial activity. Often this compromise is unsuccessful, although some useful work has been done in this area.

If anaerobic continuous-flow culture systems can be developed which simulate large bowel metabolism *in vitro* and which yield results which correlate reasonably well with observations in man or gnotobiotic animals then there is considerable scope for decreasing the time necessary to complete experiments in this subject area and to reduce the overall use of laboratory animals. The present technology in this field is limited by a lack of understanding of the complex requirements of many gastro-intestinal microorganisms, some of which still cannot be cultured outside living systems.

6. SUMMARY AND CONCLUSIONS

Classically, gnotobiotic animals have been used for a wide variety of fundamental studies of nutritional physiology. The ability of the gut bacteria to synthesize vitamins has been demonstrated and such contributions to nutritional science have facilitated a fuller understanding of the dietary requirements of GF animals, so that they can now be maintained for prolonged periods. Fundamental research continues but there is an increasing demand for systems which act as models for man in studies of food safety and nutrition.

There is no single ideal animal model but various species including rats, mice and pigs can be suitable for testing hypotheses. Gnotobiotic animals can provide information about interactions between food components and the gastro-intestinal microflora which is not readily obtainable by other means. Gnotobiotes associated with a human faecal microflora extend the range of useful models.

A fuller understanding of host-microbial interactions and the effects of diet may lead to improved health and nutrition for man and the eradication of many disorders of the alimentary canal.

REFERENCES

1. Adrian TE & Bloom SR: In 'Nutrition in health and disease and international development: Symposia XII International Congress of Nutrition (ed. Harper AE & Davis GK). New York: AR Liss Inc, p.873-882, 1981.
2. Albert MJ, Mathan VI & Baker SJ: Vitamin B_{12} synthesis by human small intestinal bacteria. Nature, 283, 781-782, 1980.
3. Andrieux C & Sacquet E: Effects of Maillard's reaction products on apparent mineral absorption in different parts of the digestive tract. The role of the microflora. Reproduction, Nutrition Developpement, 24, 379-

386,1984.
4. Bird SP: Metabolic effects of lactulose and lactitol. PhD Thesis. University of Reading, 1986.
5. Bird SP, Hewitt D & Gurr MI: Digestible and metabolizable energy values of lactulose and lactitol for the rat and miniature pig. Proc. Nutr. Soc,44, 40A, 1985.
6. Bolotskikh LA, Podoprigora GI, Kuvaeva IB, Orlova NI, Veselova OL, Kashtanova NA & Galakhova TV: Effect of different diets on blood proteins in gnotobiotic animals. Voprosy Pitaniya, 4, 44-46, 1981.
7. Borgström N, Krabish L, Lindström M, Lillienau J, Midtvedt T & Corrie M: Effects of feeding ursocholic acid to germ-free rats. Clin. Lab. Invest, 46, 1983, 1986.
8. Bustad LK & McClellan RD (eds.): Swine in biomedical research. Richland : Pacific North-West Laboratories, 1966.
9. Coates ME: Gnotobiotic animals in research: their uses and limitations. Lab. Anim, 9, 275-282, 1975.
10. Cole CB & Fuller R: The effect of dietary fat and yoghurt on colonic bacterial enzymes (β-glucosidase and β-glucuronidase) associated with colon cancer. Food Microbiology, 4, (in press), 1987.
11. Daft FS, McDaniel EG, Harman LG, Romine MK & Hegner JR: Role of coprophagy in utilization of B vitamins synthesized by intestinal bacteria. Federation Proceedings, 22, 129-133, 1963.
12. Demaine Y, Corring T, Pihet A & Sacquet E: Fat absorption in GF and CV rats artificially deprived of bile secretion. Gut, 23, 49-57, 1982.
13. Dodds WJ: The pig model for biomedical research. Federation Proceedings 41, 247-256, 1982.
14. Ducluzeau R, Raibaud P, Lauvergeon B, Gouet P, Riou Y, Griscelli C & Gnassia JC: Immediate post-natal decontamination as a means of obtaining axenic animals and human infants. Can. J. Microbiology, 22, 563-566, 1976.
15. Ducluzeau R, Lactine M & Raibaud P: Effects of bran ingestion on the microbial faecal floras of human donors and of recipient gnotobiotic mice and on the barrier effects exerted by three floras against various potentially pathogenic microbial events. Annls. Microbiol, A135, 303, 1984.
16. Fettman MJ: Endotoxemia in Yucatan miniature pigs: metabolic derangements and experimental therapies. Lab. Anim. Sci, 36, 370-374, 1986.
17. Forsum E, Goranzon H, Rundgren M, Tuilen M & Hambraeus L: Ann. Nutr. Metab, 25, 137-150, 1981.
18. Gadelle D, Raibaud P & Sacquet E: β-glucuronidase activities of intestinal bacteria determined both *in vitro* and *in vivo* in gnotobiotic rats. Appl. Environ. Microbiology, 49, 682-685, 1985.
19. Gärtner H, Zunft HJ, Schulze J, Miller W & Grutte IK: The use of gnotobiotic rats for characterization of digestibility of maltitol. Folia Microbiologica, 24, 44-45, 1979.
20. Glauser EM: Advantages of piglets as experimental animals in pediatric research. Experimental Medicine and Surgery, 24, 181-190, 1966.
21. Goodlad RA, Lenton W, Ghatei MA, Adrian TE, Bloom SR & Wright NA: Effects of an 'elemental' diet, inert bulk and different types of dietary fibre on the response of the intestinal epithelium to refeeding in the rat and the relationship to plasma gastrin, enteroglucagon and PYY levels. Gut, (in press), 1937.
22. Gordon HA & Bruckner G: Anomalous lower bowel function and related phenomena in germ-free animals. In 'The germ-free animal in biomedical research' (eds. Coates ME & Gustafsson BE)London: Lab. Anim. Ltd, 1984.

23. Gustafsson BE: Germ-free rearing of rats. General technique. Acta Path. Microbiol. Scand. Suppl. 73, 1-130, 1948.
24. Gustafsson BE: Lightweight stainless steel systems for rearing germ-free animals. Ann. N.Y. Acad. Sci, 78, 17-28, 1959.
25. Gustafsson BE, Angelin B, Bjorchem I, Einarsson K & Gustafsson JA: Effects of feeding chenodeoxycholic acid on metabolism of cholesterol and bile acids in germ-free rats. Lipids, 16,228-233, 1981.
26. Hazenberg MP, Bakker M & Verschoor-Burggraaf A: Effects of the human intestinal flora on germ-free mice. J. Appl. Bacteriology, 50, 95-106, 1981.
27. Hill MJ: Bile, bacteria and bowel cancer. Gut, 24, 871-875, 1983.
28. Høverstad T, Bohmer T & Fausa O: Absorption of short chain fatty acids from the human colon measured by the $^{14}CO_2$ breath test. Scandinavian J. Gastroenterology, 17, 373-378, 1982.
29. Høverstad T, Midtvedt T & Bohmer T: Short chain fatty acids in intestinal contents of germ-free mice mono-contaminated with *Escherichia coli* or *Clostridium difficile*. Scandinavian J. Gastroenterology, 20, 373-380, 1985.
30. Hughes HC: Swine in cardiovascular research. Lab. Anim. Sci, 36, 348-350, 1986.
31. Jadidi N, Edwards-Webb JD, Owen RW & Gurr MI: Effect of dietary saturated and polyunsaturated fats on faecal excretion of cholesterol and its catabolites in rabbits. Proc. Nutr. Soc, 46,10A, 1987.
32. Juhr NC: Intestinal enzyme activity in GF and CV rats and mice. Zeitschrift für Versuchstierlande, 22, 197-203, 1980.
33. Komai M, Takehisha F & Kimura S: Effect of dietary fibre on intestinal epithelial cell kinetics of germ-free and conventional mice. Nutrition Reports International, 26, 255-261, 1982.
34. Kovaru H, Kovaru F, Plasek J, Fisar Z & Mandel L: In 'Lectins-Biology, Biochemistry, Clinical Biochemistry. Vol.2,p.315, 1982.
35. Landy JJ & Ledbetter RK: Delivery and maintenance of the germ-free pig. In 'Swine in biomedical research' (ed. Bustad LK). Richland WA: Pacific North-West Laboratory, p.619-632, 1966.
36. Lian Loh R, Birch GG & Coates ME: The metabolism of maltitol in the rat Br. J. Nutr, 48, 477-481, 1982.
37. McNeil NI: The contribution of the large intestine to energy supplies in man. Am. J. Clin. Nutr, 39, 338-342, 1984.
38. McNeil NI, Cummings JH & James WPT: Short chain fatty acid absorption by the human large intestine. Gut, 19, 819-822, 1978.
39. Mallett AK, Bearne CA, Rowland IR, Farthing MJG, Cole CB & Fuller R: The use of rats associated with a human faecal flora as a model for studying the effects of diet on the human gut microflora. J. Appl. Bacteriology, (in press), 1987.
40. Mason VC & Just A: Bacterial activity in the hind-gut of pigs.1.Its influence on the apparent digestibility of dietary energy and fat. Z. Tierernahrg.u.Futtermittelkde, 36, 301-310, 1976.
41. Metchnikoff E: The prolongation of life. London: Heinemann, 1907.
42. Mickelsen O: Intestinal synthesis of vitamins in the non-ruminant. Vitamins Hormones, 14, 1-82, 1956.
43. Miniats OP: Production of germ-free animals.Part 2. Farm animals. In 'The germ-free animal in biomedical research' (eds. Coates ME & Gustafsson BE). London: Lab. Anim. Ltd, 1984.
44. Mount LB & Ingram DL: The pig as a laboratory animal. London: Academic Press, 1971.
45. Nuttal GHF & Thierfelder H: Thierisches leben ohne bakterien im ver-

dauungskanal.II. Mitt Hoppe Seyler's Z. Physiol.Chem, 22, 62-73, 1896.
46. Panepinto LM & Phillips RW: The Yucatan miniature pig: characterization and utilization in biomedical research. Lab. Anim. Sci, 36, 344-347, 1986.
47. Pasteur L: Observations relatives a la note precedente de M. Duclaux. CR Acad. Sci, Paris, 100, 68, 1885.
48. Peyraud JL, Thivend P & Gouet P: Role of digestive microflora on starch digestion in the pre-ruminant lamb. Recent Advances in Germ-free Research, Tokai University Press, p.285-288, 1981.
49. Phillips BP & Zierdt CH: *Blastocystis hominis*: pathogenic potential in human patients and gnotobiotes. Experimental Parasitology, 39, 358-364, 1976.
50. Pond WG & Houpt KA: Biology of the pig. New York: Comstock Publ. Assoc, 1978.
51. Raibaud P, Ducluzeau R, Dubos F, Hudault S, Bewa H & Muller MC: Implantation of bacteria from the digestive tract of man and various animals into gnotobiotic mice. Am. J. Clin. Nutr, 33, 2440-2447, 1980.
52. Ratcliffe B & Fordham JP: A technique for rearing germ-free piglets obtained without surgery. Lab. Anim, 21, 53-59, 1987.
53. Ratcliffe B & Low AG: Studies on the effect of the gut microflora on dietary nitrogen digestibility in pigs. In 'Germ-free Research: Microflora control and its application to the biomedical sciences' (ed. Wostmann BS). New York: AR Liss Inc, p.115-118, 1985.
54. Ratcliffe B, Smith MW, Miller BG & Bourne FJ: The effect of soya protein on intestinal morphology, absorption and enzyme activity in gnotobiotic pigs at 3 weeks of age. Proc. Nutr. Soc, 46, (in press), 1987.
55. Reyniers JA, Trexler PC & Ervin RF: Rearing germ-free albino rats. Lobund reports no.1. Notre Dame, Indiana: University of Notre Dame Press, p.2-84, 1946.
56. Rød TO & Midtvedt T: Origin of intestinal β-glucuronidase in germ-free monocontaminated and conventional rats. Acta Pathol. Microbiol. Scand, B, 85, 271-276, 1977.
57. Ruppin H, Bar-Meir S, Soergel CM, Wood CM & Schmidt MG: Absorption of short chain fatty acids by the colon. Gasterenterology, 78, 1500-1507, 1980.
58. Sacquet EC, Gadelle DO, Riottot MJ & Raibaud PM: Absence of transformation of β-muricholic acid by human microflora implanted in the digestive tracts of germ-free male rats. Appl. Environ. Microbiology, 47, 1167-1168, 1984.
59. Sacquet E, Leprince C & Riottot M: Dietary fibre and cholesterol and bile acid metabolisms in axenic and holoxenic rats.1.Effect of wheat bran. Reproduction Nutrition Developpement, 22, 291-305, 1982a.
60. Sacquet E, Leprince C & Riottot M: Dietary fibre and cholesterol and bile acid metabolisms in axenic and holoxenic rats.2. Effect of pectin. . Reproduction Nutrition Developpement, 22, 575-581, 1982b.
61. Sacquet E, Leprince C & Riottot M: Effects of amylomaize starch on cholesterol and bile acid metabolisms in axenic and holoxenic rats. Reproduction Nutrition Developpement, 23, 783-792, 1983.
62. Sacquet E, Leprince C, Riottot M & Raibaud P: Dietary fibre and cholesterol and bile acid metabolisms in axenic and holoxenic rats.3. Effect of non-sterilized pectin. Reproduction Nutrition Developpement, 25, 93-100, 1985.
63. Saito M & Nomura T: Production of germ-free animals. Part 1. Small mammals. In 'The germ-free animal in biomedical research' (eds.

Coates ME & Gustafsson BE). London: Lab. Anim. Ltd, p.33-48, 1984.

64. Sandaradura SS & Bender AE: Effect of oral chlortetracycline on faecal DNA excretion and mucosal cell turnover in rats fed on legumes. Proc. Nutr. Soc, 45, 118A, 1986.

65. Saxerholt H & Midtvedt T: Intestinal deconjugation of bilirubin in GF and CV mice. Clin. Lab. Invest, 46, 341, 1986.

66. Schottelius M: Die bedeutung der Darmbakterien für die Ernährung.II. Arch. Hyg, 42, 48-70, 1902.

67. Spiller GA & Kay RM: Medical Aspects of Dietary Fibre. New York/ London : Plenum. p.299, 1980.

68. Stasse-Wolthuis M: Influence of dietary fibre on cholesterol metabolism and colonic function in healthy subjects. World Review of Nutrition and Dietetics, 36, 100-140, 1981.

69. Tsuda M, Ohkubo T, Sase M & Katsunuma T: Effect of intestinal flora on the metabolism of amino acids and proteins in mouse. J. Nutritional Science and Vitaminology, 28, 315-319, 1982.

70. Umehara K, Yamanaka M, Hashimoto K & Sasaki S: Nitrogen retention in GF and CV mice reared on an antigen-free diet and the significance of the appearance of protein nitrogen in their faeces. Exp. Anim, 33, 77-84, 1984.

71. Vossen JM & Van der Waaij: Recolonization after decontamination: clinical experiences. In 'Airborne transmission and airborne infection' (eds. Hers JFP & Winkler KC). Utrecht: Oosthoek, p.549-553, 1973.

72. Ward FW, Coates ME & Walker R: Nitrate reduction, gastro-intestinal pH and N-nitrosation in gnotobiotic and conventional rats. Food Chem. Toxicology, 24, 17, 1986.

73. Weidema WF, Deschner EE, Cohen BI & DeCosse JJ: Acute effects of dietary cholic acid and methylazoxymethanol acetate on colon epithelial cell proliferation; metabolism of bile salts and neutral sterols in CV and GF SD rats. J. natl. Cancer Institute, 74, 665, 1985.

74. White FC, Roth DM & Bloor CM: The pig as a model for myocardial ischemia and exercise. Lab. Anim. Sci, 36, 351-356, 1986.

75. Young GA, Underdahl NR & Hinz RW: Procurement of baby pigs by hysterectomy. Am. J. Vet. Res, 16, 23-31, 1955.

A.C. Beynen & H.A. Solleveld (Eds), New developments in biosciences: their
implications for laboratory animal science, ISBN 978-94-010-7973-0
© 1988 Martinus Nijhoff Publishers, Dordrecht.

ANIMAL MODELS FOR CHOLESTEROL METABOLISM STUDIES

A.C. BEYNEN[1,2],

[1] Department of Laboratory Animal Science, State University, Utrecht
[2] Department of Human Nutrition, Agricultural University, Wageningen,
(The Netherlands)

ABSTRACT
The concentration of serum cholesterol is a major risk factor for
coronary heart disease in humans; the higher this concentration the
greater the risk for heart disease. This cause-and-effect relationship
has stimulated the investigation of cholesterol metabolism and its
regulation. In case animals are to be used to gain further insight into
human cholesterol metabolism, the proper choice of animal model is
crucial.
The amount of cholesterol in the body is regulated by the balance of
input and output, essentially by cholesterol absorption from the diet,
endogenous cholesterol synthesis, and by the fecal excretion of neutral
steroids and bile acids. Obviously, the amount of cholesterol in the
serum is also regulated by these processes, and in addition, it can be
influenced by cholesterol uptake and release by tissues. An increase in
cholesterol intake will trigger changes in one or more of the homeostatic
mechanisms of whole-body cholesterol metabolism. What compensatory
mechanisms gather momentum will depend on the animal species studied, and
probably also on the extent of the dietary cholesterol load.
In humans, who have limited capacity for absorption of dietary cholesterol, increased intakes of cholesterol can usually be compensated for
by inhibition of cholesterol synthesis, resulting in only a slight
increase, if at all, in serum cholesterol. Like rabbits, humans do not
display increased rates of excretion of bile acids after cholesterol
feeding. However, in rabbits cholesterol absorption is very efficient,
and even relatively low intakes cannot be compensated for by inhibition
of cholesterol synthesis, leading to hypercholesterolemia. If compensatory mechanisms are overwhelmed, cholesterol accumulates in the body,
especially in the liver. This leads to liver damage and diminished liver
function, which in turn may cause biased interpretation of experimental
results. Rats have a large capacity for depressing cholesterol synthesis
as well as to enhance bile acid synthesis after cholesterol consumption.
This makes the rat almost insensitive, with respect to the concentration
of serum cholesterol, to cholesterol loading.
Thus there are various quantitative and qualitative differences in
cholesterol metabolism between species, which should be considered in
selecting the appropriate animal model. Furthermore, the amount of
dietary cholesterol should be tailored to the characteristics of cholesterol metabolism of the animal model under study.

1. INTRODUCTION

The concentration of cholesterol in the blood of man is an important risk indicator for coronary heart disease. The higher this concentration the higher is the risk for cardiovascular disease. This notion is derived from various types of studies such as epidemiological and case-control studies in man, but also from controlled studies with animal models. Serum cholesterol may be a causative factor in the development of coronary heart disease. Lowering of serum cholesterol levels has been shown to decrease the incidence of cardiovascular morbidity (1).

Since cardiovascular disease is the leading cause of death in westernized countries, there is a great deal of interest in the regulation of cholesterol metabolism. Knowledge of those factors involved in the control of serum cholesterol concentrations might contribute to means of optimal prevention and therapy of cardiovascular disease in man. Animal models are being used in order to gain further insight into the mechanisms controlling human serum cholesterol concentrations. The value of animal models is critically dependent on their degree of similarity with man. Thus progress of cholesterol metabolism research is closely associated with the selection of appropriate animal models. I am well aware that ultimately, all hypotheses generated from animal experiments can only be proved if tested in man. In this communication I shall attempt to establish criteria which should be followed in selecting animal models in order to obtain maximum information as to human cholesterol metabolism.

2. GENERAL ASPECTS OF CHOLESTEROL METABOLISM

As a structural component of most cell membranes, cholesterol is an essential compound for all animals, including man. In addition, cholesterol is used as the substrate for the synthesis of steroid hormones, vitamin D and bile acids. In order to fulfill the cholesterol needs of the various cells, a continuous supply of cholesterol is required.

During growth, cholesterol is being accreted with new tissues, and hence the body pool of cholesterol increases. However, at steady-state, the amount of cholesterol in the body is determined by the balance of input and output. Cholesterol enters the body pool from only two sources: absorption from the diet and synthesis within the various organs of the body. Cholesterol leaves the body by various routes. Hepatic cholesterol can be transferred as such to the bile fluid and then to the gastrointestinal tract. Alternatively, cholesterol can be converted first into bile acids. Cholesterol and bile acids that escape from re-absorption eventually leave the body in feces. Cholesterol is also lost from the body through the sloughing of both intestinal mucosa and skin. The cholesterol being converted into steroid hormones is excreted from the body in urine. Quantitatively, output through fecal excretion of cholesterol, its bacterial metabolites and bile acids are most important.

At steady-state, cholesterol synthesis in the body can be assessed as the sum of fecal excretion of neutral steroids and bile acids minus cholesterol intake. Rates of cholesterol synthesis can also be determined directly by measuring the incorporation of ^3H from tritiated water into whole-body cholesterol.

The concentration of serum cholesterol is determined by the same factors as is the body pool of cholesterol. In addition, serum cholesterol can be influenced by changes in the distribution of cholesterol between serum and tissues and by the capacity of tissues to store cholesterol. However, uptake and release of cholesterol by tissues cannot play a role in the long-term homeostasis of serum cholesterol. Tissues cannot infinitely have a net influx or efflux of cholesterol. Nevertheless, in short-term studies in which serum cholesterol is perturbed, changes in tissue cholesterol is certainly a phenomenon to be on the alert for.

3. MECHANISMS OF CHOLESTEROL HOMEOSTASIS

To maintain the balance of cholesterol accross the body a series of regulatory and compensatory mechanisms have evolved in different animal species and man. The major features of these mechanisms can be illustrated nicely by describing the cascade of events in response to increased intakes of cholesterol. The efficiency of cholesterol absorption probably remains constant after cholesterol feeding and thus absorption cannot be considered a regulatory mechanism. As will be shown below, the efficiency of cholesterol can differ markedly between animal species as well as between strains of one species.

The maximum set of compensatory mechanisms triggered by cholesterol loading consists of the following parts. An increased intake of cholesterol will cause an increased influx of cholesterol into the liver, delivered to this organ by so-called chylomicron-remnants. The increased influx tends to enlarge liver cholesterol pools. The liver responds immediately by inhibition of de novo cholesterol synthesis so that no extra cholesterol is added to the liver cholesterol pools. The liver also has pathways for removing excess cholesterol. Cholesterol can be channelled into bile fluid either as such or after conversion into bile acids. In addition, cholesterol can be secreted into serum as a component of lipoprotein particles. On the other hand, cholesterol uptake in the form of lipoprotein-cholesterol can be depressed by downregulation of the activity of the receptor (apo B,E receptor) on liver membranes. The balance between cholesterol synthesis plus uptake on the one hand and cholesterol output by the liver on the other, will be reflected in the fecal excretion of neutral steroids and bile acids. In essence, cholesterol can only leave the body through the liver. The compensatory mechanisms act in concert so that in the long-run a new steady-state will be reached. At this new steady-state the concentration of cholesterol in serum may be somewhat increased, the degree of this increase being dependent on the capacity of the compensatory mechanisms.

Below, I shall compare various animal species concerning their response to cholesterol feeding. Evidence will be presented that cholesterol metabolism and its compensatory mechanisms differ between animal species in both qualitative and quantitative terms.

4. REGULATORY MECHANISMS TRIGGERED BY CHOLESTEROL LOADING OF VARIOUS ANIMAL SPECIES
4.1. Cholesterol synthesis.

Between-species variation in cholesterol synthesis while on a low-cholesterol diet involves both the rate of whole-body cholesterol synthesis and the quantitative importance of different organs. Spady and Dietschy (2) have measured in-vivo cholesterol synthesis by the ^3H_2O-method, which is considered superior to other methods. Adjustment to organ weight of the rates of cholesterol synthesis in various organs seen 1 hour after administration of ^3H_2O, allows the importance of each organ to whole-body cholesterol synthesis to be calculated. Table 1 shows a marked between-species variation in the contribution of the liver to total-body cholesterol synthesis. In the rat, about half of newly synthesized cholesterol is found in the liver, whereas in the rabbit and guinea pig this proportion is smaller than 20%.

Absolute rates of cholesterol synthesis cannot be assessed by the incorporation of [^{14}C] substrates into cholesterol. The reason for this has to do with differential dilution of the ^{14}C-labelled precursors of cholesterol in various cells. However, the incorporation data are useful in looking at the relative rates of cholesterol synthesis in a given organ.

TABLE 1. Relative importance of different organs for in-vivo whole-body cholesterol synthesis in different animal species. *After Spady and Dietschy (2).*

	Relative rate of cholesterol synthesis (%)			
Species	Liver	Small bowel	Skin	Other organs, carcass
Squirrel monkey	40	9	26	25
Rat	51	12	12	25
Hamster	27	11	19	43
Rabbit	18	20	20	42
Guinea pig	16	23	18	43

TABLE 2. Relative effect of cholesterol feeding on cholesterol synthesis in liver and small intestine in three species. *Based on references 3-5.*

		Cholesterol ingested	
Species	Organ	Low	High
Squirrel monkey	Liver	100	4
	Intestine	100	47
Rat	Liver	100	1
	Intestine	100	78
Rabbit	Liver	100	6
	Intestine	100	34

Table 2 shows the effect of dietary cholesterol on the relative rates of cholesterol synthesis in liver and small intestine as determined by the in-vitro incorporation of [2-^{14}C]acetate or [1-^{14}C]octanoate into cholesterol by tissue slices. Cholesterol synthesis in the liver is dramatically decreased by cholesterol feeding of the squirrel monkey, rat and rabbit (Table 2). However, the magnitude of inhibition of cholesterol synthesis by the small intestine appears to be species-dependent. In the rat, cholesterol synthesis by this organ seems less sensitive to dietary cholesterol than in the rabbit. It cannot be excluded that differences in the extent of cholesterol challenge have elicited an apparent species-dependent response.

It is clear that species can differ regarding organ-specific contribution to whole-body cholesterol synthesis as well as organ-specific sensitivity of cholesterol synthesis to cholesterol feeding. Thus in order to compare the impact of dietary cholesterol on cholesterol synthesis among animal species, data on whole-body cholesterol synthesis should be used. In fact, inhibition of cholesterol synthesis as a compensatory mechanism in response to increased cholesterol intakes must refer to whole-body cholesterol synthesis.

Whole-body cholesterol synthesis can be derived from sterol-balance data, that is, calculated as the difference between cholesterol excretion in feces (sum of neutral steroids and bile acids) and cholesterol intake. Prerequisite is that the animal is in steady-state so that input equals output. In many short-term cholesterol feeding experiments such a situation is not reached because the body accumulates cholesterol. In growing animals there is, in addition to cholesterol storage, formation of new tissues which contain membrane cholesterol. This leads to the calculation of falsely low or even negative rates of cholesterol synthesis. Thus cholesterol accumulation should be corrected for.

TABLE 3. Approximate rates of whole-body cholesterol synthesis in different animal species fed low or high-cholesterol diets. *Based on references 6-12. *Synthesis derived from sterol balance data and, apart from man, corrected for cholesterol retention. [+]Synthesis derived from input-output analysis after an oral dose of tracer cholesterol*

	(mg/day/kg b.w.)			
	Cholesterol ingested		Cholesterol synthesis	
Species	Low	High	Low	High
Man*	2	10	10	7
Rhesus monkey*	0	63	18	11
Swine*	0	41	57	9
Rat*	39	255	62	27
Dog*	1	115	13	0
Guinea pig[+]	2	30	24	19
Rabbit*	0		25	

Table 3 compares the rates of whole-body cholesterol synthesis determined on low and high-cholesterol diets in different animal species, including man. It appears that there are no qualitative differences between species. Cholesterol feeding invariably reduces whole-body cholesterol synthesis. As to possible quantitative differences in the response to dietary cholesterol, no solid conclusions can be drawn because cholesterol intakes differed per animal species.

Table 3 documents that there is marked variation in the rate at which cholesterol is synthesized by different species while on a low-cholesterol diet. This could imply that species with high rates of cholesterol synthesis such as the rat and swine, have greater ability to adapt to increased intakes of cholesterol than have species with low basal rates of cholesterol synthesis such as the rhesus monkey, dog and man. However, as will be discussed below, compensatory mechanisms other than inhibition of cholesterol synthesis also determine the response of serum cholesterol.

4.2. Fecal excretion of steroids

Cholesterol feeding induces an increase in the fecal output of neutral steroids in all species given in Table 4. Most of this increase probably represents non-absorbed cholesterol rather than hepatic cholesterol excreted through bile. Stimulation of bile acid excretion upon cholesterol feeding does not occur in all species. In man, swine, rabbit, and perhaps the rhesus monkey, fecal bile acid output is not enhanced. In contrast, the rat and dog show a dramatic increase in bile acid excretion. Thus there are qualitative differences between animal species in the response of bile acid excretion, unlike that of neutral steroid excretion, to cholesterol loading. Details are beyond the scope of this communication, but it should be stressed that the relative proportions of individual neutral steroids and bile acids can differ markedly between species.

TABLE 4. Fecal excretion of neutral steroids and bile acids on low and high-cholesterol diets in different animal species. *Based on references 6-10 and 13.*

	(mg/day/kg b.w.)					
			Excretion of steroids			
	Cholesterol ingested		Neutral steroids		Bile acids	
Species	Low	High	Low	High	Low	High
Man	2	10	9	14	3	3
Rhesus monkey	0	63	4	46	4	9
Swine	0	41	14	24	28	12
Rat	7	192	9	15	28	121
Rabbit	3	100	13	70	17	19
Dog	1	115	8	64	5	48

5. COMPARATIVE CHOLESTEROL FEEDING TRIALS

The data presented suggest that upon cholesterol feeding of various animal species including man at least one or two compensatory mechanisms gather momentum. Cholesterol synthesis is depressed among all species. In species such as the rat and dog this is associated with enhanced fecal excretion of bile acids. This compensatory mechanism does not operate in man, swine, and the rabbit. These qualitative similarities and differences among species should be considered when selecting an appropriate animal model for cholesterol metabolism studies.

It is extremely difficult to compare, in quantitative terms, cholesterol metabolism in various species. If one wants to make such a comparison at least the amounts of cholesterol ingested should be the same for all species. In Tables 2 and 3 cholesterol intakes were compared on the basis of mg/day/kg body weight. It could be argued that intakes should be expressed on the basis of metabolic weight rather than body mass. Likewise, one could put forward that cholesterol intakes should be normalized to cholesterol:energy ratios in the diet.

Whatever agreement is arrived so as to express and normalize cholesterol intakes among species, it should be realized that actual cholesterol absorption is the product of both absolute intake and efficiency of absorption. The capacity to absorb cholesterol can differ between species (Table 5). The rabbit and dog very efficiently absorb dietary cholesterol. It should be realized that cholesterol absorption depends on the composition of the diet. Thus in order to equate cholesterol loads among species, differences in cholesterol absorption should be accounted for.

TABLE 5. Approximate efficiency of cholesterol absorption in various species. *Based on references 7, 8, 10-12, 14-16.*

Species	Cholesterol absorption (%)
Man	45
Rhesus monkey	51
Swine	49
Rat	52
Rabbit	77
Dog	80
Guinea pig	53
Hamster	54

Theoretically, it is possible that the between-species variation in ability to step up bile acid excretion after cholesterol loading is a quantitative feature rather than a qualitative one. It could be suggested that in certain species an increase in bile acid excretion only occurs when inhibition of cholesterol synthesis cannot compensate for the increase in the amount of cholesterol absorbed. This could mean that the rate of whole-body cholesterol synthesis should be considered for the correct choice of animal model. Alternatively, the amount of dietary cholesterol should be tailored to the rate of cholesterol synthesis. If the increase in the amount of cholesterol absorbed is much higher than the basal rate of cholesterol synthesis, probably irrespective of whether bile acid synthesis is activated, cholesterol will accumulate in the body, especially the liver. This may cause liver damage and diminished liver function (17), which in turn could lead to biased interpretation of experimental results.

It is illustrative in this context to discuss how compensatory mechanisms triggered by cholesterol feeding work in concert in different species (Table 6). This interaction determines the magnitude of change in serum cholesterol. In species with relatively limited capacity to absorb cholesterol, e.g. man, a small increase in dietary cholesterol will be compensated for by inhibition of cholesterol synthesis, and serum cholesterol will rise only slightly, if at all. If the amount of absorption exceeds the quantity synthesized by the body, and if conversion into bile acids cannot increase, the pool of serum cholesterol will rise dramatically. This situation generally occurs in cholesterol feeding studies with rabbits. If the animal species, such as the rat, has large room for depressing cholesterol synthesis as well as great ability to enhance bile acid production, serum cholesterol does not necessarily increase. Thus after cholesterol consumption serum cholesterol concentrations are controlled by the efficiency of absorption, inhibition of cholesterol synthesis and the ability to activate bile acid synthesis (Table 6).

TABLE 6. Relative sensitivity of serum cholesterol to dietary cholesterol in various animal species

Species	Increase in serum cholesterol	Cholesterol absorption	Inhibition of cholesterol synthesis	Increase in bile acid excretion
Rabbit	↑↑↑	High	↓	-
Man	↑↑	Low	↓	-
Dog	↑	High	↓	↑
Rat	-	Low	↓↓↓	↑↑

6. CONCLUSIONS

I have addressed the problem of establishing criteria for choosing one animal model above another for carrying out experiments in the service of human cholesterol metabolism. Selection of the appropriate animal model depends on the particular question under study, but knowledge of qualitative and quantitative differences in cholesterol metabolism among species is required. It is important that the method of challenging cholesterol metabolism, e.g. by cholesterol feeding, is tailored to the characteristics of cholesterol metabolism in the animal model under study. If compensatory mechanisms are overwhelmed and cannot handle the challenge, cholesterol may accumulate in the body, resulting in pathological conditions, which in turn may lead to biased interpretation of the results.

In this paper attention has been focussed only on cholesterol balance accross the whole body and on serum total cholesterol. It should be emphasized that intravascular metabolism and transport of cholesterol can vary markedly among species (18). It should also be pointed out here that not only between animal species there are pronounced differences in cholesterol metabolism, but also between gender, individual animals and strains or breeds of one species (19).

REFERENCES

1. Lipid Research Clinics Program. The lipid clinics coronary primary prevention trial results. Parts I and II. JAMA 1984;251:351-374.
2. Spady DK, Dietschy JM. Sterol synthesis in vivo in 18 tissues of the squirrel monkey, guinea pig, rabbit, hamster, and rat. J Lipid Res 1983;24:303-315.
3. Dietschy JM, Wilson JD. Cholesterol synthesis in the squirrel monkey: relative rates of synthesis in various tissues and mechanisms of control. J Clin Invest 1968;47:166-174.
4. Dietschy JM, Siperstein MD. Effect of cholesterol feeding and fasting on sterol synthesis in seventeen tissues of the rat. J Lipid Res 1967;8:97-104.
5. Anderson JM, Turley SD, Dietschy JM. Relative rates of sterol synthesis in the liver and various extrahepatic tissues of normal and cholesterol-fed rabbits. Relationship to plasma lipoprotein and tissue cholesterol levels. Biochim Biophys Acta 1982;711:421-430.
6. Beynen AC, Katan MB, Van Gent CM. Endogenous cholesterol synthesis, fecal steroid excretion and serum lanosterol in subjects with high or low response of serum cholesterol to dietary cholesterol. Clin Nutr 1986;5:151-158.
7. Eggen DA. Cholesterol metabolism in groups of rhesus monkeys with high or low response of serum cholesterol to an atherogenic diet. J Lipid Res 1976;17:663-673.
8. Kim DN, Lee KT, Reiner JM, Thomas WA. Effects of a soy protein product on serum and tissue cholesterol concentrations in swine fed high-fat, high-cholesterol diets. Exp Mol Pathol 1978;29:385-399.
9. Mathé D, Chevallier F. Effects of level of dietary cholesterol on the dynamic equilibrium of cholesterol in rats. J Nutr 1979;109: 2076-2084.
10. Pertsemlidis D, Kirchman EH, Ahrens Jr EH. Regulation of cholesterol metabolism in the dog. I. Effects of complete bile diversion and of cholesterol feeding on absorption, synthesis, accumulation, and excretion rates measured during life. J Clin Invest 1973;52: 2353-2367.
11. Green MH, Crim M, Traber M, Ostwald R. Cholesterol turnover and tissue distribution in the guinea pig in response to dietary cholesterol. J Nutr 1976;106:515-528.
12. Huff MW, Carroll KK. Effects of dietary protein on turnover, oxidation, and absorption of cholesterol and on steroid excretion in rabbits. J Lipid Res 1980;21:546-558.
13. Beynen AC, Lemmens AG, Glatz JFC, Katan MB, Van Zutphen LFM. Rabbit models hypo- or hyperresponsive to changes in diet. In: Proceedings IX International Symposium on Drugs Affecting Lipid Metabolism (Paoletti R, Kritchevsky D, eds) Springer-Verlag, Berlin, 1987, in press.
14. Zilversmit DB. A single blood sample dual isotope method for the measurement of cholesterol absorption in rats. Proc Soc Exp Biol Med 1972;140:862-865.
15. Singhal AK, Finver-Sadowsky J, McSherry CK, Mosbach EH. Effect of cholesterol and bile acids on the regulation of cholesterol metabolism in hamster. Biochim Biophys Acta 1983;752:214-222.
16. Samuel P, McNamara DJ, Ahrens Jr EH, Crouse JR, Parker T. Further validation of the plasma isotope ratio method for measurement of cholesterol-absorption in man. J Lipid Res 1982;23:480-489.

17. Beynen AC, Danse LHJC, Van Leeuwen FXR, Speijers GJA. Cholesterol metabolism and liver pathology in inbred strains of rats fed a high-cholesterol, high-cholate diet. Nutr Rep Int 1986;34:1079-1087.
18. Chapman MJ. Animal lipoproteins: chemistry, structure, and comparative aspects. J Lipid Res 1980;21:789-853.
19. Beynen AC, Katan MB, Van Zutphen LFM. Hypo- and hyperresponders: Individual differences in the response of serum cholesterol concentration to changes in diet. Adv Lipid Res 1987;22: in press.

A.C. Beynen & H.A. Solleveld (Eds), New developments in biosciences: their
implications for laboratory animal science, ISBN 978-94-010-7973-0
© 1988 Martinus Nijhoff Publishers, Dordrecht.

ANIMAL MODELS IN HEMOSTASIS AND THROMBOSIS

H.C. ROWSELL

INTRODUCTION

The hemostatic mechanism comprises a complexity of interactions
between components in the intravascular compartment, such as elements
contained in the mechanisms of blood coagulation, blood platelets, and the
vessel wall. Essential to life is the balance of activities in the
homeostatic state; otherwise, blood would be a gel rather than a fluid.
The former would not be compatible with life, which depends on the fluid
state of blood except in injury to the blood itself or the vessel wall
(1), resulting in a hemostatic plug or a thrombus.

Thrombus formation involves interaction of constituents in the vessel
wall, activation of the blood coagulation mechanism, platelet aggregation
and release of thrombogenic activators. Thrombi can result in serious
clinical problems and even death for man and animals. Conversely, without
the formation of a hemostatic plug involving the same constituents and
mechanisms, life can be endangered, as the hemophilic human or animal
exemplifies (2-10).

Hemostasis involves clotting factors in the extrinsic system or
tissue pathway and the intrinsic compartment pathway. Primary hemostasis
consists of a series of reactions following vascular trauma. Hemostasis
involves release of tissue thromboplastins, as well as activation of
Factor VII and Factor X, platelet activation - release of Platelet Factor
3 (PF3), activation of von Willebrand's Factor (VWF), adhesion of
platelets to subendothelial elements (collagen), resulting in release of
platelet constituents, adenosine diphosphate (ADP) lipozymes,
phospholipids, etc., resulting in aggregation of additional platelets (11-
13). Activation of coagulation factors, conversion of prothrombin to
thrombin and fibrinogen to fibrin, polymerization of fibrin, activation of
Factor XIII (fibrin-stabilizing factor), associated with release of
vasoconstrictor substances (epinephrine) enhancing vascular response,
coupled to the formation of the platelet or hemostatic plugs, effectively
produces cessation of bleeding and a stable hemostatic plug.

The requirements of "adequate veterinary care" and management are
essential in the utilization of spontaneous, hereditary or acquired animal
models of hemorrhagic disease, or those experimentally induced. Bleeding
animals can suffer significant pain and distress. Therefore, the study of
these diseases requires the investigator and all of those responsible for
the care and handling of these animals to be guided by sensitivity and
humane and ethical considerations, as well as an understanding of the
hemostatic mechanism.

The prevention of pain and distress can best be handled by keeping
the animal quiet and isolated from noise and extraneous, arousing stimuli.
In most cases, the use of analgesics and tranquilizers is contra-

indicated because of their effect on the reactivity of blood platelets, the aggregation and release mechanism, an essential in hemostasis.

The provision of "tender loving care" is the best treatment to assure the comfort of the animal and a return to a homeostatic balance. The provision of fresh active deficient factors, i.e. normal plasma, or platelet rich plasma, are essential for hemostatsis.

HEMOPHILIA

Congenital diseases such hemophilia are expressed in the same manner in animals as in humans (8, 14, 15).

Although it was obvious that in hemophilia an injury to the vessel wall resulted in the formation of a poor or inadequate hemostatic plug, it was not until the late 1970's that the mechanisms were revealed (8, 16, 17).

In the vessel wall, especially in the endothelium, prostaglandins are synthesized from arachidonic acid derived from phospholipid metabolism. The major end product at the site of the vessel wall is the production of prostaglandin I_2 (prostacyclin), an inhibitor of platelet aggregation and a potent vasodilator. When vascular injury occurs, platelet stimulation results. Arachidonic acid from the platelet membrane, following the same pathway as prostacyclin, is acted upon by thromboxane synthetase from the injured aggregating platelets to form Thromboxane A^2 in circulating blood platelets (11, 18, 19, 20), this accelerates platelet aggregation and release.

Recently, research has further elucidated mechanisms which explain the maintenance of the fluid state of blood although the hemostatic mechanism may be activated. Thrombomodulin (T.M.) is a cell surface protein found on endothelial cells that binds thrombin and increases its ability to activate protein C, a serine protease zymogen. Activated protein C becomes a potent anticoagulant, selectively inactivating Factors V and VIII, as well as modulating the fibrinolytic system (21).

ANIMAL MODELS IN THE STUDY OF HEMOSTASIS

The importance to research in finding an animal model is how accurately that model resembles the disease state in human subjects. As well, a model may prove useful if it provides new insights, or allows the examination of single or multiple factors involved in the production of disease (22, 23).

The search for animal models for bleeding disorders such as hemophilia has been spurred on by the needs and demands of hemophilic patients. Such individuals long for the day when the researcher will develop an agent which, when injected, (perhaps not necessarily into a vein), would prevent all bleeding and its associated pain.

Giles (24) has demonstrated the value of a dog with naturally-occurring hemophilia, reporting on the genetic engineering of Factor VIII (Hemophilia A) which could in future lessen the risk of serious problems such as impurities and blood-borne viral contamination from diseases such as hepatitis and Acquired Immunity Deficiency Syndrome (AIDS).

Giles, Mann and Nesheim (25) have developed a bypass product, Factor Eight Bypassing Activity (FEBA), for the 10-15% of hemophiliacs who form antibodies against Factor VIII and thus prevent replacement Factor VIII from forming a stable hemostatic plug. This bypass product (FEBA) consists of small amounts of Factor X combined with coagulant active phospholipids.

The blood clotting factors and synonyms are:

INTERNATIONAL   SYNONYMS
CLASSIFICATION*

FACTOR I -   FIBRINOGEN
FACTOR II -   PROTHROMBIN
FACTOR III -   TISSUE THROMBOPLASTIN
FACTOR IV -   CALCIUM
FACTOR V -   PROACCELERIN, Labile factor, Accelerator
               globulin (ACg)
FACTOR VII -   PROCONVERTIN, Serum prothrombin conversion
               accelerator (SPCA), Stable Factor, Autoprothrombin I
FACTOR VIII - ANTIHEMOPHILIC FACTOR (AHF), Antihemophilic globulin
               (AHG), Platelet cofactor I, Plasma thromboplastic
               factor A
FACTOR IX -   CHRISTMAS FACTOR, Plasma thromboplastin component (PT)
FACTOR X -   STUART FACTOR, Stuart-Prower factor
FACTOR XI -   PLASMA THROMBOPLASTIN ANTECEDENT (PTA)
FACTOR XII -   HAGEMAN FACTOR
FACTOR XIII - FIBRIN-STABILIZING FACTOR (FSF), Fibrinase, Laki-
               Lorand Factor

*As recommended by the International Committee for the Nomenclature
of Blood Clotting Factors (1962). The most commonly used synonym for each
factor is capitalized.

The hereditary, acquired animal models may be associated with a
deficiency of one of the thirteen blood clotting factors, abnormalities in
platelets, the vessel wall or fibrinogen. Deficiencies have been observed
in Factors VII, VIII, IX, X, XI, and XII, platelets and fibrinogen.
Coagulation factors have been described to act in a "waterfall" or
"cascade" sequence. Those which are involved in the latter portion of the
cascade, such as Factor VIII and Factor IX which convert Factor II or
prothrombin to thrombin, produce the more serious bleeding disorders.

Factor VIII Deficiency (Hemophilia A) and Factor IX Deficiency
(Hemophilia B) are both inherited through the X chromosome. As is the
case with colour blindness, the females carry the disease, passing it on
to their male offspring. However, through selective breeding of
hemophilic dogs, Double Hemophilia AB, an extremely dangerous disease, has
been produced.

Hemophilia A and B both result from the deficiency of proteins which
have essential but distinct roles in blood coagulation. The two X-linked
hemophilias result from changes at two distinct alleles of the X
chromosome. The Hemophilia A allele (Factor VIII) is one of the cluster
of genetic loci which include protan and deutan colour blindness, and are
linked to the polymorphic marker glucose - 6 - phosphate dehydrogenase
(G6/PD). The Hemophilia B (Factor IX) allele is not close to the Factor
VIII allele, but is close to the tip of X chromosome long arm, which is
close to the site of mental retardation. The lack of close linkage of
Factor VIII and Factor IX may reflect local distortion of recombinational
activity around this site in normal females.

The hereditary bleeding disease animal models have been described as
follows:

Factor VII Deficiency - Beagle.

Factor VIII Deficiency (Hemophilia A, Classic Hemophilia) - Cat-
domestic shorthair; Dogs-Beagle, Collie, Labrador Retriever, Greyhound,
Irish Setter; ten mutant lines and crossbreeds.

Factor IX Deficiency (Hemophilia B, Christmas Disease) - Cat-British Shorthair; Dogs-Cairn Terrier, St. Bernard, Old English Sheepdog, Alaskan Malamute, crossbreeds.

Double Hemophilia AB - Factor VIII and Factor IX Deficiencies, crossbred dogs, a very severe disease.

Von Willebrand's Disease - Autosomal incompletely dominant; Variable Factor VIII deficiencies, Von Willebrand's Factors - platelet, vessel wall. Disease severe in 15% of Schnauzer-Golden Retriever, 60% of Doberman Pinscher. Autosomal recessive - Scottish Terrier, Chesapeake Bay Retriever. Severe to mild disease is found in swine, both in Poland China and Yorkshire-Hampshire crossbreeds. It is also found in Flemish Giant-Chinchilla rabbits.

Factor X Deficiency (Stuart Factor) - Severe or lethal bleeding diathesis in newborn, young dogs. Fading puppy syndrome. Mild forms survive. Autosomal inheritance.

Factor XI Deficiency. Plasma thromboplastin antecedent (PTA) Deficiency is found in the dog (Kerry Blue) and cow (Holstein). Autosomal interitance. Mild disease.

Factor XII (Hageman Trait) is found in the cat (mixed breeds). Autosomal recessive. Mild disease. Species normally lacking Factor XII (Hageman Trait) include marine mammals, birds, and reptiles.

An animal model of fibrinogen deficiency is the dog (Borzoi). Autosomal inheritance. Mild disease.

The animal models with hereditary disease platelet models are:

Thrombasthenic Thrombopathia (Glanzmann's disease) is found in the dog (Otterhound). Platelet membrane, glycoprotein defect, autosomal inheritance.

Familial Thrombopathia, autosomal inheritance reduced platelet retention and aggregation especially in concentrations of Adenosine Diphosphate (ADP) in: the cow-(Simmental), dog-(Basset hound), mouse, pale ear pigment, mutant (ep/ep) on a C57BL/6J background, rat-(fawn hooded (FH)). The Chediak-Higashi Syndrome is found in the mouse, cat, cow, and mink.

Platelet defects have been described with defects in platelet Arachidonate Pathway in the dog, cat, horse, cow, pig, and mink. Platelet dysfunction secondary to abnormal response to Thromboxane A^2 has been found in the dog (mongrel).

There are also some animal models of the miscellaneous hereditary diseases of hemostasis such as familial hypersensitivity to warfarin in the dog (Boxer), warfarin resistance in rats, congenital C-4 complement deficiency in the guinea pig (C_4D stock). Congenital C-6 complement deficiency is found in the rabbit (Freileung strain).

Experimentally-induced disease animal models in hemostasis and thrombosis may be produced by the intravascular stimuli which produce platelet aggregation and thrombosis. These would include immunological reactions, bacteria, and the viruses.

Induced bleeding disorders may be caused by an ever increasing number of drugs such as Heparin, Aspirin, Sulfinpyrazone, Indomethacin, Hydroxychloroquine, anabolic steroids, stanoxolol, Phenylbutazone, most promazine tranquillizers and anaesthetics.

CONCLUSION

As in the study of other animal models of human disease, a knowledge of the condition is essential, as is a continued re-evaluation of our conventional and customary attitudes towards the animals themselves.

Through such means, we will contribute not only to the well-being of man, but also to that of animals. Animal and human species suffering from defects in hemostasis whether spontaneous, hereditary, acquired, or induced, will benefit from the acquisition of knowledge through research in the study of hemostasis and thrombosis. Thus, animal models have made and will make, important contributions.

REFERENCES

1. Mustard JF and Packham MA: The reaction of blood to injury. In: Inflammation, immunity and hypersensitivity. (ed.) Movat HZ. New York: Harper & Row, pp. 557-664, 1979.
2. Dodds WJ: Hereditary and acquired hemorrhagic disorders in animals. In: Progress in hemostasis and thrombosis, Vol. 2. (ed.) Spaet, TH. New York: Grune and Stratton, p. 215, 1974.
3. Dodds WJ: Blood coagulation: Hemostasis and thrombosis. In: CRC handbook of laboratory animal science. Vol. 2. (eds.) Melby EC, Altman NH. Cleveland: Chemical Rubber Company, p. 85, 1974.
4. Dodds WJ: Bleeding disorders. In: Textbook of veterinary internal medicine. Vol. 2. (ed.) Ettinger SJ. Philadelphia: WB Saunders, p. 1679, 1975.
5. Dodds WJ: The diagnosis, management and treatment of bleeding disorders. Parts 1 and 2. Mod Vet Pract 58: 680-684 and 756-762, 1979.
6. Dodds WJ: Inherited hemorrhagic defects. In: Current veterinary therapy VI Ed. Philadelphia: WB Saunders, p. 438, 1977.
7. Dodds WJ: Inherited bleeding disorders. Canine Pract 5: 49-58, 1978.
8. Dodds WJ: Hemorrhagic disorders. In: Spontaneous animal models of human disease. Vol. 1. (eds.) Andrews EJ, Ward BC and Altman NH. New York: Academic Press, p. 266, 1979.
9. Spurling NW: Hereditary disorders of hemostasis in dogs: A critical review of the literature. Vet Bull 50: 151-173, 1980.
10. Landi MS and Higson JE: Factor VII deficiency in colony bred mongrels. Lab Anim Sci 32: 429, 1982.
11. Kinlough-Rathbone RL, Packham MA and Mustard JF: Platelet aggregation. In: Methods in haematology. Measurements of platelet functions. (eds.) Harker LA and Zimmerman TS. Edinburgh: Churchill Livingstone, pp. 64-91, 1983.
12. Kinlough-Rathbone RL et al: Factors influencing the deaggregation of human and rabbit platelets. Thromb Haemostas 49(3): 162-167, 1983.
13. Craig IW and Goodfellow PA: Biology of disease: Molecular genetics and the X chromosome and x-linked diseases. Lab Invest 54: 241-253, 1985.
14. Badylak SF, Dodds WJ and Van Vleet JF: Plasma coagulation factor abnormalities in dogs with naturally-occurring hepatic disease. Am J Vet Res 44(12): 2336-2340, 1983.
15. Johnston IB and Crane S: Hemostatic abnormalities in equine colic. Am J Vet Res 47(2): 356-358, 1986.
16. Mustard JF: Function of blood platelets and their role in thrombosis. The Gordon Wilson Lecture. Trans Am Clin Climatol Assoc 87: 104-127, 1976.
17. Mustard JF: The role of platelets and thrombosis in the development of atherosclerosis and its complications. Ann R Coll Phys Surg Can 14(1): 22-28, 1981.

18. Packham MA and Mustard JF: Normal and abnormal platelet activity. In: Blood platelet function and medicinal chemistry. (ed.) Lasslo A. New York: Elsevier, pp. 61-128, 1984.

19. Mustard JF, Kinlough-Rathbone RL and Packham MA: Platelets, endothelium and vessel injury. In: Advances in prostaglandin, thromboxane and leukotriene research. Vol. 13. (eds.) Neir Seineri GG et al. New York: Raven Press, pp. 235-245, 1985.

20. Langille BL, Reidy MA, and Kline RL: Injury and repair of endothelium at sites of flow disturbances near abdominal aortic coarctations in rabbits. Arteriosclerosis 6(2): 146-154, 1986.

21. De Bault LE, Esmon NL, Olson JR and Esmon CT: Distribution of the thrombomodulin antigen in the rabbit vasculature. Lab Invest 54: 172-178, 1986.

22. Institute of Laboratory Animal Resources (ILAR): Animal models of thrombosis or hemorrhagic diseases - Workshop on animal models of thrombosis and hemorrhagic diseases. DHEW Publication No. (NIH) 76-972, 1975. Available from: Institute of Laboratory Animal Resources (ILAR), National Academy of Sciences, 2101 Constitution Ave. N.W., Washington, D.C. 20418, U.S.A.

23. Dodds WJ: Second international registry of animal models of thrombosis and hemorrhagic diseases. ILAR News, 1981. Available from Institute of Laboratory Animal Resources, National Academy of Sciences, 2101 Constitution Ave. N.W., Washington, D.C. 20418, U.S.A.

24. Giles AR, Nesheim ME and Mann KG: Studies of Factor V and VIII:C in an animal model of disseminated intravascular coagulation. J Clin Invest 74: 2219-2225, 1984.

25. Giles AR, Mann KG and Nesheim ME: A combination of Factor Xa and phosphatidylcholine - phosphatidyl serine vesicles bypass Factor VIII in vivo. In press.

A.C. Beynen & H.A. Solleveld (Eds), New developments in biosciences: their
implications for laboratory animal science, ISBN 978-94-010-7973-0
©1988 Martinus Nijhoff Publishers, Dordrecht.

What's your diagnosis?
MICE WITH ULCERATING LESIONS ON THE NOSE, LIMBS AND/OR TAIL.

Authors: van Herck, H. [1], Baumans, V [2].

1 Div. of Lab. and Special Animal Diseases,
  Department of Vet. Pathology,

2 Department of Laboratory Animal Science,

  University of Utrecht,
  Yalelaan 1,
  3584 CL Utrecht,
  The Netherlands.

1 HISTORY AND PHYSICAL EXAMINATION:

An adult male mouse (Strain Sc/CpbU), was brought for
euthanasia
and necropsy. The animal showed a ruffled fur,
coniunctivitis of the left eye and superficial ulcerating
lesions on the nose and on digits of the paws.
The animal originated from a conventionally housed colony
of mice (C57Bl/U, Zo:Wk, Sc/CpbU).
Anamnesis mentioned also a sick male C57Bl/U mouse in the
same colony which showed an ulcerating lesion on the distal
part of the tail. A wild mouse with an ulcerating tail was
seen in the same room.
Thorough examination of the colony revealed several mice
with lesions on the nose, the limbs and/or tail.

Because of the possibility of an ectromeliavirus infection
it was advised to isolate the colony and to take
appropriate hygienic precautions.

2 ECTROMELIA:

Ectromelia (mousepox) is a devastating disease of mice
caused by an orthopoxvirus. The name ectromelia is derived
from the grecian words ectroma= abortion
                        melos = limb indicative for the
amputation of tail and/or limbs due to the necrosis and
inflammatory reaction occurring in the course of an
epidemic ectromeliavirus infection.

Clinical course:
The clinical course of an infection with ectromeliavirus varies between an epidemic and an endemic form in a colony of mice, depending on the susceptibility of the host and the virulence of the virus. Resistant mice are C57Bl/6 and AKR. Known to be susceptible are DBA, Balb/C, A and C3H mice.
The epidemic clinical course is characterized by acute lethal infections without any clear clinical signs. In milder cases poxlesions may develop in the skin. Severe infection may lead to amputation of the tail and/or feet.
The endemic course shows in ultimate form no clinical signs at all. Laboratory diagnostics reveal carriers of the virus or animals who have antibodies against it.

Pathogenesis:
The virus enters the body through small skin lesions or through lesions in the upper respiratory tract. Further multiplication takes place in the regional lymphnode. Having passed the efferent lymph vessels the virus enters the blood stream and causes a (primary) viremia. Liver and spleen become infected. Here the virus is multiplicated again, which results in a secundary viremia.
The secondary viremia may cause acute death. In less severe cases several other organs become infected resulting in edemas, skin lesions and coniunctivitis. Predelection sites are the limbs and tail.

Postmortem:
In sick animals major findings are necrosis in spleen, liver and skin. Necrosis in thymus, lymphnodes, Peyer's patches, intestinal mucosa and genital tract also have been observed. A histological examination can reveal inclusionbodies in the cytoplasm of cells of lymphnodes, spleen, liver, pancreas and skin.

Diagnostics:
Clinical signs, autopsy and histology are useful help to set the diagnosis of ectromeliavirus infection in a colony of mice. Proof is the demonstration of antibodies against the virus or the demonstration of the virus itself. For this several laboratory diagnostics are available. A number of these methods are based on the fact that ectromeliavirus, being an orthopoxvirus, is closely related to vacciniavirus.

3 LABORATORY DIAGNOSTICS:

Because of the fact that in the colony mentioned several mice showed lesions on the nose, limbs or tail, diagnostics were primarily headed for ectromeliavirus:
Two mice were used for autopsy. Blood from the tail was collected from 9 mice. Sera were checked against vacciniavirus in an Immuno Fluorescence Assay (IFA).

Two mice which showed lesions of the limbs, nose and tail were used for an attempt of virusisolation using Mouse Embryonic Cells (MEC).
Cryostat sections from the liver, lungs, nose, front- or hindlimb and tail of 5 mice showing lesions were used in an IFA against vaccinia antiserum.
Nine days later blood was collected for the second time from 8 mice this time for serodiagnosis (IFA).

4 RESULTS:

Necropsy of the two mice showed an infestation with mites (Myobia musculi and Myocoptes spp.) and inflammation of digit II of both the front legs. Macroscopically no other abnormalities were found. Histology showed proliferative inflammatory reactions of the subcutis. No inclusionbodies were found. In one mouse, in an ulcerating lesion of the distal part of the leg several starchgranules were observed. Liver, lungs, heart, kidneys and pancreas appeared to be normal microscopically.

Both series of sera were negative against vaccinia in the IFA. After two passages on MEC-cells, no cytopathogenic effect was seen, so virusisolation was negative.
The IFA on the cryostat sections of the tissues mentioned also apperared negative: no ectromeliavirus antigens could be demonstrated.

5 DIAGNOSIS AND DISCUSSION:

During the laboratory investigations, the number of mice showing clinical problems increased. All of them showed lesions on the nose, tail and/or digits of front or hind limb(s). In case digits were involved, often digit II was affected.
None of the animals died spontaneously. There was no preference for one of the strains of mice in particular.
Only males appeared to be involved. Both males housed individually and in groups were affected.
As soon as the investigations concerning a mouse-pox infection appeared negative, three affected mice were housed individually and placed in another building. They were checked twice a week. In about three weeks the lesions had healed and the mice appeared healthy again.
At the same time on a thorough search of the room where the colony was housed, several wild mice with litters were found. The suspicion of fights between wild mice and the male laboratory mice arose. A video-set was used for observation of the mice during the night. The video-tapes showed a wild mouse on the cages of the laboratory mice.

298

In the animal room concerned a rodenticide was placed. In about one week the wild mice were eradicated.
No more laboratory mice were affected then and in about three weeks the lesions of the animals already affected had healed.

The localisation of the lesions on the laboratory mice, the fact that only males were affected both housed individually and in groups, the presence of wild mice in the animal room and the healing of the laboratory mice after eradication of the wild ones, all these findings together add up to the diagnosis: mice with ulcerating lesions on the nose, limbs and/or tail due to fights with wild mice through the grids on the cages.

Acknowledgement: Authors wish to thank dr. A.D.M.E. Osterhaus (institute for Public Health and Environmental Hygiene) for his cooperation in the virological part of the laboratory diagnosis.

A.C. Beynen & H.A. Solleveld (Eds), New developments in biosciences: their implications for laboratory animal science, ISBN 978-94-010-7973-0
© 1988 Martinus Nijhoff Publishers, Dordrecht.

Spontaneous Hyperplasia of the Endometrium in the rabbit

Zwart,P. D.V.M., PhD.
Dept. Vet. Pathology, Division Laboratory and Special Animal Diseases, Yalelaan 1, 3584 CL UTRECHT, The Netherlands

1. INTRODUCTION

Only limited studies have been performed to analyse the causes of a decline in fertility with age of rabbits. Adams (1970) presented an overview of aging and reproduction in the female mammal and the rabbit in particular. In laboratory rabbits reproductive performance stops at 16 - 52 monthes (Adams, 1960). The number of ovulations as well as the implantation rates reveal only minor variation; there is a relatively constant level of prenatal mortality in the first three pregnancies and an increasing level of prenatal mortality in later pregnancies. By the eleventh pregnancy the number of foetusus surviving to term was significantly reduced. Rabbits can live for many more years after reproductive performance has come to an end.
Cystic endometrial hyperplasia is known in many species of domestic animals such as cattle, sheep, horses, swine, dogs and cats. In cattle, the condition is attributed to hyperoestrinism. In sheep, certain legumes with oestrogenic activity are known as causes of endometrial hyperplasia, while in swine oestrogenic mycotoxins have been incriminated (Jones and Hunt, 1983)
In rabbits, cystic endometrial hyperplasia has been in association with pseudopregnancy due to retained corpora lutea (Jones and Hunt, 1983). Oestrogen induced glandular hyperplasia of the endometrium has been used as a model in the study of uterine carcinoma (Baba et al., 1970, Baba and Haam, 1967).
More detailed morphological studies about endometrial hyperplasia to the best of our knowledge are lacking. This study aims at giving a contribution in this area.

2. MATERIALS and METHODS

Cases investigated originated from several sources. 10 Animals were send for consultancy. 37 more cases of endometrium hyperplasia originated from two groups of laboratory rabbits. One of female White Viennese x Alaska rabbits used for a lifespan feeding experiment. These

animals had never been mated. A second group were Alaska and White Viennese female rabbits which had reproduced up to the age of 2 to 3 years. Materials from these two colonies had been collected with the aim of studying endometrial carcinoma. 81 Carcinomas had been found and the frequency of coinciding endometrial hyperplasia was recorded (Elsinghorst, 1987).
After post mortem materials were fixed in 10 % buffered formalin, processed according to usual techniques; slides were stained with Haematoxylin and Eosin and in part with the van Gieson stain.

3. RESULTS

The normal endometrial mucosa was covered with a single layer of columnar cells. The epithelium of the tuba was ciliated. In the uterus groups of ciliated cells were present. Both in the tuba and the uterus, the mucosa contained relatively few glands. Normal glands were rounded or slightly elongate. The endometrial stroma was sparce and consisted of elongated fibrocytes with a moderate number of blood vessels.
The coinciding frequency of endometrial hyperplasia was recorded from 81 animals belonging to the two groups of laboratory rabbits and suffering from endometrial carcinoma. Animals varied in age from 4 years and 1 month to 9 years and 7 monthes. The animals were grouped according to their age in years. A survey is presented in Table 1.

Table 1
Hyperplasia of the endometrium in relation to age of rabbits.
===

Age(Yrs)	Number	Endometr.hyperpl.
4	10	4
5	11	5
6	30	12
7	20	7
8	5	3
9	5	5

===

It appeared that the frequency of endometrial hyperplasia was age related. In addition, 55 % of rabbits suffering from adenocarcinoma of the endometrium had also developed endometrial hyperplasia.

Macroscopy of endometrial hyperplasia revealed a wide variation in aspect. In all cases the uteri were enlarged. This however could reach extremes to such an extend that in one case an animal of 1100 gms had an uterus of 400 gm. After opening the uteri, individual cysts varyiing in size could be seen connected to the mucosa with a narrow stalk, while in other cases they had an broad basis or presented themselves as conglomerates of cysts rendering the lumen the aspect of swiss cheese. In many cases there also were more compact carcinomas bulging into the lumen.

Microscopically, hyperplasia of the endometrium was roughly divided into two different aspects, depending on the predominant feature. I.e. 1) cystic hyperplasia of the endometrium and 2) glandular hyperplasia.

ad 1) Cystic hyperplasia of the endometrium was characterised by the presence of dilated glands, which formed cystic structures demonstrating a great variety in size. There also was a great variety in complexity. Occasionally one isolated cyst was present. Generally however the changes were diffuse, leading to the presence of a great number of predominantly rounded cysts which showed a wide range in size. The internal epithelial lining of cysts mainly consisted of flattened cells; in some cases interspaced with groups of ciliated cells. Incidentally cysts were completely lined by a thin layer of squamous epithelium. In one case all cysts were covered with a columnar epithelium of mucus producing cells. In this particular case the cysts were filled with mucus.
Complications were the finding of bloodclots in the cysts and thrombosis of bloodvessels in the stroma.
In the tubae identical changes occurred. The aspect differed in that the cysts were covered with a 100 % ciliated epithelium.
Polypoid proliferations could be observed. Larger polyps with a loose structured, well vascularised stroma prominated between cystic areas.

ad 2) Glandular hyperplasia was characterised by an increase in the number of glands which were closely packed with little stroma between the glands. In one case it was noticed that a gradual increase of glands occurred over a short distance, intensifying to a situation where several layers of glands could be found. Irregularly distributed in the area, some of these closely packed glands showed cystic dilatation.
In glandular hyperplasia; the individual glands may have a high columnar epithelium. The nuclei principally beiing located basically in the cells though at different

heights. Infolding of the epithelial lining as well as outgrowth of the glandular circumference occurred, making the glands often complex in structure. In several cases it was noticed that nests of glands which were either of normal size or distinctly hypertrophic, could be found in areas of cystic dilated glands indicating the focal proliferation of glands. The glands were embedded in a well vascularised loose structured propria. Only incidentally additional pathologic changes such as gland abscesses and small precipitations of calcium were found in glandular lumina.
Glandular hyperplasia also was a feature in the tubae. In such cases glands were provided with a ciliated epithelium.

Endometrial carcinoma existed in all 81 cases of the laboratory rabbits and in 1 of the 10 derived from other sources.

4. DISCUSSION

From our results it was obvious that hyperplasia of the endometrium occured preferably at an age of 4 years and onwards.
A differentiation in cystic hyperplasia of the endometrium and glandular hyperplasia seemed to be debatable because this mainly indicated a gradual difference. There were many intermedate stages where larger cysts could be found in addition to proliferation of glands.
There has been much discussion about the carcinogenicity of oestrogen in the rabbit (Weisbroth, et al. 1974, Baba and Haam, 1967). There is little experimental evidence to support the role of oestrogen as a factor in spontaneous endometrial carcinoma in the rabbit. On the other hand glandular hyperplasia can be induced experimentally with oestrogens with sufficient certainty to be used as a model in research (Baba and Haam, 1967).
The observations of Elsinghorst (1987) incorporated in part in this study, indicate that in about 55 % of uterine adenocarcinomata of rabbits endometrial hyperplasia was present. This too was not in favour of endometrial hyperplasia as a precancerous lesion. Further arguments were that electron microscopy had revealed evidence of functional differences between hyperplastic and neoplastic cells (Baba and Haam, 1967). The hyperplastic cells were rhich in lysosomes. In addition neoplastic cells lacked both alkaline and acid phosphatases, while in glandular hyperplasia only alkaline phosphatase was lacking in the epithelial cells.
Our observations support the fact that endometrial hyperplasia is age related and may contribute to the

decline in reproductive performance of rabbits.
A separation between cystic hyperplasia of the endometrium
and glandular hyperplasia is only of limited value, this
depending on the predominant feature to be recognised at
microscopy. The pathological phenomenon described herein
could be designated as endometrial hyperplasia.

5. LITERATURE

Adams, C.E.(1970): Studies on prenatal mortality in the
rabbit; Oryctolagus cuniculus: the amout and distribution
of loss before and after implantation. J. Endocr. **19**, 325.

Adams, C.E.(1970): Ageing and reproduction in the female
mammal with particular reference to the rabbit. J. Reprod.
Fert. Suppl. **12**, 1-16.

Baba, N. and von Haam, E.(1967): Experimental carcinoma of
the endometrium. Progr. Exp. Tumor Res.**9**, 192-260.

Baba,N., Vidyarthi, S., von Haam, E. (1970): Nonspecific
phosphatases of rabbit endometrial carcinoma. Arch.
Pathol. **90**, 65-71.

Elsinghorst, Th.A.M. (1987): Pers. comm.

Jones, T.C., Hunt, R.D.(1983): Veterinary Pathology. Lea
and Febiger, Philadelphia, p 1'526.

Weisbroth,S.H. et al. (1974): The biology of the
laboratory rabbit. Academic Press New York, pp 337-338.

A.C. Beynen & H.A. Solleveld (Eds), New developments in biosciences: their implications for laboratory animal science, ISBN 978-94-010-7973-0

MONKEY GRAFFITY (environmental deficiency, boredom or artistic drive?)

D. Wesseling*, V. Baumans*, C. Goosen**.
*Department of Laboratory Animal Science, University of Utrecht,
**Primate Center TNO, P.O. Box 5815, 2280 HV Rijswijk, The Netherlands

1. INTRODUCTION
    Monkey graffity is an invented name for strange behaviour of rhesus monkeys in the Primate Centre TNO in Rijswijk. These monkeys spread their own faeces on the walls of their cages and on all other places within their reach. This is not only annoying for the animal technicians, who have to clean the cages every day, but it can also be regarded as abnormal behaviour, because rhesus monkeys normally avoid contact with their faeces. Abnormal behaviour is an indication for discomfort (Fox, 1984), and for this reason a study of this behaviour was made.
The reason for this behaviour was unknown. It was suggested that the development of this behaviour was related to boredom. The impression was that over the years the activity had become more widely spread among colony members, but that wildborn monkeys, monkeys housed in groups and female with infant did not show this behaviour at all.
The study was focussed on the following questions:
- How many rhesus monkeys show this behaviour?
- Which rhesus monkeys show this behaviour?
Animals of different age, number of years in captivity and source were studied.

2. MATERIAL AND METHODS
    Rhesus monkeys (Macaca mulatta) originally come from South East Asia. The Primate Centre TNO in Rijswijk owns a few thousand rhesus monkeys. These animals are partly wildborn animals and partly animals of the Primate Centre's own breeding unit, who are born and raised in the Centre itself. Wildborn animals arrived at the Primate Centre at ages varying from about one year to full adulthood. The monkeys are housed either individually in complete wire cages (separated by solid metal partitions), or in groups. Most adults are housed individually in small cages (60x60x80 cm) together with other monkeys in one room. Some adults are housed in groups for breeding purposes. Pregnant females and females with infants/babies are housed individually. Young monkeys are weaned at the age of three months and are housed together in groups until they are two years old. A number of 538 cages were inspected. From 31 animals the information was not complete. Therefore, the results of the remaining 507 cages were used for this study. That means 17 groups and 490 individually housed rhesus monkeys, among which 43 females with babies. Identification of the animals was possible because of a figure and/or letter combination, which was tattooed on the chest of the monkeys. If the animals were housed in breeding groups, only the name of the male was noted. If a group consisted of young monkeys, their year of birth was noted. Every morning between 9.30 and 11 a.m. the cages were cleaned.

So the cages were inspected before 10 a.m. or in the afternoon. All 507
cages were observed once during three days successively. The absence or
presence of "paintings" in or in the neighbourhood of the cage was noted.
Absence was noted as "negative" (= monkey does not show this behaviour),
and presence was noted as "positive" (= monkey shows this behaviour).
Only the clean, negative cages of lab born animals and of groups were
inspected for a second time, because of the possibility of a false
negative result in these cages.
The age and number of years in captivity of the wildborn animals were
compared to the age (= the number of years in captivity) of the lab born
animals. These results are depicted in a drawing (fig. 1). It shows that
the wildborn monkeys are not only much older than the lab born monkeys,
but also much longer in captivity.

FIG.   AGE AND NUMBER OF YEARS IN CAPTIVITY OF RHESUS MONKEYS

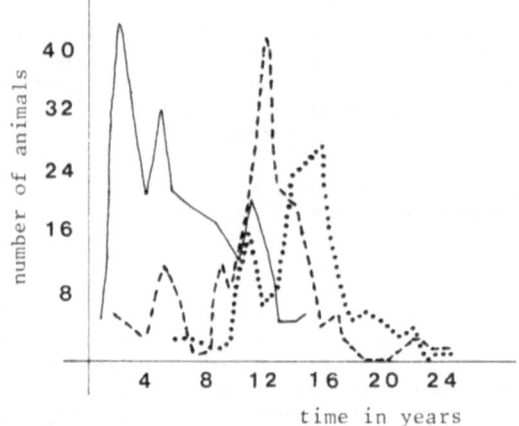

_____ age of lab born animals

_ _ _ .. _   number of years in captivity of wild born animals

••••••••••• age of wild born animals

3. Results

3.1. The wildborn rhesus monkeys.

After inspecting the cages of the wildborn animals for the first time, it appeared that the wildborn animals were largely negative. From 179 individually housed animals 3 were positive and 176 negative, among which 17 females with infant. The three positive wildborn rhesus monkeys were 16, 11 and 9 years old, and were 12, 8 and 5 years in captivity. Because of these results, inspection of the cages of the wildborn animals for a second time was not carried out.

3.2. The lab born rhesus monkeys.

The results regarding to the cages of the lab born rhesus monkeys were completely different. From 311 individually housed animals, 10 rhesus monkeys were born and weaned in safari park "De Beekse Bergen", and arrived at the Primate Centre when they were older than one year. These "semi" wildborn animals were all negative. From the remaining 301 animals were 156 positive, among which 14 females with infant, and 145 negative, among which 12 females with infant. The years of birth of the lab monkeys were compared to the results of the inspection of their cages:

	156 positive	145 negative
till '71	2 (13%)	13 (87%)
71 - 75	19 (35%)	35 (65%)
76 - 80	36 (40%)	55 (60%)
81 - 85	99 (69%)	42 (31%)

The difference between the group 1976-1980 (40% positive) and the group 1981-1985 (69% positive) is statistically significant.

A closer look at the group shows the significance:

	positive	negative
1981	25 (74%)	9 (26%)
1982	16 (76%)	5 (24%)
1983	31 (86%)	5 (14%)
1984*	25 (56%)	20 (44%)
1985	2	3

*From 1984 not all animals were housed individually; There were still 19 animals in a group.

3.3. The rhesus monkeys in groups.

Seventeen groups of rhesus monkeys were present. Five groups consisted of young, lab born animals. The groups 1984 (19 animals) and 1985 (52 animals) were positive at the time of the study (October 1986). "Positive" means that in these groups at least one monkey makes "paintings". The group 1986 (28 animals) was negative at the time of the study. The animals in this group were 3-11 months old at that time. "Negative" means that not one rhesus monkey in this group is positive. The remaining 12 groups were breeding groups. There were 7 positive and 5 negative breeding groups. Composition of the groups of monkeys were as far as the members' age and rearing background is concerned given in Table 1. If lab-born monkeys when housed in a group have the same likelihood of making graffiti as have singly housed monkeys, then the likelihood that the group makes graffiti increase as the number of lab-born animals in the group increase.

For the adult monkeys (age > 5) the expected frequency of making graffiti was .36.

Table 1.
The occurrence of graffity in different groups of rhesus monkeys.

				frequency of graffity	
	N	number of monkeys per group	number of monkeys per group	expected	observed
adult	10	2 - 6	3 (1-5)	.69	.40
	1	2	0	0	0
juvenile	5	18 (15-20)	18 (15-20)	1	.80
mixed	1	22	20	1	1

So far, groups with 1-5 (average 3) lab-born members' the likelihood of graffiti varies between .36 and .87 (average .69). The observed frequency is slightly but not significantly, less. This indicates that group housing conditions probably have little effect. That the one pair of wild-born animals also given in table 1 had no graffiti is in agreement with expectation. One expects all five juvenile groups which were relatively large in membership to be negative. The one group which was negative was the one with animals of around 1 year of age.

4. Conclusions/discussion
   From the results of this study the following can be concluded:
- Rhesus monkeys, who are born and raised in a wild or "semi" wild situation (= wildborn animals and animals from safari park "De Beekse Bergen") and who arrived at the Primate Centre at ages varying from about one year to full adulthood, do not show this behaviour.
- Lab born rhesus monkeys, who are younger than one year (group 1986), do not show this behaviour.
- Monkey graffiti is mainly shown by young, lab born rhesus monkeys (1-5 years old).
- Monkey graffiti is shown by animals in groups.
- Monkey graffiti is shown by females with babies.
Whether females with babies show this behaviour, or not, seems to be depending on the source of the females, not on the presence of the infant. Fig. 1 illustrates that for the wildborn animals the time periods over which the animals experienced captive housing conditions were quite extended, often well over 10 years. Since almost none of the wildborn animals practiced graffiti, the making of graffiti is not just due to experiencing a laboratory environment for a long period of time. Experiencing captive conditions from an early age onward seems to be a much more important factor. It appears that monkey graffiti is already shown by rhesus monkeys between one and two years old, when the animals are still housed in groups. If rhesus monkeys are born and raised "outside", they do not develop this behaviour. Apparently there is something "outside", which is missing "inside" for the young monkeys

(3 months -2 years). The presence of fellow monkeys or housing in groups does not seem to play an important role in this case. Probably the cause of this behaviour is found in the housing conditions of the young, lab born monkeys. Perhaps these young animals have a great need for explorative behaviour. The conventional wire cages do not meet this need, so that providing, for example, straw in their cages, may have a favourable effect in preventing the development of monkey graffiti.

A.C. Beynen & H.A. Solleveld (Eds), New developments in biosciences: their implications for laboratory animal science, ISBN 978-94-010-7973-0

Marking of African clawed toads (Xenopus laevis). Improvement of a skin autograft technique.

Loopstra, J.A.*, Zwart, P.*, Verhoeff-de Fremery, R.**
Vervoordeldonk, F.J.M.**
* Dept. of Vet. Pathology, Division Laboratory and Special Animal Diseases, Yalelaan 1, 3584 CL UTRECHT. Netherlands.
** Hubrecht Laboratory, Netherlands Institute for Developmental Biology, Uppsalalaan 8, 3584 CT UTRECHT.

INTRODUCTION

Since the beginning of this century many attempts have been made to elaborate the ideal method for marking the African clawed toad (Xenopus laevis). This was important because individual housing systems required a lot of work and space and therefore were too expensive. Individual identification was necessary in relation to biological experiments.

Most techniques were characterized by the fact that marks were not distinct throughout the entire lifespan of the toad. In 1981 the experiences of Verhoeff-de Fremery and Vervoordeldonk with marking Xenopus laevis by means of a skin autograft have been described. However, with this technique 25% of the autografts were lost. In addition some animals died within a few days after surgery. The research presented here had the purpose of investigating whether loss of the autograft and mortality of the toad could be reduced.

Literature about marking by means of autografts is very scarce. However, some parallels could be drawn between an autograft and an allograft technique. Du Pasquier et al. (1975) reported that the allograft was fixed within 48 hours. He used a method which greatly resembled our modified method (to be discussed later). Before the allografts were fixed he found a loss of 30% of the allografts. Clothier (1985) reported that vascularisation of the skin allograft had taken place after 4 to 6 days, 12 to 24 hours after the fixation of the allograft which took place after three days. This meant that the epidermis of both host and donor were continuous at that time and that fibroblasts had formed a new collagenous layer. He recommended not to mark animals older than 2 years (thick collagenous layer; transplantation is more difficult).

MATERIAL AND METHODS.

In an experiment with 14 animals it was tried to obtain an impression of:

1. the loss of the graft and the mortality related to the skin autograft technique.
2. the process of the growing together of the back skin and autograft.
It was expected that the large surface of skin, removed by the original method had consequences for the welfare of the Xenopus laevis. Movement and greater inconvenience may have caused early losses due to a slow healing of the wound.
Thirteen female toads of 1-3 years old were marked using two different skin marking techniques.
One female toad marked 7 years ago was also used in the experiment.
Seven animals were marked according to the original method. The original method is a standardized version of the method of Verhoeff-de Fremery and Vervoordeldonk.
This original technique is described as follows.
1. The toads were anaesthetized by immersing them in a solution of 750 mg tricaine methane sulphonate (MS 222, Sandoz) and 25 ml 0.5 M sodium hydrogen carbonate in 500 ml tap water for about 7 min. After the anaesthetization the toads were weighed and the nose-anus length was measured (table 1).
2. After anaesthesia, the toad was placed on its back on a board covered with a clear plastic sheet.
3. The skin was desinfected with 70% alcohol by means of a sterile cotton tab.
4. Scissors and forceps were sterilized in 70% alcohol and a square piece of belly skin (11x11 mm) was removed and placed in distilled water.
5. The wound was dried with a cotton tab as much as possible and covered with Histoacryl-blue[R] (Braun Dexon GmHB). The glue had to be allowed to dry before the toad was turned on its belly.
6. After desinfection a piece of skin of about 4x4 mm was cut out of the back or right shoulder. This could be thrown away.
7. The white piece of belly skin was taken out of the water and placed into the wound under the wound edges on the dorsal side. This was easy because of the subdermal lymph sacs.
8. With a cotton tab, water, blood and mucus of the dermal glands were removed. Then the dry wound was covered with Histoacryl-blue[R]. After thorough drying (3 min) the toad was placed in a small plastic tank containing tap water to recover. The tank was covered with a screen and tilted, to keep the toad's head out of the water to prevent it from drowning.
9. After recovery (1/2 - 2 hours), two toads were put together in an aquarium of 50x40x40 cm. By varying the marking site it was possible to house 6 animals in one aquarium; five of them were marked (fig 2).

Table 1.

Nr Anim	Weight (g)	Length nose-anus (cm)	Technique of marking	Site of implant	Post Mortem after grafting
1	87		M	B	3 weeks
2	79		O	B	3 weeks
3			O	B	7 years[1]
4	127	9,5	M	S	4 days
5	85	8,5	O	S	4 days
6	111	9	M	B	4 days
7	100	8,5	O	B	4 days
8	85		M	B	6 weeks[2]
9	98		O	B	6 weeks
10	101		O	B	1 week [3]
11	79	8,5	M	B	10 days
12	145	10	O	B	10 days
13	100	8,5	M	S	10 days
14	120	9	O	S	10 days

M = modified technique   B = back
O = original technique   S = right shoulder

1. Old toad that is marked 7 years ago with the unstandarized original technique.
2. Bluish film on the wound.
3. Found dead; Aeromonas was isolated.

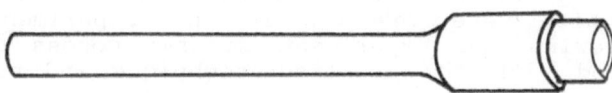

Fig.1. Biopsy punch

Six toads were marked according to the modified technique. This technique was comparable to the original one except for the part where the pieces of skin were cut out. In the alternative technique, biopsy punches were used instead of scissors and forceps. The belly skin was taken out with a biopsy punch of 8 mm in diameter (fig 1). The holes drilled in the back (4 toads) and the right shoulder (2 toads) were made by means of a biopsy punch of 4 mm in diameter.

Histoacryl-blue^R (butyl-cyanoacrylate) has been developed for use in man. It sticks the wound edges together without the need for sewing. It also has an antibacterial activity. A dry surface is prescribed. Therefore thorough drying of the wound before the Histoacryl-blue^R is put on, is very important.

In the experiment the animals are euthanized after 4 days, 10 days 3 weeks or 6 weeks (table 1) by means of a 1/2 cm^3 T 61 (Hoechst) i.m. in the hind leg.

T61=(N-2-(m-methoxyfenyl)-2-ethylbutyl(gamma)hydroxy-butyr-amide 20%;4,4'methylene-bis-(cyclohexyltrimmethyl-ammoniumjodide) 5%; Hydrochloric p-butylaminobenzoyl-dimethylaminoethanol 0,5%.

It causes no excitation and a minimum of postmortem alteration.

RESULTS

The results of the investigation can be divided into the following categories:
1. Loss of the autograft.
2. Mortality of the toad in recently marked animals.
3. Macroscopic aspects of healing of the wounds.

ad 1. No loss of the autograft has been seen with the 14 animals which were involved in the experiment. In the year following the experiment several dozens of animals were marked with the modified technique and no problems have become manifest.

ad 2. In one case a toad died 1 1/2 weeks after autografting with the original technique. This animal showed a good condition. Aeromonas hydrophila, the most important cause of red leg, was isolated.

ad 3. Macroscopic aspects of the wound:
2 days old: Histoacryl was seen as remnants on the wound.
4 days old: No Histoacryl was seen anymore. Hyperaemia was visible esp. on the edges of the wound. This was a result of vascularisation.The graft was not firmly fixed.
10 days old: Slight redness was to be seen. The graft was

obviously fixed better.

3 weeks old: No redness, the growing together of the graft and dorsal skin was advanced but not complete.

6 weeks old: The autograft had grown together firmly. A distinctly visible white marking spot was the ultimate and satisfying result.

DISCUSSION

The problems described in the introduction have not been solved completely during the experiment described here. It could be said that the loss of toads after autografting was not due to principal differences between the original and the modified technique. Although tested with a small amount of toads, both techniques could be carried out without any loss of animals. However, an advantage of the modified technique was that the belly wound was reduced twice in size compared to the original technique. This meant a reduced chance of wound infection.

It seemed that loss of the autograft did not occur if working carefully and with accuracy in surgery and narcosis. Careful drying of the wound and the autograft before the application of Histoacryl-blue^R was important. No toad involved in the experiment lost its graft within the period of observation. One month after the experiment was finished only one animal (original technique) lost the graft from the shoulder location. An individual housing system for the first two days after grafting might be considered. Two mm overlap of the skin over the graft seemed sufficient to prevent loss. One of the advantages of the modified technique is the use of two round biopsy punches ensuring an equal overlap all around.

CONCLUSIONS

It seemed to be significant that a carefully performed and standarized technique was more important than which of the two techniques was used. Histoacryl-blue^R and perhaps the method of narcosis and recovery could be of considerable importance. By following the modified technique carefully, marking Xenopus laevis with the autograft technique is a method with almost 100% success. The white autograft is clearly visible during the remaining lifespan of the toad (fig. 2). This distinguished the technique from many other marking techniques. The modified technique was preferential because the belly wound was more than twice as small compared with the original technique. In addition there was an overlap of only 2 mm of skin all around. Marking toads older than 2 years was not recommendable because of

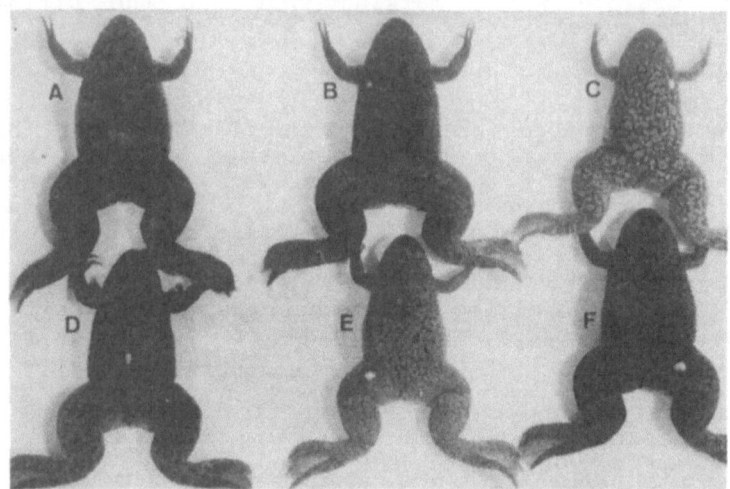

Fig. 2. A group of 6 animals coded by graft location.

the thick collagenous layer that is present in these animals. This could delay the implantation and increased the risk of losing the graft and of wound infection.
The death of recently marked toads in the period before the observations presented here might have other reasons than a failure of the technique.

REFERENCES

1.Verhoeff - de Fremery, R., F.J.M. Vervoordeldonk, Skin autografts as markers in the toad (Xenopus laevis). Laboratory Animals **16**, (1982), 156-158.
2.Du Pasquier, L., X. Chardonnens, V.C. Miggiano, A major histocompatibility complex in the toad, Xenopus laevis. Immunogenetics **1**, (1975), 482-494.
3.Clothier, R.H., personal statement (1985).
4.Gabrisch, K., P. Zwart, Krankheiten der Heimtiere. Hannover, Schlütersche Verlag (1985).

A.C. Beynen & H.A. Solleveld (Eds), New developments in biosciences: their
implications for laboratory animal science, ISBN 978-94-010-7973-0

CHANGES IN ENERGY INTAKE, BODY WEIGHT GAINS, AVERAGE FEED
EFFICIENCY AND SOME PLASMA HORMONE LEVELS IN DOGS FROM DIFFE-
RENT ENVIRONMENTS

G. Kuhn and W. Hardegg
Institut für Versuchstierkunde der Universität Heidelberg, Im
Neuenheimer Feld 347, D-6900 Heidelberg, FRG.

1. INTRODUCTION
   Protection and comfort of the laboratory animal are major requirements in research. The health and well-being of laboratory
animals are the prime factors adressed in the Animal Welfare Act
in the USA (1) as well as the guidelines for accomodation and
care of animals of the European Communities (5). These standards serve as minimum criteria for environmental control.
Many researchers have reported that differences in enclosures
can affect significantly experimental results. Variables such
as environmental temperature and relative humidity, as well as
ammonia and carbon dioxide concentrations in the air may have
a great influence on the outcome of an experiment (12). Other
factors which must be considered include lighting and noise
levels to which the animals are exposed (7,11).
The objective of the present study was to determine food intake, body weight gain, feed efficiency and circulating levels
of cortisol and thyroid hormone in plasma of dogs during indoor
and outdoor maintenance. We were interested in defining assessing changes in these parameters during indoor housing when the
dogs were maintained under typical laboratory conditions and
outdoors under comfortable natural weather conditions that
followed guidelines for accomodation and care of animals of the
European Communities.

2. MATERIALS AND METHODS
2.1. Dogs and Housing Conditions
   The experiments were conducted in Heidelberg, FRG, between
June 30 and September 1, 1986 with 20 female Foxhounds approximately 1 year of age from own breeding. During the pre-examination period all animals were housed outdoors. The dogs were
randomly divided in four groups for the experiment and were
housed in kennels, with 5 individuals in each. They were not
isolated visually from the other animals. Two groups were kept
in pens 4.00 m x 4.00 m under laboratory conditions, where the
temperature was maintained at 20 ± 2°C and the photoperiod was on
a 12L:12D light:dark cycle (photophase from 0700-1900 h). Relative humidity was not regulated. Two groups were housed outdoors in kennels 8.00 m x 9.00 m located in an undisturbed
courtyard. The fluctuations of ambient temperature and relative
humidity ranged from 7-30°C and 30-100%, respectively, and the
animals were exposed to natural daylight. During the adaptation
and experimental period, food (ssniff dog maintenance diet,
12.7 MJ/kg; Ssniff, Soest, FRG) and water were freely available

317

with the exception of Sundays, when food was given only until 0900 h. Food intake (corrected for spillage) and body weight were measured once a day and once a week, respectively. Feed efficiency (grams of body weight gain per kcal eaten) of 10 animals housed indoors and outdoors was estimated .

2.2. Blood Sample Collection

The dogs had been adapted for 4 weeks to their groups and the techniques used for examining the animal. Blood samples were collected from the Vena cephalica antebrachii with heparinized disposable syringes on Mondays between 0800 and 1000 h throughout the study. Samples were placed in ice or a refrigerator until centrifugation for 14 min at 3500 g at 10°C. Plasma samples obtained were divided into several aliquots and stored in plastic vials at ⁻20°C until each was thawed for the assay.

2.3. Hormone Assays

In plasma samples obtained at the beginning and after 3, 6 and 9 weeks of the examination, circulating concentrations of cortisol, T_3, T_4, TBK and FT_4I were determined. Circulating concentrations of cortisol were measured in duplicate by specific radioimmunoassays (RIA) using rabbit antiserum. The detection limit of the RIAs was 1.0 ng/ml of plasma. Because of the low concentrations of cortisol in the dog, additional standards for low concentrations were prepared and sample volumes were doubled.
Plasma triiodothyronine (T_3), thyroxine (T_4) and thyroxine-binding capacity (TBK as TBI or T_4-uptake) were measured with the Enzymun-Test System ES 600 (Boehringer GmbH, using enzyme-immunoassay kits purchased from Boehringer GmbH, Mannheim, FRG). The kits have been designed for the determination of the total concentrations of circulating thyroid hormones in human serum with the enzyme-linked immunosorbent assay principle (ELISA-principle) and also can be used with dogs (9).
Calculations of free thyroxine index (FT_4I) were made by dividing total circulating thyroxine by the T_4-uptake: $FT_4I = \dfrac{T_4}{TBI}$

2.4. Statistical Analyses

The data from both groups are shown in Figures 1-3 as mean ± SEM. Statistical analyses employed Wilcoxon 2-sample tests and Wilcoxon matched-pairs signed-rank tests. A sequential rejecting procedure, after Bonferroni-Holm (13) was made subsequently with Wilcoxon-tests ($p < 0.05$).

3. RESULTS
3.1. Energy Intake, Body Weight Gain, and Feed Efficieny

Weekly energy input of 10 dogs showed a significant decrease ($p < 0.01$) in indoor maintenance (663.7 ± 58.8 MJ indoors vs. 778.2 ± 61.4 MJ outdoors). Body weight increased in both housing conditions significantly ($p < 0.05$) during the study, but no significant ($p < 0.05$) differences were noted between both groups. Weight gains of dogs maintained outdoors were slightly greater than those of dogs kept indoors. Furthermore, efficiency of food utilization was comparable in both environments

(0.022 ± 0.014 indoors vs. 0.023 ± 0.008 outdoors).

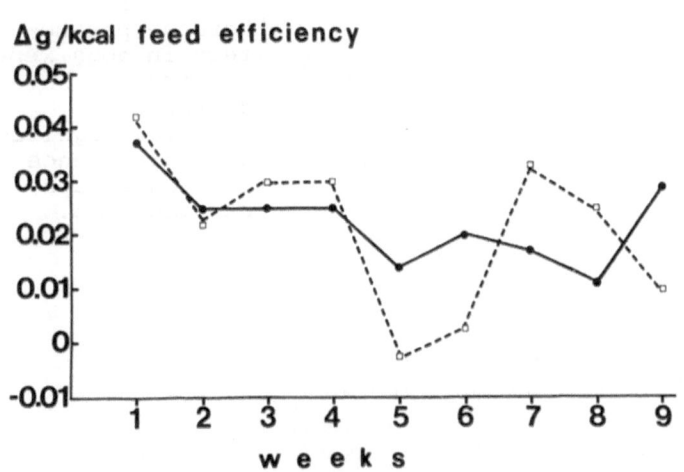

Fig. 1. Feed efficiency of dogs housed indoors (□------□) and outdoors (●———●).

3.2. Cortisol Concentrations

During the pre-examination period, levels of circulating cortisol in dogs maintained indoors throughout the experiment were near the lower limits of measurements and displayed a marked and significant increment ($p < 0.05$) within three weeks of indoor maintenance. (Fig. 2). Following the initial sharp increase, circulating cortisol concentrations stabilized at a level between those two extreme values. On the other hand, there were no significant changes in circulating cortisol concentrations in dogs maintained outdoors, and there also were no significant differences between both groups at any time during the study.

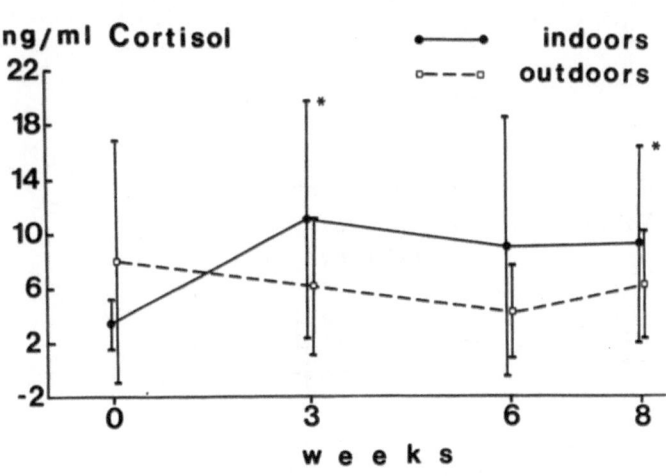

Fig. 2. Circulating cortisol activities in dogs housed indoors and outdoors.
*significant increase of circulating cortisol activity in dogs housed indoors vs. beginning of the study.

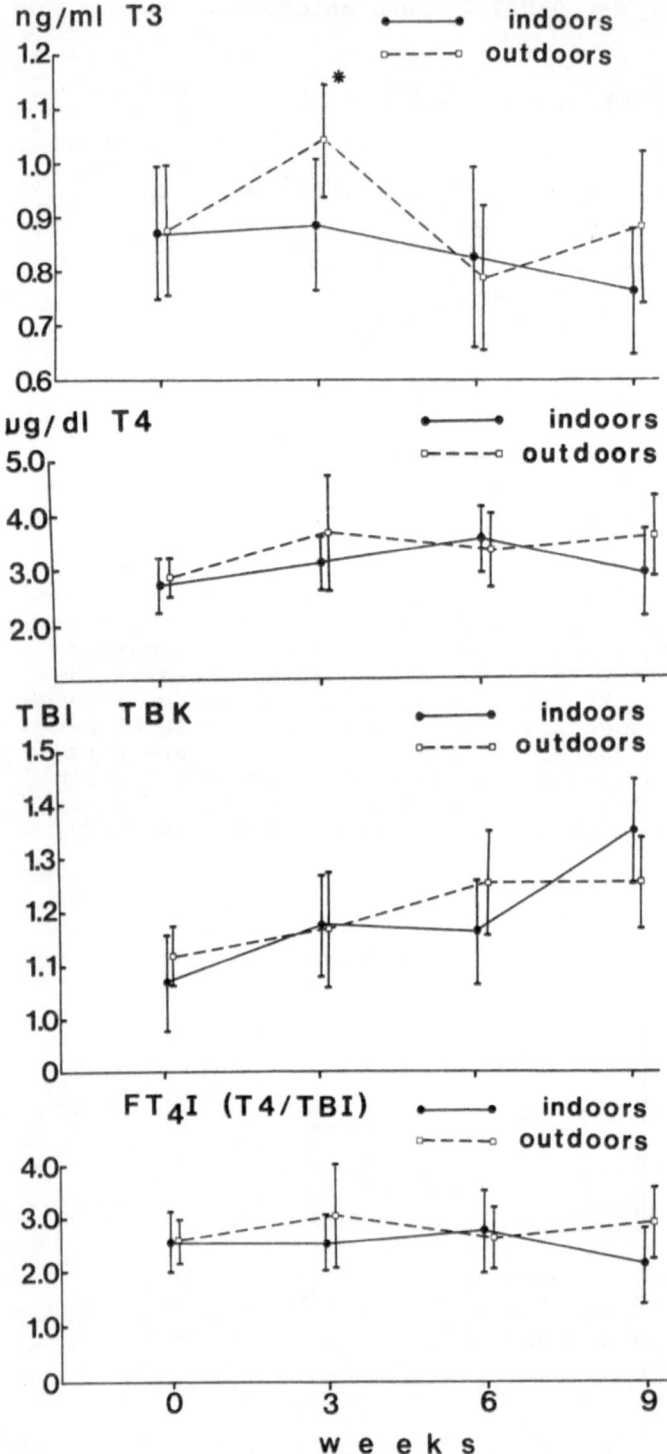

Fig.3. Circula-
ting thyroid
hormone parame-
ters in dogs kept
indoors and out-
doors.
*p < 0.05 signifi-
cant difference
between groups
at same time.

3.3. T_3, T_4, TBK, and FT_4I values

In dogs kept outdoors, a significantly higher ($p < 0.05$) plasma T_3 level was seen after 3 weeks of the study compared to dogs maintained indoors, whereas T_3 concentrations were comparable in both maintenance conditions in the other examination weeks. Dogs housed indoors showed an initial slight increase of circulating T_3 levels and thereafter, a constant and significant decrease ($p < 0.05$) towards the end of the study, while changes of plasma T_3 concentrations in dogs maintained outdoors were greater over time (Fig. 3). No significant differences occured in plasma T_4, TBK and FT_4I values between both environments, whereas TBK values significantly ($p < 0.05$) raised in both maintenance conditions.

4. DISCUSSION

This experiment yielded two noteworthy findings. Energy intake responded differently to maintenance conditions that we examined, while body weight gain was slightly different and feed efficiency was comparable in both maintenance conditions. Previously reported data suggest a complex interaction of environmental temperature and light/dark cycle for energy intake, body weight gain and hormonal factors such as serum thyroid and adrenal gland hormone levels (2,3,4,6,8,12). External factors also play an important role in controlling the energy intake of dogs. Findings of Ozon et al (1986) indicate that energy intake accompanied by increasing body weights is a function of photoperiod. Korhonen et al (1986) examined the influence of nest availability on growth of farmed raccoon dogs. Availability of a nest for the animals in that study led to the highest final body weights. The nest reduced locomotor activity and seemed to save energy, perhaps by providing a refuge from cold temperatures. Our energy intake findings also suggest that nonhomeostatic factors, for example, lower ambient temperatures outdoors, especially at night, and longer photoperiod outdoors, play an important role in controling appetite of dogs.
The time course of circulating hormone levels emphasizes the importance of looking at multiple points since, as our results demonstrate, hormone concentrations can be greater, less or comparable in dogs housed indoors or outdoors in enclosures. Hormonal rhythms in animals and man are tied to fluctuations in environmental conditions, such as ambient temperature and photoperiod. The increasing plasma cortisol levels of dogs undergoing indoor housing may result from blood samples being obtained during the initial hours of the light cycle. With termination of the dark part of the cycle, animals react with a marked peak in 17-hydroxycorticosterone secretion. If the nocturnal dark phase is extended beyond the time of awakening, then the matutinal peak do not occur until it is light, not at the usual time of awakening in darkness (7). Also, the stress of higher population density and elevated noise levels in dogs housed indoors may have produced a further elevation (11). Specifically, the decibel levels in dog rooms may be sufficient to initiate neuroendocrine responses associated with stress in

laboratory animals and caretakers (11). Moreover, the combined
physical factors of ambient temperature, relative humidity, air
movement, and thermal radiation modify heat loss from the body
and in so doing constitute an applied heat stress (2).
Our studies on circulating thyroid hormone parameters in dogs
housed indoors under laboratory conditions and outdoors under
comfortable, natural weather conditions demonstrate that circulating profile of T_3 is influenced by maintenance conditions
and that the examined conditions within certain limits don't
seem to affect T_4 and TBK levels or FT_4I values. Possible
causes of these findings are differences in ambient temperature
and photoperiod as well as stressful situations. Acute exposure
to low or high temperatures modifies thyroid activity in homeo-
thermic vertebrates and also peripheral metabolism of thyroid
hormones may be a major component of the homeorhetic response
elicited during acclimation to cold or hot environmental conditions (6). Changes of blood content of thyroid hormones and
allied substances, namely protein bound iodine, under conditions
of exposure to low and high environmental temperatures are
difficult to interpret. The observed levels represent a dynamic
balance between output of hormones from the thyroid gland on
the one hand, and its utilization and excretion on the other.
Some authors have concluded from their data at high environmental temperature that there is a reduced utilization of the
thyroid hormones reflected in an increased protein-bound level
(4). Stressful situations and glucocorticoids also reduce circulating T_3 and T_4 in mammals by their inhibitory influence at
several sites of the hypothalamo-pituitary-thyroid axis (3).
These results in animals confined to climatic chambers may help
to explain the results observed under natural weather conditions
and under constant laboratory conditions. In accordance with
these findings, results in the present study show that Foxhounds housed outdoors in summer 1986 present a significant increase of circulating T_3 levels according to lower ambient
temperature registered outdoors in the preceeding experimental
weeks. Following a period of very fine weather slightly lower
T_3 concentrations were observed in the outdoor animals relative
to dogs in enclosures indoors.
However, the real situation of this study is more complex and
there are many other factors that could have influenced our results. In conclusion, the results suggest that the examined
maintenance conditions are external factors that participate in
controlling of both energy intake and circulating T_3 and
cortisol levels.

5. ABSTRACT

We measured plasma concentrations of cortisol, thyroxine
(T_4), triiodothyronine (T_3), thyroxine-binding capacity (TBK),
free thyroxine index (FT_4I) and changes in energy intake as
well as body weight gains and feed efficiency of female Foxhounds approximately one year of age from own breeding. In
summer 1986 the dogs were maintained indoors under constant
ambient temperature and simulated natural daylight, 12:12
light:dark regimes for 9 weeks or outdoors exposed to daily
variations in temperature and summertime photoperiod at Heidel-

berg, FRG. All animals had free access to food and water.
Measurements of energy intake and body weight gain were made
daily and every week, respectively. Hormone and TBK levels
were determined in time intervals of 3 weeks.
Studies on weekly energy intake and feed efficiency of 10 dogs
as well as on circulating cortisol levels demonstrate that in
animals undergoing indoor maintenance, energy intake decreases
and circulating profile of cortisol increases. Moreover, environmental conditions induce differences in plasma T_3 levels,
while within certain limits circulating levels of T_4 and TBK
or FT_4I values don't seem to be affected. In conclusion there
may occur differences in biomedical research between dogs maintained indoors and outdoors according to GLP guidelines.

6. ACKNOWLEDGEMENTS

The authors wish to acknowledge the employees of Boehringer
Mannheim GmbH (Mannheim, FRG) for performing thyroid hormone
assays and the Institut für Pharmakologie der Universität Heidelberg, for the possibility to determine plasma cortisol levels.

7. REFERENCES

1. Anonymus: Animal Welfare Act, 1966 (Public Law 89-544),
   Amended 1970 (Public Law 91-579), Amended 1976 (Public Law
   94-279), Code of Federal Regulations, US Government Printing
   Office, Washington, 1979.
2. Barlow G, Agersborg, HP, and Keys HE: Blood levels of 17-
   hydroxycorticosteroids in hyperthermic dogs. Proc. Soc.
   Exptl. Biol. Med. 93, 280-284, 1956.
3. Bauman TR, Anderson RR, and Turner CW: Thyroid hormone secretion rates and food consumption of the hamster (Mesocricetus auratus) at 25.5. and 4.5 °C. Comp. Endocrinol.
   10, 92-98, 1968.
4. Collins KJ and Weiner JS: Endocrinological Aspects of Exposure to High Environmental Temperatures. Physiol. Reviews
   48 (4), 785-839, 1968.
5. Council of the European Communities: Council Directive of
   24 November 1986 on the approximation of laws, regulations
   and administrative provisions of the Member States regarding
   the protection of animals used for experimental and other
   scientific purposes, Official Journal of the European
   Communities, 1986.
6. Galton VA and Nissula BC: Thyroxine metabolism and thyroid
   function in the cold adapted rat. Endocrinology 85, 79-86,
   1969.
7. Hollwich F: The Influence of Ocular Light Perception on Metabolism in Man and in Animal. Springer-Verlag, New York,
   Heidelberg, Berlin, 64-98, 1979.
8. Korhonen H, Harri M, and Nurminen L: Effects of Social Competition for Feed on Growth of Farmed Raccoon Dogs. Growth
   50, 340-350, 1986.
9. Kraft W: Enzymimmunologische Schilddrüsentests beim Hund.
   Kleintierpraxis, 25, 217-222, 1980.
10. Ozon C, Dolisi C, Crenesse D, Ardisson JL: Effects of Feeding
    Time upon Daily Intake and Body Weight. Physiology & Be-
    haviour, 36, 583-586, 1986.

11. Peterson EA: Noise and Laboratory Animals. Lab. Anim. Sci., 30 (2), 422-439, 1980.
12. Serrano LJ: Carbon dioxide and ammonia in mouse cages: Effect of cage covers, population and activity. Lab. Anim. Sci., 21, 75-85, 1971.
13. Sonnemann E, Berchier P, Schiller K, Ferner K, Maurer W, Victor N: Simultane Hypothesenprüfungen. Biometrisches Seminar der Region Österreich-Schweiz der internationalen biometrischen Gesellschaft. Bad Ischl (A), 1981.

A.C. Beynen & H.A. Solleveld (Eds), New developments in biosciences: their implications for laboratory animal science, ISBN 978-94-010-7973-0
© 1988 Martinus Nijhoff Publishers, Dordrecht.

SERUM CONCENTRATIONS OF VITAMINS A, D AND E OF GROWING BEAGLES FED COMMERCIAL DOG AND FOX DIETS

H-M. VOIPIO*, T. NEVALAINEN* AND P. MÄENPÄÄ**
*National Laboratory Animal Center and **Department of Biochemistry, University of Kuopio, Finland

1. INTRODUCTION
Fox diets are generally considered unfit for dogs because of their high concentrations of vitamins A, D and E. Since serum vitamin concentrations of dogs fed with dog and fox diets were not available, this study was designed to compare a commercial dog diet with a commercial fox diet. The main differences between the diets were in their vitamin concentrations, but there were some differences also in their protein, fat and carbohydrate concentrations. In addition to concentrations of vitamins A, D and E in serum of the dogs, the effects of the diets were estimated by various blood parameters, growth, radiology, and quality of hair and feces.

2. MATERIALS AND METHODS
2.1 Animals
Twenty laboratory-bred beagles, 10 males and 10 females, were used. The dogs were taken into the experiment at the age of six weeks directly after weaning. Their weights varied from 1200 g to 2500 g, 1300 g to 2300 g in males and 1200 g to 2500 g in females.

2.2 Grouping
The dogs were divided into two feeding groups, five males and five females in each. Only sister or brother pairs were used.

2.3 Animal Care
2.2.1. Before the experiment
Before the experiment dogs were kept in pens with their mothers. Fox diet (Ketun Syystaru, Hankkija, Finland) was fed ad libitum.
At the age of three and five weeks the dogs were treated against toxocariasis with piperazine, 110 mg/kg (Oksyran[R], Lääkefarmos, Finland). At the beginning of the experiment, a clinical examination was performed.
2.2.2. During the experiment
During the experiment, the dogs were kept in single cages, 90x90x150 cm (breadth x height x depth). Temperature in animal rooms was 21±1°C and humidity mainly 50 - 75%. The cages and the room floors were washed daily. Vaccination against canine distemper and canine infectious hepatitis (Delcine-H, Tech America) was given when the dogs were 12

weeks old. The total length of the experiment per dog was 14 weeks.

2.2.3. Feeding

The dogs had an adaptation period of one week after weaning, during which the amount of the new diet was increased gradually. From the beginning of the second week, they got only the experimental diet twice daily. The amount of food was calculated to be equicaloric based on the need of 838-1365 kJ/kg depending on the age of the dog (1). Total need of energy and food was calculated per every 100 g of dog weight and the amount of food given was weighed with an accuracy of 1 g. Tap water was available ad libitum.

2.4. Diet

The experimental diets consisted of a dog diet Pentu Serti (Vaasanmylly Oy, Finland) and a fox diet Ketun Syystaru (Hankkija, Finland). The main differences between the diets were in their vitamin concentrations, which in the dog diet were 10 000 IU/kg, 1000 IU/kg and 60 mg/kg and in the fox diet 22 500 IU/kg, 6000 IU/kg and 150 mg/kg for vitamins A, D and E, respectively. The diets differed also in their protein, fat and carbohydrate contents. The amounts of calcium and phosphorus were similar in both diets, and their ratio (Ca:P) were 1.22 in the dog diet and 1.16 in the fox diet.
Further descriptions of the diets are given in table 1.

TABLE 1. Nutrients, energy, Ca, P and vitamins A, D and E in the fox diet (Ketun Syystaru) and the dog diet (Pentu Serti).

	Fox diet	Dog diet
moisture %	8.5	9.0
raw protein %	35.0	28.0
raw fat %	14.8	7.0
carbohydrate %	29.9	47.0
fiber %	2.8	3.0
ash %	9.0	6.0
total energy	2070 kJ/100g	1810 kJ/100g
calcium	1.06%	1.1%
phosphorus	0.91%	0.9%
vitamin A	22 500 IU/kg	10 000 IU/kg
vitamin D_3	6 000 IU/kg	1 000 IU/kg
vitamin E	150 mg/kg	60 mg/kg

2.5. Observations

The dogs were observed daily in the mornings and weekly at noon on the same day of the week. If some food was left over, the residual amount was weighed.The quality of feces was estimated using a scale yellow-light-normal dark-black: the amount of feces normal-subnormal-greater than normal, and the consistency of feces diarrhea-normal-hard. The dogs were weighed weekly, the quality of their hair was estimated and the thickness of a skin fold was measured above the hip. The statistical analysis of weight and growth was done with the analysis of covariance and the analysis of skin folds with the Mann-Whitney test.

2.6 Sampling
2.6.1. Blood
Blood samples were taken from the jugular vein at the beginning and at the end of the experiment. Blood leucocytes (B-leuk), erythrocytes (B-ery), hemoglobin (B-Hb), hematocrit (B-Hkr), erythrocyte mean cell volyme (E-MCV), erythrocyte mean cell hemoglobin (E-MCH) and erythrocyte mean cell hemoglobin concentration (E-MCHC) were determined from all samples. The concentrations of vitamins A (retinol), D (25-hydroxyvitamin D) and E (alphatocopherol) were determined from the samples collected at the end of the experiment. Vitamins A and E were determined by HPLC according to de Leenheer et al. (2) and vitamin D by a RIA-kit supplied by the Radiochemical Centre, Amersham, England. The statistical analysis of blood parameters was done with the analysis of variance.
2.6.2. Feces
The amount of total feces per 24 hours was measured in the beginning and after four and eight weeks from the beginning of the experiment. The statistical analysis was done with the analysis of variance.
2.6.3. Radiology
Bone mineralization was estimated radiologically based on the ossification centers. Radiography was done three weeks from the beginning of the experiment, when ossification of patella (6-9 weeks of dogs age) and the proximal epiphysis of fibula (7-13 weeks of dogs age) should have begun and eight weeks from the beginning of the experiment when the ossification of fabella (12-16 weeks of dogs age) should have begun (3).

3. RESULTS
3.1 Hematology
The blood parameters (B-leuk, B-ery, B-Hb, B-Hkr, E-MCV, E-MCH, E-MCHC) showed no statistical differences between the two feeding groups. Significant differences were seen in many parameters when comparing the samples taken in the beginning and at the end of the experiment. The amount of B-leuk was significantly higher ($p < 0.001$) in the first sample. There was a 2-way interaction in erythrocyte concentration between sex and age ($p < 0.05$); the amount of erythrocytes in females increased more than that in males between the two samples.

There was a significant difference between sex ($p < 0.01$) and age ($p < 0.001$) in Hb. Hemoglobin concentrations were lower in males than in females, also concentrations were lower in the first sample than in the second.

Hkr was significantly lower ($p = 0.005$) in males than in females and hematocrit was lower ($p = 0.001$) in the first sample than in the second sample.

E-MCV was significantly higher (p<0.001), E-MCH significantly (p<0.005) higher and E-MCHC significantly (p=0.001) lower in the first sample than in the second sample.

3.2 Serum vitamins

Serum vitamin D (25-hydroxyvitamin D) and vitamin E (alpha-tocopherol) concentrations showed no significant differences between the feeding groups. However, vitamin A (retinol) differed significantly (p<0.001) between the feeding groups: serum concentrations in the dog diet group were 628±194 ng/ml and in the fox diet group 1065±144 ng/ml. These are shown in table 2.

TABLE 2. Serum concentrations (means±SD) of vitamins A, D and E of beagles fed a dog or a fox diet for 14 weeks after weaning.

VITAMIN	Dog diet	Fox diet
Retinol(ng/ml)	628 ± 194	1065 ± 144 ***
25 OHD (ng/ml)	33.7 ± 16.1	28.3 ± 4.7
Tocoferol(µg/ml)	7.7 ± 2.2	11.2 ± 3.1

*** = p<0.001

3.3 Feces

Both diets were eaten well. Only during the adaptation period was some food either left or dropped to the floor. The feces of both feeding groups was evaluated as normal, with dark brown color and normal consistency. In the total amount of 24-hour feces there was, however, a significant difference (p<0.01) between the feeding groups. The amount of feces was smaller in the dog diet group than in the fox diet group.

3.4 Growth

The growth of dogs showed a 2-way interaction between diet and sex. The females grew significantly faster (p<0.001) with the fox diet than the males. The growth of dogs is shown in figure 1.

3.5 Other observations

No differences were observed in the thickness of skin folds in the two feeding groups. The hair of all dogs was considered as normal in each group during the experiment. In the radiological examination, no differences in ossification between the groups could be observed.

4.DISCUSSION

Two commercial diets were compared in this study, one aimed for dogs and the other for foxes. The main differences between the diets were in their concentrations of vitamins A, D and E. Despite these differences, only vitamin A concentrations in serum differed significantly between the diet groups. This difference was somewhat surprising in view of the transport system of retinol in the serum (retinol-binding protein). On the other hand, no toxic effects were seen in the fox diet group that could be associated with higher levels of vitamin A in serum.

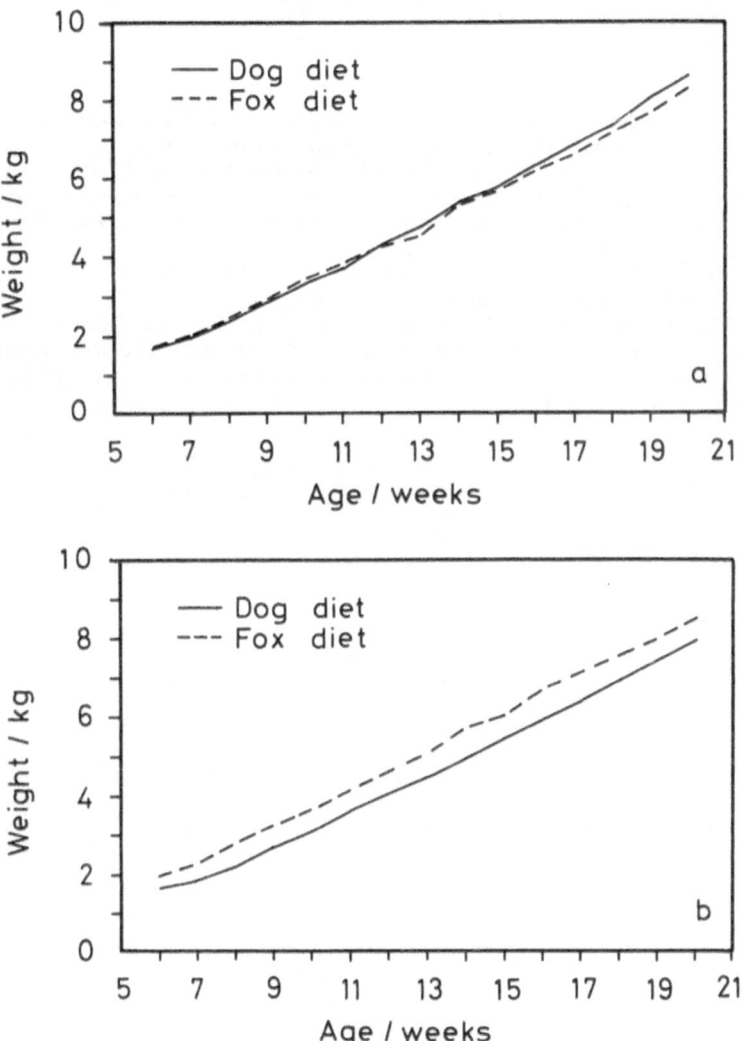

FIGURE 1. The growth of male (a) and female (b) beagles with a dog and a fox diet.

According to the manufacturer's declaration, the fox diet contained six times more vitamin D than the dog diet. However, no significant differences in serum levels of 25-

hydroxyvitamin D were observed in the two groups and no adverse effects on growth or calcification occurred during the experiment in the fox diet group. This suggests that the vitamin D contents of the diets were not as accurate as given in the declarations or that serum levels of 25-hydroxyvitamin D in dogs are not as sensitive indicators of vitamin D status as in humans or in many farm animals (4).

REFERENCES

1. Paatsama S, Ahonius B. Koiran hoito. In Paatsama S, Nurminen M. (ed). Suomalainen koirakirja. Keuruu:Otava, 1978.
2. De Leenheer AP, de Bevere VORC, de Ruyter MGM, Clayes AE. Simultaneous determination of retinol and alphatocopherol in human serum by high-performance liquid chromatography. J. Chromatogr. 162, 408-413, 1979.
3. Gillette EL, Thrall DE, Lebel JL. Carlson's veterinary radiology. Philadelphia: Lea & Febiger, 1977.
4. Horst RL, Littledike ET, Riley JL, Napoli JL. Quantitation of vitamin D and its metabolites and their plasma concentrations in five species of animals. Anal. Biochem. 116,189-203,1981.

A.C. Beynen & H.A. Solleveld (Eds), New developments in biosciences: their implications for laboratory animal science, ISBN 978-94-010-7973-0
©1988 Martinus Nijhoff Publishers, Dordrecht.

INTEGRATION OF HGH GENE IN TRANSGENIC MICE AND TRANSMISSION TO NEXT GENERATION

B. BRENIG AND G. BREM
Institut für Tierzucht und Tierhygiene der Ludwig-Maximilians-Universität München

1. SUMMARY

Transgenic mice carrying the mouse metallothionein-I promoter (mMT-I)/human growth hormone (hGH) fusion gene (MThGH) were produced by microinjection into the pronucleus of fertilized one-cell eggs.
A total of eight transgenic mice (28 %) were obtained which all contained detectable MThGH-gene construct sequences in their genome, ranging from 1 or less (mosaics) to 4 copies/cell.
One male transgenic mouse exhibiting high, metal-inducible levels of human growth hormone in serum and growing twice as large as control litter-mates was bred to three normal females. Eleven of twenty offspring reared were shown to be transgenic by Southern- and dot blot analyses. They had heritated the MThGH-gene construct from their father.

2. INTRODUCTION

In the recent years several methods have been developed, which allow stable and transmissible introduction of molecularly cloned genetic material into the mammalian genome (Gordon, 1983). At present the most effective technique, which was first established in mice, is the direct microinjection of a few hundred copies of a linear DNA-fragment into the pronuclei of fertilized one-cell eggs. This leads to a random integration of the exogenous DNA into the host genome.

The production of transgenic animals has rapidly gained importance and affords the opportunity to study the factors of gene regulation, tissue-specific expression and embryo-genesis.

Till now a variety of genes and gene constructs have been successfully introduced not only into mice (Goldberg *et al.*, 1983; Spradling and Rubin, 1983; Palmiter and Brinster, 1985; Hammer *et al.*, 1985; Brem *et al.*, 1985).

Among different others, gene constructs consisting of the mouse metallothionein-I promoter (mMT-I) and the human growth hormone structural gene have been used to produce transgenic mice (Palmiter *et al.*, 1982; Hammer *et al.*, 1985; Brem *et al.*, 1985). Most transgenics harbouring such gene constructs commonly exhibit high, metal-inducible levels of foreign mRNA in different organs, as well as detectable quantities of the human growth hormone in serum. Most animals grew significantly larger than control mice. In addition, the foreign genes became stably integrated into the germ line, where they are heritated as Mendelian loci and transmitted to progeny.

3. MATERIALS AND METHODS

3.1. The structure of the MThGH-fusiongene construct

Zygotes were microinjected with a 4.0 kb linear *Eco*RI/*Eco*RI DNA-fragment containing the mouse metallothionein-I promoter or regulatory region (1.8 kb) and the structural gene coding for the human growth hormone (1.7 kb) (Fig. 1).

The human growth hormone gene had been isolated from a size fractionated library of a human placental DNA cloned in λgt11$_{wes}$. It was ligated to the mMT-I promoter in a *Bgl*II/*Bam*HI-fusion site, about 60 bp upstream the amino acid sequence start point of hGH.

This gene construct was subcloned into the hybridvector pBR327/BPV-1, giving the 9.31 kb plasmid pXGH-1 (Goodman and Selden, 1985). Restriction enzyme recognition sites of *Pvu*II, *Bgl*I, *Bgl*II, *Sac*I and the length of restriction fragments are shown in Fig.1.

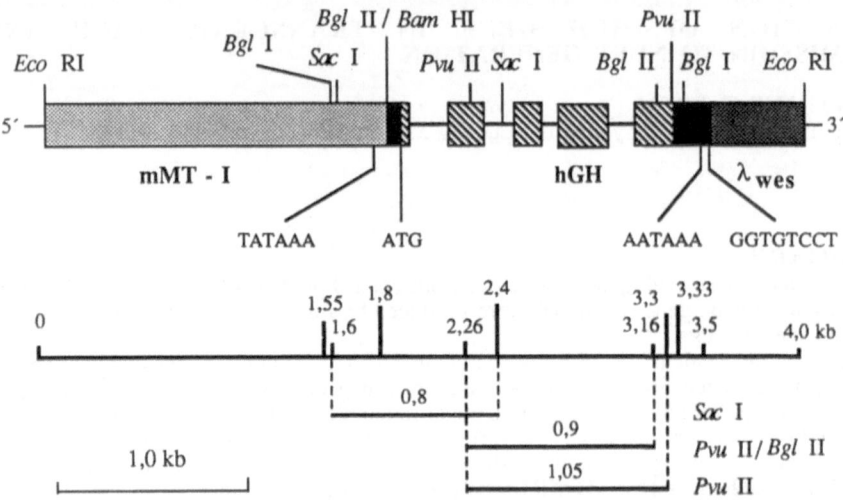

FIGURE 1. Structure of the MThGH-fusion gene construct (Goodman and Selden, 1985) For microinjection there was used a 4.0 kb *Eco*RI/*Eco*RI-linear DNA-fragment, consisting of the mouse metallothionein-I promoter (mMT-I; white shaded box) fused to the human growth hormone structural gene (hGH; coding regions: hatched boxes; non-coding regions: solid lines; 5´-, 3´-untranslated regions: black boxes). At the 3´-end of the construct there is a 0.5 kb fragment of λgt11$_{wes}$ (black shaded box). Restriction enzyme recognition sites and restriction fragment lengths are indicated.

3.2. Preparation and purification of DNA for microinjection

The 4.0 kb *Eco*RI/*Eco*RI-DNA fragment was isolated and separated from pXGH-1 on a 0.8 % agarose gel and recovered with a DEAE-cellulose membrane. Because the DNA-solution for microinjection has to be free of any contaminants, like particulate matter, all tubes and pipets had been rinsed with filtered water prior to use. After precipitation of the DNA-fragment it was resuspended in TE buffer and diluted to 1.000 copies/pl.

3.3. Extraction of high-molecular-weight DNA from small tissue samples

Isolation of high-molecular-weight DNA from small tissue samples was done according to a method described by Blin and Stafford (1976) with the following modifications.
Tail samples were directly transferred into a buffer containing 50 mM Tris-HCl (pH 8.0)/ 100 mM EDTA (pH 8.0)/100 mM NaCl/1 % SDS and 0.5 mg/ml proteinase K and were incubated at least 8 h (55 °C). After extracting twice with 1 vol. phenol/chloroform and subsequently with 1 vol. chloroform/isoamyl alcohol (24:1), samples were precipitated and resuspended in TE buffer (10 mM Tris-HCl/1 mM EDTA, pH 7.5). The DNA was then subjected to agarose gel electrophoresis and analyzed by Southern and dot blotting.

3.4. Southern blot- and dot blot analysis

For Southern blot analysis 30 μg of DNA were digested (5-10 h) in a total volume of 50 μl with a 2-3fold excess of the indicated restriction enzymes. After agarose gel electrophoresis the DNA was transferred to nitrocellulose filters (Southern, 1975) and hybridized with a [α-^{32}P]dCTP labeled 0.9 kb *Pvu*II/*Bgl*II internal fragment (hGH$_{900}$) of hGH (Feinberg

and Vogelstein, 1983; 1984), which is a smaller part of the hGH-specific probe (Palmiter *et al.*, 1983). The nitrocellulose filters were then autoradiographed for 16-48 h at -80 °C.
Dot blot analysis was done by denaturing 1, 2, 4, 8 and 10 μg of DNA in 30 μl TE buffer for 5 minutes at 100 °C. The samples were then quickly chilled on ice and treated with 30 μl 1 M NaOH for 20 minutes at room temperature and 30 μl of neutralizing-buffer 1 M HCl/ 1 M NaCl/0.5 M Tris-HCl (pH 7.5)) in succession.
After application to a nitrocellulose filter using the Bio Rad Bio-Dot™ apparatus, hybridization was done as described above.
The dots were scanned with a Berthold Automatic TLC-Linear Analyzer and copy numbers of MThGH-sequences/cell calculated by the chromatography system LB 512 (Berthold) compared to standards of MThGH assuming a haploid genomic size of mice of 2.3×10^9 bp (1.5×10^{12} daltons).

4. RESULTS

Detection of transgenic mice was done by Southern blot and dot blot analyses. The integrated genes were characterized with respect to copy number and mode of integration. Eight transgenic mice were identified from a total of 28 animals, which had been generated by microinjection directly. Seven of these mice had integrated 1 or less copy/cell of the MThGH-fusion gene construct (data not shown). One male transgenic mouse No. 211 expressed high levels of human growth hormone in serum and grew faster. Therefore it was bred and a total of 20 offspring were reared (Brem and Wanke, 1987). Fig. 2 shows the autoradiogram of Southern blotted DNA from No. 211 (Fig. 2, **1**) and No. 211-I-4 (Fig. 2, **2**), one male offspring, digested with *Eco*RI and hybridized with hGH$_{900}$. There are two prominent signals at 4.0 kb and > 20 kb in both animals detectable. The signal at 4.0 kb is corresponding to the length of a completely integrated copy of MThGH.

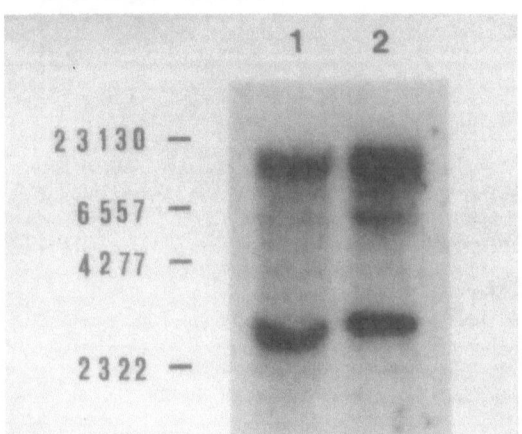

FIGURE 2. Southern blot analysis of animal No. 211 and No. 211-I-4
30 μg of DNA (No. 211, **1**; No. 211-I-4, **2**) were digested with *Eco*RI (2-3U/μg DNA), electrophoresed on a 1.0 % agarose gel and transferred to a nitrocellulose filter. Hybridization was done as discribed above. Sizes of λ-marker (bp) are indicated.

*Eco*RI-digestion of the DNA of further offspring results in three different restriction patterns (Fig. 3 and Fig. 4).
Five mice had only one signal. Mouse No. 211-I-3 (Fig. 3, **1a**) and No. 211-I-7 (Fig. 3, **3a**) at 4.0 kb and No. 211-I-5 (Fig. 3, **2a**), 211-II-7 (Fig. 4, **2a**), and 211-III-1 (Fig. 3, **6a**) at > 20 kb. Whereas animal No. 211-II-3 (Fig. 3, **4a**), No. 211-II-4 (Fig. 4, **1a**) and No. 211-I-4 (Fig. 2, **2**) had two bands at 4.0 kb and > 20 kb.

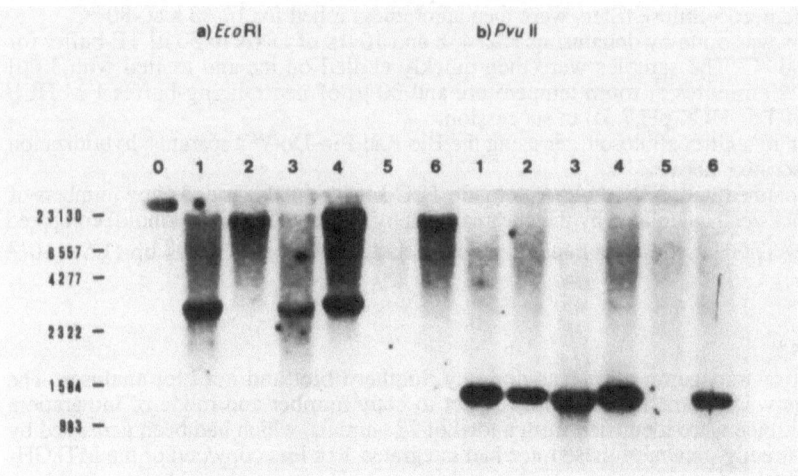

FIGURE 3. Southern blot analysis of the offspring from mouse No. 211
30 μg of DNA of each offspring (No. 211-I-3, **1**; No. 211-I-5, **2**; No. 211-I-7, **3**; No. 211-II-3, **4**; No. 211-III-1, **6**) and control (**5**) were digested with *Eco*RI (a) or *Pvu*II (b), electrophoresed on a 1.0 % agarose gel and transferred to a nitrocellulose filter. Sizes of λ-marker (**0**) are indicated (bp).

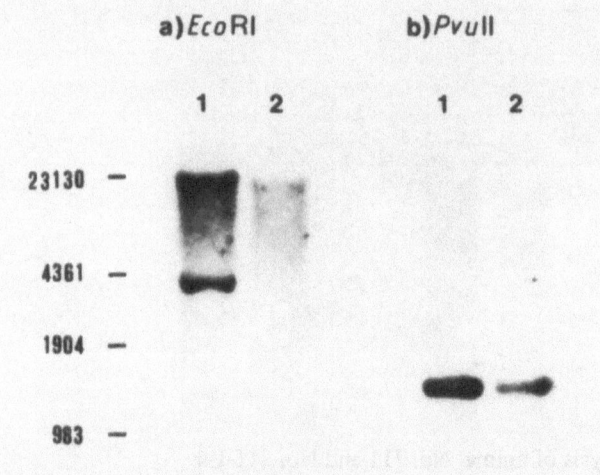

FIGURE 4. Southern blot analysis of offspring No. 211-II-4 and No. 211-II-7
Respectively 30 μg of DNA (No. 211-II-4, **1**; No. 211-II-7, **2**) were digested with *Eco*RI (a) or *Pvu*II (b), electrophoresed on a 1.0 % agarose gel and transferred to a nitrocellulose filter. Sizes of λ-marker (bp) are indicated.

*Pvu*II-digestion (Fig. 3 b and Fig. 4 b) leads to a single signal in all transgenic mice at 1.05 kb differing in intensity in respect to the number of integrated MThGH copies/cell. Fig. 5 shows the Southern blot restriction patterns of the transgenic offspring, when the DNAs are digested with *Bgl*II. Animals which had two bands at 4.0 kb and > 20 kb and those which had only one signal at 4.0 kb when digested with *Eco*RI, have two signals at 4.0 kb and at about 5.2 kb. In contrast, mice which had a single band at > 20 kb, only show the upper signal at > 5.2 kb.

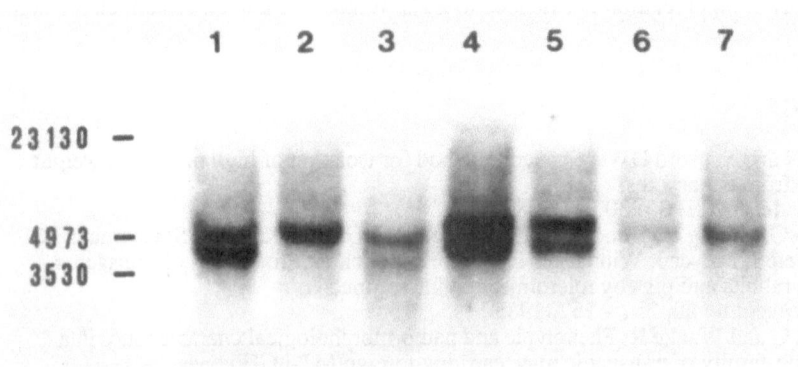

FIGURE 5. Southern blot analysis of the offspring from animal No. 211
Respectively 30 µg DNA (No. 211-I-3, **1**; No. 211-I-5, **2**; No. 211-I-7, **3**; No. 211-II-3, **4**; No. 211-II-4, **5**; No. 211-II-7, **6**; No. 211-III-1, **7**) were digested with *Bgl*II, electrophoresed on a 1.0 % agarose gel and transferred to a nitrocellulose filter. Hybridization was done as described above. Sizes of λ-marker (bp) are indicated.

TABLE 1. Summary of dot blot analyses

Animal	Sex	Copies/cell	Animal	Sex	Copies/cell
211	m	4	211-II-3	m	4
211-I-2	m	ND	211-II-4	m	4
211-I-3	m	3	211-II-5	f	ND
211-I-4	m	4	211-II-7	f	1
211-I-5	m	1	211-III-1	m	1
211-I-7	f	3	211-III-2	f	ND
m: male	f: female		ND: not determined		

To determine the integrated number of MThGH-copies/cell, the DNA of all transgenic mice were analyzed by dot blotting (Tab. 1).

5. CONCLUSION

We have analyzed transmission of a foreign DNA sequence from a male transgenic mouse to its progeny. The transmission data suggest that there are two different integration sites in mouse No. 211. The offspring reared from this transgenic mouse can be devided into three groups in respect to the mode of integration. The first group, identical to the founder mouse No. 211, has integrated 4 copies of the MThGH-gene construct per cell. One copy is localized at a single integration site, with the lost of the 5′- and 3′-*Eco*RI-sites of the construct. Three copies have been integrated tandemly arrayed in an additional chromosomal locus. The two remaining groups have either integrated the first locus with only one copy or the second one with three copies per cell.

Our data demonstrate the stable integration of a microinjected MThGH-gene construct into the germ line of transgenic mice and propagation as Medelian loci.

ACKNOWLEDGEMENTS

We gratefully acknowledge H. M. Goodman and R. C. Selden for providing us with the plasmid pXGH-1. Appreciation is extended to E.-L. Winnacker and H. Kräußlich for helpful comments.

REFERENCES

1. Blin N and Stafford DW: A general method for isolation of high molecular weight DNA from eukaryotes.
    Nucl. Acids Res. **3**, 2303 - 2308, 1976.
2. Brem G, Brenig B, Goodman HM, Selden RC, Graf F, Kruff B, Springmann K, Hondele J, Meyer J, Winnacker E-L and Kräußlich H: Production of transgenic mice, rabbits and pigs by microinjection into pronuclei.
    Zuchthygiene **20**, 251 - 252, 1985.
3. Brem G and Wanke R: Phenotypic and patho-morphological characteristics in a half-sib-family of transgenic mice carrying foreign MT-hGH genes.
    3rd FELASA Symposium, Amsterdam, 1987.
4. Feinberg AP and Vogelstein B: A technique for radiolabeling DNA restriction endonuclease fragments to high specific activity.
    Anal. Biochem. **132**, 6 - 13, 1983.
5. Feinberg AP and Vogelstein B: A technique for radiolabeling DNA restriction endonuclease fragments to high specific activity.
    Addendum Anal. Biochem. **137**, 266 - 267, 1984.
6. Goodman HM and Selden RC: pers. com., 1985.
7. Goldberg DA, Posakony JW and Maniatis T: Correct developmental expression of a cloned alcohol dehydrogenase gene transduced into the Drosophila germ line.
    Cell **34**, 59 - 73, 1983.
8. Gordon JW: Studies of foreign genes transmitted through the germ lines of transgenic mice.
    J. Experimental Zool. **228**, 313 - 324, 1983.
9. Hammer RE, Pursel VG, Rexroad jr. CE, Wall RJ, Bolt DJ, Ebert KM, Palmiter RD and Brinster RL: Production of transgenic rabbits, sheep and pigs by microinjection.
    Nature **315**, 680 - 683, 1985.
10. Palmiter RD, Norstedt G, Gelinas RE, Hammer RE and Brinster RL: Metallothionein-human GH fusion genes stimulate growth of mice.
    Science **222**, 809 - 814, 1983.
11. Palmiter RD and Brinster RL: Transgenic Mice.
    Cell **41**, 343 - 345, 1985.
12. Palmiter RD, Brinster RL, Hammer RE, Trumbauer ME, Rosenfeld MG, Birnberg NC and Evans RM: Dramatic growth of mice that develop from eggs microinjected with metallothionein-growth hormone fusion genes.
    Nature **300**, 611 - 615, 1982.
13. Spradling AC and Rubin GM: The effect of chromosomal position on the expression of the Drosophila xanthine dehydrogenase gene.
    Cell **34**, 47 - 57, 1983.
14. Southern EM: Detection of specific sequences among DNA fragments separated by gel electrophoresis.
    J. Mol. Biol. **98**, 503 - 517, 1975.

A.C. Beynen & H.A. Solleveld (Eds), New developments in biosciences: their implications for laboratory animal science, ISBN 978-94-010-7973-0
© 1988 Martinus Nijhoff Publishers, Dordrecht.

ATTEMPTS TO PRODUCE TRANSGENIC RABBITS CARRYING MTI-HGH RECOMBINANT GENE

K. ROSS, B. BRENIG, J. MEYER, G. BREM
Institut für Tierzucht
Veterinärstr. 13
D-8000 München 22

1. SUMMARY

Transgenic rabbits were produced by microinjection of a DNA-fragment consisting of the structural gene of the human growth hormone and the promoter region of the mouse metallothionein-I gene.

From 576 rabbit zygotes microinjected 120 (21 %) degenerated within two hours of culture. After transfer of the surviving injected zygotes 38 (8.3 %) offspring were born. 9 (24 %) of these offspring had integrated the MTI-hGH into their genome.

2. INTRODUCTION

The possibility to integrate foreign genes into the genome of mammals offers an interesting tool for microbiological basic research. Transgenic animals would be an excellent model to examine the mechanism of regulation and expression of recombinant gene constructs in living mammal organism.

For modern livestock production gene transfer will perhaps involve traits of economic importance like growth, disease resistance and efficiency of reproduction or feed utilization.

The first reports of the production of transgenic rabbits have been published by HAMMER et al. (1985) and BREM et al. (1985).

3. METHODS AND MATERIALS
3.1. Animals

The animals were cross-breedings of big rabbit strains. At least three weeks before using they have been caged singly. The stable had a lighting system with 10 hours light per day and air condition with regulated temperature of 17° C. The rabbits got 75 g concentrate per day and water and hay ad libitum.

3.2. Gene construct

For microinjection a 4.0 kb linear Eco RI/Eco RI-DNA fragment was used consisting of the structural gene of the human growth hormone (hGH) and the promoter/regulatory region of the mouse metallothionein-I gene (mMT-I) fused in a Bgl II/Bam HI site.
For details see BRENIG and BREM (1987).

3.3. Recovery of zygotes

Donor rabbits were superovulated with 30 I.U. PMSG per kg

body-weight and mated naturally 72 hours later. For induction of ovulation they received 90 I.U. HCG. Donors were slaughtered 17 to 22 hours after breeding and ova were collected by flushing the oviducts with BSM II (KANE and FOOTE, 1970) + 20 % FCS. The ova were assessed on microscope and judged by
- morphology: membranes, granula and mucin layer
- fertilization: presence of sperms in the perivitelline space and the second polar body
- presence of visible pronuclei.
Morphologically intact zygotes with visible pronuclei were used for microinjection.

3.4. Microinjection

Microinjection was carried out under X 400 magnification using a ZEISS Inverted Microscope IMC 405 and Nomarski interference contrast optics. Zygotes were placed on a depression slide in a drop of medium. Zygotes were attached to a holding pipette by suction, and the injection pipette (diameter 1-2 μm) was inserted into one of the two pronuclei by piezo electric power. Injection was accomplished by using compressed air. The expansion of the pronuclei (Fig. 2) demonstrated the successful injection of several picoliters of DNA-solution (4 μg DNA/ml).

3.5. Transfer of zygotes

At the same time as the donor rabbit the recipients were ovulated with 90 I.U. HCG. The microinjected zygotes which were still alive after two hours of culture were transferred to the oviducts of the pseudopregnant foster-mothers. The transfer was made under anesthesia by midline incision. In average 8 zygotes were transferred per oviduct. Some recipient rabbits received untreated control zygotes.

3.6. Analyses of gene integration

From the offspring born after gene injection the following samples were analyzed for integration of the MTI-hGH:
- Tissue probe
In the second month of life of the offspring born after microinjection a tissue probe was taken form the tail-end. Isolaton of total nucleic acid was done by powdering small samples of tail tissue in liquid nitrogen. Powdered tissures were homogenized and incubated in a 0.5 M EDTA 0.5 % N-Lauroylsarkosin 0.01 % proteinase K-buffer (pH 8.0) for 2-3 hours (50° C, 230 U/min.) in a water bath shaker. After incubation the homogenates were extracted twice with TE (10 mN Tris-HCl, 1 mM EDTA, pH 7.5) saturated phenol, twice with phenol/chloroform and once with chloroform/isoamyl alcohol (24:1). All samples were then dialyzed against 10 mM Tris-HCl, 1 mM EDTA (pH 7.8) over a period of 15-24 hours at 4° C changing the buffer several times. This procedure resulted in 1-1.8 ml of high-molecular DNA-solution with a concentration of at least 90-300 μg/ml.

FIGURE 2. Zygote before (A) and after (B) microinjection

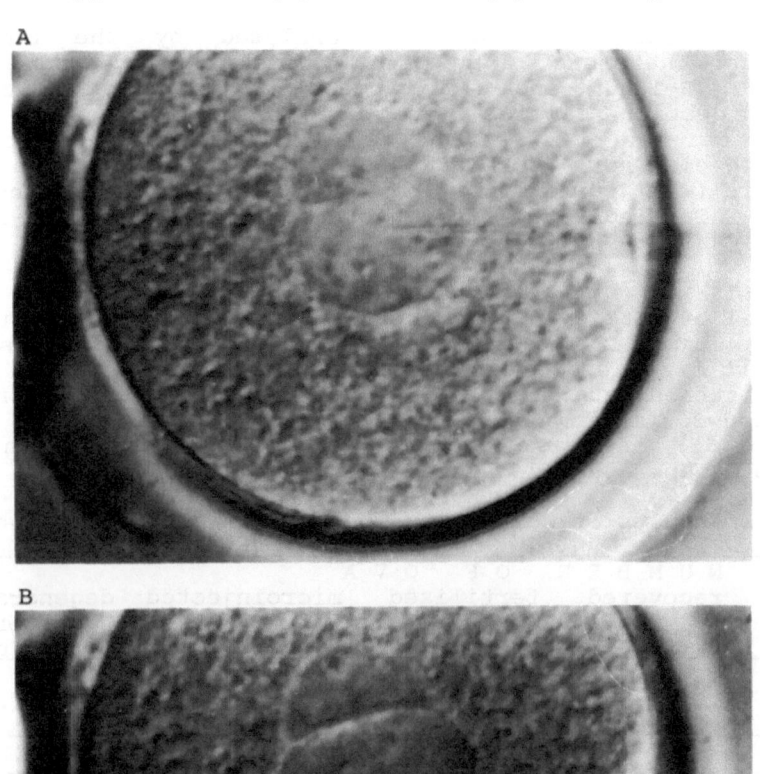

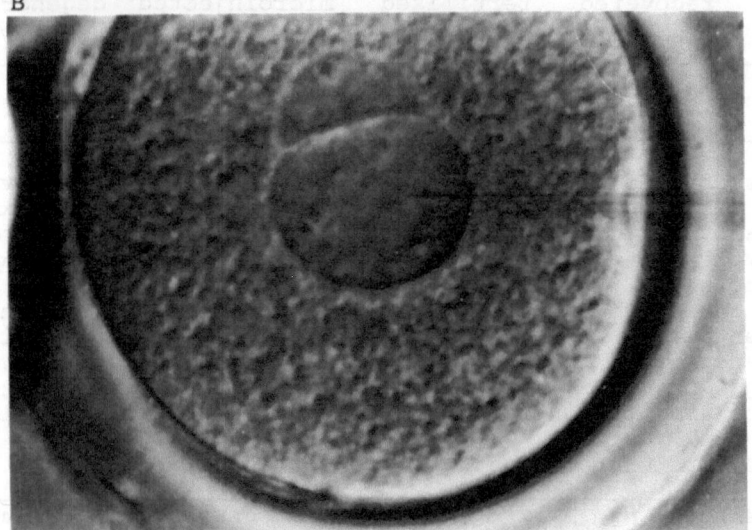

- 5 ml blood sample
These samples were collected at the age of 5 to 6 months.
Freshly taken blood was centrifuged through a Ficoll Paque
gradient to separate white blood cells which were collected
and treated as described above.
    Rabbits carrying MTI-hGH sequences were first detected by
dot blotting. 10 µg of DNA were heated to 100° C for 2 min. in
a NaOH-NaCl buffer, quickly chilled on ice, neutralized and
then spotted directly onto nitrocellulose filters, which have

been soaked in H_2O bidest. and 20 X SSC in succession prior to use.

Positive rabbits were then analyzed by the Southern technique. Therefore 30 μg of DNA were digested with <u>Eco</u> RI (60-90 U, 8-10 h) and simultaneously treated with RNase A. After digestion samples were electrophoresed on a 1.0 % agarose gel and transferred to a nitrocellulose filter as described elsewhere (SOUTHERN, 1975).

Filters were hybridized with 1 ug of pGH-1, nicktranslated with 32-P labeled -dCTP and autoradiographed for 15-30 hours at -80° C after stringent washing.

4. RESULTS

A total of 1683 ova were recovered from the 96 superovulated donor rabbits (18 ova per doe in average). 957 of these ova were fertilized and showed sperms in the perivitelline space (fertilization rate: 57 %). 576 zygotes showing both pronuclei were used for microinjection.

79 % (456) of these zygotes survived microinjection. 120 zygotes degenerated (degeneration rate 21 %) (Table 1.).

TABLE 1. Results of collection and microinjection of zygotes

| donors | N U M B E R   O F   O V A | | | |
	recovered	fertilized	microinjected	degenerated after microinjection
96	1683	957(57 %)	576	120(21 %)

456 zygotes microinjected were transferred into the oviducts of 28 recipients. 7 out of these recipients came to term (pregnancy rate: 25 %) and a total of 38 offspring were born (survival rate: 8.3 %). 24 (63 %) of these offspring were still born.

158 untreated control-zygotes were transferred into 12 recipient does. In 3 litters out of these control-recipients (pregnancy rate: 25 %) 16 offspring were born (survival rate: 10.1 %). The results of transfers with microinjected zygotes were compared to control-transfers and are shown in Table 2.

TABLE 2. Transfer of injected and control-zygotes

Treatment	recipients n	ova transferred n	Pregnancy rate	off- spring born	(Survival rate)
Microin- jection	28	456	7 (25 %)	38	(8,3 %)
Control	12	158	3 (25 %)	16	(10.1 %)

9 offspring (24 %) had integrated the MT-hGH into their

genome. 5 transgenic rabbits had integration of less than one copy/cell, presuming to be mosaics. 3 of these transgenic rabbits developed to adults. 20 offspring of one male transgenic rabbit carrying less than one copy/cell were analyzed by Southern blotting without showing integration of MT-hGH. The remaining 4 transgenic rabbits had integrated between 1 and 69 MTI-hGH copies/cell.

5. CONCLUSIONS
Direct microinjection can be used to produce transgenic rabbits. 24 % integration is comparable to other reports about the production of transgenic rabbits. HAMMER et al. (1985) described 13 % integration after microinjection of MTI-hGH into pronuclei of rabbit zygotes.

REFERENCES

1. Brem G, B Brenig, HM Goodman, RC Selden, F Graf, B Kruff, K Springmann, J Hondele, J Meyer, EL Winnacker, H Kräußlich: Production of transgenic mice, rabbits and pigs by microinjection into pronuclei. Zuchthygiene 20, 251-252, 1985.
2. Brenig B, G Brem: Integration of hGH gene in transgenic mice and transmission to next generation. 3rd FELASA Symposium, Amsterdam, 1.-5.6.1987.
3. Hammer RE, KG Purcel, CE Rexroad Jr., RJ Wall, DJ Bolt, KM Ebert, RD Palmiter, RL Brinster: Production of transgenic rabbits, sheep and pigs by microinjection. Nature 315, 680-683, 1985.
4. Kane MT, RH Foote: Culture of two- and four-cell rabbit embryos to the expanding blastocyst stage in synthetic media. Proc. Soc. exp. Biol. 133, 921-925, 1970.
5. Southern EM: Detection of specific sequences among DNA fragments separated by gel electrophoresis. J. Mol. Biol. 98, 503-517, 1975.

ACKNOLEDGEMENTS:
Project supported in part by Dr. Dr. Karl Eibl - Stiftung, Neustadt/Aisch

A.C. Beynen & H.A. Solleveld (Eds), New developments in biosciences: their implications for laboratory animal science, ISBN 978-94-010-7973-0
© 1988 Martinus Nijhoff Publishers, Dordrecht.

ELECTROFUSION OF EARLY MAMMALIAN EMBRYO CELLS

A.CLEMENT, J.MEYER, G.BREM
Institut für Tierzucht und Tierhygiene
Veterinärstr. 13
D-8000 München 22

1. SUMMARY

Electrofusion parameters for mouse, rabbit and pig two-cell embryos were found by use of a cell fusion instrument. The yield of fusion and the viability of the embryos depends on the number of fusion pulses, the pulse length (μs range) and the field strength (kV/cm range).
Tetraploid one-cell embryos were produced by electrofusion and were cultivated in vitro and in vivo.

2. INTRODUCTION

During the last decades different techniques for cell fusion were developed such as the use of polyethylenglycol(1), Sendai virus(2), lysolecithin(3) etc.. The shortcomings of these methods were(4):
1. unphysiological circumstances for the cells with decrease of their viability
2. not synchronous fusion process in treatment of two or more cells
3. low yield of fused cells
It is possible to avoid these disadvantages by use of the electrofusion method, even in fusion of early mammalian embryo cells (5,6,7).
We report here that tetraploid one-cell-stages in mice, rabbits and pigs can be produced by electrofusion of two-cell embryos. After fusion the embryos do start cleavage and can be cultivated until blastocyst stages.

3. MATERIALS AND METHODS

Mice, rabbits and pigs were superovulated with PMSG/HCG, and the two-cell embryos were collected by flushing oviducts (mouse: modified Whitten's medium, rabbits: BSM 2, pig: Whitten's medium with 15 mg/ml BSA or PBS with 20% lamb serum). Before manipulation embryos were washed in a medium with low conductance (ZIMMERMANN Cell Fusion Medium) to avoid heatening of the medium by friction. Embryos were set between two parallel stainless steel electrodes ($\emptyset$100μm), which were fixed on a slide, into a drop of fusion medium. The distance between the electrodes varied from 180 μm to 350 μm and their length was 5 cm. The electrodes were connected with a pulse generator (ZIMMERMANN Cell Fusion Instrument, CGA Corporation).

At first alternating current with 6V for all shown electrode distances was applied for several seconds (5-30) to turn the two-cell-stages into the right position for fusion (Dielectrophoresis)(4). After embryos had reached this position (cleaving line parallel to the electrodes) fusion pulses were started (direct current). Time between the fusion pulses was 0,1 s. The number of pulses was varied from 1 to 8, the pulse length from 70 to 80 μsec and the field strength from 0,52 to 1,90 kV/cm.

After pulsation the embryos were washed immediately in culture medium and were incubated (5% CO_2 in air, 37°C). One hour later they were classified under a microscope. If fusion had occured the two-cell embryo had transformed into one round cell showing both nuclei (Fig. 1).

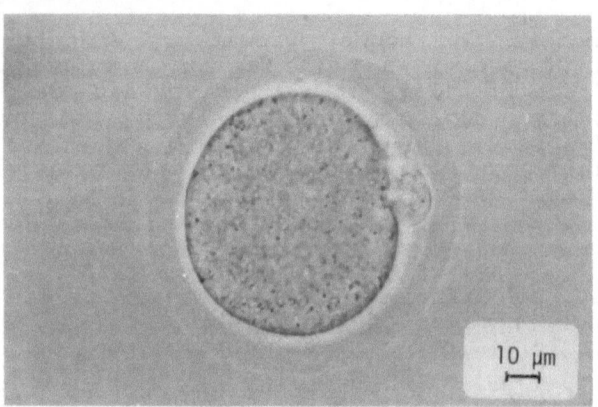

Fig. 1: Fused two-cell-embryo showing
both nuclei in one round cell

The fused embryos were cultivated in vitro (mice: Whitten's medium, rabbits: BSM 2, pigs: Whitten's medium with 15 mg/ml BSA; all in 5% CO_2 in air, 37°C) or in vivo in pseudo-pregnant foster mothers (mice, rabbits).

4. RESULTS

4.1. Fusion

Mouse: 655 two-cell embryos were treated by electric field pulses. 526 of them (81%) fused to tetraploid one-cell embryos.

TAB.1: Fusion of two-cell embryos

species	n	fused n	fused %	not fused n	not fused %	degenerated n	degenerated %
mouse	655	526	81	88	13	41	6
rabbit	66	38	57	11	17	17	26
pig	46	34	74	10	22	2	4

88 two-cell stages (13%) did not fuse. 41 embryos (6%) degenerated after electric field manipulation.
The highest fusion rate in mice (90%) was achieved by 4 pulses,
each 70 μs long at a field strength of 0,52 -0,94 kV/cm.

TAB.2: Fusion rates in different parameters in mice

pul-ses	pulse-length [μs]	treated n	fused n	fused %	not fused n	not fused %	degen. n	degen. %	voltage [kV/cm]
1	80	37	31	84	5	14	1	2	0,64
2	70	126	108	86	15	12	3	2	0,58-0,64
3	70	266	231	87	16	6	19	7	0,52-0,64
4	70	41	37	90	4	10	0	0	0,52-0,94
5	70	63	30	48	22	35	11	17	0,80-1,80
6	80	122	89	73	26	21	7	6	1,40-1,90
		655	526	81	88	13	41	6	

Rabbits: of 66 embryos treated with electric field pulses 38
fused (57%), 11 (17%) did not fuse and 17 (26%) degenerated.
More than 50% of the embryos degenerated when they were treated
by 6 pulses, each 80 μs long. More but shorter pulses give
higher yields in fused embryos (71% fused by 7 pulses, each 50
μs long).

TAB.3: Fusion rates in different parameters in rabbits

pul-ses	pulse-length [μs]	treated n	fused n	fused %	not fused n	not fused %	degen. n	degen. %	voltage kV/cm
6	80	21	6	29	4	19	11	52	1,00-1,55
7	50	45	32	71	7	16	6	13	0,58-1,00

Pigs: 34 two-cell embryos (74%) fused when 46 two-cell stages
were treated by electric field pulses. 10 embryos did not fuse
(22%). Two embryos (4%) degenerated after manipulation.

TAB.4: Fusion rates in different parameters in pigs

pul-ses	pulse-length [µs]	treated n	fused n	%	not fused n	%	degen. n	%	voltage kV/cm
≤5	70	15	12	80	1	7	2	13	0,64-1,10
6	80	43	32	74	9	21	2	5	0,75-1,88

4.2. In vitro culture of tetraploid embryos

Mouse: 219 tetraploid one-cell embryos were cultivated in vitro; 42% started cleavage. Among them, 90% developed to 2-8-cell stage, 4% to morulae and 6% to blastocysts.

TAB.5: In vitro culture of tetraploid one-cell embryos in mice

fused total	no cleavage	-----------cleaved to-------------		
		2-8-cell stage	morulae	blastocysts
219	150 (58%)	62 (90%)	3 (4%)	4 (6%)
controls total				
258	113 (44%)	96 (66%)	9 (6%)	40 (28%)

4.3. In vivo culture of tetraploid embryos

Mouse: two, three or four days after 103 fused embryos were transferred into 5 pseudo-pregnant foster mothers these animals were killed, their uteri were flushed and the embryos were collected. 50 embryos (49%) were found in four mice. 32 embryos (46%) were arrested at one-cell stage, 5 (10%) had developed to morulae (10%) and 13 (26%) to blastocysts.

TAB.6: In vivo culture of fused two-cell embryos in mice
(in five pseudo-pregnant foster mothers; time period of
2, 3 or 4 days)

controls transferred total	collected total	2-8-cell stage	morulae	blastocysts
42	13(31%)	7	5	1
fused two-cell embryos transferred total	collected total	2-8-cell stage	morulae	blastocysts
103	50(49%)	32(64%)	5(10%)	13(26%)

Eleven transfers were made in which foster-mothers were allowed
to go to term. Altogether, 163 fused embryos (white colour-
marker) and 104 control embryos (two-cell stages with black
colour-marker) were transferred. 21 black offspring (controls)
were born in 4 litters. The pregnancy rate was 36%. 42% of the
transferred embryos survived.

Rabbits: 14 fused two-cell embryos were transferred into a
pseudo-pregnant foster mother. After 72 hours, the animal was
slaughtered, and the embryos were collected. 9 embryos were
found of which 2 were degenerated, 4 were late morulae and 3
were blastocysts. These embryos were cultivated in BSM 2 for
the next 24 hours, in which they developed to 7 blastocysts.

5. CONCLUSION AND DISCUSSION

Two-cell embryos of mice, rabbits and pigs can be fused by
electric-field pulses. Provided optimal conditions, a fusion
rate up to 90% can be reached.
After fusion the tetraploid embryos do start cleavage and are
able to develop until blastocysts. The critical period for
tetraploid embryos seems to be the time after implantation.
Diploid/tetraploid-mosaics could be produced by fusion of two
blastomeres of 4-,8-,16-cell embryos or morulae. 2n/4n-aggrega-
tion chimeras are able to survive (8).
Electrofusion of enucleated cells with nuclei will be a
powerful technique for different applications of genom manipulation.

6. ACKNOWLEDGEMENTS

We thank the Dr.Dr.Karl Eibl-Stiftung, Neustadt/Aisch for
supporting this project in part.

7. REFERENCES

1. Kao K.N., Michayluk M.R.: Planta 115, 355-367, (1974)
2. Harris H., Watkins J.F.:Nature 227, 640-646 (1965)
3. Lucy J.A.: Nature 227, 815-817 (1970)
4. Zimmermann U., Vienken J.: J. Membrane Biol. 67, 165-182
   (1982)
5. Berg H.: Bioelectrochemistry and Bioenergetics 9, 223-228
   (1982)
6. Berg H., Kurischko A., Freund R.: Studia biophysica 94,
   103-104 (1983)
7. Kubiak J.Z., Tarkowski A.K.: Experimental Cell Research 157,
   561-566 (1985)
8. Bauernfeind M., Brinkmann B., Leipoldt M.: Zuchthygiene
   21.(4), 150 (1986)

A.C. Beynen & H.A. Solleveld (Eds), New developments in biosciences: their
implications for laboratory animal science, ISBN 978-94-010-7973-0
© 1988 Martinus Nijhoff Publishers, Dordrecht.

PASSIVE INFRARED MOVEMENT DETECTOR, A NEW EQUIPMENT TO MONITOR MOTOR ACTIVITY OF SMALL RODENTS IN NORMAL CAGES

Hans SIGG and Pierre TAMBORINI
Nestlé Research Center, Nestec Ltd., Vers-chez-les-Blanc,
1000 Lausanne 26, Switzerland.
Institut für Toxicologie, ETH und Universität Zürich,
8603 Schwerzenbach, Switzerland.

INTRODUCTION
Activity patterns of small mammals have proved to be very
sensitive to physiological states as well as environmental
changes. The measurement of motor activity in normal cages
without disturbing the animals is not yet possible or else, like
videosystems, too expensive for routine use. Therefore we
attempted to develop a new system with the following properties:
easy to handle; suitable for standard macrolon cages with grid
covers; possibility for continuous recordings; simultaneously
monitoring of all cages in an assay; no interference between
recording units of neighboring cages; reliability and accuracy
that permits comparisons; dataprocessing on available computer
systems; low price.

METHOD
The body of warm-blooded animals is warmer than the environment
under laboratory conditions and therefore it may be detected by a
heat radiation sensor. A variety of commercially available
Passive Infrared Movement (PIM) detectors -developped for
security purposes and automatic door opening- are based on this
principle: The area surveyed is split up into different sectors,
either using lenses or mirrors. If a warm body moves between
different sectors, the heat radiation intensity of these sectors
changes, and this is detected by the PIM sensor. If the change
exceeds a certain limit, the detector turns to a on-position. In
our system this on-position is held for 1.5 sec before it turns
off again.
An EPSON HX-20 hand-held computer supplied with a PORTBOX-inter-
face checks the on/off-positions of up to 32 PIM detectors every
second. For every on-position, the counter of the respective PIM
detector is increased by one point. After 5 min the values of
all 32 counters are transferred to the RAM-file memory, and the
counters are reset to zero. Every 24 h the values in the RAM-file
are transferred to a host computer or to a storage device, ie. a
microcassette, a tape, or a disk.
Dataprocessing may be done by the HX-20 or a host computer. Since
the treatment of the data depends on the research questions, we
do not propose a standard treatment. However, a plot of the
activity pattern is strongly recommended: A visual survey allows
the selection of the most adequate statistical procedure.

VALIDATION OF THE METHOD

The sensitivity of the OPTEX PIM detector may be regulated by a
potentiometer. A defined IR-source which may be turned on and off
is used for calibration of the different sensors. Due to
technical limitations of the optical system, the sensitivity of
the sensor is not equal over the whole area surveyed. Since the
optical properties of all sensors are the same, the deflections
also remain the same for all the cages. However this limitation has
only minor effects because the surface of an animal normally covers
an area which includes points of different sensitivity and
therefore reduces the inaccuracy.

The reliability of the system was tested using three types of
comparable activity recordings: 1. The same cage was recorded by
two PIM detectors, surveying the cage under different angles. The
shadow produced by water bottle and food cubes as well as
the optical limitations mentioned above therefore influenced the
recordings. The mean activity values differed less than 5%, while
single 5 min values differed less than 10% at a mean. 2. A
similar comparison was made using a radar system for simultaneous
recordings of the same cage. Since the recording units were not
the same, a direct comparison is not possible. However, the plot
of the activity pattern was exactly the same! 3. Finally the PIM
records were compared with the videorecords of the cage. A rough
quantification of the tapes revealed the same pattern as did the
automatic recording system.

RESULTS

We used the PIM system for longterm recordings of mice, rats and
guinea pigs in standard MACROLON type III and type IV cages. All
recordings revealed a clear-cut circadian activity pattern which
remained the same from day to day when the animals were not
disturbed. An 36-day activity record of a female Sprague Dawley
rat (Sut:SDT, 8 weeks old) is presented in Fig. 1a. Each line
presents two days.Beginning at 8.15, every 15 min the total
activity is plotted as a black bar. Darkness was from 7.00 to
19.00.The activity pattern of that rat shows a continuous activity
during the dark period every forth day (4,8,12,16, etc), while the
activity during the other dark periods show some records of rest.
This pattern persists in other female rats and reflects oestrus
cycles. As the shift in the light cycle at day 27 (dark from
19.00 to 7.00) demonstrates, the activity pattern clearly depends
on light cycle. Changes of the environmental conditions may have a
strong impact on the activity pattern as shown the first days.
In a preliminary study to improve data acquisition in acute
toxicology, the PIM system proved to be of some value (Fig. 1b).
Four groups of female Sprague Dawley rats (Sut:SDT, 8weeks) were
given a solution of 5% methylcellulose (5ml/kg p.os). This
solution contained 0, 400, 600, and 800 mg/kg aniline. The
outprint of the activity recordings began two days before the
treatment which was done at 9.00 and continued for four days.
The plots present the results of two days per line, with a dark

36 DAYS OF A FEMALE RAT

ACUTE   TOXICITY

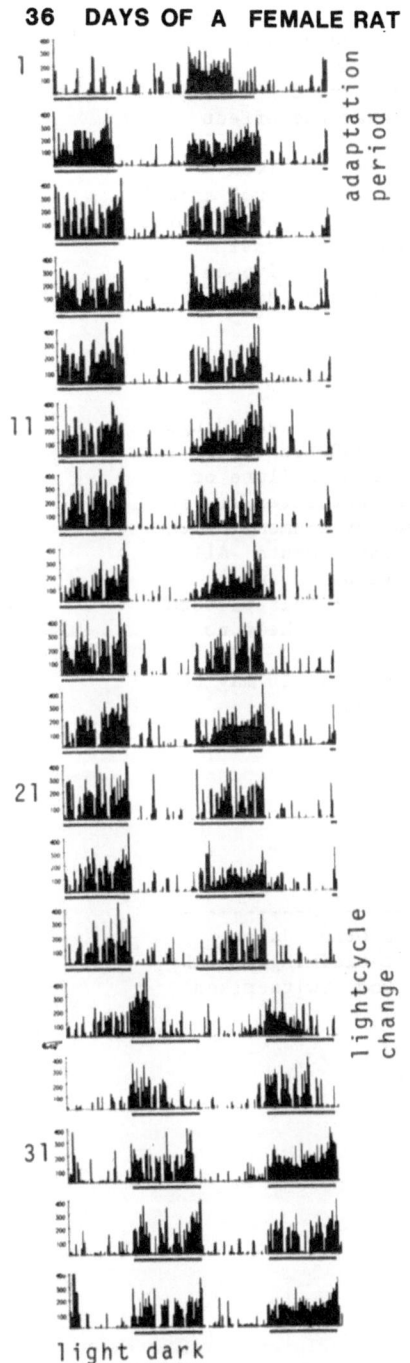

1

11

21

31

adaptation period

lightcycle change

light dark

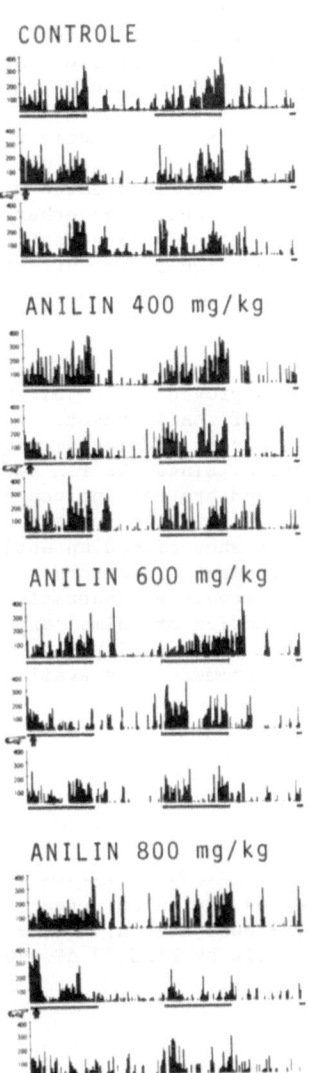

CONTROLE

ANILIN 400 mg/kg

ANILIN 600 mg/kg

ANILIN 800 mg/kg

Figure 1.
Explanations in text

period from 7.00 to 19.00. In each group, the first female was
selected for this printout. Since this study included many
other measurements which were done in the same room, the
circadian pattern is less clearcut than in undisturbed animals,
a fact to be considered in future studies. However the effect
of the treatment is clearly visible:
After the treatment the motor activity of the animals given
aniline decreases for some hours, while that in the control animals
is not impaired. The female treated with 800 mg/kg did not recover
during the four days, while the others returned to a normal
pattern. The span of an impaired activity pattern may give an
estimate of the duration of the effect and may help to define an
adequate schedule for further examinations.

CONCLUSIONS
The PIM system has proved to be suitable for longterm recordings
of motor activity of small rodents in standard MACROLON cages. In
a series of preliminary studies during 10 months no failure of
the system occurred. Since the PIM detectors do not interfere in
any physical way and are not influenced by the environment of the
cage this system is suitable for all kinds of experiments. All
laboratory animals show circadian activity patterns in an
undisturbed environment, therefore the recording of this pattern
may be proof of a completed adaptation period and may help to
improve standardization of animal experiments.
The data structure is very simple and easy to handle. It may be
processed on most commercially available software.

PIM detector: OP-04, Optex Co., Ltd., Nionohama Otsu, Japan.
Interface   : PORTBOX, W.Koch, CSM, Filderstadt-Sielmingen, BRD.
PIM-System available by ZIRELCO AG, 4657 Dullikon, Switzerland.

A.C. Beynen & H.A. Solleveld (Eds), New developments in biosciences: their implications for laboratory animal science, ISBN 978-94-010-7973-0
© 1988 Martinus Nijhoff Publishers, Dordrecht.

Establishment of a SPF Population of Cricetulus griseus

U. Märki, W. Rossbach, J. Leuenberger

Institut für biologisch medizinische Forschung AG; Wölferstrasse 4; CH-4414 Füllinsdorf

In this study different nursing methods have been studied in developing SPF Cricetulus griseus.

From a total of 192 hysterectomies, 1'182 fetus were delivered and fostered onto Ibm: MORO (SPF) mice. Animals were grouped according to treatment and conditions.

* Groups 1 - 3: Had treated foster mothers.
* Groups 4 - 6: Had untreated foster mothers but different cage conditions.
* Group 7 : Had untreated foster mothers and an early check of the nest was made with weak and dead young removal.

1. Fosters treated with the analgesics Vetalgin, kept one per cage.

2. Two fosters treated with the hypnotics Temgesic.

3. Analogous to experimental group 2 but inclusive the lactating stimulant Oxymetrin.

4. Untreated fosters with about 75 % of the own offspring, two females per cage.

5. Three fosters with 1 to 2 days old offspring and one highly pregnant foster per cage.

6. Two to three fosters without own offspring per cage.

7. One to two fosters with own offspring per cage but removing all young hamsters without milk in the stomach during the first six hours.

One hamster of the experimental group 2 and sixteen of group 7 could be raised and weaned. The treated fosters nursed all offspring very poorly or not at all and separated the mouse from the hamster offspring, which were killed. In conclusion it would appear that the success rate could be increased by using a calm, untreated good foster mother which is experiencing the second day of lactation. In addition little disturbance of the nest and removal of 'poor doers' before 6 hours of age improves success rate.

At the time we keep a small SPF-Chinese hamster population (25 pairs in a SPF unit and 3 pairs in an isolator) with a good reproduction performance of five to seven offspring per litter.

1.2 INTRODUCTION

In contrast to mice, rats and other laboratory animals the Syrian
and Chinese hamster could not be hand reared after hysterectomy
until now. It was not possible therefore to obtain gnotobiotic or
SPF hamsters apart from decontamination methods, as cross fostering onto another species had not been achieved. Since succeeding
in fostering different rodent species such as meriones, mastomys,
saccostomus and sigmodon on mice (publication in preparation),
this method was also used for the establishment of a SPF hamster
population.

The morphological-anatomical feature of tooth development in hamster babies - the incisors are covered by only a thin epithelial
layer which appears to be stretched and ruptured prenatally (Hoffmann et al., 1968) - turned out to be an extraordinary difficult
factor. For reducing the irritation of the mouse fosters, whose
own babies have no incisors at birth (Habermehl, 1980), we conceived different experimental groups either administrating an
analgesic or hypnotic drug (see table 2) or none.

1.3 MATERIALS AND METHODS

According to the availibity mature Chinese hamsters which were
kept under conventional conditions were housed together for 16
hours (= day 0). On the day 20 of pregnancy the hamster babies
were delivered by hysterectomy and brought into isolators containing the foster mice (see table 1). After 1 to 2 hours contact
with the baby mice we put the hamsters into the cage with fosters.
At this stage we rejected germfree fosters because the availibili-
ty of SPF foster mice was all the time without limits.

TABLE 1   Animals, feeding and ambient conditions

	Donor	Foster
Genus	Cricetulus griseus	Mus musculus
Stock	BestIbm: FUST (= Chi-nese hamster)	Ibm: FU albino (SPF); Syn.<Ibm: MORO (SPF)>
Temper-ature	21°C ±1°C	21°C ±1°C
Rel. humidity	50 % ± 10 %	50 % ± 10 %
LD *	12:12	12:12

	Donor	Foster
Food	NAFAG 924, ad lib	NAFAG 857 DVOD, auto-claved, ad lib
Water	ad lib	ad lib
Main-taining	Makrolon cage type II, under conventional conditions	Makrolon cage type II or III, under SPF conditions
Hygenic status	Conventional	SPF **

* LD = Light-dark with $^1/2$ h dusk-dawn

** (Checked pathogens, no positive findings: checked by G. Kouchakji)

1. Virus
   * Lymphocytic choriomeningitis (LCM)
   * Mouse Adenovirus (MAV)
   * Mouse Hepatitis (MHV)
   * Pneumonia Virus of the Mouse (PVM)
   * Polyoma
   * Sendai (parainfluenza-1)
   * SV-5
   * Theiler's Encephalomyelitis (GD-VII)
   * Toolan (H-1)
2. Mycoplasma
   * Mycoplasma pulmonis
3. Bacteria
   * Bacillus piliformis
   * Bordetella bronchiseptica
   * Clostridium perfringens
   * Corynebacterium kutscheri
   * Pasteurella spp.
   * Salmonella spp.
   * Streptobacillus moniliformis
   * Streptococcus group A
   * Streptococcus pneumoniae
   * Yersinia pseudotuberculosis
4. Fungus
   * Dermatophytes
5. Tissue protozoa
   * Encephalitozoon cuniculi
   * Toxoplasma gondii
6. Intestinal Protozoa
7. Helminths
8. Ectoparasites

The first experiments were started on the 18 of August 1983 and on th 25 of June 1986 we finished this program. The entire experiment is divided into 3 main groups.

A. Treatment of the fosters either with hypnotics or with analgesics or with hypnotics and lactating stimulant (group 1 to 3 in table 3)

B. No treatments, but different cage conditions (group 4 to 6 in table 3)

C. In contrast to main group A and B the first check after starting fostering was made after 6 hours (and not after 24 hours) and all weak and dead hamster babies were removed immediately (group 7 in table 3)

TABLE 2   Treatments of the fosters

Drug	Dosis	Application
Vetalgin (analges-ics), Veterinaria	0.02 ml	i.m.
Temgesic (hypnotics), Reckitt and Colman	0.125 or 0.25 ml	i.p. or s.c.
Temgesic and Oxymetrin (lactating stimulant)	0.25 or 0.50 ml and 0.05 ml	i.m.

1.4 RESULTS

TABLE 3   Number of fostered and weaned hamsters per group

Group	Method	Cage Type	Treatment	Fos-tered	Wea-ned
1	1 foster with lit-ter (1-2 days old) or none	II	Vetalgin	123	0
2	2 fosters with litter (1-2 days old) or none	III	Temgesic	57	0
3	2 fosters with litter (1-2 days old) or none	III	Temgesic and Oxy-metrin	24	0
4	2 fosters with litter (1 day old)	III	None	46	1

Group	Method	Cage Type	Treatment	Fostered	Weaned
5	3 fosters with litter (1-2 days old) and 1 highly pregnant foster	III	None	(460)	0
6	2-3 fosters without litter	II	None	30	0
7	1-2 fosters with litter (1 day old). First control after 6 hours, removal of dead and weak hamsters	II	None	322	16

Explanations to table 3:

- Group 1 - 6: First control after 24 hours
- Group 4: 1 female hamster
- Group 5:(460) is the estimated number of fostered hamsters

- Total hysterectomies: 192
- Delivered fetus: 1'182 (Average of 6.16 delivered hamsters per donor)

TABLE 4   Sex ratio of the weaned hamsters from group 7

Cage	No. fosters	Female hamsters	Male hamsters	Total
1	2	1	1	2
2	2	1	0	1
3	2	1	0	1
4	2	2	1	3
5	2	2	1	3
6	2	3	1	4
7	2	0	2	2

1.5 DISCUSSION

Treating of foster mice, based on the unsuccessful fostering of
hamsters younger than 3 days old (Sickel, 1978, 1987), had a
negating influence on mothering ability.   The mice did not
suckle the hamster babies properly or not at all. The babies
died during the first days.

Under different cage conditions with 1 to 3 fosteres either
with own offspring or not, the mice separated their own offspring from the hamsters and suckled the hamster rarely. Either
the hamsters died during the first 2 days or they were killed
or even eaten by the fosters.

The first check of the nest after 6 hours and the elimination
of all weak and dead hamsters increased the chance for a successful fostering. Probably weak and dead hamsters irritated
the fosters and had a deleterious effect on the mothering ability.

In conclusion the factors increasing the success rate of weaning hamsters under the described conditions seem to be:

    Calm, untreated good foster mother
    Second lactation day of foster mother
    Few handlings but first check after six hours
    Elimination of weak and dead hamsters.

At the time we keep a small hamster population of 25 breeding
pairs in a SPF unit and with 3 breeding couples in a SPF isolator. These SPF hamsters have a good reproduction performance
of five to seven offspring per litter.

1.6 LITERATURE

Hoffman, R.A., Robinson, P.F. and Magalhaes, H. (1968): The golden
     hamster, its biology and use in medical research. The Iowa
     State University Press.

Habermehl, K.-H. (1980): Die Altersbestimmung bei Versuchstieren,
     Verlag Paul Parey.

Sickel, E. (1978): Jahresbericht 1978, Zentralinstitut für Ver-
     suchstiere, Hannover.

Sickel, E. (1987): Personal communications.

May 19th, 1987

A.C. Beynen & H.A. Solleveld (Eds), New developments in biosciences: their implications for laboratory animal science, ISBN 978-94-010-7973-0
© 1988 Martinus Nijhoff Publishers, Dordrecht.

CHRONIC PHLORIZIN INTOXICATION IN ADIPOSE MUTANT MICE C57BL/KS db/db AND IN NORMAL CONROLS

I. ORTLEPP, K. GAERTNER, H. SCHOETZ, A. ZOGBAUM

1. INTRODUCTION

In a test, using phloricin to inhibit tubùlar glucose reabsorption in the kidney, obese mice showed a loss of body weight and many died during the experiment. Non adipose mice of the same strain do not show any lasting effects when treated with a similar dose of phloricin. (For details see 1, 2)
Results of further experiments on this subject are presented in the following sections.

2. METHODS

2.1. Animals:
2.1.1. Mice: 16 female C57BL/6J/Han (B6) (Series I) and 74 C57BL/KS$^{db/db}$ mice (Herberg, Düsseldorf) of both sexes (db/db) (Series II) from own breeding colony, bred in repulsion model with the misty-gene (m) as marker, were used in the studies.
2.1.2. Animal housing: The mice were housed in makrolon cages Type II on woodchips at $22 \pm 2°C$, illuminated from 06:00 to 18:00.
Food(Altromin 1324; 18% raw protein, 3,8% fat, 6% crude fibre) and water were provided ad libitum.

2.2. Experimental technics:
2.2.1. Phloricin application: Chronic phloricin intoxications were induced by subcutaneous depots of phloricin (Phloretin-2-β-glucoside; MG 472.5) (Serva Heidelberg) (Series I 60 mg; series II 50 mg) contained in polyvinyl tubes of 2 cm length and 2mm $\emptyset$. Controls are treated with similar tubes containing starch. In the test series II the implants were changed every second week.
2.2.2. Blood collection: Blood was collected from the retro orbital sinus using heparinized "Delbrück" tubes (volume 20 μl; Blaubrand^R)
2.2.3. Urine collection: Urine was collected in metal metabolism cages (Jackson model) for 24 h (Series I) or in glass cylinders on perforated metal sheets for 8 h (Series II) under paraffin seal. Bladders were massaged before and after the collection period.
2.2.4. Urine and blood glucose test: In preparation the blood samples were deproteinized and the urine samples were diluted

1:100. The prepared samples were analysed by the "Glucoseoxi-
dase - Perid method" (Boehringer) using an Eppendorf photometer
at an wave length of 578 nanometer in double tests. The variation is less than 5% of mean.

2.2.5. Weight measure: Weigths were measured on an electrical
scale with damping and 0,01 g indication steps (Sartorius).

3. RESULTS and DISCUSSION

Survival rate, body weight growth, blood glucose levels and the
renal glucose excretion in healthy (B6) and in obese (db/db)
mice with or without phloricin treatment were examined. The results are shown in the figures 1 - 6.
The B6-mice were randomly divided into two groups of eight animals each(Series I); 51 of the db/db-mice were treated with
phloricin, 23 served as control (Series II).

3.1. Survival rate under the influence of chronic phloricin in-
toxication:
All the healthy (B6) and the non treated obese mice (db/db)
survived this experiment, but 45 of 51 treated obese mice died
within 68 days. According to survival time the animals were di-
vided into four groups:
Group A   00 - 29 days (n = 16)
Group B   30 - 49 days (n = 20)
Group C   50 - 68 days (n = 09)
Group D   killed after 68 days (n = 06)

3.2. Body weight under the influence of chronic phloricin intoxication:

Figure 1:   Figure 2:

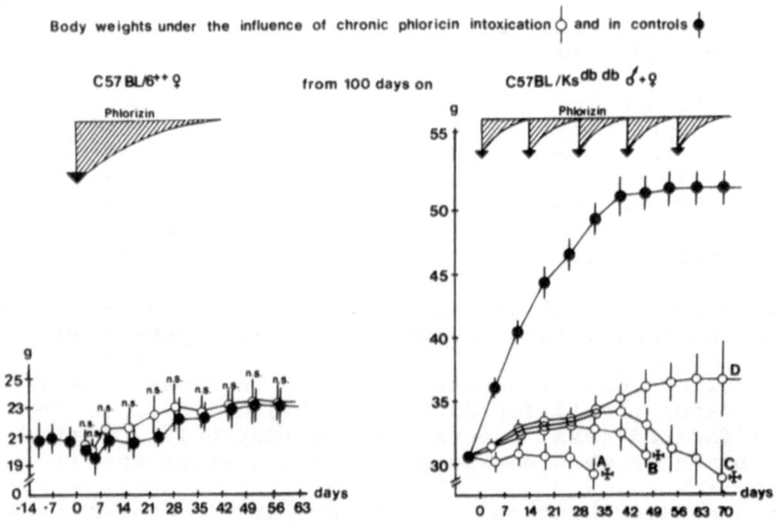

Body weights under the influence of chronic phloricin intoxication ⊘ and in controls ●

Phloricin has no influence on the body weight in the healthy
mice (See figure 1) Obese mice (db/db) however showed greatly
decreased growth rates under the influence of phloricin treatment. Therefore the weights of the treated mice are more than
30% below the weights of the untreated group (see figure 2). It
is remarkable, that weight increase is highly correlated with
the survival. Loss of weight for more than one week is a sign
of impending death.

3.3. Blood glucose levels under the influence of chronic phloricin intoxication:

Figure 3:   Figure 4:

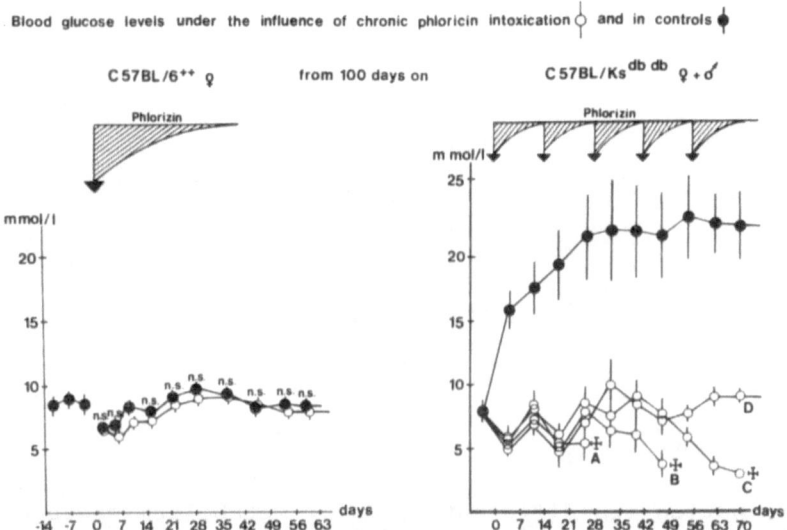

Blood glucose levels under the influence of chronic phloricin intoxication ○ and in controls ●

Phloricin causes in the B6-mice a slight decrease of blood glucose levels for a short time after application (see figure 3).
In the db/db-mice the decrease of blood glucose levels is
higher and the recompensation lasts longer. If the application
of phloricin implants every second week starts before blood
glucose levels rise in the onset of diabetes, like the untreated group in figure 4 shows, blood glucose levels are kept in
the range of normal blood glucose levels in healthy mice (7 - 8
mmol/l).
The comparison of figures 2 and 4 shows the high correlation
between blood glucose levels and body weight. The lack of ability to compensate the influence of phloricin treatment seems
to lead via decrease of blood glucose level to prefinal weight
loss.
The influence of phloricin on the membranes of fat cells might
be responsible for the effects, shown in the figures 2 and 4.
Phloricin inhibits the active transport of glucose into the fat
cells and at low blood glucose levels lipogenesis decreases because of lack of substrate. On the other hand phloricin increases lipolysis and leads to raising blood levels of fatty acids.

So the fat depots become exhausted and this decreases gluconeogenesis on the long term because no more substrate is available.
The glucose metabolism of obese mice seems not to be able to
handle both renal glucose excretion and the disturbance of gluconeogenesis in spite of increased food intake.

3.4. Renal glucose excretion under the influence of chronic
phloricin intoxication:

Figure 5:   Figure 6:

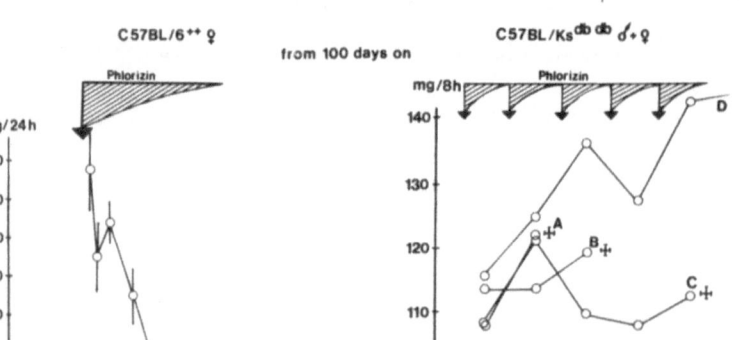

Renal glucose excretion under the influence of chronic phloricin intoxication ⬦ and in controls ●

In this examination two different types of metabolism cages and
and a shorter urine collection period for the obese mice. were
used. Both alterations in the experimental design have the tendency to lower results. Therefore the results of this part of
the experiment can not be interpreted absolutly. However
qualitative comparision of the experimental groups with each
other is possibel.
The healthy (B6) mice showed considerable renal glucose excretion for 35 days without lasting influence on blood glucose
levels and body weight.
In obese mice (db/db) phloricin causes an increase of 10 - 20%
in renal glucose excretion. It is astonishing, that the mice of
group D, which showed the smallest phloricin effect on blood
glucose levels and body weight, had the highest glucose excretion rate of the four groups of treated obese mice.

4. CONCLUSIONS

Phloricin treated obese mice which had died spontaneously
showed remainig fat layers at autopsy. Therefore exhaustion of
fat depots can not be the only cause of death.
Two further theories can be discussed:
1.) Phloricin is able to block glucose transport through the

blood - brain - barrier and so the brain, poorly provided with glucose, sustains damage, leading to death by malfunction.
2.) The disturbance of glucose metabolism hinders the synthesis of glucogene amino acids and leads to breakdown of protein metabolism.
These theories are subjects to running experiments.

5. SUMMARY

Phloricin inhibits the glucose intake into fat cells and increases lipolysis. This leads to exhaustion of fat depots. The exhaustion of substrate reserves for gluconeogenesis or other unknown effects of phloricin prevent hyperglycemia but lead to death within 68 days in 45 from 51 treated fat mice (db/db).
In healthy, non adipose mice (B6) long term phloricin treatment can be maintained in similar doses without influence on survival, body weight and blood glucose level.

1. Ortlepp I: Untersuchungen zur langfristigen Phlorizinbelastung bei Mäusen der Diabetesmutante C57BL/KS db/db,+/+ Hannover Tierärztl. Hochschule, Diss. 1986
2. Schötz H: Die chronische Phlorizinintoxikation und ihre Einflussnahme auf die renale Natrium und Glucoseausscheidung sowie den Glucosestatus bei gesunden und diabetischen Mäusen Hannover Tierärztliche Hochschule, Diss. 1983
3. Zogbaum A: Untersuchungen an der Maus zur Hemmung der glomerulären Hyperfiltration bei Hyperglycämie durch Phlorizin Hannover Medizinische Hochschule Diss. 1982

A.C. Beynen & H.A. Solleveld (Eds), New developments in biosciences: their
implications for laboratory animal science, ISBN 978-94-010-7973-0
©1988 Martinus Nijhoff Publishers, Dordrecht.

PLASMA ⍺-AMYLASE AND LIPASE ACTIVITIES IN DOGS WITH VARIATIONS
IN FOOD COMPOSITION AND AVAILABILITY

G. Kuhn and W. Hardegg

Institut für Versuchstierkunde der Universität Heidelberg, Im
Neuenheimer Feld 347, D-6900 Heidelberg, FRG.

1. INTRODUCTION
    Early in this century digestion was assumed to lack the re-
gulatory sophistication of internal metabolism. Neither composition of a meal nor end products of its digestion are thought
to modulate the rate of specific digestive hydrolyses. Only the
amount of enzyme that is secreted can be varied, and not the
composition of the mixture. The term parallel secretion or parallel transport has been applied to this concept and implies
that various enzymes are secreted at parallel rates (1). Grossman et al (6) were the first to show directly an adaptation of
pancreatic enzymes to diets rich in starches or proteins. They
observed increased amylase and trypsin activities in rat pancreatic tissue after feeding high starch and high protein diets.
Other investigators have demonstrated subsequently that not
only storage of enzymes, but also synthesis and secretion of
pancreatic enzymes depend on the composition of the diets that
were fed (2,3,7). These changes are usually attributed to long-
term alterations in the rate of synthesis of different enzymes
produced by specific content of the diet in an as yet unclear
manner (7). Enzyme activities in canine pancreatic juices also
are modified by the composition of the diet. Behrman and Karl
(2) showed that pancreatic lipase of dogs adapt to dietary
lipids. Furthermore, responsibility for pancreatic lipase induction does not seem to be the amount of dietary lipid, but
the percentage of energy provided by dietary fat.
Exocrine pancreatic secretion takes place both in the absence
of food (basal secretion) and in response to a meal. Basal or
interdigestive secretion occurs when all food has emptied from
the stomach and has been digested and absorbed by the small
intestine. In the dog, pancreatic secretion has been found to
reach basal secretory levels not earlier than 16 to 17 hours
after feeding (8).
The goal of our study was to determine the influence of a high
and low caloric intake and food composition upon circulating.
⍺-amylase and lipase activities of clinically normal dogs
during phases of interdigestive secretion. Circulating ⍺-
amylase and lipase concentrations were estimated 24 hours after
the last feeding. The research was designed to answer several
questions. Are plasma ⍺-amylase and lipase activities influenced by food availability or food composition? Are there
comparable changes in both enzymes? What is the temporal
sequence of circulating ⍺-amylase and lipase changes in a
starvation period over 72 hours, and are there influences of

energy intake in the prefasting period?

2. MATERIALS AND METHODS
2.1. Animals and Housing Conditions
Twenty obese female Foxhounds, approximately one year of age, bred in our animal facility and ranging in body weight from 19 to 30 kg were used for the experiments. The animals were individually maintained in enclosures indoors according to guidelines of good laboratory practise.

2.2. Experimental Design and Diets
In two separate experiments the dogs were fed with a commercial dog maintenance diet (ssiff HH; Ssniff, Soest,FRG) or a dog breeding diet (ssniff HZ; Ssniff, Soest, FRG) for 21 days, respectively (see Table 1 for the composition of the diets). In both experiments one group of 10 dogs was fed at a high caloric intake and the other group of 10 dogs at a low caloric intake. After the first experiment, dogs with high and low caloric input were exchanged, and at the end of the second experiment, all dogs were fasted for 72 h. Dogs were weighted in 1-week intervals, and, then, food intake per animal each day was calculated for one week. Amount of diet was 70 x body weight$^{0.75}$ = fed diet in grams per animal each day for dogs fed with a high caloric intake, as well as for all dogs in the 14-day preexamination periods used in both experiments. Dogs from the low caloric input conditions were fed at the rate of 23 1/3 x body weight$^{0.75}$ = fed diet in grams per animal each day. At the end of the second experiment, dogs of both groups were food deprived for 72 h. Water was available ad libitum ,and dogs were fed once daily during the morning hours.

2.3. Collection of Blood Samples, Determination of Enzyme Activities and Quality Control
Blood samples were drawn in 1-week intervals from the cephalic vein in experiment I and II and every 24 h when dogs were food deprived for 72 h. Blood samples were heparinized and centrifuged for 14 min at 3000 g at 10°C. Plasma was obtained and frozen at -20°C until assays were performed. We used automated methods for plasma α-amylase (EC 3.2.1.1) and lipase activities (EC 3.1.1.3) of Boehringer GmbH, Mannheim, FRG. Reagents were used as suggested by the test manufacturer. Enzyme assays were optimized according to the German Society of Clinical Chemistry. The evaluation method was implemented with an Abbott Bichromatic Analyzer ABA-100 (Abbott, Langen, FRG). Measurements were performed whilst corresponding samples of both groups were determined together. Internal quality control was assured by Precinorm E^R (Boehringer GmbH, Mannheim, FRG).

TABLE 1. Composition of diets and energy content

Dietary constituent	Dog maintenance diet	Dog breeding diet
crude protein	22.0 %	26.0 %
carbohydrate	30.5 %	26.8 %
crude fat	5.8 %	7.8 %
fibre	4.3 %	4.2 %
ash	7.2 %	7.0 %
water	12.0 %	12.0 %
energy content	12.7 MJ/kg	13.5 MJ/kg

2.4. Statistical Analyses

Data from each group are shown in Figures 1 and 2 as the mean $\pm$ SEM. Statistical analyses employed Wilcoxon 2-sample tests and Wilcoxon matched-pairs signed-ranks tests. Thereafter, a sequential rejecting procedure, after Bonferroni and Holm (9) was made with Wilcoxon tests. Results were considered statistically significant if $p < 0.05$.

3. RESULTS

Effects of the two diets and different amounts of food intake on circulating α-amylase and lipase activity are illustrated in Figures 1 and 2. Maintenance and breeding diet yielded similar plasma α-amylase and lipase activities in both situations, when the animals were fed a high caloric intake. However, none of the alterations in circulating α-amylase activity was significantly different ($p < 0.05$) between animals fed a high or a low amount of the maintenance diet or the breeding diet or even during total food deprivation. When the animals were fed the dog breeding diet, a significant decrease ($p < 0.05$) of plasma α-amylase occured after 14 days of low food ingestion, while a significant increase ($p < 0.05$) in circulating activity of α-amylase was seen in the same animals during starvation for 72 h. Moreover, values of circulating α-amylase ranged from 500 to 2500 U/l. In contrast with plasma α-amylase levels, circulating lipase activity variations showed a significant decrease ($p < 0.05$) when dogs were fed a low caloric intake of the dog maintenance diet after 14 days of the experiment, but no significant differences ($p < 0.05$) were observed between the groups . Feeding the dogs with the breeding diet, a significant lower lipase activity ($p < 0.05$) was noted in animals fed a low energy intake after 21 days and after 48 h of starvation. As compared to the lipase activity of dogs fed a low amount of the breeding diet in the prestarvation period, that showed no significant change ($p < 0.05$) during starvation, circulating lipase activity decreased significantly ($p < 0.05$) in dogs fed a high energy input in the prestarvation period and there were seen similar activities in both groups after 72 h of starvation.

368

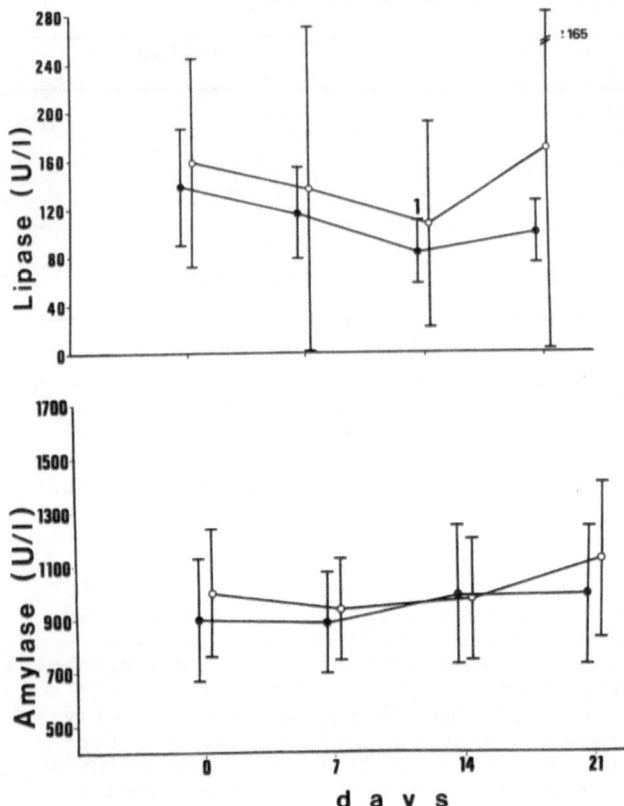

Fig. 1. Circulating ∝-amylase and lipase activity in dogs fed a dog maintenance diet. ●—● dogs fed a low caloric intake; O—O dogs fed a high caloric intake. [1]p < 0.05 significant decrease of circulating lipase activity in dogs fed a low amount of food after 14 days of the experiment.

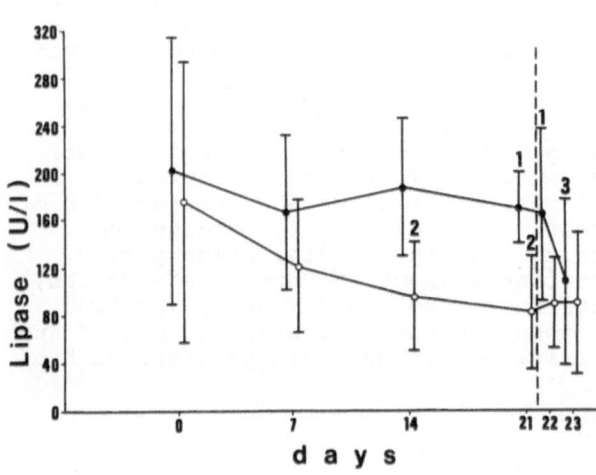

Fig. 2. Circulating ∝-amylase and lipase activity in dogs fed a dog breeding diet. ●—● dogs fed a high caloric intake for 21 days; O—O dogs with a low caloric intake for 21 days. [1]p < 0.05 significant difference of circulating lipase activity between groups at same time [2]p < 0.05 significant decrease of circulating lipase activity after 14 and 21 days of low energy input [3]p < 0.05 significant decrease of circula-

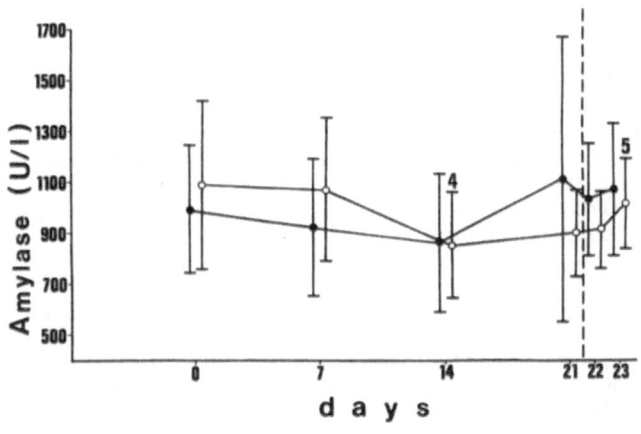

ting lipase activity after 72 h of starvation in dogs fed a high energy amount in the prestarvation period. [4]p < 0.05 significant decrease of circulating α-amylase activity after 14 days of low energy input [5]p < 0.05 significant increase of plasma α-amylase activity after 72 h of starvation.

4. DISCUSSION

In the present investigation we were able to show that a reduction in daily input (breeding diet and maintenance diet) as compared to animals with high caloric intake resulted in significant decreases in circulating lipase activity in phases of interdigestive pancreatic secretion. In contrast, plasma α-amylase activity wasn't significantly different between both groups in all situations. The results suggest that adaptations of circulating lipase and α-amylase activities to food amounts are not parallel. There are many factors responsible for these data. Hiatt (1961) observed that total extirpation of the pancreas causes a surprisingly small change in serum amylase levels of dogs. Dogs studied for one and a half years after total pancreatectomy have serum amylase activities well within normal limits. These findings suggest that dogs have amylase producing organs, others than the pancreas that are capable of maintaining the serum amylase level within normal limits (5). Stickle et al (1980) observed amylase activities greater than serum activities in tissue extracts from pancreas, duodenum, ileum, ovaries and testicles. Hiatt (1961) found that the mucosa of the small intestine of dogs secretes an amylase apparently identical with that of the pancreas. The production of amylase in the small intestine of dogs varies widely between and within animals. This could explain the variation in plasma α-amylase activities in the present study. In clinically normal dogs two isoamylase groups were identified by electrophoresis in the serum. Most of the serum amylase activity is the more cathodal isoamylase. The anodal isoamylase is rarely observed in serum from normal dogs and, when present, accounts for little of the enzymatic activity. The anodal isoenzyme was found in uterus and pancreas (10). Results of Challis et al (1957) suggest that ACTH and cortisone play an important role in the elevation of serum amylase activity. This may explain elevating α-amylase levels during starvation in dogs fed a low caloric intake in the prestarvation period. We noted a great disturbance in the animals after 60 h of starvation.

The pancreas is the primary source of serum lipase (6). Investigations of lipase activities in other organs than the pancreas also showed activities in the liver, spleen, kidney and heart, but only pancreatic lipase is effective in liberating C_{14}, C_{16} and C_{18} fatty acids from their triglycerides at appreciable rates. The lipolytic enzymes from the duodenum and liver are much less effective (6,11). Therefore, serum lipase results supplement the information available from amylase tests for determination of the exocrine pancreatic activity.

5. CONCLUSION

In contrast to reports by other investigators that pancreatic enzymes adapt to diet composition, circulating α-amylase activity was not influenced by either the amount nor contents of the diets that we used. On the other hand, reduction of food intake decreased circulating lipase levels. Only pancreatic lipase is effective in liberating C_{14}, C_{16} and C_{18} fatty acids from their triglycerides at appreciable rates and, therefore, our results indicate that circulating pancreatic lipase activities decrease in dogs with decreasing food intake. Determination of total circulating amylase levels is a nonspecific test for pancreatic function. Thus, determination of total plasma α-amylase activity is an inappropriate diagnostic tool for pancreatic function. It is necessary to evaluate isoenzymes of canine serum amylase by electrophoresis.

6. ABSTRACT

Circulating levels of α-amylase and lipase were examined after manipulations to the diet of 20 obese female Foxhounds. The animals were divided randomly to a high or a low energy intake with either a commercial dog maintenance diet (30.5 % carbohydrate, 22 % protein and 5.8 fat) or a commercial dog breeding diet (26.8 % carbohydrate, 26 % protein and 7.8 fat). After 21 days of feeding the animals with the commercial dog breeding diet, all animals were food deprivated for 72 h.

Measurements of plasma α-amylase and lipase were made following collection of blood each week, 24 h after the last meal was consumed. Blood samples were collected at 24-h intervals during the food deprivation period.

Results suggest that varying the diets of dogs did not influence plasma levels of α-amylase. There was substantial between and within individual variability between 500 to 2500 U/1, but the changes showed no clear pattern in association with food availability. Dogs switched to the low caloric diet experienced gradually declining lipase levels but no further declines during the food deprivation period. Dogs maintained on high caloric diets experienced a rapid decrease in plasma levels of lipase during the food deprivation period and reached levels of the animals previously fed the low caloric diets.

Conclusions are that circulating α-amylase activities were not influenced by either food availability or composition whereas lipase levels depend on food availability. Moreover, only pancreatic lipase is effective in liberating C_{14}, C_{16} and C_{18} fatty acids from their triglycerids at appreciable rates and therefore, we observed an adaptation of lipolytic exocrine pancreas function to the amount of food intake, whereas determination of total circulating $\measuredangle$-amylase is a nonspecific test for the function of several organs and further separation of isoamylases by electrophoresis and other techniques is necessary to evaluate the activity of a specific organ like the pancreas.

7. REFERENCES

1. Babkin BP: Einige Grundeigenschaften der Fermente des Pankreassaftes. Zentralbl. Gesamte Physiol. Pathol. Stoffwechsels 1, 98-108, 1906.
2. Behrman HR, Karl MR: Adaptation of Canine Pancreatic Enzymes to Diet Composition. J. Physiol. 205, 667-676, 1969.
3. Challis TW, Reid C, Hinton JW: Study of Some Factors which Influence the Level of Serum Amylase in Dogs and Humans. Dogs and Humans. Gastroenterology 33(5), 818-822.
4. Grossman MI, Greengard H, Ivy AC: The Effect of Dietary Composition on Pancreatic Enzymes. Am. J. Physiol. 138, 676-682, 1942.
5. Hiatt N: Investigation of the Role of the Small Intestine in the Maintenance of the Serum Amylase Level of the Dog. Annals of Surgery 154 (5), 864-873, 1961.
6. Lott JA, Patel ST, Sawkney AK, Kazmurczak SC, and Love JE: Assays of Serum Lipase: Analytical and Clinical Considerations. Clin. Chem. 32 (7), 1290-1302, 1986.
7. Marchis-Mouren G, Paséro L, and Desnuelle P: Further Studies on Amylase Biosynthesis by Pancreas of Rats Fed on a Starch-Rich Diet. Biochem. Biophys. Res. Comm. 13, 262-266, 1963.
8. Singer MV: Neurohormonal Control of Pancreatic Enzyme Secretion in Animals. In the Exocrine Pancreas: Biology, Pathobiology and Diseases, ed. by VLW Go et al,315-331, New York: Raven Press, 1986.
9. Sonnemann E, Berchier P, Schiller K, Ferner K, Maurer W, and Victor N: Simultane Hypothesenprüfungen. Biochemisches Seminar der Region Österreich-Schweiz der internationalen biometrischen Gesellschaft, Bad Ischl (A), 1981.
10.Stickle JE, Carlton WW, and Boon GD: Isoamylases in Clinically Normal Dogs. Am. J. Vet. Res 41 (4), 506-509, 1980.
11.Zieve L, Vogel WC, and Kelly WD: Species Difference in Pancreatic Lipolytic and Amylolytic Enzymes. J. Apll. Physiol. 18 (1), 77-82, 1963.

A.C. Beynen & H.A. Solleveld (Eds), New developments in biosciences: their
implications for laboratory animal science, ISBN 978-94-010-7973-0
© 1988 Martinus Nijhoff Publishers, Dordrecht.

LOCOMOTOR ACTIVITY OF A BEHAVIOR MUTANT IN NMRI MICE

C. HECKL-ENSSLIN, I.v.BUTLER

1. INTRODUCTION
   In an NMRI-population an activity mutant was observed that
is similar to " Dreher ". Homozygotes are hyperactive and manifest rotations in small and/or wide circles.
Preliminary genetic studies indicated a rezessive inheritance
at a single locus.
The purpose of this experiment was to quantify differences in
the 24-hour-activity cycles between the mutants and normal
mice.

2. MATERIAL AND METHOD
   Locomotor activity of 17 mutants and 62 phenotypic normal
NMRI mice was investigated. Age at testing was about 80 days.
Activity of single animals in their cages was measured indirectly. When the animal is active it sets the cage's walls
into vibrations. They are transmitted via electric signals,
which can be recorded. Therefore vibrations are a measure for
locomotor activity.

3. RESULTS
   Locomotor activity was tested continuously over three days.
Figure 1 shows typical 24-hour-activity cycles of a mutant and
a phenotypic normal mouse.

Figure 1. 24-hour-activity cycles of a
a) Mutant
b) Normal mouse

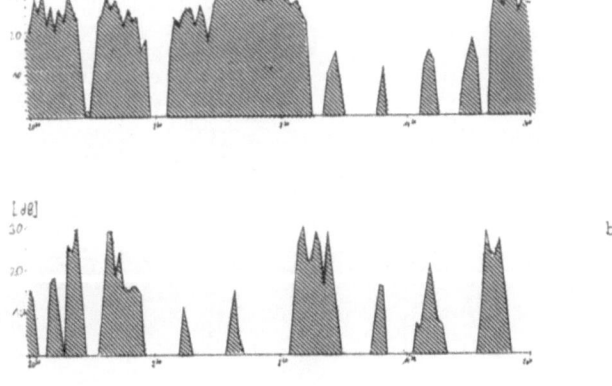

We tried to quantify activity by the following measures (mean of three cycles):

- Duration of locomotor activity within 24 hours
- Number of phases of activity within 24 hours
- Intensity of locomotor activity within 24 hours.

TABLE 1. Means x and standard deviations s of activity traits duration (tA), number of phases (Ai) and intensity (Lm)

Trait	Mutant n=17		Control n=62	
	x	s	x	s
tA (h)	15.0	1.2	12.7	1.4
Ai	24.7	15.4	31.6	16.0
Lm (dB)	23.8	3.3	16.7	2.2

In comparison to the controls, mutants are longer active (t=6.8; p 0.01) and their activity is more intense (t=8.6; p 0.001). No significant differences in number of activity phases appeared between mutants and normal mice.

A.C. Beynen & H.A. Solleveld (Eds), New developments in biosciences: their
implications for laboratory animal science, ISBN 978-94-010-7973-0
© 1988 Martinus Nijhoff Publishers, Dordrecht.

The application of embryo transfer and cryopreservation to commerciel
laboratory animal breeding.

CRYOPRESERVATION OF MOUSE STRAINS BY A QUICK
FREEZING METHOD.

Frederik Dagnæs-Hansen, Bomholtgaard Breeding and Research Center Ltd.

Bomholtvej 10,8680 Ry, Denmark.

Introduction.
    The recent progress in the development of ·cryopreservation of the mamma-
lian embryo, now permits embryos to be frozen rapidly with a minimum of e-
quipment and to be thawed in an equally rapid and efficient way(1,7,12,13,
16,20,21,25).
    Furthermore aspects of preventing disease transmission by use of embryo
transfer has come into focus in the field of laboratory animal science a-
gain (2,3,4,6).
    The investigations which is reported, represent some of the preliminary
results in a research program designed to explore the applicability of embryo transfer and cryopreservation to large scale laboratory animal production. Some of the results concerning superovulation, embryo transfer and
synchronization of donor and recipient have been presented previously (5).
    Cells generally are injured by being cooled rapidly by intracellular ice
formation during cooling and crystal grain growth during warming(11). The
occurence of intracellular ice formation during freezing and thawing can be
minimized by dehydration of the cells either using slow cooling and thawing
or dehydrating the cells using hyperosmotic solutions. The early success in
cryopreservation of mouse embryos   depended on the use of cryoprotectants and
on the partial dehydration of cells by slow cooling and warming(24,26).
    Since 1984 several reports on cryopreservation by the use of hyperosmotic
solutions of glycerol and impermeable sugars have been published (7,12,13,
21,25). The ability of highly concentrated solution to undergo solidifica-
tion without ice crystal formation (a process referred to as VITRIFICATION
) is an important element in the partial dehydrated cytoplasm as well as in
the extracellular solution during freezing by the conventional method(14,
15). Attempts to optimize the quick freezing of mammalian embryos by use of
vitrification solutions was first proposed by Rall et al.(1985), who reported high survival rate provided cryoprectectant toxicity and osmotic events
could be properly controlled.
    The previously mentioned reports on quick freezing mouse embryos are all
dealing with cryopreservation of eight-cell embryos, morulae and blastocysts
from outbred or hybrid mice.
    The aim of this study was to examen the possibilities of u-
sing a procedure based on that of Scheffen et al.(1986) for
the quick freezing of different inbred strains, and the possibilities of adapting this procedure to the cryopreservation of
earlier developmental stages of mouse embryos than the eight-
cell embryo.
Material and Methods.
    Experiments were performed with the outbred stock Bom:Sencar, the inbred
strains CBA/JBom,NZW/NBom,NZB/NBom,B1OA(2R)/Bom and hybrids C3D2F1/Bom ob-

tained from Bomholtgaard Breeding and Research Center Ltd. All females were 8-10weeks old and virgins, the males were 10-11 weeks old.

The animal were maintained under condition of uniform temperature (21 $\pm$ 2°C) and humidity (50 $\pm$ 5 %). Light were on from 0600 to 1800 h. Drinking water and feed (Standard,Altromin) were available through the studies ad libitum.

Females were maintained in cages (12x20x30cm) in groups of 10-15 per cage, while the males were housed individually in smaller cages (13x15x25 cm)

Superovulation of the females were induced by an i.p. injection of 10 I.U. of PMSG (Antex^{R} Leo,Denmark) followed 48 hours later by an i.p. injection of 5 I.U.hCG (Physex^{R}Leo,Denmark). Two hours after the hCG injection the females were placed with the males (one female per male) of the same genotype. Morulae and early blastocysts were collected from the uterus 84 hours after the hCG injection, while two-,four- and eight-cell embryos were collected from the oviduct 41,68 an 78 hours after the hCG injection respectively.

The oviducts and uterus were flushed using modified PBS (23) + 5 % FCS and normal developed two-,four-,eight-cell embryos and morulae were washed in PBS + 5% FCS before freezing.

The freezing was performed by placing the embryos for 10 minuts in the equilibration solution consisting of 10 % glycerol and 10 % 1,2-propandiol in PBS. After equilibration time the embryos were loaded quickly in 0.25 ml transparent French straws (I.M.V.France) in a drop (10-20 ul) of the vitrification medium separated by two air bubbles from the dilution medium filling the rest of the straw. The dilution medium consisted of 1 M sucrose in PBS. The vitrification medium was 25 % glycerol + 25 % 1,2-propandiol in PBS. Within 30 seconds the straws were plunged into liquid nitrogen (6). The straws were stored 1-14 days in liquid nitrogen.

In experiment B freezing of two-,four- and eight-cell embryos were either performed like previously described or the embryos were first equilibrated in 1.5 M glycerol for 10 minuts to fascilitate the permeation of water(13,20). Two-cell embryos were equilibrated in gradually increasing solutions of glycerol (0.5 M,1.0 M and 1.5 M) or the glycerol solutions + 0.5 M DMSO. DMSO was also added to the freezing medium, when used in the equilibration solutions.

Thawing was performed by placing the straws in a waterbath approximately 30°C until the ice had disappeared in the sucrose solution. The freezing medium and sucrose diluent were mixed immediately after the thawing by shaking the straws like a clinical thermometer and the embryos were left in the sucrose solution for 5 minuts before dilution out of the sucrose solution using PBS + 5% FCS.

The embryos were evaluated by microscopic examination before transfer to culture medium M16 (23). The embryos of the morulae stage were cultured for 48 hours while the earlier embryonic stages were cultured for 72 to 96 hrs. Embryos developing into morphologically normal blastocysts were considered as having survived.

A proportion of the embryos that had developed normally in vitro were transferred to the oviduct of day 1 pseudopregnant recipient Bom:SENCAR, using the method described by Zeilmaker (1981). The recipient females were allowed to litter.

Results and discussion.

The survival rate of embryos from inbred strains after conventional cryopreservation is described to be lower than for embryos from hybrid or outbred mice(17,18,27).

The present study indicates that embryos form inbred mice can be cryo-

preserved using the vitrification method and that the results (in vitro
cultivation **survival**) are comparable with other reports in which the embryos
have been frozen using timed prefreezing (table A)

TABLE   A.

STRAIN	NO. OF EMBRYOS (MORULAE)			
	FROZEN	RECOVERED after thawing	SURVIVED*	(%)**
NZW/NBom	26	26	22	(85)
NZB/NBom	73	68	48	(71)
ĊBA/JBom	20	20	10	(50)
B1OA(2R)/Bom	10	9	7	(78)
C3D2F1/Bom	100	82	73	(89)

* evaluated by cultivation in vitro   ** Percent survived of recovered.

Frozen-thawed embryos from hybrids (C3D2F1/Bom) have been transferred to
pseudopregnant females and yielded ·normal live offspring. Results from
transfer of frozen-thawed embryos obtained from inbred strains still awaits.

The suitability of various preimplantation developmental stages of embryos
to freezing is probably due to differences in the membrane permeability (8).
Although mouse embryos at all the preimplantation developmental stages have
been successfully frozen by slow cooling (24) ·no stages earlier than eight-
cell embryos have been frozen successfully when quick freezing procedure have been used (12,13).

The present results indicate that earlier developmental stages of embryos
can be frozen and thawed successfully by the vitrification (quick freezing)
method, although the survival rate for two- and four-cell embryos is lower
than for eight-cell embryos and morulae. To obtained the same survival rate
for these embryonic stages the procedure have to modified even more.

TABLE   B.

EMBRYONIC STAGE PROCEDURE (C3D2F2/Bom embryos)	FROZEN	RECOVERED after thawing	SURVIVED*	(%)**
8-cell embryos				
direct equilibration	40	38	8	(21)
permeation in 1.5 M glycerol before eq.	40	31	24	(77)
permeation in 1.5 M glycerol + 0.5 M DMSO before equilibration	40	38	36	(95)
4-cell embryos				
direct equilibration	40	39	9	(23)
permeation in 1.5 M glycerol before eq.	40	35	23	(66)

TABLE B. (continued)

EMBRYONIC STAGE PROCEDURE	FROZEN	RECOVERED after thawing	SURVIVED [*] (% [**])	
2-cell embryos				
permeation in 1.5 M glycerol before eq.	20	17	0	
permeation in 0.5 M, 1.0 M and 1.5 M glycerol before eq.	20	16	0	
permeation in the glycerol solutions + 0.5 M DMSO	20	19	3	(16)

SUMMARY:

Cryopreservation of embryos from various inbred strains of mice by a quick freezing method is described. The survival rate evaluated by in vitro cultivation of the embryos, is varying from 50 to 85 % depended on the genotype. The survival rate for embryos from hybrids is 89 %.

In experiment B it is shown that by modifying the procedure of quick freezing,earlier developmental stages than eight-cell embryos can be frozen and thawed successfully.

Keywords: cryopreservation, mouse embryos, direct plunging, quick freezing inbred strains.

REFERENCES:

(1) Biery,K.A.,G.E.Seidel and R.P.Elsden:Theriogenology 23,140,1986.
(2) Bittner,J.J. and C.C.Little:J.Heredity 128,117-122,1937.
(3) Carthew,P.,M.Wood & C.Kirby:J.Reprod.fert.69,253-357,1983.
(4) --   -- & --   :J.Reprod.fert.73,207-213,1985.
(5) Dagnæs-Hansen,F.:Scand.J.Lab.An.Sci.14,no 2.suppl.1987.
(6) Fekete,F.& C.C.Little:Cancer research 2,525-530,1942.
(7) Krag,K.T.,I.-M.Koehler & R.W.Wright:Theriogenology 23,199,1985.
(8) Leibo,S.P.: from"Frozen storage of Lab.An." G.H.Zeilmaker(ed)1-21,1981.
(9) Leibo,S.P.: from"Genetic engineering of animals"J.Warren Evans & A.Hollander(eds):Basic Life Science 37,251-272,1986.
(10)Massip.A.,P.van Der Zwalmen & F.Leroy:Cryobiology 21,574-577,1984.
(11)Mazur,P.:from"Freezing of Mammalian Embryos".Ciba Found Symposium 52,pp-19-42,1977.
(12)Miyamoto,H.&T.Ishibashi:J.Reprod.fert.78,471-478,1986.
(13)Miyamoto,H.,Y.Miyamoto & T.Ishibashi:Jpn.J.Zootech.Sci.57,250-256,1986.
(14)Rall,W.F.& G.M.Fahy:Nature 313,573-575,1985.
(15)Rall,W.F.& G.M.Fahy:Theriogenology 23,220,1985.
(16)Scheffen,B.,P.van Der Zwalmen & A.Massip:Cryoletters,7,260-266,1986.
(17)Schmidt,P.M.,C.T.Hansen & D.E.Wildt:Biol.Reprod.32,507-514,1985.
(18)Schmidt,P.M. & D.E.Wildt:Theriogenology 23,195,1986.
(19)Szell,A. & J.N.Shelton:J.Reprod.Fert.76,401-408,1986.
(20)Szell,A. & J.N.Shelton:J.Reprod.Fert.78,699-703,1986.
(21)Takahashi,Y. & H.Kanagawa:Jpn.J.Vet.Res.33,140-144,1985.
(22)Takeda,T.,R.P.Elsden & G.E.Seidel:Theriogenology 21,266,1984.
(23)Whittingham,D.G.:J.Reprod.Fert.suppl.14,7-21,1971.
(24)Whittingham,D.G.,S.P.Leibo & P.Mazur:Science 178,411-414,1972.
(25)Williams,T.J. & S.E.Johnson:Theriogenology 23,235,1985.
(26)Wilmut,I.:Life Science 11,part 2,1071-1079,1972.
(27)Yokoyama et al.(28)Zeilmaker G.H.fr.Frozen storage of Lab.An.101-119,1981

A.C. Beynen & H.A. Solleveld (Eds), New developments in biosciences: their implications for laboratory animal science, ISBN 978-94-010-7973-0
© 1988 Martinus Nijhoff Publishers, Dordrecht.

SEROLOGICAL FOLLOWING OF A LABORATORY RAT BREEDING
CONTAMINATED WITH RESPIRATORY VIRUSES DURING 1981-1986.

M.PŘIBYLOVÁ, T.SVOBODA, P.KLÍR

1.INTRODUCTION
Nowadays diagnostics of viral diseases in the breedings of small laboratory rodents is an unseparable part of "Health quality control of animals"(Heine 1982). Among the most frequent detected contaminants of laboratory rat breedings belong pneumotropic paramyxoviruses Sendai and PVM. That is why we wellcomed the possibility of observing, from the beginning,the development of the infections with these viruses in natural conditions. We focused at similarities and differences in the way of spreading and maintaining the infections from the serological point of view. By observing the dynamic of antibodies in the young we wanted to point to the fact, that some age groups of animals are not suitable for the serological diagnostics, even if the breeding is considerably contaminated,being more suitable for the other diagnostic methods (histopathology, virus isolation).

2.PROCEDURE
2.1.Materials and methods
2.1.1.Characteristics of the breeding colony: semibarrier breeding unit of laboratory rats Ipcv:Wist, II category (Bondy et al.1987), established in 1978, average production 20 000 animals per year.
In June 1981 contamination by Sendai virus was ascertained and in September of the same year contamination by virus PVM. Besides the breeding was controlled for the presence of antibodies against the following viruses: Theiler GDVII, Reo 3, Kilham rat, Toolan's H-1, SDA x MHV, Mouse adenovirus and LCM results being negative.
2.1.2.Methods used: For determining the titers of antibodies against the pneumotropic viruses micromethods of hemagglutination inhibition tests were used. The description of the method including used puffers and preparation of tested sera according to Diagnostic Procedures for Viral Infections of Laboratory Rodents (1978). Antigens used: Microbiological Associates, Bethesda, Maryland USA.
Tissues for histological investigation were stained with hematoxylin and eosin.
2.1.3.Blood collection: in light ether narcoses. In repeating collection blood was gained in the amount of 0.3-0.5 ml by cutting the tail end. One time collections were performed by a. carotis and v. jugularis bleeding. Newborn young were decapited.

380

2.1.4.<u>The arrangement of serological observations</u>: serological observations were divided into four basic points given in
the following list:
A. Longitudinal observing of the breeding in total in 1981-
1986. 903 randomly chosen animals at the age of 90-180 days,
both sexes (each being of a different cage) were tested.
B. Following time of antibody lasting in chosen seropositive
females in 1982-1983. We chose 14 bred 5 month old females positive for both observed infections. In the course of 21 months we performed 267 blood samplings. During the tested period
the females were mated 3 times and gave birth also 3 times.
The third litters were serologically investigated together
with the mothers. The young were investigated at the 1,14,28
and 42 days after birth.
C. Following seroconversion in young animals (from the weaning to 90 days of age) in 1983. Totally were tested 318 young
animals divided according to the age into the following groups:
4-6 weeks, 6-8 weeks, 8-12 weeks and 13 weeks.
D. A long time observation of antibody dynamics in the young
from the mothers with the high or low antibody titers (high
titer=80 and more, low titer=40 and less; for both infections).
24 litters from 6 months old females (secundiparae) with a determined antibody level, were followed from birth to 8 months
of age. Individual groups were formed on the basis of the titer
level of the mothers directly before partums (June 1983-Feb.
1984).

3.RESULTS
A. In the proved anti Sendai and anti PVM positivity the
percent of seropositive animals increased till 1984, when in
the control of breeding we noted 6,52%anti Sendai negative
animals only and all controlled animals were anti PVM positive
or suspective (positive titer 1:20 and more, suspective 1:10-
1:$^+$ 20, negative lower than 1:10, for both infections). The
number of animals investigated in single years including per-
cents of seropositivity see in tab.1. The next two years showed
a considerable decrease of anti Sendai positivity, while anti PVM
antibody kept on being very frequent.
The infections in the breeding had the subclinical course.
Neither any clinical symptoms, nor production decrease after
contamination were observed. Gross pathology did not show any
extend changes in the lung tissues. Nevertheless there were
proved changes in histological lung sections, best evident
between 2-3 months of age. See pict.1.
B. In 14 bred females followed from 5-26 months of age,
positive for both infections from the beginning of the observation, in total 267 samplings divided into 20 terms we did not
note a single case of the anti PVM titer decrease under the
significant level (1:20), even when the level of the titer was
not always the same in the individual animal. The titer level
oscilated in both directions. The same being for anti Sendai
positivity with the exception of a temporary decrease in four
animals totally in 15 samples (5,62 %)under significant level.

TAB.1

			number of animals					
RESULTS OF THE SEROLOGICAL CONTROL OF THE SEMIBARRIER BREEDING UNIT								
year	virus	tested	negative		suspective		positive	
			n	%	n	%	n	%
1981	SENDAI	156	134	85.89	8	5.12	14	8.97
	PVM	"	140	89.74	7	4.49	9	5.77
1982	SENDAI	97	46	47.42	10	10.31	41	42.27
	PVM	"	28	28.86	18	18.56	51	52.58
1983	SENDAI	240	44	18.33	57	23.75	139	57.97
	PVM	"	34	14.17	5	2.08	201	83.75
1984	SENDAI	138	9	6.52	4	2.90	125	90.58
	PVM	"	0	0	6	4.35	132	95.65
1985	SENDAI	– 138	105	76.09	3	2.17	30	21.74
	PVM	"	7	5.07	8	5.79	123	89.13
1986	SENDAI	134	102	76.12	0	0	32	23.88
	PVM	"	11	8.21	2	1.49	121	90.30

PICT.1.

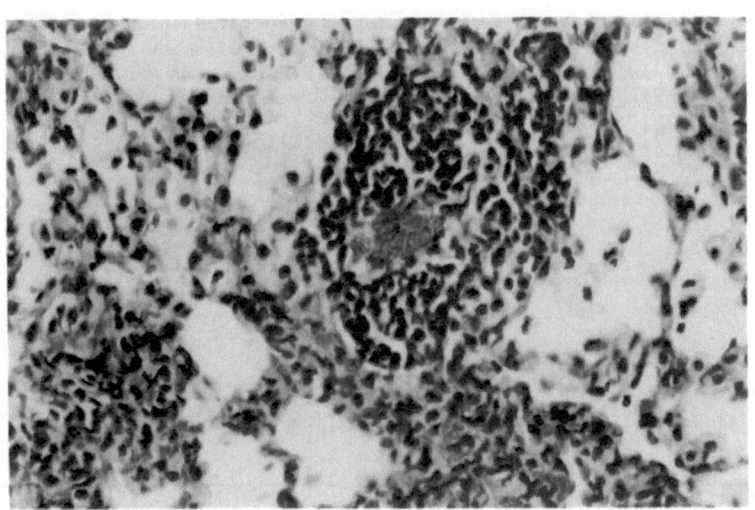

Picture 1 shows a lung tissue preparate of 90 days
old animal with the typical finding of perivascular
infiltration by lymphocytar and plasmatic cells.
Septa containing mononuclear cells are thick and in
alveoli are aggregates of macrophages (foam cells)
ascertained.

Average values gained in this group of animals are in Pict.2.
It is evident that the curves of average values of 14 adult
females kept their significant positivity for both antigens
during the whole time. PVM used to elicite a higher antibody
response than the Sendai virus did in most observed animals.

Investigation results in the third young litters, tested in
4 terms from birth to 6 weeks are expressed by two single peak
curves between the 9-11 months 1982. In the young tested the
first day after birth we noted 2 x positive, 4 x suspective
and 8 times negative anti Sendai titers, while in PVM we ascertained 13 x positive and only 1 suspective titer. The titer
height in the young reached its top in both infections the
14th day after birth, anti Sendai antibody decreased under
significant level the 28th, PVM antibody the 42nd day.

PICT.2.

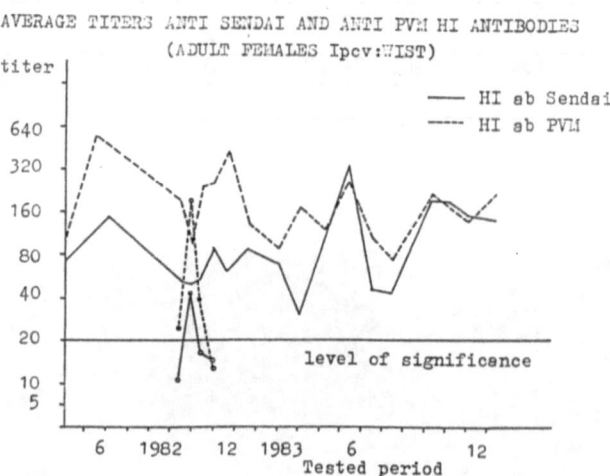

AVERAGE TITERS ANTI SENDAI AND ANTI PVM HI ANTIBODIES
(ADULT FEMALES Ipcv:WIST)

C. In animals 4-6 weeks old a considerable difference was
found in seropositivity between both infections, see tab.2.

TAB.2.

			RESULTS OF SEROLOGICAL INVESTIGATION IN ANIMALS UNDER 90 DAYS OF AGE					
age (in weeks)	virus	tested	number of animals					
			negative		suspective		positive	
			n	%	n	%	n	%
4-6	SENDAI	50	37	74	6	12	7	14
	PVM	"	14	28	15	30	21	42
6-8	SENDAI	91	55	60	21	23	15	17
	PVM	"	43	47	16	18	32	35
8-12	SENDAI	122	54	46	27	22	41	34
	PVM	"	8	7	13	10	101	83
13	SENDAI	55	8	15.5	8	15.5	39	71
	PVM	"	1	2	2	4	52	94

The highest number of anti PVM neg ative cases were not found in the lowest age category, however it was ascertained in the group of 6-8 weeks.

   D. From 24 chosen females 9 were put into a group with high and 14 with low anti PVM titers. In anti Sendai antibodies a high titer was ascertained in 17 and a low in 7 animals. Average female values with high and low titers PVM and Sendai antibodies are shown on pict.3 with horizontal lines. From the curves on the graph it is evident that the ascertained level of antibodies in young passes two phases. The immediate increase after birth is followed by a sudden decrease the beginning of which falls into the period before weaning, after which follows the time when even suspective antibodies are not detectable, than the titers gradually increase in the young animals.

PICT.3.

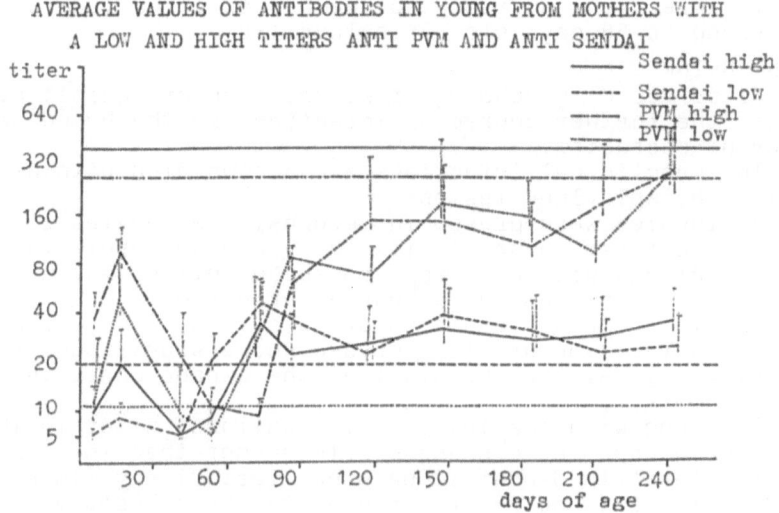

AVERAGE VALUES OF ANTIBODIES IN YOUNG FROM MOTHERS WITH
A LOW AND HIGH TITERS ANTI PVM AND ANTI SENDAI

Both phases were noted with a higher increase of titers in anti PVM and the animals of 8 months achieved the values of mothers with high titers, in anti Sendai the dynamics was similar having a difference in values,especially in the 8 months old animals when the titers were closely above the level ascertained in mothers with low positivity.

4.DISCUSION

   In a six year serological following the subclinical infections of the breeding colony with PVM and Sendai viruses we have found, after contamination, a rapid increasing number of seropositive animals in both infections. Further phases, however, showed considerable differences. While the contamination with PVM remains high, the breeding infection with the Sendai virus seems to gradually disapear or to pass into another form not detectable by this method. The fact that both infections after reaching the maximum of contamination, behave completely

differently is necessary to be explained, as well as the fact
that in D following, no single anti Sendai antibody titer in 8
months old animals achieved the antibody level which mothers
with high titers had. The course of two peak curves of antibodies in the young with the first peak being passive and the
second active response, offers the consideration that the infection of the young in the breed appears in the time when the
titer of passive antibody falls not under the significant but
the protective level. The precise time is dependent on the
mother titers. Groups of young of low positive mothers i.e.
with lower titer in the first phase of life have an earlier increase in the second phase (average 12 days Sendai; 21 days
PVM). In our observation, it was in all cases lower in Sendai
antibodies, the beginning of the second phase being also earlier against that of the anti PVM groups. The young seems to
have encountered the infection with Sendai virus earlier than
the PVM infection. We isolated the Sendai virus from seronegative young 56-58 days old (Přibylová 1984).

5.CONCLUSION
    1. After rapid increasing of seroconversion against PVM and
Sendai, the further course of infections in the breed had a
different character.
    2. The subclinical infections are accompanied with histolo-
gical detectable lung lesions.
    3. Antibodies were proved in seropositive females 21 months.
    4. In the first phase of life the young are protected against
both infections with passive transfered antibodies.The first
phase reaches maximum on 14[th]day after birth.
    5. The titer height achieved by passive antibody transport
influences the time of its decrease under significant (better
protective) level and the start of the second phase in the
young.
    6. In young with the low positive antibody titer in the first
phase the second curve phase starts sooner than in young with
higher titers (in Sendai 12 days earlier; 21 days in PVM).
    7. Passive antibodies are transported by colostrum and
mother milk better against PVM than against Sendai.

REFERENCES

1.Bondy R,Klír P,Přibylová M,Svoboda T,Jelen P,Pravenec M:
   Health Quality Control of Laboratory Animals of the Czechoslovak Academy of Sciences (ČSAV) Scand.J.Lab.Anim.Sci.14,
   1987:15-24
2.Diagnostic Procedures for Viral Infections of Laboratory
   Rodents: Murine Virus Diagnostic Laboratory, Microbiological
   Associates Bethesda, Maryland 1978:4-8
3. Heine W: Importance of Quality Standards and Quality Control
   in Small Laboratory Animals for Toxicological Research.In:
   Animal in Toxicological Research, Eds Bartošek I, Guitani A,
   Pacei E, Raven Press, NY 1982:1-12
4.Přibylová M: (unpublished data) 1984.

A.C. Beynen & H.A. Solleveld (Eds), New developments in biosciences: their implications for laboratory animal science, ISBN 978-94-010-7973-0
© 1988 Martinus Nijhoff Publishers, Dordrecht.

ERADICATION OF ENCEPHALITOZOONOSIS IN RABBIT BREEDING COLONIES BY CARBON IMMUNOASSAY

T. WALLER, Pharmacia AB, Uppsala, Sweden

INTRODUCTION

Encephalitozoonosis, caused by the microsporidian parasite Encephalitozoon cuniculi, is a common protozoan disease in rabbits which influences negatively on results in research and meat production. Since 1971 different immunological tests have been developed for diagnosis of encephalitozoonosis. The carbon immunoassay (1) has proved to be a reliable serological test which is easy to perform without use of expensive equipment.

Rabbits from several breeding colonies in Sweden have been tested for the presence of antibody to E. cuniculi. This paper presents a successful health monitoring program for encephalitozoonosis based on the carbon immunoassay.

MATERIALS AND METHODS

Animals

287 does and bucks from 8 rabbit breeding units were tested monthly and yearly for antibody to E. cuniculi by carbon immunoassay (CIA).

Blood sampling and preparation

Blood was collected on filter paper and dried. The dried blood was processed in the laboratory as described by Waller and Bergquist (1). Approximately 1 drop of dried blood was extracted with 1 ml phosphate buffered saline solution, pH 7.2, for 30 min. and heat treated at 56°C for another 30 min. After centrifugation at 400 g for 5 min. the supernatant was used for testing.

Carbon immunoassay (CIA)

5 ul of an E. cuniculi antigen suspension (Testman, Box 9020, S-75009 Uppsala, Sweden) was mixed with equal amounts of prepared blood extract and carbon suspension (Testman) on a microscope slide. A coverslip was applied and the result was read after 5 min. using an ordinary light microscope at 400 magnifications. With positive extracts the E. cuniculi organisms were stained grayish-black while with negative extracts the organisms appeared white against the dark background of carbon.

Seropositive animals

Seropositive animals and their youngs were regarded as infected and were removed from the breeding unit.

Testing intervals

Each breeding animal was tested monthly until all animals in the colony were seronegative in 2 following tests. After that, yearly tests were performed.

Introduction of new breeding animals

Before introduction of new breeding animals into a colony the new animals were tested twice with one month interval. If both tests were negative the animals were allowed to be incorporated in the stock.

RESULTS

The results are presented in Table 1. All breeding units had seropositive animals at the first test. The frequency of infected breeding animals varied between 3.3 and 30.5 %. One unit was seronegative from the second test, 4 colonies were seronegative from the 3rd test and 3 colonies were seronegative from the 4th test. Yearly tests since all breeding animals were seronegative have not revealed any reinfection after 6 years in 2 colonies and after 3 years in the other units.

Table 1. Number of rabbits seropositive to E. cuniculi in 8 breeding units as detected by monthly carbon immunoassay tests.

Breeding unit No.	No. of breeding animals at start	No. of seropositive animals				
		1st test	2nd test	3rd test	4th test	5th test
1	30	1 (3.3%)	0	0	–	–
2	42	4 (9.5%)	2	0	0	–
3	28	6 (21.4%)	1	0	0	–
4	51	5 (9.8%)	1	0	0	–
5	36	11 (30.5%)	2	0	0	–
6	30	7 (23.3%)	4	1	0	0
7	44	8 (18.2%)	2	2	0	0
8	26	7 (26.9%)	3	2	0	0
Total	287	49 (17.1%)	15	5	0	0

– = Not done

DISCUSSION

The testing intervals of 1 month was chosen as CIA only detects IgG and measurable IgG levels will be reached in rabbits within 1 month after infection (2). The laboratory processing of dried blood ends in a dilution of the blood equivalent to a serum dilution of about 1:40 (1). Many years of experience have shown that infected rabbits always reach a CIA titer of 1:160 or more during the course of infection and against this background collection of blood on filter paper is an easy and safe way to get blood samples for testing.

As infected does transmit the disease to most of their youngs during the time from birth to weaning, youngs from infected does must be regarded as infected and removed from the colony.

The result of this monitoring program for detecting rabbits seropositive to E. cuniculi is very promising. Since the last seropositive animal found by monthly tests was removed from each colony no more infected rabbits were found. It was also easy to prevent reinfection by testing new animals before introduction into a colony. This indicates that encephalitozoonosis might be a disease which is rather easy to avoid in rabbit breeding units.

REFERENCES

1. Waller, T. and Bergquist, N.R.: Rapid simultaneous diagnosis of toxoplasmosis and encephalitozoonosis in rabbits by carbon immunoassay. Lab.Anim.Sci. 32, No. 5, 515-517, 1982.

2. Waller, T., Morein, B. and Fabiansson, E.: Humoral immune response to infection with Encephalitozoon cuniculi in rabbits. Lab. Anim. 12, 145-148, 1978.

A.C. Beynen & H.A. Solleveld (Eds), New developments in biosciences: their
implications for laboratory animal science, ISBN 978-94-010-7973-0
© 1988 Martinus Nijhoff Publishers, Dordrecht.

SPONTANEOUS MURINE HEMOSIDEROSIS, A MODEL FOR HUMAN HEMOCHROMATOSIS?

T.S. Veninga and H. Morse*
Dept. of Radiopathology, State Univ. of Groningen, Bloemsingel 1, 9713 BZ
Groningen, The Netherlands
*Hope Farms, PO Box 85, 3440 AB Woerden, The Netherlands

1. INTRODUCTION
    The study of animal diseases resembling those of man may justify use of
laboratory animals. Alternatives for organic derailments in living organisms are as yet not available. We studied the natural occurrence of iron
loading in the spleen of C57Bl mice in order to find characteristics
similar to human hemochromatosis, which is of genetic origin. In animals
the disorder is commonly indicated as hemosiderosis. The blood pigment,
hemosiderin, accumulates in macrophages notably of the spleen. Other organs
can be involved especially the liver. In human hemochromatosis this is the
principle organ which is heavily loaded with iron pigment. A certain amount
of hemosiderin is regularly present in the spleen of mice. The amount increases with age. However, as far as we know, it does not lead to an abnormal colour at the surface of the spleen. At the age of 2 months fifteen
per cent of our C57Bl mice may show a spleen with a black anterior part.
There appears to be no apparent sex preference for the disorder. We studied
mice primarily at the age of 3 months by histological techniques, thereby
paying attention to some endogenous factors.

2. METHODS
    The spleens of 209 male and 41 female C57Bl mice of our institute inbred
colony were screened for the presence of a black anterior part as a co-
activity in other running experiments. Most of the animals were 3 months
old, but individuals of 2 months of age were also included. Thirty-one
males (14.8%) and five females (12.2%) possessed a spleen with a black part
of varying size.
    The animals received food and acidified tap water (pH 2.3) ad libitum.
The diet (RMH-B, Hope Farms, Woerden, The Netherlands) was composed as
previously described (Veninga and Morse 1984).
    The normal iron content of the diet varied from 180 to 200 mg/kg, however
a group of 20 males and 16 females received the same diet with a moderately
reduced iron content, 134 mg/kg, for 14 weeks from weaning. To trace an
eventual effect of the acidified water 10 males and 10 females were
supplied with alkaline tap water (pH 8.3) for 15 weeks after weaning at 3
weeks of age.
    Preparation of 2 μm sections of affected spleen parts and their staining
was described in detail elsewhere (Veninga, Wieringa and Morse, submitted).
Staining concerned hematoxilin-eosin (HE), Perls' Prussian Blue (PPB),
silver-methenamine (SM) and the enzymatic activity of acid phosphatase (AF)
and α-naphthylbutyrate esterase (BE).
    Tissue sections of liver, lung, pancreas, heart and kidney were inciden-
tally included and stained in the same way. Smears of peritoneal and lung
cells obtained by lavage with physiological saline were stained by the May-
Grünwald Giemsa and Perls' Prussian Blue dye technique. Approximately 0.2

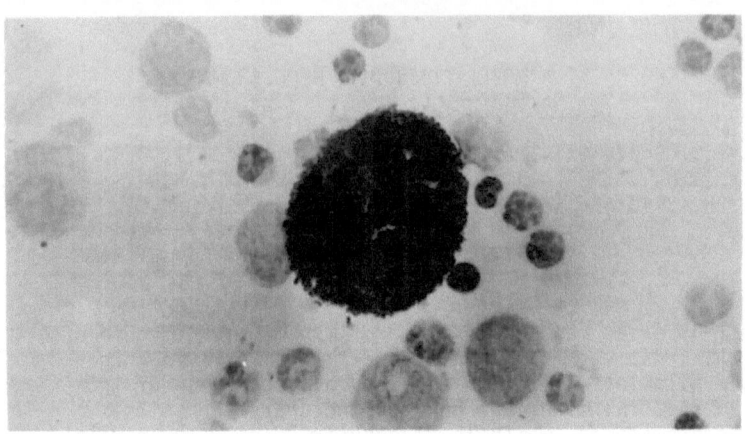

FIGURE 1. Macrophage isolated from the spleen of a mouse with hemosiderosis. In the cell hemosiderin granules. Stain: α-naphthylbutyrate esterase. Magnification 1200x.

ml urine of 17 males and 13 females was collected in a small disc provided with excess reduced ascorbic acid and mixed. Thereafter the fluid was tested for the occurrence of iron employing Merckoquant iron strips (Merck, Darmstadt, Germany).

Isolated spleen cells were obtained by the introduction of syngeneic lymphosarcomatous spleen cells into living mice. This procedure enlarges the spleen 4-fold in 5 days, loosening its cells. At the same time the apex of the spleen remains black. Black apical cells suspended in murine blood plasma were transferred to glass slides and stained as indicated.

3. RESULTS

Light microscopic observations of HE-stained sections of the black portion of a spleen revealed several agglomerations of yellow-brown granules in cells situated primarily in the red pulpa. Several granules stained blue with PB indicating hemosiderin accumulation. Hemosiderin laden cells stained positive for AF and BE which indicated the cells to be macrophages (fig. 1). Isolated spleen cells laden with granules had the typical morphologic appearance of macrophages. The granules stained heavily with SM.

With exception of the liver that contained very small spots of hemosiderin pigment after staining with SM. Organs other than the spleen did not demonstrate hemosiderin pigment.

Moderate diminution of the iron content in the animals' food from 196 to 134 mg/kg did not significantly influence the incidence of affected mice. Of the 20 males 3 had a spleen with a black tip, of the 16 females it was one. Neither did the omission of acidification of the drinking water exert an effect. One of the 10 males and 2 of the 10 females had the disorder.

Macrophages from the lungs or the peritoneal cavity of affected mice did not contain significant amounts of pigment. Similarly, no divalent iron was found in the urine.

4. DISCUSSION

The disorder described in this report concerns hemosiderosis, the accumulation of hemosiderin in macrophages principally in the spleen. The pigment laden macrophages are found to a greater or lesser degree in the anterior part of the spleen. With exception of the liver hemosiderin granules have not been detected in the other organs investigated. The liver contains a few dispersed hemosiderin bearing hepatocytes. In this respect, the murine disorder differs from human chromatosis where the liver is the primary organ showing iron overload in the parenchyma. Other organs like the pancreas can also be affected. Liver cirrhosis may also occur. The disease in humans is of genetic origin, which becomes manifest in middle-aged man (Kreeftenberg 1986). The defect may be sought at the level of the intestinal mucosa, where a disturbance in the regulation of iron absorption occurs.

Regarding the murine hemosiderosis the occurrence of the hemosiderin in macrophages of the spleen points to senescent erythrocytes as the source of the iron. This iron is collected by the reticulo-endothelial system (RES), particularly, the spleen. However, liver parenchymal cell accumulation of iron via the RES has been observed in mice frequently injected with erythrocytes in combination with a chelating agent (Corden 1986). We studied the disorder in young adult mice. The involvement of aged individuals may permit a more adequate comparison to human hemochromatosis, and lead to a more fitting model than is now presented. So far, hemosiderotic mice do not apparently suffer from the excess amount of iron in the RE cells of their spleen. This agrees with the observation that iron in RE cells is relatively innocuous (Hershko 1978). The lack of an obvious superficial marker prevents recognition in the intact mouse. We therefore looked for an adequate determinable factor for the disorder, but did not succeed. Pulmonary and intraperitoneal macrophages that can be harmlessly obtained by lavage, do not contain pigment granules. Furthermore, mice with demonstrable amounts of iron in their urine, have not been observed. Since "only" 15% of our C57Bl mice possess a hemosiderotic spleen a macroscopic recognizable characteristic is a prerequisite. The employment of X-ray photography may be a possible solution to this problem and should be tested in the future.

Although highly unlikely with laboratory animals can external factors cause iron overload. One should not forget the classic example of South African native males imbibing beer produced in iron utensils.

Exogenous factors of nutritional origin can influence the bioavailability of heme and non-heme iron in the diet. Theoretically, increased absorption can lead to iron deposition. Although it is not within the scope of this short report to explore all the potential nutritional factors which might affect absorption, a few should be mentioned. (Extensive reviews can be found in the Annual Review of Nutrition Vol. 1, 1981 and Vol. 6, 1986).

Besides the iron status of the animal and the quantity of the metal in the diet, heme iron is better absorbed than non-heme iron which is the principle form consumed by most animals. Ascorbic acid increases non-heme but not heme iron absorption. The presence of non-milk animal products such as meat and fish in the diet will facilitate the absorption of both kinds of iron. On the other hand tannins, phosphates and phosphate complexes, fiber and zinc reduces absorption. Phytates have been suspected as inhibitors as in the case of zinc. However, the isolated monoferric phytate of wheat bran appears to be highly available, although wheat bran as a totality is an inhibitor. In the complex natural ingredient diet composed of many ingredients fed to our animals promoters as well as inhibitors were present. (This is comparable to the situation with the human). Although the

one usually mitigates the other we examined two obvious possibilities.
    A low pH facilitates mucosal iron absorption (Kreeftenberg 1986). Since
we normally supply our mice with acidified tap water, we performed an
experiment in which we replaced it by alkaline water. Although treated for
a period of 5 months, this did not significantly alter the incidence of
hemosiderotic animals. Finally, the drinking water of the animals contains
3-4 μg/l iron as communicated by the municipal watersuppliers. This
quantity is too low to deliver a significant contribution to the iron state
of the animals.
    Since dietary factors are highly unlikely a defect of genetic origin as
the cause of the disease may form a real possibility (Veninga, Wieringa and
Morse, submitted). If so, this would be a characteristic in common with
human hemochromatosis.

REFERENCES

1. Veninga TS and Morse H: Influence of low protein diet on the life span
   of male and female C57Bl mice. Nutrition Research 4, 709-717, 1984.
2. Kreeftenberg H: Liver iron: pathology and diagnostics. Thesis (Dutch),
   University of Groningen, 1986.
3. Corden BJ: Hemosiderosis in rodents and the effect of acetohydroxamic
   acid on urinary iron excretion. Exp. Hematol. 14, 971-974, 1986.
4. Hershko C: Determinants of fecal and urinary iron excretion in des-
   ferrioxamine-treated rats. Blood 51, 415-423, 1978.

A.C. Beynen & H.A. Solleveld (Eds), New developments in biosciences: their implications for laboratory animal science, ISBN 978-94-010-7973-0
©1988 Martinus Nijhoff Publishers, Dordrecht.

IN VITRO ADHESION OF K88 POSITIVE E. COLI TO INTESTINAL VILLI OF JUST-WEANED PIGLETS.

E. Cox, V. Cools and A. Houvenaghel.

Laboratory of Veterinary Physiology, State University Centre of Antwerp, Slachthuislaan 68, 2008 Antwerp, Belgium.

1. INTRODUCTION

Enterotoxigenic E. coli (ETEC) strains are an important cause of diarrhoea in young animals and children. These strains produce enterotoxins which stimulate small intestinal electrolyte and fluid secretion. Most ETEC strains possess long, threadlike protein polymers (fimbriae) on their surface, which mediate attachment to specific receptors or receptor sites on villous enterocytes. Specific adherence plays two important roles in the pathogenesis of diarrhoea. It allows ETEC to resist the natural flushing mechanisms in the intestine and to colonize the small gut. But it also facilitates the interaction between ETEC elaborated enterotoxins and small intestinal epithelial cells. Although a wide variety of fimbrial adhesins have been reported (1), K88 fimbriae possessing ETEC strains are most commonly isolated from cases of diarrhoea in just-weaned piglets. The K88 antigen exists in 3 serological variants: K88ab, K88ac and K88ad (2). In piglets, presence or absence of receptors or receptor sites for these variants is genetically determined (3). The genetic susceptibility of piglets to adhesion of K88 ETEC to their villi can only be identified using a jejunum resection or biopsy technique in combination with an in vitro adhesion assay (4,5). In this study we investigated whether the in vitro villous adhesion could be replaced by an easier to perform buccal cell adhesion assay.

In just-weaned piglets, in neonatal animals and in children in developing countries, ETEC infections are often combined with enteric viral infections. These viral infections induce small intestinal villous atrophy (6). We have tried to evaluate the effect of virus induced villous atrophy on adhesion of K88ac ETEC. Several in vitro adhesion assays can be used to study bacterial adhesion to intestinal epithelial cells. We used a method described by Girardeau (7).

2. MATERIALS AND METHODS
2.1.Bacterial strains

ETEC strains used are listed in Table 1. Strains were inoculated over-night on blood agar base (Oxoid) plus 5 % (v/v) whole defibrinated sheep blood at 37 °C, examined for production of K88ab, K88ac, K88ad fimbriae with a slide agglutination test using specific antisera (RIVM, Bilthoven, The Netherlands), reinoculated on tryptone soy broth (TSB)(Oxoid) or tryptone soy agar (TSA)(Oxoid) during 24 hours at 37 °C and recontroled for production of fimbriae. The strains on TSA were suspended and diluted in sterile physiological saline. The number of viable bacteria per ml culture or suspension was determined by counting colony forming units (CFU).
2.2.Animals

Just-weaned female piglets (Pietrain X Belgian Landrace), three to four weeks old, were individually housed and were allowed to drink UHT sterilized whole cow's milk ad libitum during the experiment.

2.3. Experimental procedure

On day one of the experiment, a catheter was implanted in the left A. carotis communis. Then, piglets were divided into 3 groups. In the first group (group 1), piglets were infected with 2 ETEC strains on day 4 (GIS 25 and 156KP83; 0.7×10^{10} CFU/strain), as previously described (8). Piglets of the second group (group 2) were treated with chloramphenicol(1875 mg/kg/day) (suppression of the coloniza-tion resistance) during 3 days, whereafter they were infected with transmissible gastro-enteritis (TGE) virus on day 4

Table 1: Strains and serotypes of ETEC used in this study

Strains	Serotype	Fimbriae
GIS 25	0149:K91:K88ac	K88ac
156KP83	0149:K91:K88ac	K88ac
G 7	08:K87:K88ab	K88ab
H 56	08:K87:K88ad	K88ad
19KP74	0149:K85:K88ad	K88ad
E 57	0138:K81	none

and with both ETEC strains on day 5. The remaining piglets (group C) were kept as uninfected controls. The control piglets (n=6) and piglets of group 1 (n=15) were euthanatized 5 to 10 days after catheterization. In group 2, 6 piglets were killed 2 to 3 days (group 2.1), 3 piglets 4 to 5 days (group 2.2) and 6 piglets 7 days (group 2.3) after the TGE virus challenge. Piglets were euthanatized with an overdose methomidate (Hypnodil® , Janssen).

2.4. Samples

Heparinized arterial blood samples were taken through the implanted catheter. These samples were tested for mannose resistant haemagglutination. Immediately after euthanasia, the abdomen was opened and 15 cm long small intestinal segments were excised at the caudal jejunum and ileum. In the control group and group 2, specimens of the caudal end of the excised small intestinal segments were fixed in 10 % (v/v) phosphate-buffered formalin for histological examination. In all 3 groups, the segments were opened, gently washed in Krebs buffer pH 7.2 at 4 °C (NaCl 0.12 M, KCL 0.014 M, NaHCO$_3$ 0.025 M and KH$_2$PO$_4$ 0.001 M) supplemented with 1 % (v/v) formaldehyde. Then the villi were gently scraped from the mucosa with a glass slide, washed 3 times in the Krebs buffer and stored in this buffer until assayed in the villous adhesion assay within 3 weeks. Simultaneously, buccal mucosal epithelial cells were scraped from the oral mucosa of the piglets, washed and maintained in the same buffer for not longer than 3 weeks. These cells were used in the buccal cell adhesion assay.

2.5. Villous adhesion assay

This assay was based on the in vitro technique described by Girardeau (7). The stored villi were placed in Krebs buffer without formaldehyde for one hour, after which the buffer was renewed. This was repeated 3 times. Approximately 50 villi were placed into 1 ml 0.1 M phosphate buffer (pH 6.7) containing 1 % D-mannose (Difco), and 0.1 ml of a bacterial suspension (approximately 4.2×10^9 CFU/ml) was added. D-mannose inhibits adhesion mediated by type 1 pili. These pili are present on many ETEC strains. After gently mixing villi and bacterial suspension for one hour at roomtemperature, some villi were removed with a pipette, placed under a cover slip and examined by light microscopy at magnification x 600. In a first part of this study, in which villous adhesion was compared with buccal cell adhesion, the adhesion was scored : + = adhesion, + = weak-adhesion, - = no adhesion . In the second part of this study, in which we evaluated the effect of virus induced villous atrophy on K88ac ETEC attachment, adhesion was quantified by counting the number of visible bacteria adhering along 250 μm villous length.

2.6.Buccal cell adhesion assay

The same technique was used as for the villous adhesion assay. Adhesion, was scored (see villous adhesion assay) .

2.7.Mannose resistant haemagglutination(MRHA)

The porcine erythrocytes (RBC) were washed in phosphate buffered saline (PBS) and centrifuged at 1.4 x g for 5 minutes. Washing and centrifugation were repeated 3 times. Then the RBC were suspended in PBS containing 2.5 % (w/v) α-methyl-D-mannopyranoside (a mannose analogue)(Janssen Chimica). One ml of the erythrocytes suspension was supplemented with 1 ml of a 24 hours TSB culture (approximately 1.9×10^9 CFU/ml) (in glass tubes). The reaction was read after an incubation period of 3 hours at 4 °C. When MRHA occurred, erythrocytes formed a sheet at the bottom of the glass tube. In case of a negative MRHA, the erythrocytes formed a red spot.

2.8.Histological examination

This examination was partly performed in the Laboratory of Veterinary Pathology, State University of Ghent, Belgium (Dr. H. Thoonen and Prof. Dr. J. Hoorens). The in formalin fixed small intestinal specimens (group C and 2) were dehydrated, embedded in paraplast, cut at 5 μm and stained with hematoxylin and eosin. Villous height was measured, using an ocular micrometer for ten well-oriented villi per small intestinal segment.

2.9.Statistical procedures

Data were presented as means ± SEM and were analysed by a two-level nested ANOVA with unequal sample sizes. Significance was accepted at the $p < 0.05$ level. Pearson product moment correlation was calculated between mean villus-height of an intestinal segment and the mean number of bacteria adhering in vitro to the villi of the same segment.

3. RESULTS AND DISCUSSION

The results of the in vitro adhesion of K88 ETEC to villi and buccal epithelial cells are summarized in Table 2. Adhesion to villi was piglet-specific. Fourteen piglets were susceptible to adhesion of the 3 K88 variants to their villi, one piglet was susceptible to adhesion of K88ab and K88ac ETEC, but not to K88ad ETEC. Two piglets showed only adhesion of the K88ad strains to their villi. Four piglets were totally resistant to K88 adhesion to their villi. These findings are in agreement with the results of Bijlsma et al (3). They, however, observed piglets which were susceptible to adhesion of K88ab and K88ad ETEC, but not to K88ac ETEC. We did not. As can be seen in Table 2, no relationship was observed between adhesion to villi and to buccal epithelial cells. In this study porcine erythrocytes always showed haemagglutination with the K88ad ETEC strains, but not with the K88ab and K88ac strains. So, neither the buccal cell adhesion assay, nor haemagglutination can replace in vitro adhesion to intestinal villi.

The results of TGE virus-ETEC infection on K88ac ETEC adhesion are summarized in Table 3. No adhesion was observed with strain E 57 (no fimbriae). In piglets euthanatized 2 to 3 days after TGE virus infection (1 to 2 days after ETEC inoculation; group 2.1) a significant villous atrophy occurred in the caudal jejunum in comparison with control piglets (group C). We observed a significantly decreased in vitro adhesion of K88ac ETEC to villi obtained from the same segments (segments with atrophic villi), in comparison with adhesion to villi in group C. In groups 2.2 and group 2.3 villous height and in vitro adhesion of ETEC were near normal. The correlation between mean villous height per intestinal segment and the mean number of bacteria adhering along 250 µm villous brush border of the same segment was R = 0.66 (p < 0.001).

Table 2: In vitro adhesion of 3 K88 ETEC variants to villi and buccal cells of just-weaned piglets.

N	Villous adhesion			Buccal cell adhesion			
	K88ab	K88ac	K88ad	K88ab	(N)	K88ac	K88ad
14	+	+	+	±	(12),−(2)	−	+
1	+	+	−	±		−	+
2	−	−	+	±		−	+
4	−	−	−	±	(3),−(1)	−	+

N = number of piglets; + = adhesion; ± = weak adhesion; − = no adhesion.

Table 3: Mean (± SEM) villous height (jejunum and ileum) and mean number of K88ac ETEC adhering along 250 µm villous brush border of villi isolated from jejunal and ileal segments in group C (not infected piglets) and in groups 2.1, 2.2 and 2.3 (infected with TGE virus and ETEC).

Groups	Villous height (µm)		K88ac ETEC adhesion (number/250µm)	
	Jejunum	ileum	jejunum	ileum
group C	339 ± 15	238 ± 8	84 ± 9	50 ± 4
group 2.1	185 ± 8*	164 ± 14	35 ± 5*	31 ± 3
group 2.2	234 ± 12	239 ± 11	95 ± 16	60 ± 16
group 2.3	304 ± 9	251 ± 14	86 ± 6	69 ± 5

* p < 0.05 in comparison with group C.

Several in vitro techniques are described for studying attachment of ETEC to small intestinal epithelial cells using isolated brush borders (9) or enterocytes (10) or villi (7). In this study a villous adhesion assay was used. This method proved to be a reliable method to determine the genetic susceptibility of piglets for K88 ETEC (Table 1) and allowed us to quantify ETEC adhesion by bacterial count along a defined villous lenght (Table 2). So we observed that in vitro adhesion to intestinal villi was decreased after an in vivo infection of piglets with TGE virus and ETEC, probably as a result of the virus induced villous atrophy.

4. ACKNOWLEDGEMENTS

This work was supported by grant 4640A from the Belgian Institute for the Encouragement of Research in Industry and Agriculture (IWONL). The technical assistance of Mr J. Bollé, Mrs C. De Schepper and Mrs R. Hens are gratefully acknowledged.

5. REFERENCES

1. Parry SH and Rooke DM, 1985 : Adhesins and colonization factors of Escherichia coli. In: M Sussman (ed): The Virulence of Escherichia coli. Academic Press Inc., London, p.79-155.
2. Guinee PAH and Jansen WH, 1979: Behaviour of Escherichia coli K antigens K88ab, K88ac and K88ad in immunoelectroforesis, double diffusion and hemagglutination. Inf. Immunity, 23, 700-705.
3. Bijlsma IGW and Bouw J, 1985: Inheritance of K88-mediated adhesion of Escherichia coli to jejunal brush border in pigs: A genetic analysis. In IGW Bijlsma: Adhesion of K88-positive Escherichia coli to porcine intestinal epithelium: Immunological and genetic aspects. Proefschrift, Faculteit der Diergeneeskunde, Rijksuniversiteit Utrecht, p.45-56.
4. Bijlsma IGW, Rijkshuizen ABM and Frik JF, 1985: Determination of adhesive phenotypes in live pigs: The use of jejunum resection and haemagglutination by means of the different variants of Escherichia coli K88-antigen. In: IGW Bijlsma: Adhesion of K88-positive Escherichia coli to porcine intestinal epithelium: Immunological and genetic aspects. Proefschrift, Faculteit der Diergeneeskunde, Rijksuniversiteit Utrecht, p.35-44.
5. Snodgrass DR, Chandler DS and Makin TJ, 1981: Inheritance of Escherichia coli K88 adhesion in pigs: Identification of nonadhesive phenotypes in a commercial herd. Vet. Rec., 109, 461-463.
6. Moon HW, 1978: Mechanisms in the pathogenesis of diarrhoea: a review. JAVMA, 172, 443-447.
7. Girardeau JP, 1980: A new in vitro technique for attachment to intestinal villi using enteropathogenic Escherichia coli. Ann. Microbiol., 131 B, 31-37.
8. Cox E, Schrauwen E and Houvenaghel A, 1986: Experimental infection of piglets with enterotoxigenic E. coli: effect of oral chloramphenicol pretreatment. IRCS Med. Sci., 14, 374-375.
9. Sellwood R and Lees D, 1980: Adhesion of Escherichia coli pathogenic in pigs, calves and lambs to intestinal epithelial cell brush borders. In: PW de Leeuw and PAM Guinee (ed): Laboratory diagnosis in neonatal calf and pig diarrhoea. Kluwer Academic Publishers Group, AH Dordrecht, p.171-174.
10. Thorns CJ and Morris JA, 1985: In vitro techniques for the attachment of Escherichia coli to Epithelial cells. In: M Sussman (ed): The Virulence of Escherichia coli. Academic Press Inc., London, 333-338.

A.C. Beynen & H.A. Solleveld (Eds), New developments in biosciences: their implications for laboratory animal science, ISBN 978-94-010-7973-0
© 1988 Martinus Nijhoff Publishers, Dordrecht.

THE EFFECTS OF GENTLING ON OPEN-FIELD BEHAVIOUR OF RATS

Hirsjärvi, Paula A. and Junnila, Mirja A.
Department of Applied Zoology, University of Kuopio, Box 6,
70211 Kuopio and Institute of Occupational Health, Helsinki,
Finland

ABSTRACT
The fear experienced by the animal in an experimental
situation may distort the results of the experiment. Although
handling is known to influence the animals' reactions, this
fact has been largely ignored in behavioural studies. We
studied the effects of gentling on behaviour of rats in three
fundamentally different open-field situations: stressful,
indifferent and frightening.
90 male Kuo:Wistar rats, aged 95 days, were tested on open-
field for 5 min. on 4 successive days. 3 weeks before testing
the rats were randomly divided into two groups: one to be
individually gentled and the other to be left with routine
care only.
The behaviour in all the groups was typical to the test
situation. A basic difference was found in the behaviour of
the gentled and nongentled rats. While all the rats showed a
slight and transient fear of the new situation, the nongentled ones also showed a different type of fear, suggesting
fear towards the experimenter.

INTRODUCTION
It is well known that the stress experienced by the animal
influences its responses. All the stress-eliciting factors,
however, are not too well recognized. The human-animal
relationship as well as the animal's experiences of handling
have a great influence on several biochemical, physiological
and behavioural parameters (Corda et al, -80, Hughes &
Beveridge, -80, Meaney et al, -85, Williams & Russell, -72,
Uphouse, -81, Uphouse et al, -82).
As a rule, the effects of gentling have been found positive:
handling-habituated animals are more resistant to infections
and their recovery from stress and operations is better than
that of nonhandled animals (Bustad, -83, Meaney et al, -85).
In open-field, handling has been observed to decrease
emotionality and increase exploration and activity in rats
(Hughes & Beveridge, -80). On the other hand, standardized
stroking can be used as a punishment, just like an electric
shock (Candland et al, -60, -61).
The reason for these contradictory observations may be the
differrring conception of handling. When the aim of handling
is to decrease stress, handling must be gentling rather than
standardized stroking or mere lifting from cage to cage. The
individuality of the animals has to be considered; the fear-

399

less one reacts towards human contact in a different way than his timid cagemate.

We studied the effects of gentling on open-field behaviour of Wistar rats in three fundamentally different test situations: stressful, indifferent and frightening.

MATERIALS AND METHODS

The animals were 90 male Kuo:Wistar rats housed in stainless steel wiremesh cages, 5 rats per cage. They were given pelleted food for rats and tap water, both ad libitum.

3 weeks before testing, at the age of 10 weeks, the animals were transferred into experimental quarters and randomly divided into 2 groups: one to be individually gentled twice a day and the other to be left with routine care with minimal handling once a week. The animals of the two groups were separated into separately air-conditioned cubicles in the experimental room. After that the experimenter took care of the animals, no one else was admitted into the quarters.

The temperature in the animal and experimental quarters was maintained at 21-22 ^{O}C, the relative humidity at 50-80 %. The light/dark cycle was converted to enable working with the rats at their activity period.

The open-field used was a circular grey plastic arena, $\emptyset$ 83 cm. Each rat was individually tested for 5 min. on 4 successive days towards the end of the dark period. The rats were observed directly, the experimenter standing by the field. An assistance who had made acquintance with the gentled animals for 4 days before testing was standing on the opposite side of the field to avoid the animal's preference for either side (McCall et al, 69).

The test situations were 1) stressful: 78 dB white noise with 1500 lx light (N=15/15) 2) indifferent: 1500 lx light with background noise of 55 dB (N=15/15) 3) dark with background noise (N=15/15).

The parameters scored were: ambulation (total and middle field), rearing (total and middle field), motionlessness (frequency and duration), grooming and defecation.

Qualitative estimation was also made: rigid movements, teeth chattering, fluffyness, moving flat with ears backwards, and proportion of loose stools were observed. Motionlessness was divided into 'active' (sniffing, exploring) and 'passive' (fluffyness, teeth chattering, backward turned ears, flat or lying).

Analysis of variance for repeated measurements was employed to evaluate the results.

RESULTS

The behaviour in all the test groups was typical to each test situation. In the stressful situation ambulation is low, groomings frequent and the proportion of loose stools high. In the frightening situation ambulation is extremely high, defecation low; rigid movements and freezing are frequent (Hirsjärvi & Junnila, -86).

In each situation the behaviour on the whole of the gentled and nongentled animals was different. As to quantitative parameters ambulation and rearing were higher in the gentled animals on the first trial. The differences decreased on repeated trials.

In the stressful situation behaviour on the whole changed as a function of trials in the gentled as well as nongentled animals; the scores of the nongentled rats changed towards those of the gentled ones.

In the indifferent situation the behaviour on the whole on the first trial differed from that of later trials in both the gentled and nongentled rats. It may be noted that defecation decreased on repeated trials in the gentled animals but remained high in the nongentled rats throughout the trials.

In the frightening situation all the rats had extremely high ambulation, especially on the first trial. The difference between the gentled and nongentled rats was marked on the first trial. Ambulation decreased as a function of trials in both the gentled and nongentled rats while decrease in defecation was only seen in the gentled rats.

Qualitative observation revealed fluffyness and teeth chatteringwith ears backwards in all the groups. These features vanished on repeated trials, more quickly in the gentled rats. The most marked qualitative differences between the gentled and nongentled rats were the high frequencies and durations of passive motionlessness and higher proportions of loose stools and rigid movements in the nongentled rats. These features decreased on repeated trials and the difference between the gentled and nongentled rats diminished.

DISCUSSION

Our results are in accord with those of Hughes (-78) who observed decreased ambulation in the presence of the experimenter.

In the stressful situation the gentled animals showed mainly stress, expressed as low ambulation and high defecation (Broadhurst, -57, Walsh & Cummins, -76), and loose stools. The stress was combined with exploratory tendency which appeared as increasing activity (Hughes, -78, Williams & Russell, -72) and decreasing signs of fear (Walsh & Cummins, -76).

In the behaviour of the nongentled rats stress, and on later trials,exploration were mixed with yet another component, which could be interpreted as prey-like behaviour with passive motionlessness and rigid movements. This behaviour decreased with habituation but did not disappear altogether.

In all the groups slight and transient signs of fear, evidently caused by the new situation, were seen.

In the indifferent situation explorativity and the initial fear in the new situation soon subsided with habituation. In the nongentled animals the other type of fear - passive motion-lessness and rigid movements - was also observed.

The present results support the conception put forward in our previous study (Hirsjärvi & Junnila, -86), that the rat, although a nocturnal species, experiences the dark open place frightening, presumable because of fear of nocturnal predators. All the rats showed escape behaviour with extremely high ambulation and rearing (Williams & Russell, -72), and the gentled animals also had very low defecation. Ambulation remained high in spite of some habituation. Passive motion-lessness and rigid movements were, again, typical to the non-gentled rats - as if they had already had contact with the predator. Also the fact that their defecation remained high throughout the trials enhances the picture of fear (Williams & Russell, -72).

In conclusion, there seems to be several competing components in the open-field behaviour of rats: the behaviour caused on purpose by the test situation itself (stress, fear...), the transient fear of the new situation, the explorativity typical to the rat and, finally, the often unintended fear elicited by man - a predator. This disturbing component can be easily minimized by gentling.

REFERENCES

1. Broadhurs PL: Determinants of Emotionality in the Rat: I. Situational Factors. British Journal of Psychology, 48, 1-2, 1957.
2. Bustad LK: Laboratory Animal Scientists and Their New Role in Human-Animal Bonding. The 8th ICLAS/CALAS Symposium, Vancouver, 1983.
3. Candland DK, Faulds B, Thomas DB and Candland MH: The Reinforcing Value of Gentling. Journal of Comparative and Physiological Psychology, 53, 55-58, 1960.
4. Candland DK, Horowitz SH and Culbertson JL: Acquisition and Retention of Acquired Avoidance with Gentling as Reinforcement. Journal of Comparative and Physiological Psychology, 55, 1062-1064, 1962.
5. Corda MG, Biggio G and Gessa LG: Brain Nucleotides in Naive and Handling-habituated Rats: Differences in Levels and Drug-sensitivity. Brain Research, 188, 287-290, 1980.
6. Hirsjärvi PA and Junnila MA: Effects of Light and Noise Test Stimuli on the Open-field Behaviour of Wistar Rats. Scandinavian Journal of Psychology, 27, 311-319, 1986.
7. Hughes CW: Observer Influence on Automated Open-field Activity. Physiology and Behavior, 20, 481-485, 1978.
8. Hughes RN and Beveridge IJ: Effects of Experimental and Test Location Novelty on Nonspesific Activity in Rats and Its Modification by Metamphetamine. New Zealand Psychologist, 9, 15-18, 1980.
9. McCall RB, Lester ML and Corter CM: Caretaker Effect in Rats. Developmental Psychology, 1, 771, 1969.
10. Meaney MJ, Aithen DH, Bodnoff SR, Iny LJ, Tatarewicz JE and Sapolsky RM: Early Postnatal Handling Alters Gluco-corticoid Receptor Concentrations in Selected Brain Regions. Behavioral Neuroscience, 99, 765-770, 1985.
11. Uphouse LL: Interactions between Handling and Acrylamide on the Striatal Dopamine Receptor. Brain Research, 221, 421-424, 1981.
12. Uphouse LL, Nemeroff CB, Mason G, Prange AJ and Bondy SC: Interactions between 'Handling' and Acrylamide on Endocrine Responses in Rats. Neurotoxicology, 3, 121-125, 1982.
13. Walsh RN and Cummins RA: The Open-field Test: A Critical Review. Psychological Bulletin, 83, 482-504, 1976.

14. Williams DJ and Russell PA: Open-field Behaviour in Rats: Effects of Handling, Sex and Repeated Testing. British Journal of Psychology, 63, 593-596, 1972.

A.C. Beynen & H.A. Solleveld (Eds), New developments in biosciences: their implications for laboratory animal science, ISBN 978-94-010-7973-0

BLASTOMERE KARYOTYPING: A DIRECT METHOD FOR PRODUCING MOUSE TRISOMY 16 ↔ DIPLOID AGGREGATION CHIMERAS AS AN ANIMAL MODEL OF HUMAN DOWN'S SYNDROME

C. Bacchus and W. Buselmaier
Institut für Humangenetik und Anthropologie der Universität, Im Neuenheimer Feld 328, D-6900 Heidelberg. FRG

INTRODUCTION

Chromosome 21 (HSA21) is the smallest human chromosome and contains about 1.8% of genomic DNA (1). The decisive process for manifestation of trisomy 21 (Ts21) is the triplication of the distal band 21q22 (2), which is known to carry the genes for a proto-oncogene sequence (ETS2) and the enzymes super-oxide dismutase (SOD1) and phosphoribosylglycinamide synthetase (PRGS) (3,4). A regional localization has been described for the gene for phosphofructokinase (PFKL) (5,6). It is not yet known which genes are responsible for manifestation of Down's syndrome. However, up to now, eight individual genes have been localized, which - besides those located on band 21q22 - are distributed over the entire chromosome (for appropriate surveys see 1).

At present, work is currently in progress to elucidate this syndrome via an animal model (7,8). Many aspects of causal interaction can be resolved by means of the experimentally inducible trisomy 16 (Ts16) of the mouse. Murine Ts16 suggests itself since the previously named genes (SOD1, ETS2, and PRGS) and a gene for an interferon receptor protein (IFRC) have been localized on both HSA21 and the distal part (segment C3→ter) of mouse chromosome 16 (MMU16), which contains 3.86% of genomic DNA (1,9,10).

Since murine Ts16 is not viable beyond term, analysis of basic mechanisms responsible for the trisomic phenotype is limited to the pre- and perinatal phase. Thus, fundamental studies on the causation of mental retardation and its physiological and morphological basis, which is often associated with human autosomal trisomies, can only be performed in aggregation chimeras with a longer lifespan. Conventional aggregation methods, however, lead to extremely low proportions of Ts16 ↔ diploid chimeras (11). The Ts16 component is induced by a breeding procedure, which generates only 33.3% trisomic indivi-duals together with 66.6% chromosomally balanced litter mates. The aim of this study was, therefore, to develop a direct method for the specific production of mouse Ts16 ↔ diploid aggregation chimeras by embryo splitting, blastomere karyotyp-ing and subsequent implantation of selected embryos.

MATERIALS AND METHODS

Ts16 is induced by mating males (Rb8Lub^{WLub3}/2H+), doubly heterozygous for two metacentric translocation chromosomes Rb(11.16) and Rb(17.16) with the laboratory strain NMRI outbred.

In the male meiotic prophase I this chromosome constellation leads to the formation of quadrivalents and therefore to the induction of nondisjunction events, which result in the generation of trisomic embryos (Fig.1) carrying two (2Rb), and chromosomally balanced litter mates carrying one metacentric chromosome (1Rb). Superovulated pregnant females are sacrificed 48 hours after appearance of a vaginal plug. 4-cell embryos are recovered by flushing the oviduct with M2 medium. For identification of the chromosomal status the zona pellucida is enzymatically digested followed by mechanical blastomere disaggregation in calcium-free M16 medium (Fig.2).

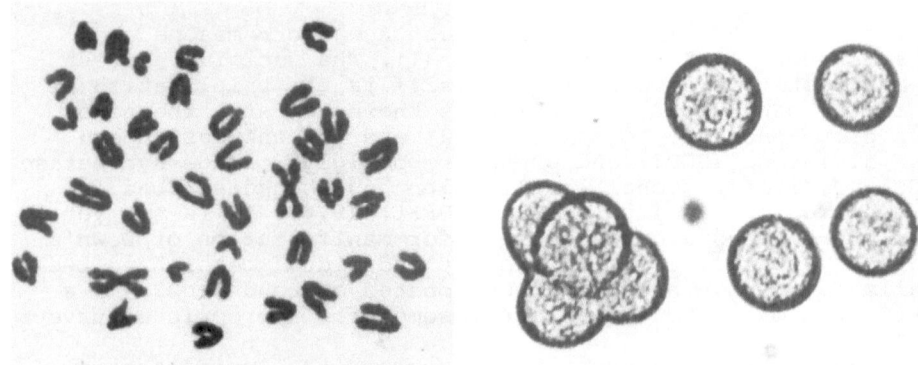

FIGURE 1. Karyotype Ts16 embryo.   FIGURE 2. 4-cell embryo before and after embryo splitting.

Three of the isolated 1/4 blastomeres are transferred into calcium-containing M16 medium and incubated in a 37°C humidified, 5% CO_2 incubator until they have developed to 8-cell embryos. Parallel to this procedure the remaining 1/4 blastomeres are incubated in Colcemid (final conc. 25µl/ml) for 3-5 hours. Embryos are then individually transferred to a slide in a microdrop of 1.5% Na-citrat (37°C) and left at room temperature for approx. 2 minutes. The chromosomes are fixed in methanol: glacial acetic acid (3:1). After staining (10% Giemsa) Ts16 embryos can easily be distinguished (2Rb) from their balanced litter mates (1Rb). Aggregation chimeras are then produced by aggregating the selected trisomic 8-cell embryos with diploid embryos, derived from unselected 1/4 NMRI outbred blastomeres. After aggregation (Fig.3) the chimeric embryos are incubated until they have developed into a single blastocyst which is transferred to pseudopregnant recipients.

RESULTS AND DISCUSSION
Using the described technique one is able to analyze more than 70% of the embryos successfully. Furthermore, an in vitro triplication of the trisomic embryos is achieved by embryo splitting. Theoretically it should be possible to establish

a trisomic cell line through repeated embryo splitting. However, during first cell divisions no protein biosynthesis occurs which leads to mini-blastocysts (Fig.4).

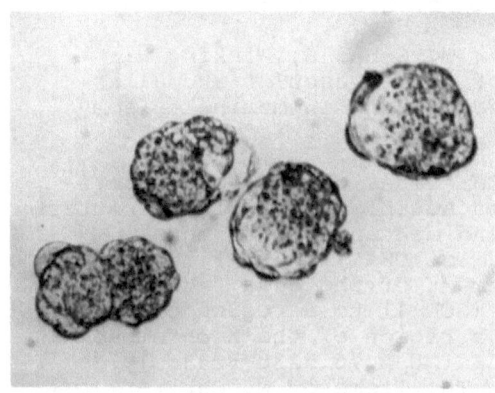

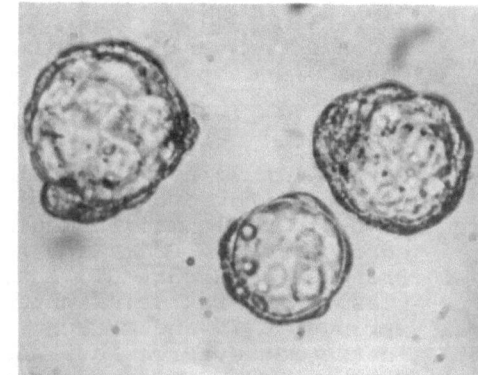

FIGURE 3. Aggregation of 8-cell embryos and chimeric blastocysts.

FIGURE 4. Reduction of blastocyst size after repeated embryo splitting.

The application of this method is therefore limited. According to our experience half the cell mass of a normal blastocyst is necessary for successful reimplantation.
This design can be applied for the specific production of any trisomic chimera of interest by using appropriate mouse strains. The method can also be used for embryo sexing in stock-farming via embryo transfer.

REFERENCES

1. Reeves RH, Gearhart JD, Littlefield JW: Genetic basis for a mouse model of Down syndrome. Brain Res Bull 16:803-814, 1986.
2. Summitt RL: Chromosome 21: Specific segments that cause the phenotype of Down Syndrome. In: de la Cruz FF and Gerald PS (eds): Trisomy 21 (Down Syndrome): Research Perspectives. Baltimore: University Park Press, 225-236, 1981.
3. Epstein CJ: Mouse monosomies and trisomies as experimental systems for studying mammalian aneuploidy. Trends Genet 1: 129-134, 1985.
4. Watson DK, Smith MJ, Kozak C, Reeves R, Gearhart J, Nunn MF, Nash W, Fowle JR, Duesberg P, Papas TS, O'Brien SJ: Conserved chromosomal positions of dual domains of the ets protooncogene in cats, mice, and humans. Proc Natl Acad Sci USA 83:1792-1796, 1986.
5. Chadefaux B, Rethore MO, Allard D: Regional mapping of liver type 6 phosphofructokinase isoenzyme on chromosome 21. Hum Genet 68:136-137, 1984.

6. Cox DR, Kawashima H, Vora S, Epstein CJ: Regional mapping of SOD-1, PRGS, and PFK-L on human chromosome 21: Implications for the role of these genes in the pathogenesis of Down syndrome. Am J Hum Genet 35:188A, 1983.
7. Epstein CJ: The consequences of chromosome inbalance: principles, mechanisms and models. New York: Cambridge University Press, 1986.
8. Bacchus C, Sterz H, Buselmaier W, Sahai S, Winking H: Genesis and systematization of cardiovascular anomalies and analysis of skeletal malformations in murine trisomy 16 and 19 - two animal models for human trisomies. Hum Genet, in press, 1987.
9. Cox DR, Epstein LB, Epstein CJ: Genes coding for sensitivity to interferon (IfRec) and soluble superoxide dismutase (SOD-1) are linked in mouse and man and map to mouse chromosome 16. Proc Natl Acad Sci USA 77:2168-2172, 1980.
10. Francke U, Taggart RT: Assignment of the gene for cytoplasmic superoxide dismutase (SOD-1) to a region of chromosome 16 and of Hprt to a region of the X chromosome in the mouse. Proc Natl Acad Sci USA 76:5230-5233, 1979.
11. Gearhart JD, Singer HS, Moran TH, Tiemeyer M, Oster-Granite ML, Coyle JT: Mouse chimeras composed of trisomy 16 and normal (2N) cells: Preliminary studies. Brain Res Bull 16:815-824, 1986.

A.C. Beynen & H.A. Solleveld (Eds), New developments in biosciences: their implications for laboratory animal science, ISBN 978-94-010-7973-0
© 1988 Martinus Nijhoff Publishers, Dordrecht.

A PROPOSED SPECIES DIFFERENCE IN THE RENAL EXCRETION OF PERFLUORO OCTANOIC ACID IN THE BEAGLE DOG AND RAT

H. Hanhijärvi[*], M. Ylinen[**], T. Haaranen[**], T. Nevalainen[**]
* Oy Star Ab Pharmaceutical Co., P.O. Box 33, 33721 Tampere, Finland
** University of Kuopio, Kuopio, Finland

INTRODUCTION
Perfluorinated fatty acids has been used as corrosion inhibitors, wetting agents, fire extinguishers and surface active agents (Olsen and Andersen 1983). On the other hand, there is not much information available on their biological effects. Griffith and Long (1980) showed that in a 90-day feeding study male albino ChR-CD rats were more susceptible to the toxicity of perfluoro octanoic acid (PFO, $CF_3(CF_2)_6 COOH$) than the females. The organic fluoride concentrations in the pooled serum were also considerably higher in the male than in the female animals. With radiolabelled PFO it was demonstrated that female Holzman rats actively secrete PFO in the kidney (Hanhijärvi et al. 1982). In the male rats only filtration could be demonstrated. From the data presented by Ophang and Singer (1980) it can be calculated that the plasma half life for PFO in the female Holzman rats is roughly 10 h. This far the plasma half life of PFO for the male rats has not been calculated. Indirectly it has been suggested that the plasma half life probably is longer than 7 days, because steady state could be reached only in the females by this time during a subchronic feeding study (Hanhijärvi et al. 1987).
As well known, the rat and beagle dog are among the most widely used species for the nonclinical safety evaluation studies. It is very important to know the possible species differences in the renal excretion to be able to induce and predict toxicity in a proper manner. Therefore, the aim of the present study was to study the excretion kinetics of PFO in the beagle dog, and when possible, to compare the results with those previously demonstrated in the rat.

MATERIALS AND METHODS
Altogether 10 laboratory bred beagle dogs (5 males and 5 females) were used during the present study. Six dogs (3 males and 3 females) were anaesthetized with methoxyflurane. Catheters were fixed in both ureters after laparototomy and cystotomy.
The animals were first given an intravenous dose of 30 mg/kg of PFO whereafter a continuous infusion with 5 % mannitol solution 1.7 ml/min was maintained. Urine was collected at 10 min intervals for 60 min. A 5 ml blood sample was collected in the middle of each urine sampling period. After that probenecid was given 30 mg/kg i.v. After the injection more urinary and blood samples were collected equally as before the injection. Renal clearances for PFO were calculated using the following equation:

$$Cl \ (ml/min/kg) = \frac{U \times V}{P \times W}$$

where U = urinary PFO concentration (ug/ml)

V = urine flow rate (ml/min)
P = plasma concentration of PFO (ug/ml)
W = animal weight (kg)

Four more dogs (2 males and 2 females) were given 30 mg/kg of PFO intravenously. After the injection the animals were kept in metabolic cages, and blood samples were collected intermittently for 30 days. From the conscious dogs plasma half lives were determined for each animal separately.

PFO concentrations were analyzed with a GLC-method described in detail elsewhere (Ylinen et al. 1985).

RESULTS

The results of the clearance study are given in TABLE 1. It is quite clear that there is no difference between the renal clearances of the male and female dogs either before or after probenecid. Both sexes seem to have an active secretion mechanism for PFO, because probenecid effectively and statistically significantly reduces the PFO clearance in each dog (paired Student's t-test).

The individual PFO concentrations for each four animals are given in Figure 1. The plasma half lives are also given. The plasma half lives of PFO were longer in the male dogs (473 h and 541 h) than in the female dogs (202 h and 305 h).

DISCUSSION

In our previous study with Holzman rats the total renal clearance for PFO was 58 ml/min/kg in the females and 1.7 ml/min/kg in the males, correspondingly (Hanhijärvi et al. 1982).The PFO doses for both sexes were essentially the same 5.2-5.6 mg/kg. In the rat study with radio-labelled PFO, the free PFO concentration in the serum was used for the calculation of the clearances. This time the sensitivity of the GLC-method did not give this opportunity. When the findings in the rat are corrected using the total PFO concentrations in the plasma (97.5 % bound), the clearance figures of the female rats still remain considerably higher (1.45 ml/ min) than those of the dogs. It is most interesting to observe that after this correction the PFO clearances of the male rats before and after probenecid (0.0025-0.0043 ml/min) and those of the female rats after probenecid (0.0028 ml/ min) resemble the figures observed in the dogs after probenecid. Therefore, it is obvious that the filtration rate of PFO is quite similar in both species for this particular compound.

On the other hand, the active tubular secretion rate of PFO in the rat and dog seem to be quite different. Basing on the previous study the very effective active tubular secretion rate in the female rat enables this sex to excrete PFO about 50 times more effectively than the dog (Hanhijärvi et al. 1982). On the other hand, the slower but still active tubular secretion of PFO in the dog kidney seems to make the PFO elimination a little faster in the dog than in the male rat. It should be kept in mind that this far there is no evidence available to suggest significant metabolism of perfluorinated fatty acids (Rozner 1981, Clark et al. 1973).

Surprisingly enough, plasma half lives of PFO in the male and female dogs were somewhat different. Also in this species the female appears to be able to eliminate PFO faster than the male, although no significant difference could be observed in the acute renal excretion test (TABLE 1). As indicated in FIGURE 1 the plasma half lives of PFO in the males were 473 h and 541 h, which means about 20 and 23 days. In the females the

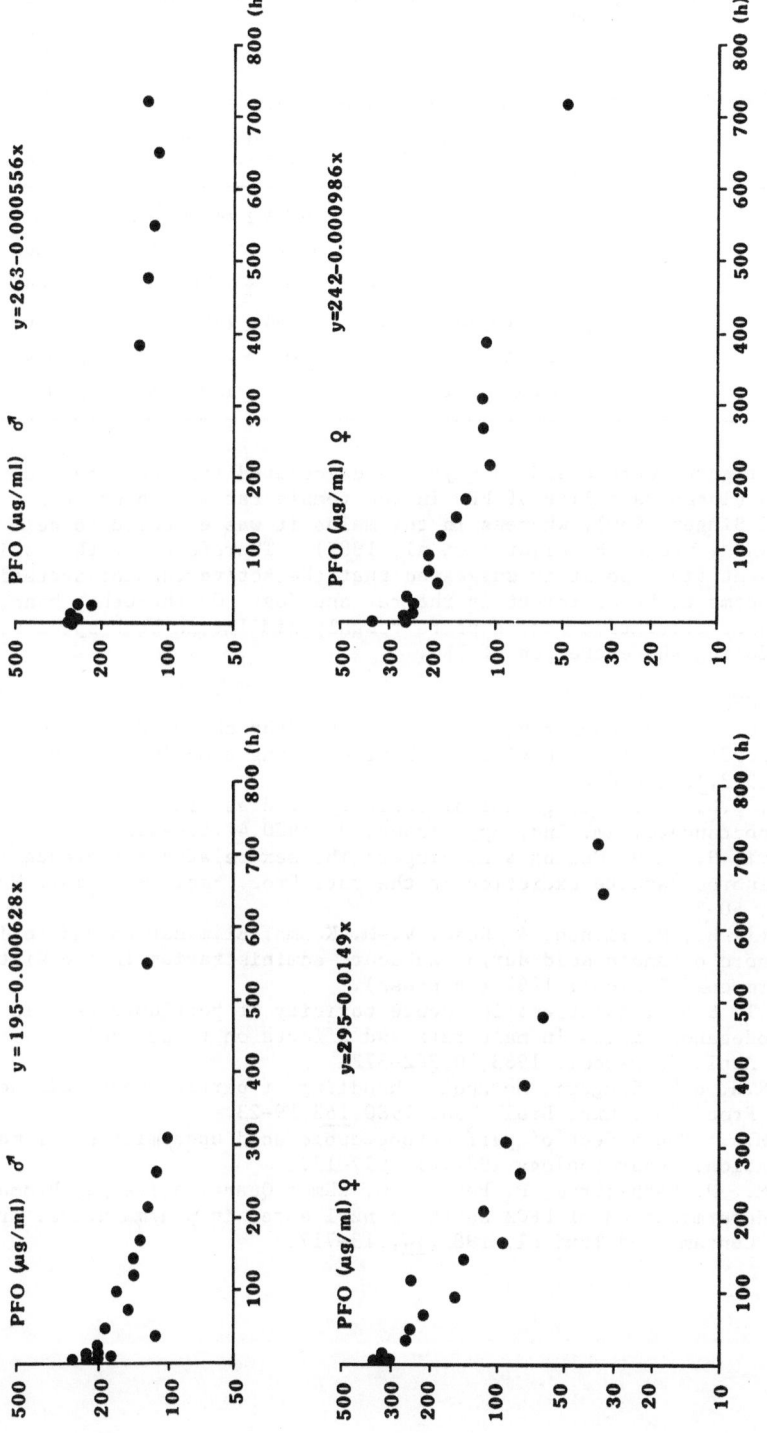

FIGURE 1. PFO concentrations in the plasma of two male and two female dogs after a single dose of 30 mg/kg i.v. administration.

TABLE 1.   The mean renal clearances (+ SD) for PFO before and after the administration of 30 mg/kg of probenecid. The number of experiments in parenthesis.

CLEARANCE OF PFO (ML/MIN/KG)

DOG NUMBER	SEX	BEFORE PROBENECID Mean + SD	(N)	AFTER PROBENECID Mean + SD	(N)	SIGNIFICANCE (paired t-test)
1	♂	0.0382 + 0.0034	(5)	0.0056 + 0.0036	(5)	p < 0.001
2	♂	0.0265 + 0.0094	(5)	0.0036 + 0.0014	(3)	p < 0.01
3	♂	0.0216 + 0.0120	(5)	0.0050 + 0.0013	(5)	p < 0.05
4	♀	0.0310 + 0.0097	(3)	0.0065 + 0.0034	(5)	p < 0.01
5	♀	0.0336 + 0.0062	(4)	0.0035 + 0.0008	(5)	p < 0.01
6	♀	0.0396 + 0.0091	(5)	0.0045 + 0.0006	(5)	p = 0.001

respective figures were 8 and 13 days. As calculated from the previous results the plasma half life of PFO in the female rat was about 10 h (Ophang and Singer 1980), whereas in the males it was expected to be between 1 and 4 weeks (Hanhijärvi et al. 1987). Therefore, on the basis of the present findings it is suggested that the active tubular secretion mechanism seems to be different in the rat and dog. On the other hand, the glomerular filtration rate appears roughly similar in both species, when considering the excretion of PFO.

REFERENCES
1.Clark, L.C. Jr., F. Becattini, S. Kaplan, V. Orbrock, D. Cohen & C. Backer: Perfluorocarbons having a short dwelling time in the liver. Science 1973,181, 680-682.
2.Griffith, F.D. & J.E. Long: Animal toxicity studies with ammonium perfluorooctanoate. Am. Ind. Hyg. Assoc. J. 1980,41,576-583.
3.Hanhijärvi, H., R.H. Ophaug & L. Singer: The sex related difference in the perfluorooctanoate excretion in the rat. Proc. Soc. Exp. Biol. Med. 1982,171, 50-55.
4.Hanhijärvi, H., M. Ylinen, A. Kojo, V.-M. Kosma: Elimination and toxicity of perfluoro octanoic acid during subacute administration in the Wistar rat. Pharmacol. Toxicol. 1987 (in press).
5.Olson, C.T. & M.E. Andersen: The acute toxicity of perfluorooctanoic and perfluorodecanoic acids in male rats and effects on tissue acids. Toxicol. Appl. Pharmacol. 1983,70,362-372.
6.Ophaug, R.H. & L. Singer: Metabolic handling of perfluorooctanoic acid in rats. Proc. Soc. Exp. Biol. Med. 1980,163,19-23.
7.Rozner, M.A.: The effect of perfluorooctanoic acid upon microsomal membrane per oxidation. Pharmacology 1981,41,2997-13.
8.Ylinen, M., H. Hanhijärvi, P. Peura & O. Rämö: Quantitative gaschromatographic determination of PFOA as the benzyl ester in plasma urine. Arch Environ. Contam. and Toxicol. 1985,14,713-717.

A.C. Beynen & H.A. Solleveld (Eds), New developments in biosciences: their implications for laboratory animal science, ISBN 978-94-010-7973-0
© 1988 Martinus Nijhoff Publishers, Dordrecht.

MEASUREMENT OF ORTHOSTATIC RESPONSES IN CONSCIOUS DOGS

E. AHONEN[1] T. NEVALAINEN[2] E. LÄNSIMIES[1] K. TAHVANAINEN[1] J. HARTIKAINEN[3] and M. HAKUMÄKI[3]. 1) Department of Clinical Physiology, Kuopio University Central Hospital, 2) National Laboratory Animal Center and 3) Department of Physiology, University of Kuopio, P.O.Box 6, SF-70211 Kuopio, Finland

1. INTRODUCTION

Most acute experiments have been carried out with animals in horizontal position lying on their back or on one side, a normal prerequisite for the use of instrumentation. However, these positions are abnormal to the dog. The change from horizontal four-leg standing position to head-up tilted position causes redistribution of the circulating blood volume and tendency of blood to pool in the lower extremities. This causes a fall in central venous pressure and tends to decrease venous return and cardiac output. Accordingly, stroke volume is decreased, and an increase in heart rate is followed to maintain sufficient cardiac output.

Vertical tilting has been used to predict drugs potentially causing postural hypotension in pentobarbital anesthetized dogs (1). Tilting has been used also to measure the sympatholytic activity of drugs in anesthetized dogs (2). Few studies are available on orthostatic regulation of conscious, unrestrained animals. Baum et al. (3,4) used conscious dogs to assess the orthostatic potential due to antihypertensive therapy. A head-up tilt of 90 degrees was done by lifting the forelimbs. Tilting elevated both heart rate and arterial blood pressure. Quantitative testing of baroreceptor dynamics has been done by measuring vascular responses to head motion in intact, unanesthetized dogs (5).

The aim of the present study was to develop a trained, chronically instrumented experimental model for measurement of orthostatic responses in conscious dogs, and to compare the responses in the high and low pressure circulation systems.

2. MATERIAL AND METHODS

2.1. Animal procedures

For the experiments, a total of six laboratory bred beagles of both sexes (1 - 1.5 years old), weighing 10 - 12 kg were used. Before the surgical procedures the dogs were accustomed to stand and sit in place without restraining devices in standardized laboratory surroundings. Tge surgical procedures (Fig. 1) were done under general anesthesia. In sterile left thoracotomy, the catheters were passed to the descending aorta, left atrium and pulmonary artery by the method of Herd and Barger (6). All the catheters were exteriorized in the same route out of the thorax to the neck of the dog. The animals were allowed to recover for at least four days after the operation before the experiments. The catheters were flushed every two days with a solution of heparin in saline. Each of the dogs was trained for the measurements, and in this way no restraint was needed during the experiments.

2.2. Measurements

The experiments were began after an adaptation period of 5 min with the animal in four-leg standing position. Thereafter, the measurement was continued 2 min in standing position, then 2 min in sitting, and again 2 min in standing position. The arterial blood pressure (AP), pulmonary arterial pressure (PAP) and left atrial pressure (LAP) were measured and stored into an instrumentation tape recorder. From the AP-curve, heart rate (HR), tension time index (TTI, heart rate x systolic blood pressure), mean arterial pressure (MAP) and pulse pressure (PP) were calculated.

During the head-up tilting, the body position of the dog was changed 60 degrees from the standing position. The altitude of the heart was controlled by a liquid capillary indicator and kept constant during the experiment.

The left ventricular function was analyzed by first-pass radionuclide ventriculography (7) both in four-leg standing and in 60-degree head-up tilted positions. Left ventricular end-diastolic volume (LVEDV), ejection fraction (LVEF) and heart rate were measured from the left ventricular time-activity curve. Stroke volume (SV) was calculated as LVEDV x LVEF and left ventricular end-systolic volume (LVESV) as LVEDV - SV. Cardiac output (CO) was calculated as HR x SV. The maximal velocity of contraction was measured as left ventricular peak ejection rate (LVPER).

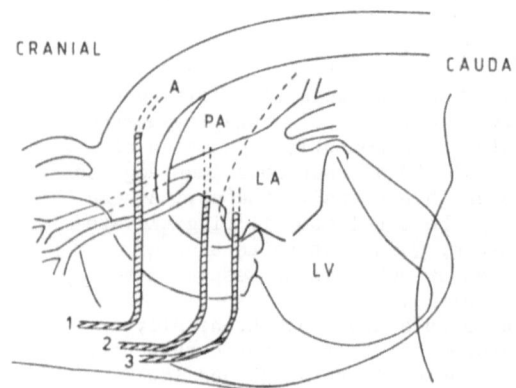

FIGURE 1. Surgical procedures as seen from the left lateral view of the dog. A = aorta, LA = left atrium, LV = left ventricle and PA = pulmonary artery. The catheters 1 – 3 were used for pressure recordings, and the catheter 3 for radionuclide ventriculography.

2.3. Statistical methods

For the measurement of initial changes in heart rate, the AP-curve was used for beat-to-beat heart interval (HI) analysis. The values 15 s prior to tilting were regarded as control state, and thereafter the acceleration and brake indices of heart interval (8) for 15 s were calculated from the AP-curve.

The constant changes of the parameters were analyzed from 30 to 60 s after tilting. The recording was transferred off-line from the instrumentation recorder to a multichannel analyzer. Sixty seconds prior to tilting were regarded as control state for the calculation of changes. The significance of changes was calculated with paired Student's t-test.

3. RESULTS
3.1. Initial heart rate changes

The initial heart rate changes occurred in 15 s after head-up tilting. Firstly, the maximal acceleration of heart rate (28 ± 14 % shortening in HI, mean $\pm$ SD) occurred 6.4 ± 1.9 s after tilting. Thereafter, a transient deceleration (52 + 25 % increase in HI, from maximal acceleration to maximal deceleration relative to the control state) followed in 11.9 + 1.9 s after tilting.

3.2. Constant changes
The control values and the constant changes due to tilting are given in Table 1. After tilting, the most marked changes occurred in PAP and in LAP. Although the initial heart rate changes were large, the constant HR response on average remained minimal. LVEDV and SV decreased significantly due to tilting.

TABLE 1. Hemodynamic orthostatic responses in six conscious dogs (mean + SD). The control level of each parameter has been measured in four-leg standing position, and the response indicates the change due to 60-degree head-up tilted position. Significance: ns = not significant, * = p < 0.05, ** = p < 0.01.

Parameter	Control level	Response	Significance
HR (bpm)	90.7 + 17.0	4.8 + 12.5	ns
TTI (kPa/min/10^2)	17.0 + 4.6	0.1 + 1.8	ns
MAP (kPa)	15.2 + 3.1	0.3 + 0.8	ns
PP (kPa)	7.6 + 2.0	0.0 + 0.7	ns
PAP (kPa)	2.0 + 0.2	-0.6 + 0.2	**
LAP (kPa)	0.6 + 0.5	-0.5 + 0.3	**
LVEDV (ml)	32.2 + 3.9	-4.5 + 3.7	*
LVESV (ml)	13.5 + 3.6	-1.4 + 3.7	ns
SV (ml)	18.8 + 3.4	-3.1 + 2.3	*
CO (1/min)	1.7 + 0.6	-0.2 + 0.3	ns
LVEF (%)	58.3 + 9.2	-2.5 + 5.9	ns
LVPER (1/min)	3.8 + 0.8	0.1 + 0.5	ns

4. DISCUSSION
4.1. Evaluation of the animal model and instrumentation
The method for catheter implantation (6) proved easy to perform in dogs. An advantage of this method is that all catheters needed can be placed through the thoracic incision. Dissecting the neck area for arterial catheterization may damage other structures, e.g. nerves, and carotid clamping abolishes the baroreceptor function in that area.

Trained standing and sitting dogs were used for the orthostatic experiments. Unexpectedly, the dogs were rather easy to handle by this method. The dogs were only lightly restrained in the arms if needed, but in all cases the recordings were obtained without problems, although the dogs were connected to the catheters 50 cm in length. In a special restraining harness (9,10) with freely hanging extremities of the dog, the orthostatic regulation can be presumed to change due to venous pooling and altered distribution of the circulation. In addition, training and chronical instrumentation play an important role in standardization of the method, since different modifying factors easily alter the sympathetic and vagal activities (11) mediated in the orthostatic responses.

416

4.2. Orthostatic responses

The orthostatic responses in the pressure and heart rate registrations could be divided into initial, transient phase and to late, constant phase. Similar characterization of orthostatic responses has been demonstrated also in man (8, 12).

In the present study the orthostatic pressure changes in the high pressure system were minimal, while in the low pressure system the head-up tilting caused hypotension. In the work of Baum et al. (3,4), a 90-degree head-up tilting in conscious dogs caused increases in both heart rate and in mean arterial pressure. In the present work, in accordance with the decreased venous return, the left ventricular size both in end-diastole and in end-systole was decreased during head-up tilting, but because the decrease in the end-diastolic volume was larger, stroke volume was also decreased.

It is obvious that the orthostatic changes induce a complex chain of events that cause adjustments in both central and peripheral circulation. The latter response to head-up tilting can be supposed to be caused by vasoconstriction, which was not possible to evaluate in the present experiments. Some hyperemic organs such as the spleen and the lungs may affect the orthostatic responses by acting as a dynamic reservoir of circulating blood volume (13). These effects could not be eliminated in these results, because the aim was to study the orthostatic changes in intact dogs.

In the orthostatic experiments the head-up tilting of 60 degrees was sufficient to induce hemodynamic changes in conscious dogs. Head-up tilting of 90 degrees has been used previously in anesthetized (1,2) and in conscious dogs (3,4). Also tilting of 30 degrees has been observed to induce hemodynamic changes in anesthetized cats (14).

From the pressure parameters studied in this work, left atrial pressure proved to be the most sensitive for indicating orthostatic regulation. This is in agreement with the finding of Furnival et al. (15) that the left atrium acts as a first link controlling heart volumes. There was no significant correlation between changes in left atrial pressure and heart rate during the orthostatic experiments. From that point it seems that the left atrial distension-induced tachycardia (10) has no significant role in the cardiovascular regulation at physiological pressure changes induced by orthostatic stress.

As a whole, a 60-degree heap-up tilting in conscious dogs causes strong transient acceleration and deceleration responses, but no significant constant changes in heart rate. In the constant phase, the most marked changes are hypotension in the low pressure system, and decrease in cardiac size.

REFERENCES

1. Constantine JW, McShane WK, Wang SC: Comparison of carotid artery occlusion and tilt responses in dogs. Am J Physiol 1971,221,1681-5.
2. Antonaccio MJ, Robson RD, Povalski HS: Modification by antihypertensive drugs of reflex circulatory responses induced by vertical tilting or bilateral carotid occlusion. Eur J Pharmacol 1973,29,23-31.
3. Baum T, Vander Vliet G, Glennon JC, Novak PJ: Antihypertensive and orthostatic responses to drugs in conscious dogs. J Pharmacol Methods 1981,6,21-32.

4. Baum T, Vander Vliet G, Glennon JC: Captopril alone and in combination with hydrochlorothiazide, on responses to upright tilt in conscious dogs. Arch Int Pharmacodyn Ther 1981,252,139–46.
5. Lamberti JJ, Urguhart J, Siewers RD: Observations on the regulation of arterial blood pressure in unanesthetized dogs. Circ Res 1968, 23,415–28.
6. Herd JA, Barger AC: Simplified technique for chronic catheterization of blood vessels. J Appl Physiol 1964,19,79–80.
7. Ahonen E, Hakumäki M, Hartikainen J, Kuikka J, Länsimies E, Nevalainen T, Tahvanainen K: First-pass radionuclide ventriculography in conscious dogs. Scand J Clin Lab Invest 1987,47,75–81.
8. Sundkvist G, Almer L–O, Lilja B: A sensitive orthostatic test on tilt table, useful in the detection of diabetic autonomic neuropathy. Acta Med Scand Suppl 1981,656,43–5.
9. Sine PN, Englers F: Restraining jacket for prolonged experiments on dogs. Lab Anim Care 1965,15,452–5.
10. Nevalainen TO, Hakumäki MOK, Hyödynmaa SJ, Närhi MVO, Sarajas HSS: Distension of pulmonary vein–left atrial junction: Heart rate responses in conscious and anesthetized dogs. Acta Physiol Scand 1980,110,47–52.
11. Kollai M, Koizumi K: Cardiovascular reflexes and interrelationships between sympathetic and parasympathetic activity. J Auton Nerv Syst 1981,4,135–48.
12. Borst C, Wieling W, van Brederode JFM, Hond A, de Rijk LG, Dunning AJ: Mechanisms of initial heart rate response to postural change. Am J Physiol 1982,243,H676–81.
13. Horton JW, Longhurst JC, Coln D, Mitchell JH: Cardiovascular effects of hemorrhagic shock in spleen intact and in splenectomized dogs. Clin Physiol 1984,4,533–48.
14. Koyama S, Ammons VS, Manning JW: Visceral afferents and the fastigial nucleus in vascular and plasma renin adjustments to head–up tilting. J Auton Nerv Syst 1981,4,381–92.
15. Furnival CM, Linden RJ, Snow HM: Reflex effects on the heart of stimulating left atrial receptors. J Physiol (Lond) 1971,218,447–63.

A.C. Beynen & H.A. Solleveld (Eds), New developments in biosciences: their implications for laboratory animal science, ISBN 978-94-010-7973-0
© 1988 Martinus Nijhoff Publishers, Dordrecht.

ORAL 65ZN LOADING TEST IN RATS FED IRI-OB DIET WITH VARIOUS ZN CONCENTRATIONS

J.P. van Wouwe[1],[2], M. Veldhuizen[2] and C.J.A. van den Hamer[2]: [1] Dep of Pediatr, Leiden State Univ, PO 9600, 2300RC Leiden, [2] Dep of Radiochem, Interuniv Reactor Inst, Mekelweg 15, 2629JB Delft. The Netherlands.

INTRODUCTION

The metal Zn is of biological importance as cofactor in enzymes, active in all major metabolic pathways. Deficiency of Zn is characterized by delay of cell multiplication and tissue growth. No specific biochemical indicator of deficiency is available. Homeostatic mechanisms in deficiency state have been proven by radiotracer whole body retention studies: the increase of biological halftime Tb is a reliable tool to diagnose (degrees of) Zn deficiency. Laboratory animals provide a suitable model to study the physiology of the homeostasis. Rats and mice react immediatelly with growth retardation when becoming Zn deficient (1). Available radiotracers for retention studies are ^{65}Zn (Tf=235 day, γ-peak 1.12 MeV) and ^{69}Zn (Tf=12h, γ-peak 0.44 MeV).

TABEL 1. Clinical features of Zn deficiency (4).

FEATURE	MILD	MODERATE	SEVERE
Skin			
Alopecia	−	−	+
Dermatitis	−	−	+
Delayed Woundhealing	−	+	+
Gastrointestinal			
Anorexia	−	+	+
Hypogeusia	−	+	+
Diarrhea	−	−	+
Endocrine			
Growth Retardation	−	+	+
Oligospermia	+	+	+
Hypogonadism	−	+	+
Insulin Hypersensitivity	−	+	+
Metabolic			
Weight Loss	+	+	+
Hyperammonemia	+	+	+
Neuropsychiatric			
Depression	−	+	+
Irritability	−	+	+
Emotional Dysfunction	−	−	+
Ophtalmologic			
Impaired Darkadaptation	−	+	+
Immunologic			
T-cel Dysfunction	−	+	+
Recurrent Infections	−	−	+

Human dietary Zn deficiency is known to occur in otherwise well nourished
children (2). Subclinical deficiency causes various health hazards (Table
1). Laboratory monitoring of deficienies is restricted to sophisticated
radiotracer research (3). Search for a specific and sensitive screening
test in nutritional Zn deficiency is desirable in order to improve health
potential. It can also lead to insight into the precise incidence and
into the causes of Zn deficiency.
Our institute provides whole body counting (WBC) facilities for lab
animals and the opportunity to produce and safely use Zn radiotracers. In
cooperation with Hope farms (PO Box 85, 3440 AB Woerden, NL) a diet apt
to study trace element physiology has been developed (5).
The aim of this study is to survey the effects of the nutritional Zn
status on the radio Zn retention in the tissues of the rat.

MATERIALS AND METHODS
Two groups of 6 male Wistar rats at weaning (CBP, POB #167, 3700 AD
Zeist, NL) were individually housed in Macrolon metabolic cages with
stainless steel lid in a climatized room with 12 h light-dark cycle. They
were pair-fed with IRI-OB-diets which contain 30 and 150 μmol Zn kg^{-1}
respectively (Tabel 2).

TABEL 2. Composition of the IRI-OB diet.

IRI-OB DIET	W%	TRACE ELEMENTS μmol kg^{-1}			
glucose	50	As	1	I	2
corn starch	15	B	10	Mn	900
ovalbumin	20	Co	2	Mo	2
sunflower oil	4	Cr	40	Ni	40
α-cellulose	5	Cu	150	Se	2
choline Cl	<1	F	130	Sn	20
minerals	4	Fe	1000	V	2
vitamin mix	1	Zn-deficient diet			30
.trace elements:	0.1.	.Zn-control diet			150.

Radiotracer Zn as $^{65}ZnCl$ was obtained from The Radioactive Center,
Amersham, UK (5×10^{10} Bq mol^{-1}). Doses were given by intraperitoneal injection on day 1 and by orogastric sonde on day 15. Body weight was daily
recorded. WBC was performed every other day by putting the animal in a
container into a tank filled with pseudo-cumene equipped with a ND60 multichannel analyser (ND, Schaumburg, Ill, USA). Counts were compared with
the standard and expressed in % of the i.p. dose given on the first day.
On day 15 after an overnight fast the animals were given 11 nmol ^{65}Zn g^{-1}
body weight per orogastric sonde. Half an hour later they were anes-
thesized by CO_2 and decapitated. The following tissues were collected:
liver, pancreas, femor, heart muscle and duodenum. The duodenum was
lavaged with saline to collect the content. All samples were counted in a
γ-counter with window setting between 960 and 1240 keV. Whole blood was
collected in heparine 5% and separated (at $4°C$, 3000 rpm during 15 min)
in plasma and red cells. The cells were thrice washed in saline. The
washed red cells were subsequently in vitro incubated with a near
physiological medium with ^{65}Zn at 37°C during 1h, washed and counted and
chased with inactive Zn (6). The tissues were freeze dried during 48 h
and weighed. The tissues were dissolved in 65% nitric acid (Suprapur,
Merck, Darmstadt, FGR) and 20% hydrogen peroxide (Aristar, BDH Chem Ltd,

Poole, UK) at 70°C. Zn content was measured by AAS (Perkin Elmer 2380)
with an acetylene flame at 213.6 nm. The tissue specific activity
(cpm/µmol total Zn) was calculated. The values in the two groups of
animals were analysed by ANOVA and judged for sensitivity and specificity
by the overlap index (7).

RESULTS
Body growth of the animals is shown in Figure 1. The results of the whole
body retention of the same animals is shown in Figure 2.

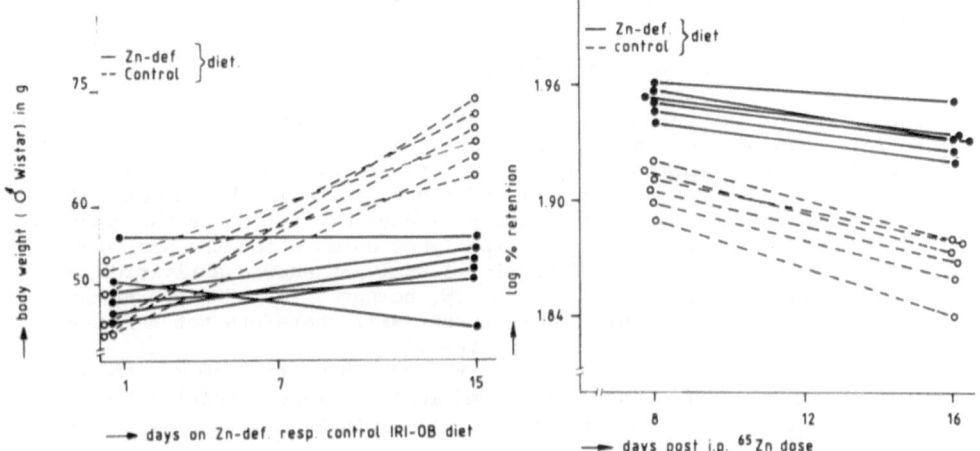

FIGURE 1. Growth of male Wistar
rats pair-fed a Zn deficient resp.
a control IRI-OB diet

FIGURE 2. Whole Body ^{65}Zn retention curves in the same animals
after < 1 nmol ^{65}Zn g^{-1} on day 1

The duodenum content contained no activity (average of 50 ncpm, with a
background fluctuating between 42 and 56 cpm). The specific activity in
plasma and tissues is given in Tabel 3. The in vivo and in vitro red cell
^{65}Zn uptake, and the specific activity after the chase experiment are
shown in Tabel 4.

TABEL 3. Specific activities (^{65}Zn/total Zn) in tissues and blood of male
Wistar rats pair-fed during 15 days a Zn deficient and a control IRI-OB
diet resp., 30 min after an oral ^{65}Zn loading, expressed in cpm ^{65}Zn
µmol^{-1}, mean ± s.d., % overlap and probability (ANOVA).

sample origin	Zn-deficient	control	[%O]	p
plasma	2710 ± 580	1980 ± 610	[83]	n.s.
bone	9010 ± 2840	4810 ± 1050	[17]	0.005
liver	53520 ± 9830	57230 ± 13060	[71]	n.s.
pancreas	34420 ± 11030	34930 ± 11130	[99]	n.s.
duodenum	305520 ±120260	264340 ± 53260	[92]	n.s.
heart muscle	29170 ± 3960	22730 ± 3390	[17]	0.008.

422

TABEL 4. In vivo (30 min after an oral ^{65}Zn loading) resp. in vitro (1h at 37°C) ^{65}Zn uptake, and chase experiments (1h at 37°C) of washed red blood cells in counts per minute, of male Wistar rats pair fed during 15 days a Zn deficient resp. a control IRI-OB diet. Mean ± s.d. (n=6), % overlap and probability (ANOVA).

ery experiment	Zn deficient	control diet	[%O]	p
in vivo	297 ± 53	222 ± 38	[25]	0.018
in vitro	1273 ± 156	1010 ± 135	[22]	0.011
chase	1134 ± 142	907 ± 122	[17]	0.014

DISCUSSION

Growth of the animals was severely impaired when fed the Zn deficient IRI-OB diet. The retention curves of the ^{65}Zn i.p. dose given at day 1 shows decreased excretion of the activity as observed before (8). Extrapolation of the retention curve (9) shows the apparant retention to be 86 resp. 77% of the dosage in the deficient or control animals. The combination of a retention study with an oral loading both of ^{65}Zn enables us to judge the nutritional deficiency state in the animals. The dose of the retention study (< 1 nmol g^{-1}) does not influence the results of the oral loading. In a previous experiment animals fed the same diets and exposed to such a similar dose of < 1 nmol g^{-1} showed activities in the tissues studied of <100 cpm at day 15, compared to >2500 cpm after the oral loading (± 4%). This latter amount will therefore not influence the results as shown in tabel 3 significantly.

Specific activities in the tissues under study show after 30 min to be only significantly different in bone and muscle, tissues holding the largest compartments of body Zn. After different time intervals this pattern changes.

The increased accumulation of activity in the red cells after 30 min, and the similar increase after in vitro incubation reflect a decrease in the exchangeable red cell Zn (cpm after chase minus cpm after in vitro incubation, multiplied by the in vitro uptake). The kinetic analysis of these phenomena are at present under study. It is desirable to reveal the physiological mechanism and homeostatic function of this increased radio Zn accumulation in red cells. Our laboratory animal data point at the possibility to use an in vitro erythrocytic test that reflects nutritional Zn deficiency. The potential for such a reliable test for diagnostic purposes is extense. The calculated overlap index showes encouraging specificity and sensitivity (17-25% overlap).

CONCLUSION

Nutritional Zn deficiency in male Wistar rats at weaning is accompanied by growth retardation, increased ^{65}Zn retention as measured by WBC, increased specific activities after 30 min in bone and muscle, and increased accumulation of ^{65}Zn in red cells. The in vitro accumulation of ^{65}Zn in red cells was also increased in nutritional Zn deficiency, and has little overlap in the groups of rats fed the two IRI-OB diets.

REFERENCES
1. Roth H.P. and Kirchgessner M. Zn metalloenzyme activities. Wld Rev Nutr Diet 34:144-160. 1980

2. Henkin R.I. and Aamodt R.L. in Inglett G.E. (ed) Nutrition Bioavailability of Zinc. American Chemical Society, Washington D.C. pp 83-107. 1983
3. Cornelisse C. Zn absorption and retention in man, with special emphasis on surgical patients, Thesis State University of Utrecht.1985
4. Garnica A.D. Chan W.Y. and Rennert O.M. Current Problems in Pediatrics volume XVI # 2 pp 61-72. 1986
5. Van Barneveld A.A. and Morse H. A new diet for trace element research with rats and mice, IRI-rapport 133-84-02, FELASA-1, Dusseldorf, FRG.1981
6. Van Wouwe J.P and Van den Hamer C.J.A. How to diagnose Zn-deficiency? Pediatr Res 20:1034A. 1986
7. Hertz A.J. Overlap index. An alternative to sensitivty and specificity in comparing the utility of a laboratory test. Arch Pathol Lab Med 108: 65-67. 1984
8. Van Wouwe J.P. et al. In preparation
9. Van Barneveld A.A. Trace Element Absorption and Retention Studies in Mice, Thesis University of Nijmegen. 1984

ACKNOWLEDGEMENTS
Ton van der Meer, Stefanie Hoogenkamp and Kees Volkers are acknowledged for their skilled technical assistence, H Morse for help to prepare the diets. JPvW is supported by Grant 28-1549 of the Dutch Prevention Fund.

A.C. Beynen & H.A. Solleveld (Eds), New developments in biosciences: their implications for laboratory animal science, ISBN 978-94-010-7973-0

MORTALITY AND TUMOUR INCIDENCE OF BDII/HAN RATS

J.KASPAREIT-RITTINGHAUSEN and F.DEERBERG

SUMMARY
    In a life-span study 400 BDII/Han rats (200 males and 200 females) were kept from weaning to their natural death . The mean life-span of male rats was 33,6 ±4,7 months , that of females 22,5 ± 4,4 months. The life expectancy was mainly influenced by neoplasms , that occurred in 83% of the males and 96,5% of the females . Out of the large tumour spectrum only few neoplasms exceeded a 10% incidence: Endometrial carcinomas (91%) in females and pheochromocytomas (26,5%), histiocytic sarcomas (22,5%), pituitary gland tumours (18%) as well as hemangiomas of the mesenteric lymph nodes (14%) in males .

INTRODUCTION
    The correct interpretation of results of carcinogenicity studies requires a precise knowledge of the spontaneous neoplastic lesions of the rat strain used. Because the incidence and the tumour spectrum of rats vary to a great extent between different strains or stocks and even between different colonies of the same strain kept under different maintenance conditions (12,15),   numerous studies on various strains of rats (4,7, 10, 13) were carried out to obtain information about their mortality, tumour frequency and tumour spectrum. The present study reports the mortality rate and incidence of spontaneous neoplasms in BDII/Han rats kept in a longevity   study.

MATERIAL AND METHODS
    A total of 400 BDII/Han rats (200 males and 200 females) were maintained in groups of five (separated according to sex) in polycarbonate-cages, in a barrier-type animal house, until their natural death. Maintenance conditions, microbiological status and procedures of examination were described in detail elsewhere (9). All rats were inspected twice daily and a complete necropsy was performed on those animals found dead or moribund. After fixation in 10% formalin tissue samples from almost all organs and from all pathological alterations were embedded in paraffin wax. Tissue sections (4 μm) were stained routinely with hematoxylin-eosin. Selected tumour sections were stained with Azan according to Heidenhain and with reticulin stain according to Gomori. Representative tumour tissue was embedded in hydroxyethyl-metacrylate and 1 μm thin sections were stained with toluidine blue.

RESULTS
Mortality
    The cumulative mortality of female and male BDII/Han rats is presented in Fig.1. The first female died during the 13th month and the first male during the 15th month of life. Thereafter, the mortality increased with the age of the animals, in females much more rapidly than in   males. The mean life-span of males was 33,6 ± 4,7, that of female rats 22,5 ± 4,4 months.

426

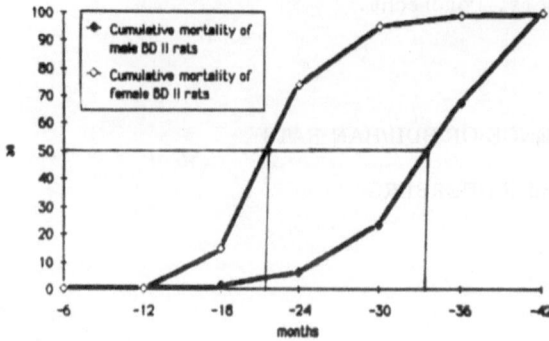

Fig 1. Cumulative mortality of male and female BDII/Han rats .

Tumour incidence

As shown in Table 1, 83,4% of the males and 96,5% of the females revealed one or more spontaneous tumours. Almost all of the male tumour-bearers were observed during the third or fourth year of life, while most of the females developed tumours already during the second year. The majority of tumour-bearing rats had a single neoplasm, but in 47,8% of the male and 18,7% of the female rats two or more neoplasms occurred (Table 2).

	Age (months)				
Sex	1-12	13-24	25-36	37-48	Total
Males	0/0	9/4,5	106/53	41/25,5	166/83,4
Females	0/0	141/70,9	49/24,6	2/1	192/96,5

Table 1. Tumour frequency in male and female BDII/Han rats (n/%).

Sex	1	2	3	4	5	6	7	tumours
Males	86/52,0	59/35,3	15/9,0	5/2,9	-	-	1/0,5	
Females	156/81,3	28/14,6	7/3,6	-	1/0,5	-	-	

Table 2 . Frequency of multiple tumours (n/%) in tumour-bearing BDII/Han rats.

427

Frequent neoplasms (>10%) in BDII/Han rats

The majority of neoplasms occurred with frequencies below 1%. Only few neoplasms exceeded a 10% incidence: Endometrial carcinomas in females and pheocromocytomas, histiocytic sarcomas, adenomas and adenocarcinomas of the pituitary gland, as well as hemangiomas of the mesenteric lymph nodes in males.

With an incidence of 91% endometrial carcinomas dominate in the female tumour spectrum (Fig.2). In almost all cases they are responsible for the death of the animals, because of their strong tendency to metastasise and their frequent association with severe purulent inflammation of the uterus. In 69,2 % of the affected females widespread metastases, often invading intraabdominal organs, were found in the abdominal cavity. In 38,5% the tumours showed distant metastases to the lungs and in 39% to the mesenteric lymph nodes. Microscopically several histological types of endometrial carcinomas could be distinguished. The majority were adenocarcinomas (86,7%). With lower frequency anaplastic carcinomas (6,6%), adenoacanthomas (5%) and squamous cell carcinomas (1,7%) occurred.

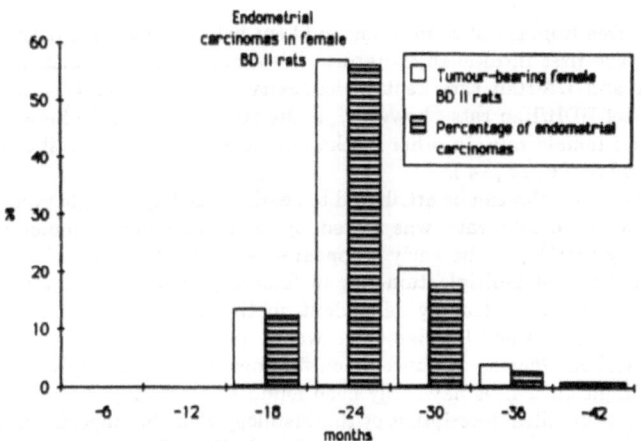

Fig.2. Frequency of endometrial carcinomas in female BDII/Han rats.

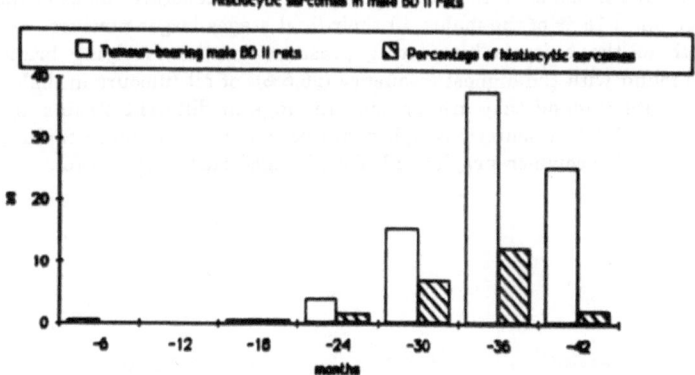

Fig.3. Frequency of histiocytic sarcomas in male BDII/Han rats.

The most frequent neoplasms in male rats were pheochromocytomas (26,5%). All were found in males older than two years. With somewhat lower incidence (22,5%) histiocytic sarcomas were diagnosed. They accounted for up to 40% of the tumours observed in some half-year periods (Fig.3). Histologically the tumours were characterized by neoplastic proliferation of histiocytic cells. Langhans-type multinucleated giant cells, palisading of histiocytes arround areas of necrosis, and occasionally a granuloma-like arrangement of cells were further common histological features of the tumours. In nearly all affected rats a multiple tumour growth in various organs was found. Most of the histiocytic sarcomas were localized in the subcutis (71%), but lungs (62%), liver (53%), mediastinal lymph nodes (36%), mesentery (29%) and mesenteric lymph nodes (20%) were also frequently involved. Adenomas and adenocarcinomas of the pituitary gland were observed in 18% of the males. They entirely developed in the anterior lobe of this endocrine gland. Hemangiomas of the mesenteric lymph nodes were found in 14 % of the male BDII/Han rats. In only one male a hemangioma of another lymph node (ln. mediastinalis) was diagnosed.

DISCUSSION

As expected of a rat colony free from specific infections and kept behind a hygienic barrier, the mortality was low during the first third of their natural life-span. Similiar results have been described for Han:Wist and DA/Han rats kept in longevity studies under the same conditions (7,8). A peculiarity of BDII/Han rats, however, is the striking difference between the life expectancy of male and female rats. In other stocks and strains identical or shorter life-spans of males were observed (4, 11 , 13 , 14).

The higher rate of mortality in females can be attributed to death caused by a single type of neoplasm, since the majority of female rats was killed by metastasising endometrial carcinomas. The high frequency (91%) and the early appearance of the neoplasms is also responsible for the lower incidence of multiple tumours in females, compared to males, because the multiplicity of neoplasms is strongly dependent on the mean life-span of the animals. With the exception of Han:Wist and DA/Han rats, where incidences of 40% and 60% were reported (7,8), endometrial carcinomas are uncommon findings in rats. Among various species of laboratory animals high incidences have only been found in aged rabbits (1,5) and aged Chinese hamsters (3,16). A detailed description of the histology and the importance of hormonal influences on the development of the tumours in female BDII/Han rats is given elsewhere (9).

An uncommon high frequency (22,5%) of histiocytic sarcomas was found in male BDII/Han rats. This incidence considerably exceeds previously reported in Wistar (2%) and F344 (0,4% in females and 1,5% in males) rats (2,13). Adenomas and adenocarcinomas of the pituitary gland were observed in 18 % of the males. In their final stages larger tumours were responsible for the death of the animals by putting pressure on the overlying brain. Pheochromocytomas were found with the highest frequency (26,5%) of all tumours in males. Just as tumours of the pituitary gland they are common findings in different strains and stocks of rats. Hemangiomas of the mesenteric lymph nodes occurred as secondary findings in 14% of the males. They are also common neoplastic lesions in aged Han:Wist rats (6).

REFERENCES

1. Baba N.E., von Haam E. : Animal model : Spontaneous adenocarcinoma in aged rabbits. Am.J.Path. 68, 653-656, 1972 .

2. Barsoum N.J., Wedead H., Gough A.W., Smith G.S., de la Iglesia F.A. : Histiocytic sarcoma in Wistar rats. Arch. Pathol. Lab. Med. 108, 802-807, 1984 .

3. Brownstein D.G., Brooks A.L.: Spontaneous endomyometrial neoplasms in aging Chinese hamsters. J.Nat. Cancer Inst. 64, 1209-1214, 1980 .

4. Burek J.D. : Pathology of aging rats. CRC Press, Palm Beach , Florida , 1978 .

5. Cotchin E. : Spontaneous uterine cancer in animals. Br.J.Cancer 18 , 209-227 , 1964 .

6. Deerberg F., Pittermann W., Rapp K.G. : Longevity study in Han:Wistar rats : Experience in maintaining aging rats for gerontological investigations. Interdisipl. Topics Geront. 13, 66-74, 1978 .

7. Deerberg F., Rapp K.G., Pittermann W., Rehm S. : Zum Tumorspectrum der Han: Wist-Ratte. Z. Versuchstierk. 22, 267-280, 1980.

8. Deerberg F., Rehm S., Rapp K.G. : Survival data and age-associated spontaneous diseases in two strains of rats. In: Archibald J., Ditchfield J., Rowsell H.C. (Eds.). The contribution of laboratory animal science to the welfare of man and animals. 8th ICLAS/CALAS Symposium, Vancouver, 1983. Gustav Fischer Verlag, Stuttgart, New York , 171-181.

9. Deerberg F., Kaspareit J. : Endometrial carcinoma in BDII/Han rats: A model of a spontaneous hormone-dependant tumour. J.Natl.Cancer Inst. 78,1987 (in press).

10. Goodman D.G., Ward J.WM., Squire R.A., Paxton M.B., Reichhardt W.D., Chu K.C. ; Linhart M.S. : Neoplastic and nonneoplastic lesions in aging Osborne-Mendel rats . Toxicol. and appl. Pharmacol. 55 , 433-447, 1980 .

11. Hollander C.F. : Current experience using the laboratory rat in aging studies. Lab.Anim.Sci. 26 , 320-328, 1976.

12. McKenzie F.W., Gamer F.M. : Comparison of neoplasms in six sources of rats. J.Nat.Cancer Inst. 50, 1243-1257, 1973.

13. Solleveld H.A., Haseman J.K., McConnell E.E. : Natural history of body weight gain, survival, and neoplasia in the F344 rat. J.Nat. Cancer Inst. 72, 929-940, 1984.

14. SuterP., Luetkemeier H., Zakowa N., Christen P., Sachse K., Hess R. : Lifespan studies on male and female mice and rats under SPF-laboratory conditions. Arch.Toxicol. Suppl. 2 , 403-407 , 1979 .

15. Tarone R.E., Chu K.C., Ward J.M. : Variability of the rates of some common naturally occurring tumours in Fischer 344 rats and (C57Bl/6NxC3HeN)F1 (B6C3F1) mice. J. Nat. Cancer Inst. 66, 1175-1181, 1981.

16. Ward B.C., Moore W. : Spontaneous lesions in a colony of Chinese hamsters. Lab.Anim.Care 19, 516-521, 1969.

A.C. Beynen & H.A. Solleveld (Eds), New developments in biosciences: their implications for laboratory animal science, ISBN 978-94-010-7973-0
© 1988 Martinus Nijhoff Publishers, Dordrecht.

ASSESSMENT OF DISCOMFORT INDUCED BY ORBITAL PUNCTURE IN RATS

A.C. BEYNEN[1,2], V. BAUMANS[1], J.W.M. HAAS[3], K.K. VAN HELLEMOND[3], F.R. STAFLEU[1] and G. VAN TINTELEN[3]

[1]Department of Laboratory Animal Science, State University, Utrecht
[2]Department of Human Nutrition, Agricultural University, Wageningen
[3]Small Animal Center, Agricultural University, Wageningen
(The Netherlands)

ABSTRACT
    Possible discomfort induced by orbital puncture in rats was assessed by assigning scores per punctured and non-punctured eye of each animal to the following parameters: photophobia, enophthalmos, discharge, superficial damage, internal damage and reduced withdrawal upon approaching with a cotton tip. Scoring was performed independently and blind (the rats were punctured either at the left or right eye) by 4 assessors. Out of the 6 parameters only enophthalmos tended to occur more frequently in punctured than control eyes. There was a considerable between-assessor variation in the assignment of scores to enophthalmos. It remains to be established whether enophthalmos is associated with discomfort or pain in rats.

1. INTRODUCTION
    Orbital puncture is a technique which is carried out frequently to take blood samples from rats. In the course of our studies on the assessment of discomfort in laboratory animals (1), we have now attempted to assess possible discomfort induced by orbital puncture.

2. PROCEDURE
2.1. Animals, housing and diets
    For this study we used rats from a current experiment concerning the effects of dietary proteins on cholesterol metabolism (2). Male rats from a random-bred Wistar Cpb/WU colony were used. At the age of 5 weeks the animals were transferred from a commercial diet (RMH-B[R], Hope Farms, Woerden, The Netherlands) to the pre-experimental semipurified diet containing soybean protein. After another 14 days, on day 0 of the experiment, the rats were put on the experimental, semipurified diets. The composition of the diets can be found elsewhere (2). As described below, diet was not a variable in this study because each animal served as its own control.
    The rats were housed individually in cages (24 x 17 x 17 cm) constructed of stainless steel with wire mesh bases. The cages were placed in a room with air conditioning (20 °C), controlled lighting (light: 06.00-18.00 h; dark: 18.00-06.00 h) and humidity (55-65%).

2.2. Orbital puncture

Blood samples were taken in the non-fasting state between 08.00 and 10.00 h on days -2 and 15 of the experiment. Each rat was invariably sampled on either the left or right eye. The site of sampling was determined per rat on the basis of random selection. Blood was drawn by orbital puncture under light diethyl-ether anesthesia. A half-way broken, heparinized capillary tube was inserted with its sharp end into the nasal corner of the eye, and subsequently about 0.5 ml of blood was collected (Fig.1).

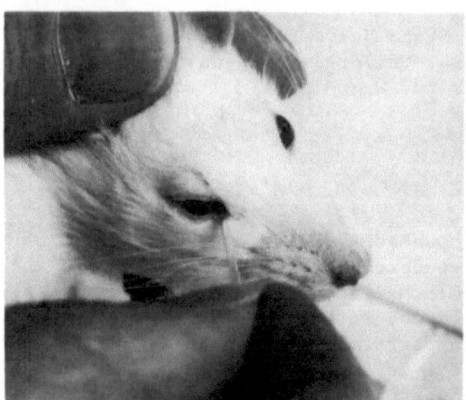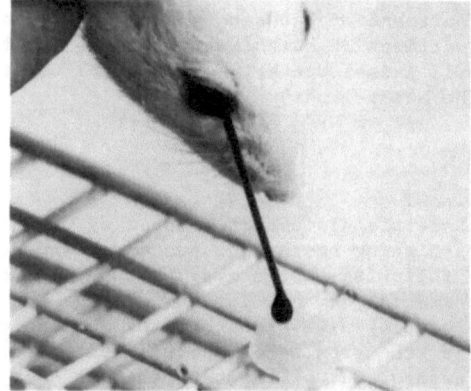

FIGURE 1. Puncturing of the right eye and collection of blood.

2.3 Assessment of discomfort

Four persons (VB, FRS, KKvH and JWMH), who are familiar with laboratory animals, assessed discomfort on days 4 to 6, 12 to 14 and 21 to 23 of the experiment (Fig. 2). Assessors no. 1 and 2 scored in the morning, whereas assessors no. 3 and 4 did it in the afternoon. The assessment scheme was discussed beforehand among all authors. Concerning the animals the assessors knew only that they would be dealing with rats which had been punctured either at the left or right eye. The animal technician (GvT) and ACB were the only ones acquainted with the puncturing code.

Prior to scoring, the rats were divided into 4 groups consisting of 9 animals each; the formation of groups was independent of the diet code. The 4 groups were scored by the 4 assessors at the 3 time intervals (Fig. 2) according to a partial 4x4 Latin square. During the course of the experiment there was no deliberation among the assessors concerning the outcome of their scores. Only after the completion of the experiment the puncturing code was broken, and the score distribution calculated.

433

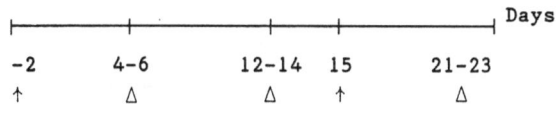

↑ = orbital puncture

Δ = assessment of discomfort

FIGURE 2. Experimental design.

   Assessment of the magnitude of discomfort was performed by measuring 6
parameters per eye of each individual animal. Thus each animal served as
its own control, one eye being punctured, and the other not. Scores of 0 to
3 were assigned to each variable, with 0 given if no abnormal variation was
detected. Depending on the severity of the abnormality, scores ranging from
1 to 3 were assigned. The procedures and variables measured successively,
and the criteria for scoring were as follows.
   The barren cage was taken from the shelf and placed on a table. The rat
was now assessed while in its cage.
Photophobia. The rat was examined facially for possible photophobia of one
of the eyes. If present, this was considered abnormal.
Enophthalmos. The rat was examined from a dorsal view for possible
recession of the eyeball. This procedure implied that the left and right
eye were compared directly.
   Subsequently, the rat was handled, and its eyes were assessed more
closely. The following variables were measured.
Discharge from the eyes.
Superficial damage of the eye and conjunctiva
Internal damage of the eye.
Withdrawal of the eye. The eye was approached with a moist cotton tip.
Lack of withdrawal was considered abnormal.

3. RESULTS
   Tables 1 to 3 show that most of the selected variables did not
discriminate between the control and punctured eyes. These variables are
photophobia, discharge, superficial damage, internal damage and withdrawal
of the eye. In essence, these variables scored normal for both eyes.
   The assessment of enophthalmos displayed a considerable between-assessor
variation. As shown in Tables 1 to 3, enophthalmos tended to occur more
frequently in punctured than control eyes as indicated by assessors no. 1
and 2 (P < 0.01, Chi-square test). However, assessors no. 3 and 4 gave
similar scores to both eyes.
   The results for assessment of enophthalmos were averaged for the 4
assessors. It is clear from Table 4 that frequency distributions for both
eyes differed systematically. However, the difference between the
distributions did not reach statistical significance (Chi-square test).

TABLE 1. Absolute frequency distribution of discomfort assessment scores on days 4 to 6.

		Score							
		Control eyes (n=36)				Punctured eyes (n=36)			
Parameter	Assessor	0	1	2	3	0	1	2	3
Photophobia	1	36	–	–	–	36	–	–	–
	2	36	–	–	–	36	–	–	–
	3	36	–	–	–	36	–	–	–
	4	36	–	–	–	34	1	1	–
Enophthalmos	1	32	4	–	–	21	14	1	–
	2	26	9	1	–	19	16	1	–
	3	28	8	–	–	36	–	–	–
	4	27	8	1	–	25	8	3	–
Discharge	1	36	–	–	–	35	–	1	–
	2	35	1	–	–	32	3	1	–
	3	36	–	–	–	36	–	–	–
	4	34	2	–	–	34	2	–	–
Superficial damage	1	36	–	–	–	36	–	–	–
	2	36	–	–	–	36	–	–	–
	3	36	–	–	–	36	–	–	–
	4	36	–	–	–	35	1	–	–
Internal damage	1	36	–	–	–	36	–	–	–
	2	36	–	–	–	36	–	–	–
	3	36	–	–	–	36	–	–	–
	4	36	–	–	–	36	–	–	–
Withdrawal	1	35	1	–	–	34	2	–	–
	2	35	1	–	–	28	8	–	–
	3	32	4	–	–	34	2	–	–
	4	27	6	3	–	22	12	2	–

4. DISCUSSION

Out of 6 selected variables only enophthalmos tended to be different between control and punctured eyes, in that the incidence was higher in the latter. There was however a considerable between-assessor variation in the assignment of scores to this parameter. Part of this variation could lie in the fact that assessors no. 1 and 2 operated in the morning and no. 3 and 4 in the afternoon. Between-assessor variation was also noted in previous work (1).

Whether enophthalmos in rats is associated with discomfort or pain remains to be established. It is important to note that the outcome of the present study cannot be extrapolated readily because it must have been influenced strongly by the technique of blood sampling and the skill of the animal technician.

TABLE 2. Absolute frequency distribution of discomfort assessment scores on days 12 to 14

Parameter	Assessor	Control eyes (n=36)				Punctured eyes (n=36)			
		0	1	2	3	0	1	2	3
Photophobia	1	36	–	–	–	36	–	–	–
	2	36	–	–	–	35	1	–	–
	3	36	–	–	–	36	–	–	–
	4	31	5	–	–	30	5	1	–
Enophthalmos	1	32	4	–	–	24	11	1	–
	2	27	9	–	–	20	14	2	–
	3	36	–	–	–	36	–	–	–
	4	31	5	–	–	31	5	–	–
Discharge	1	36	–	–	–	36	–	–	–
	2	35	1	–	–	34	2	–	–
	3	36	–	–	–	36	–	–	–
	4	36	–	–	–	36	–	–	–
Superficial damage	1	36	–	–	–	36	–	–	–
	2	36	–	–	–	36	–	–	–
	3	36	–	–	–	36	–	–	–
	4	36	–	–	–	36	–	–	–
Internal damage	1	36	–	–		36	–	–	–
	2	36	–	–	–	36	–	–	–
	3	36	–	–	–	36	–	–	–
	4	36	–	–	–	36	–	–	–
Withdrawal	1	35	1	–	–	35	1	–	–
	2	35	1	–	–	36	–	–	–
	3	35	1	–	–	36	–	–	–
	4	33	3	–	–	32	4	–	–

TABLE 4. Mean frequency distributions of scores from assessment of enophthalmos, expressed as percentages

	Control eyes (n=36)				Punctured eyes (n=36)			
	0	1	2	3	0	1	2	3
days 4 – 6	78	20	1	0	70	26	3	–
days 12 – 14	88	13	0	0	77	21	2	–
days 21 – 23	91	9	–	–	76	23	1	1

TABLE 3. Absolute frequency distribution of discomfort assessment scores on days 21 to 23

Parameter	Assessor	Control eyes (n=36)				Punctured eyes (n=36)			
		0	1	2	3	0	1	2	3
Photophobia	1	36	–	–	–	36	–	–	–
	2	35	1	–	–	36	–	–	–
	3	36	–	–	–	36	–	–	–
	4	34	2	–	–	35	1	–	–
Enophthalmos	1	36	–	–	–	25	11	–	–
	2	35	1	–	–	24	10	1	1
	3	35	1	–	–	36	–	–	–
	4	25	11	–	–	24	12	–	–
Discharge	1	36	–	–	–	36	–	–	–
	2	36	–	–	–	36	–	–	–
	3	36	–	–	–	36	–	–	–
	4	35	1	–	–	36	–	–	–
Superficial damage	1	36	–	–	–	36	–	–	–
	2	36	–	–	–	36	–	–	–
	3	36	–	–	–	36	–	–	–
	4	36	–	–	–	36	–	–	–
Internal damage	1	36	–	–	–	36	–	–	–
	2	36	–	–	–	36	–	–	–
	3	36	–	–	–	36	–	–	–
	4	36	–	–	–	36	–	–	–
Withdrawal	1	36	–	–	–	36	–	–	–
	2	36	–	–	–	36	–	–	–
	3	36	–	–	–	36	–	–	–
	4	30	5	1	–	28	7	1	–

REFERENCES

1. Beynen AC, Baumans V, Bertens APMG, Havenaar R, Hesp APM, Van Zutphen LFM. Assessment of discomfort in gallstone-bearing mice: a practical example of the problems encountered in an attempt to recognize discomfort in laboratory animals. Lab Anim 1987; 21:35–42.
2. Beynen AC, Lemmens AG. Dietary glycine and cholesterol metabolism in rats. Z Ernährungswiss 1987; in press.

A.C. Beynen & H.A. Solleveld (Eds), New developments in biosciences: their implications for laboratory animal science, ISBN 978-94-010-7973-0

EFFECT OF MORPHINOMIMETICS IN DIFFERENT PAIN TESTS

Dhasmana KM, Banerjee* AK and Rating W
Departments of Anaesthesiology and Laboratory Animal Centre*,
Erasmus University, 3000 DR Rotterdam, The Netherlands

1. Introduction

Despite recent development in pain treatment morphine is still used as first line of defence against the pain. In order to acheive better understanding of the mechanism and control of pain, reliable and valid methods for measurement of pain are needed and which can be performed in animals. Various methods for measurement of pain in animals are described (Beecher, 1957; Chapman et al., 1985). On a dose to dose basis narcotic analgesic can be classified into three groups - stronger (sufentanil), moderate (morphine) and weak (mepridine) narcotic analgesics. In order to analyse the analgesic activity of morphinomimetics we used three most commonly known pain tests - tailflick, hot plate and formalin test. These tests were chosen because they are considered more suitable than others for demonstrating a particular drug effect.

2. Materials and methods

Wistar rats (175-300 g) were used through out. A 12 hr light - 12 hr dark schedule was imposed and the experiments were conducted during light hours.

2.1. Surgical preparation

Implantation of intrathecal catheter in the subarachnoid space was done according to the method of Yaksh and Rudy (1976) and as described by Dhasmana et al.(1986, 1987).The animals were anaesthetized with halothane (0.5 - 1.0%) and nitrous oxide in oxygen (2:1). A polyethylene catheter (external diameter 0.61 mm) was introduced into subarachnoid space through a slit made on atlanto-occipital membrane. The catheter was advanced towards the root of the tail for 8 cm. The total length of the catheter was 16 cm and its volume 10 ul. The wound was closed. The catheter was filled with sterile saline and sealed by heat. After recovery from anaesthesia, the animals were housed individually with free access to food and water. Animals with neurological abnormality were discarded from the experiment.

2.2. Nociceptive tests

All the morphinomimetic drugs were tested on tailflick, hot plate or formaline test and were either by subcutaneously or intrathecally administered.

2.3. Tailflick test

Rats were restrained with a towel and radiant heat was applied to the root of the tail. The reaction time to the first movement of the tail was recorded. The heat source was adjusted to give a reaction time between 3-5 sec in a group of rats. In the absence of response cut off time was kept at 30 sec in order to prevent damage to the tissues.

2.4. Hotplate test

A fibreglass cylinder (30 cm height with a circumference 30 cm) and the basement plate of a metal aluminum was used. for this experiment. The cylinder was placed in a hot bath and the temperature of the metal plate was maintained at 55 +/- 1 degree centigrade. The test animal was placed in the cylinder and the reaction time was noted when the animal licked its hind paw or jumped. The cut off time was kept at 60 sec in order to prevent the tissue damage. If the animal did not show any reaction within 60 sec was removed from the cylinder.

2.5. Formalin test

The procedure described by Dubuisson and Dennis (1977) was used. Before starting the test, the animals were placed in individual plastic cages for observation of behavioral count. After 30 min of adaptation to the cage, 50 ul of 5% formalin was injected into the dorsal surface of the right fore paw. The position of formalin-injected paw was assessed by noting the position at which it was placed. The following criteria was taken into consideration: - 0 - full weight is placed on the injected paw; 1- the injected paw is rested lightly on the floor or another part of the body; 2- the injected paw is elevated and in contact with no other surface while the uninjected paw bears the weight and 3- the rat licks or bites the injected paw. This behavioral count was made at every 3 min for 15 min and then every 5 min for rest of the period (60 min) for each rat.
In control animals immediately after formalin injection pain score is maximum but after several minutes (10 min) the pain score becomes very low and after the end of the period increase in pain score persisted throughout the remainder of the session.

3. Drugs- morphine hydrochloride (pharmacy department, Erasmus University, Rotterdam, The Netherlands), sufentanil tartrate, (Janssen pharmaceuticals, Belgium), mepridine hydrochloride (OPG, The Netherlands).

4. Statistics

All data were analysed as mean +/- SEM. The latencies obtained after drug administration were compared with control by two way analysis of variance followed by Duncan's new multiple range test. The significant effect was taken as $P < 0.05$.

5. Results
5.1. Effect of morphinomimetics on tailflick and hotplate tests following subcutaneous administration

Morphine was administered at three dose levels (1, 3 and 10 mg/kg. Morphine was effective in increasing tailflick latency at 10 mg/kg dose but it was more effective as analgesic in hotplate at 3 and 10 mg/kg (Fig 1). Sufentanil (3, 5 and 15 ug/kg) which is closely related to fentanyl was found to be very effective as analgesic in these nociceptive tests. It raised the pain threshold in a dose-dependent manner (Fig 2). In both tailflick and hotplate tests mepridine (10, 20, 40 and 60 mg/kg) interfered in pain reaction time but only at the highest doses. In the hotplate test however, lower doses were more effective than highest dose (Fig 3). With each drug a saline control was always taken into consideration and which did not influence the nociceptive threshold either in tailflick or hotplate tests.

5.2. Effect of morphinomimetics on tailflick and hotplate tests following intrathecal administration

Morphine was injected at three doses (5, 15 and 45 ug) in 10 ul volume followed by 10 ul of saline. At the highest dose (45 ug) morphine significantly increased the threshold of nociception in tailflick test but not at the lower doses. In addition, all three doses of morphine were quite effective in producing analgesia by increasing the reaction time in hotplate test (Fig 4). The peak effect appeared 20 min after the drug administration and lasted more than 60 min. Sufentanil (0.01, 0.25 and 0.5 ug) significantly increased the pain threshold in tailflick and hotplate tests (Fig 5). The peak effect was observed 20 min after drug administration and 30 min after the analgesic effect began to decline but was still apparent after 60 min. Mepridine was administred at two dose levels 100 and 200 ug. Mepridine has little effect in the tailflick test but did produce analgesic effect in the hotplate test lasting 20 min (Fig 6).

5.3. Effect of morphinomimetics on formalin test

Morphine (1, 3 and 10 mg/kg sc) showed prominent effect in this test by lowering the pain score. Lowest dose of morphine (1 mg/kg) however, was effective only upto 30 min but morphine at higher doses (3 and 10 mg/kg) significantly suppressed the pain score throughout the observation period (60 min). The analgesic effect of morphine begins to appear after 6 min of drug administration (Fig 7). In addition sufentanil (3, 5 and 15 ug/kg sc) exhibited an analgesic effect in a dose-dependent manner (Fig 8). The effect of sufentanil was evident within 3 min after administration of the drug. As can seen in fig 9. Mepridine is quite active when given subcutaneously (40 and 80 mg/kg sc). The onset of analgesic effect starts after 6 min and continued during observation period. However the intrathecally administered mepridine (100 and 200 ug) depressed the pain score upto 30 to 35 min. Pretreatment with saline (1 ml/kg sc) as shown in fig 7, 8 and 9 did not influence the pain score. Pain score is maximum at 3 starting immediately after formalin injection and then gradually decreases till 10 min (first phase) and subsequently rises again (second phase) lasting more than 60 min.

440

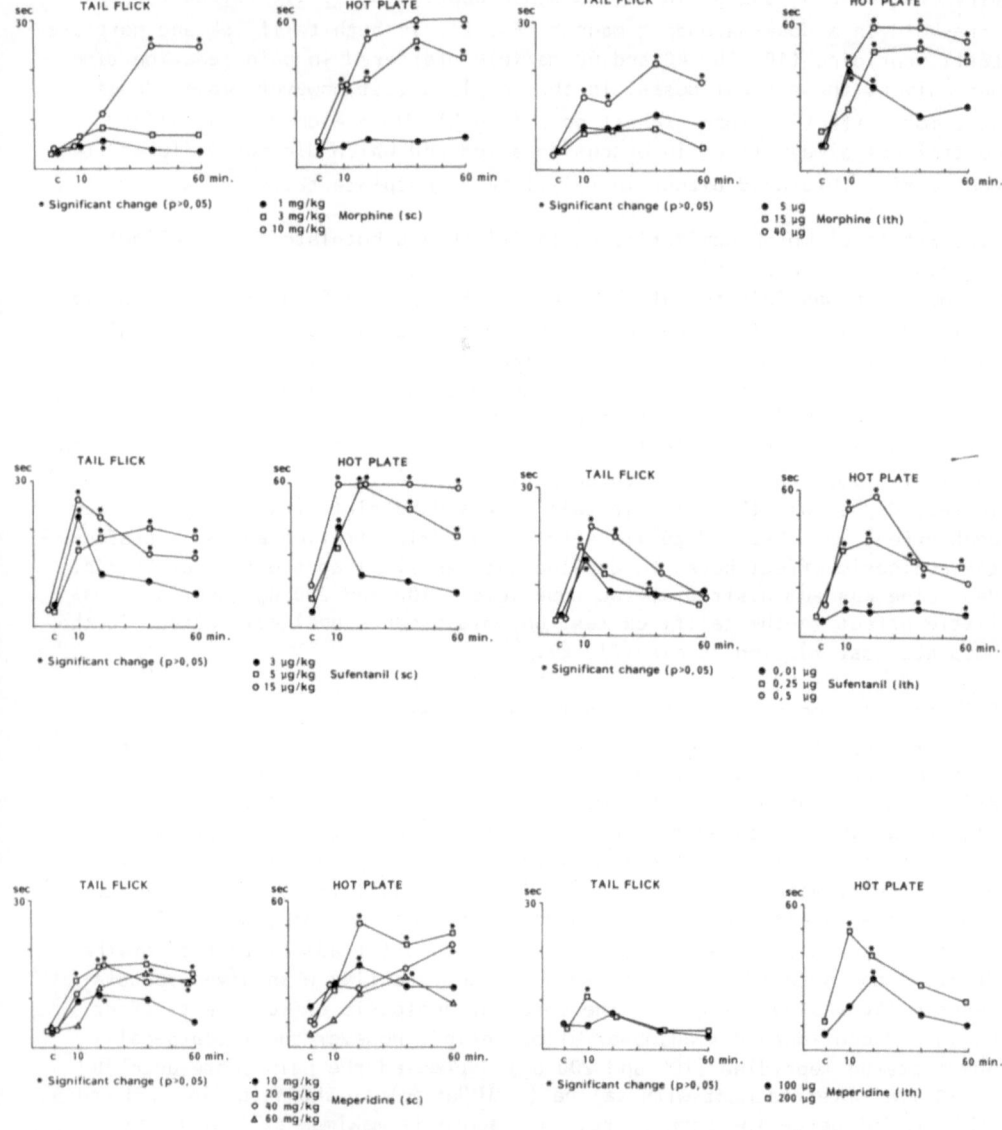

Figure (1-6) shows the effect of different morphinomimetics on latencies in tailflick and hotplate tests following subcutaneous (sc) or intrathecal (ith) administration. Verticle line indicates latencies in seconds (sec) and horizontal line indicates duration in minutes (min).

441

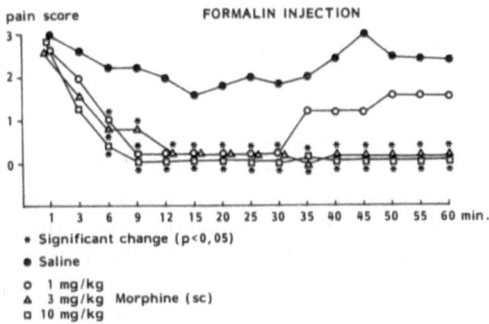

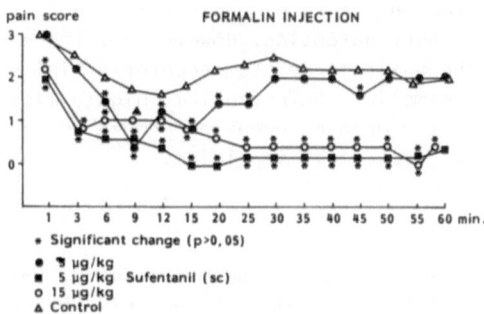

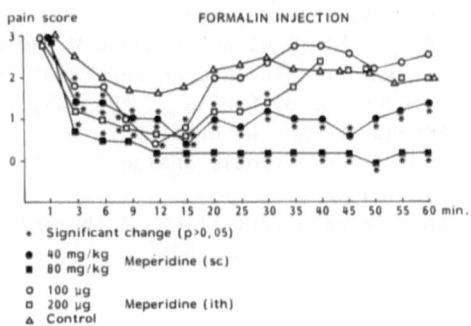

Figure (7-9) depicts the effect of morphine (fig 7), sufentanil (fig 8) and mepridine (fig 9) on pain score (verticle line) and duration (horizontal line) indicated in minutes (min).

442

6. Discussion

Pain tests (tailflick, hotplate and formalin) have been employed by
many workers to study the mechanism and nature of pain receptors. Although
tailflick and hotplate tests are the most commonly used method in pharmacological and physiological investigations, we utilized these tests to
ascertain the analgesic effect of weak, moderate and strong morphinomimetics.
Certain pain tests are more suitable than others for demonstrating particular
drug effects but the reason why one pain test varies from the others are
not precisely known. For assessing the analgesic effect of different
morphinomimetics drugs were administered either subcutaneously or intrathecally. The results, from all three tests demonstrate that strong analgesic
compound (sufentanil) is highly active in a dose-dependent manner. However
the dose-ratio between intrathecal and subcutaneous administration, there
was no difference in regards to sufentanil. For morphine the ratio between
intrathecal and subcutaneous was 1:50 and for mepridine the ratio was
1:1000. From these results two important points emerge a) the stronger the
morphinomimetic a lower intrathecal dose is required b) the ratio between
intrathecal and subcutaneous route of administration is less for stronger
opiates and more for weaker narcotics. However all these characterizations
can be explained on the basis of opiate receptor affinity and lipophilicity
of individual morphinomimetics. Sufentanil is highly lipid soluble and has
very high affinity for the opiate receptors proved to be one of the strongest
morphinomimetic and the weakest is mepridine.

References

Beecher HK: The measurement of pain. Pharmacol Rev 9, 59-209, 1957.
Chapmann CR, Casey KL, Dubner R, Foley KM, Gracely RH and Reading AE:
    Pain measurement: an overview. Pain 22, 1-31, 1985.
Dubuisson D and Dennis SG: The formalin test: A quantitative study of the
    analgesic effects of morphine, mepridine, and brain stem stimulation
    in rats and cats. Pain 4, 161-174, 1977.
Dhasmana KM, Banerjee AK, Faithfull NS and Erdmann W: Role of 5-hydroxy-
    tryptamine receptors in narcotic-induced reduction in gastrointestinal
    transit in rats. Acta Pharmacol Sinica 7, 499-503, 1986.
Dhasmana KM, Banerjee AK and Erdmann W: Gastrointestinal transit following
    intrathecal or subcutaneous narcotic analgesics. Arch Internat
    Pharmacodyn 286, 152-161, 1987.
Yaksh TL and Rudy TA: Chronic catheterization of the spinal subarachnoid
    space. Physiol Behav 17, 1031-1036, 1976.

A.C. Beynen & H.A. Solleveld (Eds), New developments in biosciences: their
implications for laboratory animal science, ISBN 978-94-010-7973-0
© 1988 Martinus Nijhoff Publishers, Dordrecht.

PREVENTION OF EARLY DEATHS IN MICE CONTAMINATED WITH GRAM NEGATIVE ENTERIC
BACTERIA AND FUNGUS FOLLOWING IRRADIATION.

A.K. Banerjee, A.F. Angulo* and J. Kong-A-San.
Laboratory Animal Centre, Erasmus University of Rotterdam, P.O. Box 1738,
3000 DR Rotterdam, The Netherlands.
* National Institute for Public Health and Environmental Hygiene, Bilthoven,
The Netherlands.

Introduction

Mice harbouring Gram-negative organisms die earlier than expected.
The predominant organisms found after whole body irradiation of mice are
Escherichia coli, Proteus mirabilis, Enterobacter cloacae, Citrobacter
freundii, staphylococcus species and enterococci (Brook et al. 1984;
Gerachi et al. 1985).

Wensink, 1961 and Flynn, 1963, reported that fatal septicaemia can
occur in mice harbouring Pseudomonas aeruginosa 3-8 days after irradiation.
After irradiation the animals become susceptible to both endogenous and
exogenous infections and also damaged enteric mucosa become permeable to
enteric bacteria which cause sepsis and endotoxaemia. Endotoxin has been
observed in the liver and blood of mice after irradiation.

It has been reported that germ-free mice survive significantly longer
than conventional mice following intestinal radiation injury. The partial
decontamination of the tract by antibiotic therapy can, however, prolong
the survival of conventional animals following irradiation. The reasons
for early death may be sepsis, which may cause severe fluid and electrolyte
loss, vascular shock or diarrhoea (Gerachi et al. 1985).

Many workers have reported that coliforms and other Gram-negative
bacteria commonly cause bacterimia following irradiation. Diarrhoea caused
by Citrobacter freundii in mice has been reported by Brennan et al. 1965.
Matsumoto (1980) found that a group of mice contaminated with Enterobacter
cloacae started to die 8 days post irradiation and at 13 days mortality
was 65-80% compared with less than 5% in uncontaminated mice.

Dubos (1968) stated that the intestinal flora of the Specified
Pathogen Free (S.P.F.) animals contain few coliforms, proteus, pseudomonas
and clostridia and they are much more resistant to endotoxin, cortison and
total body radiation than conventional animals. He further reported that
as soon as conventional animals are maintained in cages with air filters
and are provided with sterile food and water, the total count of coliforms,
enterococci and other Gram-negative bacteria progressively decreased in
numbers in the intestine.

Recently we have observed a high mortality in inbred mice following
irradiation. The normal pattern of mortality in inbred mice following 950
rad irradiation and intraperitoneal injection of bone marrow suspension
should not exceed more than 10% at 10 days post irradiation. However 70 to

80% of the animals died between 3 to 7 days post irradiation. Bacteriological investigation of the dead animals showed that the animals were contaminated with Citrobacter freundii, Enterobacter cloacae, Proteus species, E.coli and in one group of animals in addition to Gram-negative organisms fungus contamination was found. The situation was needed careful investigation. Our aim was to find a solution which would help to decrease the mortality rate of animals after irradiation either by providing antibiotic treatment or by providing adequate protection from exogenous infection.

Materials and Method

Three groups of animals were compared:

A. A group of 180 strain 129 mice, 10-12 weeks old were used. The animals were maintained in a conventional system and they were provided with autoclaved bedding material, clean cages twice a week and acidified drinking water. Bottles were changed twice a week. Hope Farms (Woerden, The Netherlands) non-sterilised pellets were given ad libitum. The animal caretakers used clean working clothes every day but did not use disposable gloves during the cleaning procedures.

B. A group of 120 strain BCBA mice (SPF), 6-8 weeks old, were housed in a conventional system for 6-8 weeks before the animals were used for experiment. Similar hygienic measures were adopted as described above.

C. A group of 120 strain BCBA mice were pretreated with oxytetracycline hydrochloride (cyclovit[R], Aesculaap, Boxtel, The Netherlands) 2 g/ltr drinking water (non acidified) for 10 days post-irradiation and 7-10 days post treatment the animals were used for irradiation experiment. Similar hygienic measures were adopted as described above.

Bacteriological examination

Bacteriological examination was conducted in animals prior to treament with oxytetracycline hydrochloride and from the dead animals following irradiation. 10 animals of each group (A, B and C) of dead mice were examined and samples from intestine, liver and heart-blood were cultured on blood agar and MacConkey agar, both under aerobic and anaerobic conditions, at 37°C for 24 hours.

Irradiation experiment

Approximately 80-120 mice were used per experiment. Animals were divided into groups of 10 per Makrolon cage, type III. The animals were subjected to whole body irradiation (950 rad). Following irradiation homogenised bone marrow suspension was injected intraperitoneally into the mice. A few animals were kept as a control. Following irradiation the animals were transferred to an animal room.

Results

A.   Following irradiation, 62 of the 80 animals used in group A died within 10 days including the control animals. E.coli, Citrobacter freundii and Enterobacter cloacae were isolated from the intestine, liver and heart-blood. There was a heavy growth on the plate.

B. In group B 91 animals died within 7 days of irradiation including control animals. On postmortem examination, multiple necrotic nodules, 0.2 to 1 mm in diameter, were observed in the liver. They were grey to white in colour. Heavy growths of Enterobacter cloacae and Streptococcus faecalis were isolated from the intestine. Histological section of the liver stained by HES revealed the presence of aspergillus species of fungi in the liver.

C. Of the 120 BCBA animals which were pretreated with oxytetracycline hydrochloride in group C, only 3 animals died on the 4th and 2 more on the 7th days following irradiation. Bacteriological examination of the dead animals revealed few colonies of Enterobacter cloacae from the intestine and liver. It appeared that after the introduction of antibiotic therapy,the mortality rate decreased considerably.

Further experiment concerning four groups of animals pretreated with antibiotic treatment.

Four more experiments were carried out with BCBA conventional mice consisting of 120 animals in each group. The animals were pretreated with oxytetracycline hydrochloride (2 g/ltr) for 10 days. The animals were irradiated two weeks after the end of the treatment in one group. In another 3 groups the period between pretreatment and irradiation was longer than 4-6 weeks. In these animals, however, the result was not constant. The mortality rate increased considerably. It appeared that the mortality rate of the mice increased considerably the longer the interval between antibiotic therapy and irradiation. The animals need to be protected from exogenous infection.

Use of S.P.F. animals for irradiation experiment.

4 weeks old BCBA of SPF category were housed in a separate disinfected animal room in our Laboratory Animal Centre. Although they were maintained in a conventional system for about 6-8 weeks before irradiation, during the holding period the animals were provided with clean cages, with sterilized bedding material, acidified drinking water and Hope Farms non-sterilized pellets. The animal caretakers used clean working clothes every day and disposable gloves during the cleaning procedures. No other animals were kept in the same room.

The decision was taken to place the animals in a Laminar flow (cross-flow) cabinet (Animal isolator Bos & Eelman type 313H90, Wieringerwerf, The Netherlands) immediately after the animals had been irradiated in a separate room. Following irradiation the animals were housed in autoclaved makrolon cages with stainless steel wire top, and sterilized bedding material. The animals were provided with sterilized drinking bottles with acidified drinking water and irradiated Hope Farms pellets. The animal caretaker used clean working clothes, facemask and disposable gloves during changing of the cages. The door of the animal room was kept closed.

Result

Following the above mentioned methods the overall mortality was less than 20%. The first animal died 13 days post irradiation. 80% of the animals survived more than 30 days post irradiation. Subsequent experiments conducted using a similar method gave constant good results.

Discussion

In the present study we found that Enterobacter cloacae, Citrobacter freundii, Escherichia coli and Streptococcus faecalis with a high count were the cause of a large number of early deaths in mice following irradiation (Matsumoto, 1980, 1982; Brook et al. 1984 and Gerachi et al. 1985). We also observed growth of fungus in the liver following irradiation.

Conventional animals are not suitable for irradiation experiments. SPF animals should be used for experimental purposes and the animals should be housed in an animal isolator (Laminar flow) following irradiation and given sterilized food, bedding material and acidified drinking water. It is essential that animals should be protected from exogenous infection following irradiation. We found dramatic decrease in mortality rate, as soon as the SPF animals were housed in a Laminar flow cabinet after irradiation. The animals were adequately protected from exogenous infection by using this method. Pretreatment with antibiotic did not provide constant results.

Refenrences
1. Brook, I., MacVillie, T.J. and Walker, R.I. (1984) Recovery of aerobic and anaerobic bacteria from irradiated mice. Infection and Immunity. 46, 270-271
2. Gerachi, J.P., Jackson, K.L. and Mariano, M.S. (1985). The Intestinal Radiation Syndrome: Sepsis and Endotoxin. Radiation Research 101, 442-450
3. Matsumoto, T. (1980). Early deaths after irradiation of mice contaminated by Enterobacter cloacae. Laboratory Animals. 14, 247-249
4. Matsumoto, T. (1982). Influence of Escherichia coli, Klebsiella pneumoniae and Proteus vulgaris on the mortality pattern of mice after lethal irradiation with Y rays. Laboratory Animals. 36-39

A.C. Beynen & H.A. Solleveld (Eds), New developments in biosciences: their implications for laboratory animal science, ISBN 978-94-010-7973-0
© 1988 Martinus Nijhoff Publishers, Dordrecht.

Detection of Mycoplasma neurolyticum in a colony of
inbred mice: clinically silent infection.

A.F.Angulo, A.K.Banerjee* and A.A.Polak-
Vogelzang.
National Institute for Public Health and Environmental Hygiene, Bilthoven,
The Netherlands.
* Laboratory Animal Centre, Erasmus University Rotterdam, The Netherlands.

Introduction
Mycoplasma neurolyticum was isolated from the brain and after intracere-
bral passage of infected tissue the "rolling disease" was evoked experimentally by Findlay et al. (1938). The isolation rate was 78% in experimentally infected animals and 58% from the naturally infected mice (Hill
1979). Following Tully (1981) M.neurolyticum remains as a latent
organism in the brain of mice but under various conditions of stress the
organism multiplies and the disease reveals itself by causing neurological
disorders which in a late phase lead to the animal rolling on the long
axis of its body. M.neurolyticum can be isolated from the nasal mucosa
and lungs of carrier animals (Adler 1965). In the present study M.neuro-
lyticum was isolated from the nasal cavity and the conjunctiva of two
inbred strains of mice that had no apparent clinical symptoms. The entire
colony was found free of M.pulmonis and M.arthritidis.

Materials and methods
Animals and housing: Various strains of inbred and congenic mice were
maintained in the experimental department for research purposes. Autoclaved cages, bedding material and drinking bottles were used. Acidified
drinking water and food pellets (Hope Farms, Woerden, The Netherlands)
were provided ad libitum. The temperature was maintained at 21°C, the
air was filtered and renewed 10 times per hour.

Microbiological examination.
A total of 134 mice were examined and swabs from the nasal cavity, trachea,
middle ear, conjunctiva, lungs, brain and ovaries were cultivated onto
blood agar plates and P.P.L.O. agar and in P.P.L.O. broth. The two last
media did not contain thallium acetate or penicillin. Incubation was made
aerobically at 37°C. The P.P.L.O. plates were examined at 100 x magnification and M.neurolyticum was further identified by immunofluorescence.

Results.
M.neurolyticum was isolated from the conjunctiva and the nasal cavity of
one B10BYR mouse and two BCBA inbred mice. A total of 134 animals were
submitted to microbiological examination.

Discussion.
The most remarkable fact in the cases described is that neither clinical
symptoms nor mortality were observed. That this infection did not cause
symptoms of disease in our colonies may be attributed to a low level of
stress due to careful management as well as to the possibility that the
concerned strain does not produce toxin. The low incidence of infected

animals indicates that this organism does not propagate so easily or so efficiently as that of other Mycoplasma species such as M.pulmonis.

References
Adler,H.E.: Mycoplasmosis in animals. Adv. Veterinary Science 10: 205-244. 1965
Findlay,G.M., Klieneberger,E., MacCallum,F.O. and MacKenzie,R.D.: Rolling disease: New syndrome in mice associated with a Pleuro-pneumonia-like-organism. Lancet 235: 1511-1519. 1938
Tully,J.G.: Mycoplasmal Toxins. Israel Journal of Medical Science 17: 604-607. 1981

A.C. Beynen & H.A. Solleveld (Eds), New developments in biosciences: their
implications for laboratory animal science, ISBN 978-94-010-7973-0
© 1988 Martinus Nijhoff Publishers, Dordrecht.

EFFECTS OF HANDLING AND TRANSPORTATION STRESS ON RODENTS

M LANDI, T BOWMAN, S CAMPBELL

Introduction

The first reported observation regarding the response of an organism to
stress was made by W B Cannon in 1929 (3). Cannon described stress as
"fright and flight" mechanism, restricting his discussion to signs which
are caused by the release of catecholamines. These compounds, released
from the adrenal medulla, cause an increase in heart rate and blood
pressure, constriction of the peripheral vasculature and an increase in
glycogen mobilization.

In 1950, Hans Selye described stress as a "specific syndrome,
nonspecifically induced". Selye, unlike Cannon, described the effects of
stress as being mediated by compounds from the adrenal cortex. He felt
that stress manifested itself in an organism as a "general adaptation
syndrome" that has three stages. The first stage is the "alarm reaction",
which involves the stimulation of the hypothalmic pituitary-adrenal axis.
After the first stage, the body may progress to the stage of
"resistance". This occurs after the organism has acquired adaptation as a
result of continuous exposure to stimuli. The third and final stage of
the system is "exhaustion". Exhaustion is the result of prolonged
exposure to stimuli, to which adaptation has developed, but can no longer
be maintained.

There are many mechanisms that can produce stress in people and animals.
Earlier workers demonstrated the reductions in body weight and rate of
weight gain and microscopic pathologic alterations as the result of
shipment. The changes were attributed to the stress experienced by the
animal during transit.

A variety of factors can produce stress during shipment - variations in
the intensity and duration of light, availability of food and water, age
and physical condition of the animal at the time of shipment. This stress
involves stimulation of the hypothalamohypophyseoadrenal axis.

Activation of the hypothalamohypophyseoadrenal axis results in
corticosterone release and this steroid is often used as a marker for
stress in rodents.

The objective of this study was to determine whether the stress of
shipment would produce a significant change in the immune response of mice
and if changes occur, how long it would take for these responses to return
to baseline.

On a small study with rats the effects of handling versus anaesthesia were also monitored by measuring increases in corticosterone levels. This was done to evaluate what was least stressful; animal restraint for a short time or anaesthesia.

METHODS AND MATERIALS

Shipping Study

A total of 285 CRL:CD mice were used. All of the experimental mice were randomly assigned to two groups at the breeding facility on the basis of transportation either by truck or place. Control animals were shipped by truck from the breeder to the research site and acclimated to the environment for one month prior to use. The mice were not born on the same calendar date but all were males weighing between 30 - 35 grams and at least eight weeks of age when used in the studies. All of the mice were shipped in commercial cardboard packing cases with not more than 15 mice per box. Water and food were provided by the presence of a moist commercial diet.

Four separate shipments of experimental groups were made; two by truck and two by plane. During transit temperature and light alterations were monitored. Experimental animals were weighed at arrival and randomly allotted to three groups for evaluation at 0, 24 and 48 hours after arrival (Table 3). The control animals shipped to the research site 1 month prior to use were randomly divided into nonstressed and experimentally stressed control groups and housed under similar conditions. To induce experimental stress the animals were placed in modified centrifuge tubes followed by placement into a cold (0 C) environment for 15 minutes. The animals were then returned to their cages for 45 minutes before use. The nonstressed control animals were handled normally.

Animals from each group were used for 1 of the following assays: (i) plasma corticosterone concentration, (ii) number of plaque-forming cells in the spleen, (iii) serum hemagglutnation titer and (iv) delayed-type hypersensitivity reaction (foot pad test). The weight of every animals was obtained at the start and end of each of the assays.

Corticosterone measurement - Plasma corticosterone concentrations were measured by a radioassay method, based on a modification of Murphy's competitive protein-binding techniques. All mice were decapitated and bled within 2 minutes of contact to avoid handling stress. Blood sample collections were done between 1100 and 1300 hours, the time of arrival at the research center.

Immune function assays - Three assays were used to evaluate the immune function of the animals; the foot pad test, the hemagglutination assay, and the plaque-forming cell assay. Each assay was done on different sets of age, sex and weight-matched animals. In the foot pad test a sensitizing dose of 1×10^8 sheep RBC was injected intradermally into the left rear foot pad. The animals were challenge exposed 1 week later in the right rear foot pad, using the same concentration of antigen. The thickness of the right rear foot pad was measured, using a dial guage before challenge exposure and again 24 and 48 hours after challenge exposure. Three measurements were taken at each period to determine the average foot pad thickness.

To quantitate the antibody level, samples of blood were collected by cardiac puncture following the 48-hour foot pad measurement. Antibody titers were determined by the modified hemagglutination procedure. Antibody levels, based on 2-fold dilutions of the serum, were reported in $\log_2$ hemagglutination titers. To further evaluate antibody production, a modification of Jerne's plaque-forming cell assay was used. For statistical analysis the number of plaques were then ranked as.

Statistical analysis - The student's t test was used to analyse differences in weight gain, temperature variation, light intensity, and for each biochemical assay at each period. A level of $P \leq 0.05$ was considered significant. To attain additional information, the values obtained for the groups transported by truck and plane were combined and a 1-way analysis of variance was done between the nonstressed controls, experimentally stressed controls, and on- 24- and 48-hour groups. Comparisons subsequent to significant overall F values were made, using the Newman-Keuls procedure. Differences at the value $P \leq 0.05$ were then calculated.

Handling/Restraint Study

A total of 4 - 6 rats (CRL:CD(SD)) were used, per time point. All rats were male and weighed approximately 400 grm. Each phase of the study was started at the same time of day. All blood was obtained by collection into heparinized tubes following decapitation.

Corticosterone measurement was done using a commercially available kit, (Radioassay Systems Lab, Inc, Carson, CA, USA).

Statistical analyses were done using a VAX RS1 data base.

Results

Shipping Study

Findings were as follows :

1   There was no significant increase in weight gain between the treatment groups.

2   There were no significant differences in light exposure between the groups transported by truck and plane. However, temperature differences were evident. The truck group had a higher minimal value (18.4 $\pm$ 2.6°C vs 15.8 $\pm$ 0.6°C) and smaller temperature differential (7.1 $\pm$ 2.2°C vs 11.4 $\pm$ 2.3°C) than the plane group. There were no differences in the maximal value between the groups shipped by truck(25.6 $\pm$ 2.7°C) and plane 27.2 $\pm$ 2.7°C). It would be noted that temperatures were within present governmental (federal) recommended temperature ranges of no less than 10°C (45°F) or greater than 29.4°C (85°F).

I. RESULTS AND STATISTICAL SIGNIFICANCE

Treatment Group	Plasma Corticosterone Concentration (ug/dl)
(Nonstressed controls)	4.3 ± 0.5 (15)*
(Experimentally stressed controls)	17.0 ± 3.7 (12)**
0 hours after arrival	26.4 ± 13.2 (16)***
24 hours after arrival	15.8 ± 5.2 (12)
48 hours after arrival	15.8 ± 5.8 (17)

*   Mean 1 SD; (No. of mice per group in parentheses).
**   Statistically significant ($P \leq 0.05$) difference when compared with all other treatments.
*** Statistically significant ($P \leq 0.1$) difference when compared with negative controls.

II. RESULTS AND STATISTICAL SIGNIFICANCE

Treatment Group	Foot Pad Thickness (mm)
(Nonstressed controls)	3.2 ± 1.9 (12)*
(Experimentally stressed controls)	1.0 ± 0.6 (15)
0 hours after arrival	1.0 ± 0.6 (18)
24 hours after arrival	2.8 ± 1.2 (18)
48 hours after arrival	3.2 ± 1.1 (19)

*   Mean 1 SD; (No. of mice per group in parentheses).
**   Statistically significant ($P \leq 0.01$) difference when comared with 24-, 48-hour negative treatment groups.

III.   RESULTS AND STATISTICAL SIGNIFICANCE

Treatment Group	Hemagglutination Titers ($\log_2$)
(Nonstressed controls)	6.58 ± 1.1 (12)*
(Experimentally stressed controls)	5.33 ± 1.8 (15)**
0 hours after arrival	4.25 ± 1.4 (18)***
24 hours after arrival	6.24 ± 1.0 (17)
48 hours after arrival	6.82 ± 1.6 (19)

*   Mean 1 SD; (No. of mice per group in parentheses).
**   Statistically significant ($P \le 0.01$) difference when compared with
     48-hour and negative treatment groups.
*** Statistically significant ($P \le 0.01$) difference when compared with
     24-, 48-hour and negative treatment groups.

IV.   Results and Statistical Significance

Treatment Group	Plaque-Forming Cell Assay (rank)		
(Nonstressed controls)	18.5	3.5	(8)*
(Experimentally stressed controls)	8.6	3.1	(8)**
0 hours after arrival	5.2	2.5	(14)***
24 hours after arrival	12.0	3.0	(14)**
48 hours after arrival	19.1	2.4	(13)

*   Mean 1 SD; (No. of mice per group in parentheses).
**   Statistically significant ($P \le 0.01$) difference when compared with
     48-hour and negative treatment groups.
*** Statistically significant ($P \le 0.01$) difference when compared with
     24-, 48 hour and negative treatment groups.

3   The plasma corticosterone concentrations (Results I) of the
     nonstressed controls were significantly lower than the other four
     groups.

4   The foot pad test (Results II) showed that animals assayed at arrival
     had a significantly smaller increase in foot pad thickness 24 hours
     after challenge exposure when compared with those sensitized at 24 or
     48 hours after arrival, however, were not significantly different from
     nonstressed control animals.

5   Several significant differences were found in the hemagglutination
     assay (Results III). Stated simply, animals sensitized at arrival
     were the least able to respond to sheep RBC antigen. Significantly
     lower antibody titers were found in those animals sensitized at
     arrival when compared with the titers at 24 hours and 48 hours and the
     nonstressed controls. The experimentally stressed control animals
     also had significantly lower antibody levels when compared with those
     sensitized 48 hours after arrival and the nonstressed controls.

6   For the PFC assay the findings (Results IV) were that the number of
     plaques for the animals sensitized at arrival were significantly less
     than those sensitized at 24 and 48 hours after arrival, as well as the
     nonstressed controls. The experimentally stressed control animals had
     significantly fewer plaques than did those sensitized at 48 hours
     after arrival and the nonstressed controls. Finally, the animals
     sensitized at 24 hours after arrival had a significant decrease in the
     number of plaque-forming cells, compared with those animals sensitized
     48 hours after arrival and nonstress controls.

Handling/Restraint Study

Results are listed in Table 1. Values for corticosterone were similar to
those reported by other investigators using RIA technique. Figure 1 is a
graph of the corticosterone values.

Discussion

In the shipping study, stress and release of adrenocorticotrophic hormone,
with the subsequent increase in plasma corticosterone, had been initiated
at some point during transit. Reduction of the immune response to an
antigen had been produced but within 48 hours after arrival, a normal
immune response was mounted. This indicated that even with elevated
steroid levels, resistance or adaptation to stressful stimuli had
developed. This paradox of high corticosterone concentrations, with a
recovering immune system is difficult to explain, however, it is possible
that the mice had progressed toward the 2nd stage of Selye's general
adaption syndrome. Again, this 2nd stage occurs after prolonged exposure
to stimuli, to which the organism has acquired adaptation as a result of
continuous exposure.

In the study comparing handling restraint of rats versus anesthetic
restraint it was found that based on CCSTG values, the manual restraint of
the rats for short time periods, had minimal effect on steroid circardian
rhythm. This can be seen in Figure I. Animals receiving pentobarbital
anesthesia had significantly higher concentrations of CCST when compared
to controls, saline injected controls and manual restraint.

TABLE I

MEAN CORTICOSTERONE VALUES

0 TIME HRS.	1 CONTROL	2 SALINE	3 MANUAL REST.	4 PENT. 30 mg/kg	5 PENT. 60 mg/kg
1 0.0	2.8 (0.4)				
2 0.5		13.3 (3.0)a	50.5 (4.0)a	318.0 (72.0)	380.0 (78.0)
3 1.5		42.5 (17.0)b	26.0 (12.0)b	369.8 (73.0)e	6.1 (4.0)
4 3.0	4.4 (1.5)d	16.7 (10.0)c	9.7 (3.0)c	161.2 (85.0)	221.0 (87.0)
5 6.0	43.0 (13.0)	112.5 (36.0)	55.8 (18.0)	54.0 (16.0)	46.0 (16.0)

ALL CORTICOSTERONE VALUES ARE IN ng/ml.

NUMBERS IN PARENTHESIS INDICATE STANDARD ERROR OF MEAN.

a STATISTICALLY SIGNIFICANT ($P \leq 0.01$) DIFFERENCE WHEN COMPARED TO PENTOBARBITAL (60 mg/kg).
b " ($P \leq 0.01$) " " " " (30 mg/kg).
c " ($P \leq 0.06$) " " " " (60 mg/kg).
d " ($P \leq 0.06$) " " " " (60 mg/kg).
e " ($P \leq 0.05$) " " " " (60 mg/kg).

Figure 1

MEAN CORTICOSTERONE LEVELS

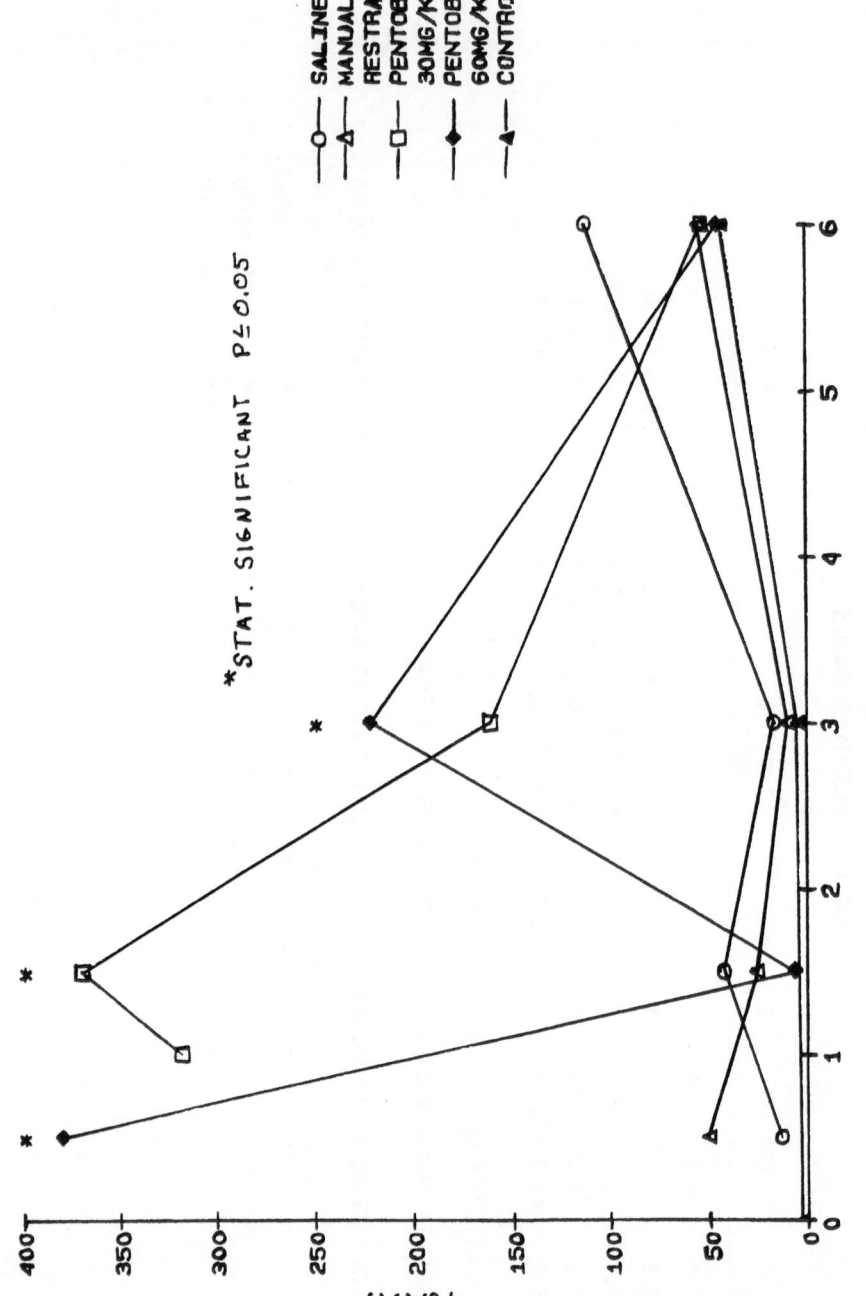

*STAT. SIGNIFICANT P≤0.05

TIME (HOURS)

C C S T

—○— SALINE IP
—△— MANUAL RESTRAINT
—□— PENTOBARB 30MG/KG
—◆— PENTOBARB 60MG/KG
—▲— CONTROLS

In the light of the findings of depressed immune function after shipping, it seems reasonable to propose that a postshipping stabilization period is necessary before the animals are moved into a research project or experiment. During this period, animals should be introduced into an environmentally controlled facility, separate from the main animal colony. Water and feed should be similar to those which the animals will be consuming during a research experiment or procedure. On the basis of the present study, a 48 hour stabilization period is the minimum necessary to allow the immune function of mice to return to normal. However, it should be noted that other considerations, such as possible bacterial or viral hazards, could lengthen the postshipping stabilization period beyond the suggested 48 hour period.

REFERENCES

1 Brodie, D., Valetski, L.: Production of gastric hemorrhage in rats by multiple stresses. Proc. Soc. Exp. Biol. Med. 113: 998-1003, 1963.

2 Cannon, W.: Organization of physiological homeostasis. Physiol. Rev. 9: 399-431 (1929).

3 Cohen, J.J., Claman, N.H.: Thymus-marrow immunocompetence. V Hydrocortisone resistant cells and processes in the hemolytic antibody response of mice. J. Exp. Med. 133: 1026-1034 (1971).

4 Dracott, B., Smith, C.E.T.: Hydrocortisone and the antibody response in mice. I. Correlation between serum cortisol levels and cell numbers in the thymus, spleen, marrow and lymph nodes. Immun. 38: 425-429 (1979).

5 Dymsza, H., Miller, S., Maloney, J., Foster, H.: Equilibration of the laboratory rat following exposure to shipping stress. Lab. Anim. Sci. 13: 61-65 (1963).

6 Jones M.T. Control of adrenocortical hormone secretion. In: The Adrenal Gland, James, V.H.T. (ed.). Raven Press, NY, 1979, pp. 93-130.

7 Kawate, T., Abo, T., Himuma S., and Kumagai, K., Studies on the bioperiodicity of the immune response. II. Co-variations of murine T and B cells and the role of corticosteroid. J. Immuno. 126: 1364-1367, 1981.

8 Perhach, J., Jr., Barry, H., III: Stress response of rats to acute body or neck restraint. Physiol. Beh. 5: 443-448 (1970).

9 Selye, H.: The general adaptation syndrome and the diseases of adaptation. J. Clin. Endoc. 6: 118-127 (1946).

10 Slanetz, C., Fratta, I., Crouse, C., Jones, S.: Stress and transportation of animals. Proc. An. Care. Panel. 7: 278-289 (1957).